WHO FOOD ADDITIVES SERIES: 50

Safety evaluation of certain food additives

Prepared by the
Fifty-ninth meeting of the Joint FAO/WHO Expert Committee on Food Additives (JECFA)

World Health Organization, Geneva, 2003

IPCS - International Programme on Chemical Safety

WHO Library Cataloguing-in-Publication Data

Safety evaluation of certain food additives / prepared by the fifty-ninth meeting of the Joint FAO/WHO Expert Committee on Food Additives.

(WHO food additives series ; 50)

1.Food additives – analysis 2.Food additives – toxicity 3.Flavouring agents – analysis 4.Flavouring agents – toxicity 5.Risk assessment I.Joint FAO/ WHO Expert Committee on Food Additives. Meeting (59th : 2002 : Geneva, Switzerland) II.Series

ISBN 92 4 166050 3 **(NLM classification: WA 712)**

CONTENTS

This publication is a contribution to the International Programme on Chemical Safety.

The International Programme on Chemical Safety (IPCS), established in 1980, is a joint venture of the United Nations Environment Programme (UNEP), the International Labour Organisation (ILO), and the World Health Organization (WHO). The overall objectives of the IPCS are to establish the scientific basis for assessing the risk to human health and the environment from exposure to chemicals, through international peer-review processes, as a prerequisite for the promotion of chemical safety, and to provide technical assistance in strengthening national capacities for the sound management of chemicals.

The Inter-Organization Programme for the Sound Management of Chemicals (IOMC) was established in 1995 by UNEP, ILO, the Food and Agriculture Organization of the United Nations, WHO, the United Nations Industrial Development Organization, and the Organisation for Economic Co-operation and Development (Partici-pating Organizations), following recommendations made by the 1992 United Nations Conference on Environment and Development to strengthen cooperation and increase coordination in the field of chemical safety. The purpose of the IOMC is to promote coordination of the policies and activities pursued by the Participating Organiza-tions, jointly or separately, to achieve the sound management of chemicals in relation to human health and the environment.

PREFACE

The monographs contained in this volume were prepared at the fifty-ninth meeting of the Joint FAO/WHO Expert Committee on Food Additives (JECFA), which met at WHO Headquarters in Geneva, Switzerland, 4–13 June 2002. These monographs summarize the safety data on selected food additives reviewed by the Committee.

The fifty-ninth report of JECFA has been published by the World Health Organization as WHO Technical Report Series No. 913. Reports and other documents resulting from previous meetings of JECFA are listed in Annex 1. The participants in the meeting are listed in Annex 3 of the present publication; a summary of the conclusions of the Committee is given in Annex 4. Some of the substances listed in Annex 4 were evaluated at the meeting only for specifications.

Specifications that were developed at the fifty-ninth meeting of JECFA have been issued separately by FAO as Food and Nutrition Paper No. 52, Add. 10. The toxicological monographs in the present publication should be read in conjunction with the specifications and the report.

JECFA serves as a scientific advisory body to FAO, WHO, their Member States, and the Codex Alimentarius Commission, primarily through the Codex Committee on Food Additives and Contaminants and the Codex Committee on Residues of Veterinary Drugs in Foods, regarding the safety of food additives, residues of veterinary drugs, naturally occurring toxicants and contaminants in food. Committees accomplish this task by preparing reports of their meetings and publishing specifications or residue monographs and toxicological monographs, such as those contained in this volume, on substances that they have considered.

The toxicological monographs contained in this volume are based upon working papers that were prepared by Temporary Advisers. A special acknowledgement is given at the beginning of each monograph to those who prepared these working papers.

Many proprietary unpublished reports are referenced. These were voluntarily submitted to the Committee by various producers of the food additives under review and in many cases represent the only data on safety available on those substances. The Temporary Advisers based the working papers they developed on all of the data that were submitted, and all of these reports were available to the Committee when it made its evaluations. The monographs were edited by E. Heseltine, Lajarthe, 24290 St Léon-sur-Vézère, France.

The preparation and editing of the monographs included in this volume were made possible through the technical and financial contributions of the Participating Organizations of the International Programme on Chemical Safety (IPCS), which supports the activities of JECFA.

Any comments or new information on the biological or toxicological properties of the compounds evaluated in this publication should be addressed to: Joint WHO Secretary of the Joint FAO/WHO Expert Committee on Food Additives, International Programme on Chemical Safety, World Health Organization, Avenue Appia, 1211 Geneva 27, Switzerland.

ALITAME

First draft prepared by
Dr Peter J. Abbott[1], Professor R. Walker[2] and Dr C. Leclercq[3]
[1]Food Standards Australia New Zealand, Canberra, Australia;
[2]School of Biomedical and Life Sciences, University of Surrey, Guildford, Surrey, England; and
[3]National Institute for Food and Nutrition Research, Rome, Italy

1. EXPLANATION

Alitame is an intense sweetener, with a sweetness potency 2000 times greater than that of sucrose. It is a dipeptide of L-aspartic acid and D-alanine, with a terminal *N*-substituted tetramethylthietanyl-amine moiety. Alitame is prepared by a multistep synthesis involving the reaction between two intermediates, (*S*)-[2,5-dioxo-(4-thiazolidine)] acetic acid and (*R*)-2-amino-*N*-(2,2,4,4-tetramethyl-3-thietanyl) propanamide. The final product is isolated and purified by crystallization of an alitame-4-methylbenzenesulfonic acid adduct, followed by additional purification steps, and finally recrystallization from water.

Alitame was considered by the Committee at its forty-fourth and forty-sixth meetings (Annex 1, references *116* and *122*). Most of the toxicological studies were examined at the forty-fourth meeting, and specifications and a toxicological monograph were prepared; however, an ADI could not be allocated because of concern about deficiencies in the studies of carcinogenicity. At its forty-sixth meeting, the Committee reconsidered this issue in the light of a general discussion of the survival rates of animals in contemporary long-term studies and a further statistical analysis of the data from the two available studies. At that meeting, the Committee concluded that there was no evidence that alitame is carcinogenic. An ADI of 0–1 mg/kg bw was allocated on the basis of the NOEL of 100 mg/kg bw per day in an 18-month study in dogs. The Committee noted that, although not specifically requested, a further study of tolerance to repeated doses of alitame in persons with diabetes was under way and that, as is customary, the results should be forwarded to the Committee when available.

2. BIOLOGICAL DATA

While consideration of the 90-day study of tolerance in diabetic subjects was on the agenda of the present meeting, it was postponed pending receipt of the final report.

3. DIETARY INTAKE

According to the Codex draft General Standard for Food Additives (GSFA), alitame is permitted for use at a maximum level of 40–300 mg/kg in a wide range of foods and beverages, including bakery wares, water-based flavoured drinks, dairy-based drinks, dairy-based desserts, cream, edible ices, jams, confectionery and some dietetic foods. It is permitted for use within good manufacturing practice in table-top sweeteners and in some sugars (brown) and syrups (maple). The permitted uses in the Joint Food Standards Code of Australia and New Zealand, which was provided to the Committee, largely overlap with those in the Codex draft GSFA but also include use in flour products (including noodles and pasta), some processed fruit and vegetables, custard, jelly, sauces and toppings. In contrast, the Joint Food Standards Code allows its use only in specific bakery wares, not including bread (Australia New Zealand Food Authority, 2002).

3.1 Screening by the budget method

The budget method was used to estimate the theoretical maximum level of use as a food additive in the proportion of food and beverages likely to contain alitame that would not result in exceedance of the ADI by the population (Table 1). The proportion of solid food and beverages that might contain alitame was set at 50%, since, according to the GSFA, it is allowed for use in food categories 7.0 (bakery wares which include a staple food, bread) and 14.1.4 (all carbonated and non-carbonated drinks.

The theoretical maximum level of use of alitame is less than the maximum level permitted in solid foods and beverages by both the GSFA and the Australia New Zealand Food Authority. A detailed assessment of use of alitame in both solid and liquid foods was therefore required.

3.2 Assessment of intake

National assessments of intake based on a dietary survey at the individual level were provided by Australia and New Zealand (Hambridge, 2002a,b). National projections of intake of alitame in the USA, prepared for the Food and Drug

Table 1. Theoretical maximum level of alitame estimated by the budget method

Category	Per cent containing alitame	Theoretical maximum level (mg/kg)	Highest maximum permitted level (mg/kg)	
			Australia New Zealand[a]	GSFA[b]
Solid foods	50	40	60	100
Beverages	50	40	300	300

JECFA ADI for alitame, 0–1 mg/kg bw

[a] Highest maximum permitted level for solid foods and beverages according to the Joint Food Standards Code of Australia and New Zealand, excluding table-top sweeteners, in which alitame is permitted within good manufacturing practice

[b] Highest maximum permitted level for solid foods and beverages according to the Codex draft GSFA (2002), excluding table-top sweeteners and 'other sugars and syrups', in which alitame is permitted within good manufacturing practice

Administration in 1986 (MRCA Information Services, 1986), were also available. The main results of these assessments are presented in Table 2.

3.3 Evaluation of estimates of intake

Estimates based on individual consumption and levels of use in the GSFA and the Joint Food Standards Code showed that the ADI is slightly exceeded by consumers at high percentiles in both Australia and New Zealand. It should be noted, however, that high levels of consumption measured in short surveys, such as 24-h recall, overestimate long-term intake. Intake of alitame from 'other sugars and syrups' (category 11.2) was not assessed, as there was no maximum permitted level of use in this food category and no data were available on the levels of use in industry. Nevertheless, this food category is unlikely to contribute much to overall intake, in view of the low percentages of consumers (6% and 8%) and their low level of intake (9 g/day and 7 g/day) observed in the surveys in Australia and New Zealand, respectively. On the basis of permitted levels in Australia and New Zealand, the main contributors to the mean estimated intake of alitame were biscuits, cakes and pastries, flour products and water-based flavoured drinks in Australia and biscuits, cakes and pastries, sauces, toppings, mayonnaises and salad dressings and flour products including noodles and pasta in New Zealand. The estimated intake of alitame from the categories biscuits, cakes and pastries and flour products were widely overestimated in both countries, as alitame was assumed to be present in all items (no data were available on the presence of artificial sweeteners). The same was true for sauces and toppings in New Zealand.

Assuming Codex draft GSFA permissions, the category breads and bakery products accounted for 89% of the overall intake of alitame from all food categories in Australia and 88% in New Zealand.

The projections of intake in the USA were based on frequency of intake over 14 days, giving a better indication of long-term intake and thus a smaller overestimate of high intake. Nevertheless, the projected use levels in 1986 may not be reliable for adequately assessing current intake. The sponsors did not provide current manufacturers' levels of use in the USA to the Committee.

4. COMMENTS

While consideration of the 90-day study of tolerance in diabetic subjects was on the agenda of the present meeting, it was postponed pending receipt of a final report.

No data on the dietary intake of alitame were available at the previous meetings of the Committee. At its present meeting, the Committee compared the maximum levels of alitame listed in the Codex draft General Standard on Food Additives (GSFA) with the theoretical maximum level calculated by the budget method (see Annex 1, reference *137*). On the assumption that alitame is used in all foods, the theoretical maximum level of alitame, based on the current ADI, was calculated to be 40 mg/kg. The maximum levels in the GSFA are up to 300 mg/kg in a wide range of foods and beverages, with no limit in 'other sugars and syrups' or in table-top sweeteners. Detailed assessments of the intake of alitame when used in foods were therefore required.

Table 2. Intake of alitame estimated from individual surveys

Country	Survey (year)	Assumptions	Intake (mg/kg bw per day) by consumers only[a]		% ADI[b]
Australia	National, 24-h recall; ≥ 2 years; sample size, 13 858; mean body weight, 67 kg (1995)	(i) Maximum permitted level (Aus NZ or GSFA) or Australian industry use level (if GMP)[c] or 0 if GMP and no use level available[d] (ii) Assume alitame is always present if permitted in food group, unless information on presence is available. (iii) Take into account presence of sweeteners in both foods and ingredient. (iv) Modified GSFA classification system (v) Adjusted for individual body weight	Mean Aus NZ Median Aus NZ 95th %ile Aus NZ Mean GSFA Median GSFA 95th %ile GSFA	0.4 0.2 1.1 0.59 0.46 1.5	40 20 110 59 46 150
New Zealand	National, 24-h recall; ≥ 15 years; sample size; 4636; mean body weight, 71 kg (1997)	(i) Maximum permitted level (Aus NZ or GSFA) or Australian industry use level, similar to NZ (if GMP)[c] or 0 if GMP and no use level available[d] (ii) Assume alitame is always present if permitted in food goup, unless information on presence is available. (iii) Take into account presence of sweeteners in both foods and ingredient. (iv) Modified GSFA classification system (v) Adjusted for individual body weight	Mean Aus NZ Median Aus NZ 95th %ile Aus NZ Mean GSFA Median GSFA 95th %ile GSFA	0.45 0.29 1.4 0.59 0.50 1.4	45 29 140 59 50 140

Table 2 (contd)

Country	Survey (year)	Assumptions	Intake (mg/kg bw per day) by consumers only[a]		% ADI[b]
USA	National, menu census, 14-day average intake (1984–86); sample size, 5435 and USDA portion sizes (1977–78); all ages; mean body weight not reported	(i) Assume presence at projected industry use level in all foods in which alitame is intended for use[e] (ii) MRCA food categories (iii) Adjusted for individual body weight	Mean 90th %ile	0.35 0.79	35 79

Aus, Australia; NZ, New Zealand; GSFA, Codex draft general Standard on Food Additives; GMP, good manufacturing practice; MRCA, Manufacturers ??

[a] Persons who consume a product that, according to assumptions, might contain the additive. With Aus NZ assumtion, 10 624 persons in the Australian survey and 4084 in the New Zealand survey; with GSFA assumptions, 13 428 persons in the Australian survey and 4559 in the New Zealand survey; with projected USA industry use level, 5310 persons

[b] JECFA ADI, 0–1 mg/kg bw

[c] Alitame is allowed in table-top sweeteners within GMP; the manufacturers' use level is 500 mg/kg.

[d] Alitame is allowed within GMP in sugars and syrups according to the GSFA, but no use level was available from industry.

[e] Alitame was intended for use in beverages, sugar substitutes, cold desserts, candy, baked goods, jams, jellies, sweet spreads and preserves.

National assessments of intake based on recent individual dietary records and GSFA maximum levels were provided by Australia and by New Zealand. These data showed that the ADI might be exceeded by some consumers, at levels representing 140% of the ADI in Australia and 140% of the ADI in New Zealand for consumption at the 95th percentile. These are likely to be overestimates, as the surveys were of short duration (24 h) and because it was assumed that alitame was systematically present in some major food categories. The intake of alitame from 'other sugars and syrups' was not assessed since no manufacturer's use levels were available. In both Australia and New Zealand, the main contributor to the mean estimated intake was sweet bakery wares, which was responsible for 89% of the overall estimated intake from all food categories in Australia and 88% in New Zealand. These estimates are based on data from only two countries, and further work is required to refine the intake estimates with recent data from other countries.

The estimates of intake were based on data from only three countries, and they should be further refined with intake estimates from other countries. The levels of use of alitame in 'other sugars and syrups' should be provided.

The Committee identified GSFA category 7 (bakery wares, including bread) as a food group that might contribute to a high intake of alitame, since this group includes foods that are staples in many countries. It therefore suggested that the Codex Committee on Food Additives and Contaminants consider reviewing standards for use of alitame.

5. EVALUATION

The ADI of 0–1 mg/kg bw allocated by the Committee at its forty-sixth meeting was retained.

6. REFERENCES

Australia New Zealand Food Authority (2002) Australia New Zealand Food Standards Code, Standard 1.3.1 Food Additives, Canberra (available at http://www.anzfa.gov.au)

Hambridge, T. (2002a) Information sheet for Australia on alitame. Canberra, Australia New Zealand Food Authority. Submitted to WHO by the Australia New Zealand Food Authority.

Hambridge, T. (2002b) Information sheet for New Zealand on alitame. Canberra, Australia New Zealand Food Authority. Submitted to WHO by the Australia New Zealand Food Authority.

MRCA Information Services (1986) Frequency distributions of the 14-day average daily intake and the sources of mean intake, of CP-54,802 in milligrams per kilogram of body weight for grand total and by 10 food subgroups for the period from July 1984 through June 1985. (Confidential) report to Pfizer, Inc. Submitted to WHO by Danisco USA, Inc., Ardsley, New York.

CROSS-LINKED SODIUM CARBOXYMETHYL CELLULOSE

First draft prepared by
Dr D.J. Benford
Food Standards Agency, Aviation House, London, United Kingdom

1. EXPLANATION

Cross-linked sodium carboxymethyl cellulose is manufactured by acidifying an aqueous suspension of sodium carboxymethyl cellulose and heating the suspension to achieve cross-linking. The product is then washed and dried. It is also produced during the manufacture of sodium carboxymethyl cellulose by lowering the pH and heating to cause cross-linking. Cross-linked sodium carboxymethyl cellulose is used in tablets of table-top sweeteners and dietary food supplements, as it facilitates disintegration in aqueous solutions, with a maximum level of use of 30 g/kg. It is also widely used as an excipient in pharmaceutical applications.

Cross-linked sodium carboxymethyl cellulose has not been evaluated previously by the Committee. The parent compound, sodium carboxymethyl cellulose, was considered by the Committee in its evaluations of modified celluloses at its fifth, seventh, tenth, thirteenth, seventeenth, twenty-sixth, twenty-seventh, thirtieth and thirty-fifth meetings (Annex 1, references *5*, *7*, *13*, *19*, *32*, *59*, *62*, *73* and *88*) and of ethyl cellulose and ethylhydroxyethyl cellulose at the twenty-sixth and twenty-seventh meetings (Annex 1, references *59* and *62*). At its thirty-fifth meeting, the Committee reviewed new data on seven modified celluloses (Annex 1, reference *89*), including sodium carboxymethyl cellulose. The data consisted of the results of studies in rats on caecal enlargement, effects on caecal flora and developmental toxicity and studies of mutagenicity in vitro. New data on humans were also available, indicating that some individuals may experience laxative effects at a dose as low as 5 g per person per day. The Committee concluded that modified celluloses are of low toxicity and allocated a group ADI 'not specified'[1] to seven modified celluloses: methyl cellulose,

[1] ADI 'not specified' is used to refer to a food substance of very low toxicity which, on the basis of the available data (chemical, biochemical, toxicological and other) and the total dietary intake of the substance arising from its use at the levels necessary to achieve the desired effect and from its acceptable background levels in food, does not, in the opinion of the Committee, represent a hazard to health. For that reason, and for reasons stated in the individual evaluation, the establishment of an ADI expressed in numerical form is not deemed necessary. An additive that meets this criterion must be used within the bounds of good manufacturing practice, i.e. it should be technologically efficacious and should be used at the lowest level necessary to achieve this effect, it should not conceal food of inferior quality or adulturated food, and it should not create a nutritional imbalance.

ethyl cellulose, hydroxypropyl cellulose, hydroxypropylmethyl cellulose, sodium carboxymethyl cellulose, ethyl cellulose and ethylhydroxyethyl cellulose. The Committee pointed out that their laxative properties should be taken into account when they are used as food additives.

Enzymatically hydrolysed sodium carboxymethyl cellulose was evaluated by the Committee at its fifty-first meeting (Annex 1, reference *137*). The Committee compared the toxicity and absorption, distribution and metabolism of enzymatically hydrolysed carboxymethyl cellulose with that of the parent compound and concluded that the similarities in the results were consistent with the absence of any toxicologically significant difference between the two. Therefore, the enzymatically hydrolysed form was included in the group ADI 'not specified' with the other seven modified celluloses.

2. BIOLOGICAL DATA

2.1 Biochemical aspects

2.1.1 Absorption, distribution and excretion

No specific information was available. Studies reviewed previously by the Committee relating to sodium carboxymethyl cellulose and its enzymatically hydrolysed form indicated that 90–99% of both forms was excreted in faeces. The carboxymethyl cellulose was partially degraded during passage through the gastrointestinal tract (Annex 1, reference *137*).

Cross-linking of carboxymethyl cellulose results in it becoming insoluble in water, and it is therefore less likely to be absorbed and degraded than the parent compound.

2.2 Toxicological studies

2.2.1 Acute toxicity

Studies of acute toxicity with cross-linked sodium carboxymethyl cellulose have been conducted in rats and mice, apparently demonstrating low toxicity, but the reports were not available to the Committee.

2.2.2 Short-term studies of toxicity

In a study that complied with good laboratory practice (GLP) and was conducted to OECD guideline 408, groups of 20 male and female Sprague-Dawley CD rats were given diets containing 10 000 or 50 000 ppm of Ac-Di-Sol SD 711 or 50 000 ppm of 'modified cellulose gum' for about 90 days. Both compounds were defined in the report as 'a mixture of polymers of internally cross-linked sodium carboxymethyl cellulose', and Ac-Di-Sol was reported to be the brand name of 'modified cellulose gum'. However, this terminology is inappropriate, since 'modified cellulose gum' should more correctly be used as a synonym for the parent sodium carboxymethyl cellulose rather than the cross-linked form. As control animals received a diet consisting of unblended basal diet, the test diets had a lower caloric value. Clinical signs were recorded daily, and body weights and food consumption were recorded weekly. Ophthalmic examinations were performed before the beginning of the study and before terminal sacrifice. Blood samples were taken before termination for haematological and clinical chemical determinations. At necropsy, brain, liver, adrenals, testes and kidneys were weighed, and a wide range of organs were taken for histopathological examination. Urine was not analysed.

Three males (one receiving 50 000 ppm Ac-Di-Sol and two receiving 'modified cellulose gum') were killed because of mechanical injury to the snout. There were no treatment-related deaths or clinical signs. The average doses received by the animals during the study were 760, 3900 and 4000 mg/kg bw per day for males and 890, 4700 and 4600 mg/kg bw per day for females given 10 000 ppm Ac-Di-Sol, 50 000 ppm Ac-Di-Sol and 50 000 ppm 'modified cellulose gum', respectively. Males receiving 50 000 ppm Ac-Di-Sol showed no change in food consumption but had a statistically significant decrease in body-weight gain during the final 3 weeks of the study, resulting in a mean terminal body weight that was 93% that of control animals. Males receiving 'modified cellulose gum' had significantly increased food consumption throughout the study, but their body-weight gain was unaffected. Females in all treated groups showed increased food consumption and no effect on body-weight gain. The authors considered that these results were not toxicologically significant. No significant effects were reported in ophthalmicl examinations, at necropsy or in haematological parameters. The only significant changes in clinical chemistry were small decreases in aspartate aminotransferase activity in males and females receiving 50 000 ppm 'modified cellulose gum'. The authors considered that these alterations were not toxicologically significant. The absolute weights of the kidney and the weight of the kidney relative to that of brain were significantly increased by 7–8% in males receiving 50 000 ppm of 'modified cellulose gum'. The relative adrenal weights were significantly decreased by 10–12% in rats of each sex receiving 50 000 ppm of 'modified cellulose gum'. These changes were not accompanied by histopathological changes and were considered not to be treatment-related.

Renal mineralization at the corticomedullary junction was observed in females in all groups but not in males. This lesion was equated to nephrocalcinosis and was observed in 4/20 control females (three minimal, one mild), 7/20 at 10 000 ppm Ac-Di-Sol (six minimal, one mild), 7/20 at 50 000 ppm Ac-Di-Sol (five minimal, one mild, one moderate) and 9/20 at 50 000 ppm 'modified cellulose gum' (five minimal, three mild, one moderate). In addition, one female receiving 50 000 ppm 'modified cellulose gum' had mineralization in the urinary space of the kidney. No statistical analysis was conducted. The study pathologist considered that, although renal mineralization at the corticomedullary junction is a common spontaneous change in control female rats, it was a microscopic lesion related to treatment in female rats fed 50 000 ppm Ac-Di-Sol or 'modified cellulose gum'. No other treatment-related lesions and no caecal enlargement were observed. (Freeman, 1997a).

The Committee recalled the results of a 90-day toxicity study in rats in which the effects of enzymatically hydrolysed sodium carboxymethyl cellulose were compared with those of the parent compound, sodium carboxymethyl cellulose, which was discussed by the Committee at its fifty-first meeting (Annex 1, reference *137*). In this study, the test substances were added to the diet at a concentration of 2.5, 5 or 10%, and 10% wheat starch was added to the diet of control animals. The monograph and report of the fifty-first meeting focused on the histopathological observations that were statistically significantly different from those in control animals, i.e. pelvic urothelial hyperplasia and pelvic nephrocalcinosis in male animals. However, corticomedullary nephrocalcinosis was noted more frequently in females than in males in all groups of animals on test diets and was not found in control animals. The increased frequency was statistically significant in females receiving a diet containing 10% sodium carboxymethyl cellulose (Bär et al., 1995).

2.2.3 Reproductive toxicity

In a study of teratogenicity that complied with GLP and was conducted according to OECD guideline 414, groups of 25 female Sprague-Dawley CD rats were fed diets containing 10 000 or 50 000 ppm of Ac-Di-Sol SD 711 or 'modified cellulose gum' on days 6–15 of gestation. The animals were then fed untreated basal diet until the end of the study on day 20 of gestation. The report defined both compounds as 'a mixture of polymers of internally cross-linked sodium carboxymethyl cellulose'; however, this is ian nappropriate use of the term 'modified cellulose gum', which is more correctly used as a synonym for the parent sodium carboxymethyl cellulose, rather than the cross-linked form. Control animals received a diet consisting of unblended basal diet. Clinical signs were recorded daily, and maternal body weights and food consumption were recorded on days 0, 6, 15 and 20. At necropsy, the dams were examined for gross lesions, the number and distribution of implantation sites, early and late resorptions, live and dead fetuses and corpora lutea. Body weight, sex and external signs were recorded for all fetuses, and half were subjected to visceral examination and half to skeletal examination.

There were no deaths, although one animal receiving 10 000 ppm 'modified cellulose gum' was killed early after premature delivery on day 20. A small number of clinical signs were ascribed to mechanical injury and were considered not to be treatment-related. The average doses received by the animals during the study were 910, 4600, 910 and 4600 mg/kg bw per day for animals receiving 10 000 ppm Ac-Di-Sol, 50 000 ppm Ac-Di-Sol, 10 000 ppm 'modified cellulose gum' and 50 000 ppm 'modified cellulose gum', respectively. Food consumption, body-weight gain and gravid uterine weight did not differ between groups. All dams appeared normal at necropsy, and there were no differences between treated groups in the numbers of implants, live and dead fetuses or corpora lutea, litter size or resorptions. Dams receiving 50 000 ppm 'modified cellulose gum' had a statistically significant increase in preimplantation loss, which was considered not to be treatment-related because implantation occurred before administration of the test material.

The fetal body weights and sex ratios were comparable between test groups. External malformations were observed in one fetus of each of three dams receiving 10 000 ppm 'modified cellulose gum' but in no other group. These were considered not to be treatment-related because the finding was not statistically significant and they occurred only in fetuses of dams at the lowest dietary concentration. Visceral malformations were observed in one fetus of one dam receiving 10 000 ppm of Ac-Di-Sol and one fetus of one dam receiving 50 000 ppm of Ac-Di-Sol. This finding was considered not to be treatment-related as it was not statistically significant. There were no visceral malformations in fetuses of control dams or those receiving 'modified cellulose gum'. Sporadic skeletal variations, but not malformations, were reported in fetuses of control dams. One fetus of one dam receiving 50 000 ppm of Ac-Di-Sol had a non-ossified lumbar vertebra. This finding was not statistically significant and therefore considered not to be treatment-related. Skeletal malformations were reported in one fetus of each of two dams receiving 10 000 ppm of 'modified cellulose gum'. These findings were considered not to be treatment-related because they were isolated occurrences in the group at the lowest dose. A statistically significant increase in the incidence of a skeletal variation (incompletely ossified thoracic vertebra centra) was noted among fetuses of dams receiving 50 000 ppm. The fetal incidence was 27% and the litter incidence was 72%. This was considered not to be treatment-related because it occurred in the absence of any statistically or biologically significant effects on other parameters in the study, and because the incidence was within the range of that in other control groups in the laboratory (16–28% and 72–84% for fetal and litter incidences, respectively). There was no evidence of maternal or developmental toxicity at a limit dose of 50 000 ppm in the diet (Freeman, 1997b).

3. COMMENTS

At its present meeting, the Committee reviewed two studies on cross-linked sodium carboxymethyl cellulose and considered its properties in the light of previous evaluations of modified celluloses. Like other modified celluloses, cross-linked sodium carboxymethyl cellulose is considered to be of low toxicity. Because cross-linking renders it insoluble in water, it is less likely to be absorbed than the parent compound. In a 90-day dietary study in rats, the only treatment-related observations were increased food consumption, decreased body-weight gain and effects on the kidneys of animals receiving cross-linked sodium carboxymethyl cellulose at 50 000 mg/kg of diet, equivalent to 3900–4700 mg/kg bw per day averaged over the study period. The changes in food consumption or body-weight gain were consistent with a 5% reduction in the nutritional content of the diet, as compared with the basal diet fed to control animals, due to addition of the test substance. In contrast to the effects of some other non-digestible substances evaluated previously by the Committee, caecal enlargement was not observed. Mineralization at the corticomedullary junction was observed in the kidneys of some female rats, with a higher incidence in treated groups than in controls. The report equated the renal mineralization to nephrocalcinosis and noted that, although this is a spontaneous change in control female rats, it was related to treatment in female rats fed cross-linked sodium carboxymethyl cellulose at 50 000 mg/kg of diet, but not at 10 000 mg/kg of diet. Pelvic nephrocalcinosis and corticomedullary nephrocalcinosis were observed in rats given sodium carboxymethyl cellulose or enzymatically hydrolysed carboxymethyl cellulose in a study evaluated by the Committee at its fifty-first meeting (Annex 1, reference *137*). In that study, it was also reported that corticomedullary nephrocalcinosis was observed more frequently in females than in males. Because of the inconsistency of the dose–response relationship, however, the Committee at its fifty-first meeting had not been convinced that these changes were of toxicological concern. The changes observed in the study evaluated at the current meeting appeared to represent a minimal increase in the incidence and severity of a spontaneous lesion, which may be within normal variation. As discussed in the report of the twenty-sixth meeting (Annex 1, reference *59*), nephrocalcinosis is a common finding in laboratory rodents. No implications for human hazard have been established. Therefore, the NOEL for cross-linked sodium carboxymethyl cellulose in the above study was 3900 mg/kg bw per day.

The study of developmental toxicity in rats showed no evidence of maternal toxicity or teratogenicity at the highest possible dose of cross-linked sodium carboxymethyl cellulose of 50 000 mg/kg of diet. The NOEL was 4500 mg/kg bw per day.

4. EVALUATION

The Committee noted that evaluation of the results of the 90-day toxicity study was not straightforward, because of nutritional differences between the basal diet given to control animals and the diets supplemented with cross-linked sodium carboxymethyl cellulose. However, taking into account evaluations of other modified celluloses, the Committee concluded that the effects of cross-linked sodium carboxymethyl cellulose were consistent with dietary overloading with a poorly absorbed substance of low toxicity. The Committee included cross-linked sodium carboxymethyl cellulose in the group ADI 'not specified'[2] with the other modified celluloses.

5. REFERENCES

Bär, A., Til, H.P. & Timonen, M. (1995) Subchronic oral toxicity study with regular and enzymatically depolymerized sodium carboxymethylcellulose in rats. *Food Chem. Toxicol.*, **33**, 909–917.

Freeman, C. (1997a) Ac-Di-Sol SD-711 and modified cellulose gum ninety-day feeding study in rats. Unpublished report No. I95-2054 from FMC Corporation Toxicology Laboratory, Princeton, New Jersey, USA. Submitted to WHO by the Organisation des Fabricants de Produits Cellulosiques Alimentaires, Leidschendam, Netherlands.

Freeman, C. (1997b) Ac-Di-Sol SD-711 and modified cellulose gum dietary teratology study in rats. Unpublished report No. I96-2090 from FMC Corporation Toxicology Laboratory, Princeton, New Jersey, USA. Submitted to WHO by the Organisation des Fabricants de produits Cellulosiques Alimentaires, Leidschendam, Netherlands.

---

[2] ADI 'not specified' is used to refer to a food substance of very low toxicity which, on the basis of the available data (chemical, biochemical, toxicological and other) and the total dietary intake of the substance arising from its use at the levels necessary to achieve the desired effect and from its acceptable background levels in food, does not, in the opinion of the Committee, represent a hazard to health. For that reason, and for reasons stated in the individual evaluation, the establishment of an ADI expressed in numerical form is not deemed necessary. An additive that meets this criterion must be used within the bounds of good manufacturing practice, i.e. it should be technologically efficacious and should be used at the lowest level necessary to achieve this effect, it should not conceal food of inferior quality or adulturated food, and it should not create a nutritional imbalance.

MINERAL OILS (MEDIUM- AND LOW-VISCOSITY) AND PARAFFIN WAXES

First draft prepared by
E. Vavasour[1] and Junshi Chen[2]
[1]Health Canada, Ottawa, Canada; and
[2]Chinese Centers for Disease Control and Prevention, Beijing, China

1. EXPLANATION

Mineral oils (medium- and low-viscosity) comprise a subgroup of the family of mineral oils and waxes. They are manufactured from crude mineral oils in various refining steps, such as distillation, extraction and crystallization, and are subsequently purified by acid treatment (oleum method) and/or hydrotreatment (catalytic hydrogenation). Mineral oils (medium- and low-viscosity) are mixtures of highly refined paraffinic and naphthenic liquid hydrocarbons with boiling-points greater than 200 °C.

In order to define the materials more clearly, the Committee at its forty-fourth meeting (Annex 1, reference *116*) prepared individual specifications for micro-crystalline wax and for two groups of mineral oils: mineral oils (high-viscosity) and mineral oils (medium- and low-viscosity). The latter group was furthermore divided into three classes. For the purpose of characterization of the different types of oils and waxes, the criteria viscosity, average relative molecular mass and carbon number at 5% distillation-point were used in the specifications (see Table 1).

Food-grade mineral oils were last evaluated by the Committee at its forty-fourth meeting (Annex 1, reference *116*), when four 90-day studies conducted in Fischer 344 rats with a range of mineral oils and waxes representative of materials in commercial use were reviewed. The materials tested included low-, medium- and high-viscosity mineral oils (N10(A), N15(H), P15(H), N70(A), N70(H), P70(H) and P100(H)); paraffin waxes (low-melting-point wax and intermediate-melting-point wax); and microcrystalline waxes (high-melting-point wax and high-sulfur wax). All the

Table 1. Classification of mineral hydrocarbons

Name	ADI (mg/kg bw)[a]	Viscosity at 100 °C (mm^2/s)	Average relative molecular mass	Carbon number at 5% distillation-point
Microcrystalline wax	0–20[a,b]	≥ 11	≥ 500	≥ 25
High-melting-point wax		Not included in present studies		
Low-melting-point wax	Withdrawn[c]	No JECFA specification		
Low-melting-point wax		3.3	380	22
Mineral oil (high viscosity)	0–20[a]	> 11	> 500	≥ 28
P100		11	520	29
Mineral oil (medium and low viscosity) class I	0–10[d]	8.5–11	480–500	≥ 25
P70		9.0	480	27
Medium-viscosity liquid petroleum		8.7	480	25
P70(H)		8.6	480	27
Mineral oil (medium and low viscosity) class II	0–0.01[e]	7.0–8.5	400–480	≥ 22
N70(H)		7.7	420	23
Mineral oil (medium and low viscosity) class III	0–0.01[e]	3.0–7.0	300–400	≥ 17
P15(H)		3.5	350	17
N15(H)		3.5	330	17

P100 oil, crude: paraffinic, viscosity (40 °C): 100 mm^2/s
P70 oil, crude: paraffinic, viscosity (40 °C): 70 mm^2/s
P70(H) oil, crude: paraffinic, viscosity (40 °C): 70 mm^2/s, hydrotreated (catalytic hydrogenation)
N70(H) oil, crude: naphthenic, viscosity (40 °C): 70 mm^2/s, hydrotreated (catalytic hydrogenation
P15(H) oil, crude: paraffinic, viscosity (40 °C): 15 mm^2/s, hydrotreated (catalytic hydrogenation
N15(H) oil, crude: naphthenic, viscosity (40 °C): 15 mm^2/s, hydrotreated (catalytic hydrogenation)

[a] Established by the Committee at its forty-fourth meeting
[b] Group ADI
[c] Previous ADI 'not specified'
[d] Established at the present meeting
[e] Temporary group ADI, extended at the present meeting

substances, with the exception of the microcrystalline waxes, appeared to accumulate in the tissues of the animals to varying degrees, depending on the material and dose. Except for P70(H) and P100(H) oils, there was evidence of accumulation of the mineral hydrocarbons and effects indicative of a reaction to a foreign body at one or more doses. The types of effects seen were similar and included focal histiocytosis, increased weights of liver, lymph nodes, spleen and kidneys, granulomas or microgranulomas of the liver, haematological changes typical of a mild, chronic inflammatory reaction and biochemical changes indicative of mild hepatic damage.

The ADI of 0–20 mg/kg bw for mineral oils with the specifications of high-viscosity oils and of high-melting-point and high-sulfur waxes was based on NOELs at the highest dose tested (2% in the diet) in 90-day studies in Fischer 344 rats. The NOELs for all the other materials except P70(H) oil (i.e. class II and III medium- and low-viscosity oils and low-melting-point wax) were based on an increased incidence of histiocytosis in the lymph nodes at the next highest dose. For P70(H) oil, the NOEL was based on an increased incidence of pigmented macrophages in male rats at a dietary concentration of 2%, a minor effect considered of doubtful biological significance. Accordingly, because effects were observed at all doses, the ADIs for low-melting-point and intermediate-melting-point waxes were withdrawn. A group temporary ADI of 0–0.01mg/kg bw was allocated for class II and III medium- and low-viscosity mineral oils, the temporary nature of the ADI being due to uncertainty about the long-term significance of the inflammatory response to accumulated dietary mineral hydrocarbons. A temporary ADI of 0–1 mg/kg bw was allocated for P70(H) oil. These temporary ADIs were extended at the fifty-first meeting (Annex 1, reference *137*).

At its forty-fourth meeting, the Committee considered that, although the types of effects seen were essentially reactions to a foreign body, it was possible that a prolonged inflammatory response of the type observed could result in functional changes in the immune system and that this aspect required further investigation. It also noted that the oils and waxes for which high NOELs were observed contained a greater proportion of hydrocarbon components of high relative molecular mass (high carbon number) and had higher viscosities than those materials with a low NOEL, which contained a greater proportion of hydrocarbon components of lower relative molecular mass (low carbon number). The Committee required information about the compositional factors in mineral oils that influence their absorption and toxicity. It also required a study of at least 1 year's duration on one of these materials in Fischer 344 rats, including an assessment of immune function at appropriate times and an investigation of the kinetics of accumulation of the material, to determine whether a plateau is reached. It recommended that a reversal period of 1 year be included, in order to determine whether the granulomatous hepatic lesions observed in rats in the 90-day studies were fully reversible.

At its present meeting, the Committee reviewed a number of new studies, including the results of a combined 2-year study of toxicity and carcinogenicity and a 1-year study of toxicity with a 1-year recovery period with P70(H) and P100(H) oils (with the specifications of class-I medium- and low-viscosity mineral oils and of high-viscosity mineral oil, respectively) conducted in parallel. In addition, the Committee received and reviewed a number of studies conducted with low- and medium-viscosity mineral oils, including: a 2-year study of the carcinogenicity of a medium-viscosity liquid petroleum (class I medium- and low-viscosity mineral oil); studies of pharmacokinetics and studies of humoral immune function after administration to P15(H) mineral oil (class III medium- and low-viscosity mineral oil) in Fischer 344 and Sprague-Dawley rats; and a 90-day study of histopathological responses and compositional analysis of absorbed hydrocarbons with N15(H), N70(H) and P70(H) oils (classes III, II and I medium- and low-viscosity mineral oils, respectively). Several studies of low-melting-point paraffin wax were also reviewed by the Committee, as they were considered to provide information relevant for the evaluation of low- and medium-viscosity mineral oils relating to the difference in response to mineral hydrocarbons in Fischer 344 and Sprague-Dawley rats.

2. BIOLOGICAL DATA

2.1 Biochemical aspects

2.1.1 Absorption, distribution and excretion

Differences in the disposition and pharmacokinetics of a low-viscosity mineral oil were assessed in rats of two strains given a representative mineral hydrocarbon component. For the study of disposition, fasted female Fischer 344 and Sprague-Dawley rats received a single dose by gavage of 2 ml/kg bw of a 4:1 mixture (v/v) of olive oil and a food-grade paraffinic white mineral oil meeting the specifications for medium- and low-viscosity mineral oil class III, containing [1-^{14}C]-1-eicosanylcyclohexane and a non-absorbable marker, [1,2-^{3}H]polyethylene glycol 4000. The dose of mineral oil was 340 mg/kg bw. The rats were placed in metabolism cages for collection of urine, faeces and expired air at regular intervals over 96 h, at which time they were killed, and liver and mesenteric lymph nodes were collected. For the study of pharmacokinetics, the rats received the same single dose of the olive oil:mineral oil mixture containing [1-^{14}C]-1-eicosanylcyclohexane by gavage, and serial blood samples were collected from in-dwelling jugular vein catheters at regular intervals for up to 96 h, in addition to the collection of urine and faeces. Three rats of each strain were killed at each time, and blood, liver and mesenteric lymph nodes and the contents of the bladder and intestines were analysed for total radioactivity.

A comparison of the values for the area under the curve of the plasma concentration *versus* time (AUC), an indicator of overall systemic exposure, indicated that the bioavailability of total ^{14}C label was greater in Fischer 344 rats than in Sprague-Dawley rats. The maximum blood concentration attained (C_{max}), and the time to reach C_{max} were also greater in Fischer 344 rats. Faecal excretion of ^{14}C-labelled material was the major route of elimination for both strains of rat, although there were strain differences in the rate of faecal excretion. The majority of ^{14}C label (70%) had been eliminated by 16 h in the Sprague-Dawley rats, which corresponded to the gastrointestinal transit time in both strains. In contrast, only 11% of the ^{14}C label had been eliminated by 16 h in the Fischer 344 rats, and the majority (75%) was eliminated within 48 h. The authors speculated that the longer delay in faecal excretion of labelled hydrocarbon in Fischer 344 than in Sprague-Dawley rats was the result of biliary excretion of absorbed ^{14}C label or sloughing of gastrointestinal endothelial cells containing the label. Urinary excretion was the second route of elimination of ^{14}C-containing products and reached near-maximal values (10%) by

24 h in Sprague-Dawley rats. By contrast, urinary excretion of radioactivity by Fischer 344 rats was linear over time during the 96-h monitoring period, attaining 7% of the administered ^{14}C label. The authors indicated that this was characteristic of a depot effect in which the rate of elimination is determined by the export of ^{14}C from a peripheral compartment before urinary excretion, in this case, probably the liver. In the livers of Sprague-Dawley rats, the maximum amount of radioactivity (2%) occurred at 8 h and decreased thereafter to reach 0.1% by 96 h. By contrast, the maximum amount of radioactivity in the livers of Fischer 344 rats (4%) was not reached until 24 h, and 3% of the administered radioactivity was still present at 96 h. The authors concluded that Sprague-Dawley rats can efficiently metabolize and excrete ^{14}C-containing compounds, whereas the inability of Fischer 344 rats to clear these compounds from the liver probably resulted in deposition and possible retention. The amount of radioactivity in the mesenteric lymph nodes was similar in Fischer 344 and Sprague-Dawley rats until the 96-h sampling time, when the percentage of the administered radioactivity increased by nearly 10-fold, from 0.002% to 0.02%, in Sprague-Dawley rats (Sipes & Halladay, undated).

The above study was repeated with a 10-fold lower dose of the olive oil:mineral oil mixture. Fasted female Fischer 344 and Sprague-Dawley rats received 0.2 ml/kg bw of the oil mixture, to deliver a single oral dose of 34 mg/kg bw of the low-viscosity oil (mineral oil, medium and low viscosity, class III), by gavage. As in the previous study, [1-^{14}C]-1-eicosanylcyclohexane (0.34 mg/kg bw) was used as a tracer. The animals were placed in metabolism cages for the collection of urine and faeces at regular intervals up to sacrifice at 96 h. Blood samples were collected from an in-dwelling jugular vein catheter, also at regular intervals. Liver and mesenteric lymph nodes were collected and analysed for radioactivity at sacrifice. In contrast to the results with the high dose, the radioactivity in blood of Fischer 344 rats showed a biphasic profile, with two C_{max} values at 4 and 16 h (one at 18 h with the high dose). The C_{max} in Sprague-Dawley rats was at 5–5.6 h in both studies. As in the previous study, the AUC was greater in Fischer 344 than in Sprague-Dawley rats, indicating greater bioavailability. Faecal excretion was the major route of excretion, with 76% (Fischer 344) and 72% (Sprague-Dawley) of the administered radioactivity eliminated by 96 h. The delay in faecal elimination of radioactivity seen in Fischer 344 rats was less apparent with the low dose. The differences in patterns of urinary excretion and liver retention of radioactivity by Fischer 344 and Sprague-Dawley rats were similar to those observed with the high dose. As in the study with the high dose, more radioactivity (approximately 2.5% of the administered dose) was retained in the liver of Fischer 344 than Sprague-Dawley rats (0.5%), and the radioactivity retained in lymph nodes in Fischer 344 rats exceeded that in Sprague-Dawley rats over the 96-h observation period by twofold (approximately 0.02% versus 0.01%) (Sipes & Halladay, 2001a).

The study of disposition of low-viscosity mineral oil, described above, was repeated with the addition of a 2-week dietary pretreatment before the single gavage dose of radiolabelled tracer. Female Fischer 344 and Sprague-Dawley rats were fed standard diet containing 2% white oil (mineral oil, medium and low viscosity, class III) or untreated diet for 2 weeks, during which time food consumption and body weight were monitored. At the end of the 2-week period, the treated animals received a single oral dose of the white oil (340 mg/kg bw) in a 1:4 ratio with olive oil

(2 ml/kg) containing [1-^{14}C]-1-eicosanylcyclohexane (2.1 mg/kg bw). The animals were placed in metabolism cages for the collection of urine and faeces at regular intervals up to 96 h. At sacrifice, liver and mesenteric lymph nodes were collected for determination of total radioactivity.

Consumption of mineral oil in the diet during the pretreatment period did not affect weight gain or food consumption, although Sprague-Dawley rats had a higher rate of food consumption and greater weight gain than Fischer 344 rats. Faecal excretion was the major route of excretion, with 84% (Fischer 344) and 94% (Sprague-Dawley) of the administered radioactivity eliminated by 96 h, compared with 92% and 88%, respectively, in un-pretreated control animals and 82% and 87% in the study with a high single dose by gavage. The delay in faecal elimination of radioactivity by Fischer 344 rats in the study with the single high dose was shorter. The differences in patterns of urinary excretion of radioactivity between Fischer 344 and Sprague-Dawley rats were similar to those observed with the single high dose (nearly complete by 16 h in Sprague-Dawley rats, linear over the 96-h observation period in Fischer 344 rats). However, more ^{14}C label was excreted in the urine of Fischer 344 rats after dietary pretreatment with mineral oil than in un-pretreated controls (19% versus 11%), and slightly less label was excreted in the urine of Sprague-Dawley rats (8% versus 11%) than in un-pretreated controls.

As the data on faecal excretion by rats that did not receive pretreatment in this study and that with the single high dose are inconsistent, no conclusion could be reached about whether the increased urinary excretion of radioactive tracer in pretreated Fischer 344 rats indicated increased absorption of mineral hydrocarbon or reduced retention. The proportion of administered radioactivity recovered in the liver was not altered by mineral oil pretreatment and, as observed previously, was greater in Fischer 344 than in Sprague-Dawley rats (about 2% versus 0.1%). In the mesenteric lymph nodes, Fischer 344 rats again retained more radioactivity than Sprague-Dawley rats (0.03% versus 0.003%). A threefold increase in retention of ^{14}C label in mesenteric lymph nodes was seen in pretreated Fischer 344 rats as compared with un-pretreated controls after 96 h, although no difference in retention was seen from that of rats given a single dose of radioactive tracer by gavage in the previous study. Inconsistencies between the results obtained for Fischer 344 rats in the control group in this study and those receiving a single high dose by gavage in the previous study make it difficult to draw any conclusion about the possible effects of mineral oil pretreatment on the disposition of ^{14}C-labelled tracer in the mesenteric lymph nodes. These results confirm the observation in the previous studies, that Fischer 344 and Sprague-Dawley rats differ in terms of the amount of mineral hydrocarbon retained in liver and mesenteric lymph nodes (Sipes & Halladay, 2001b).

A qualitative investigation was conducted to determine whether low-melting-point paraffin wax accumulated in the liver and Kupffer cells of female Fischer 344 and Sprague-Dawley rats fed 0 or 2% in the diet for 14, 30 or 60 days. Extracts of livers from Fischer 344 rats fed 2% in the diet for 60 days and of Kupffer cells from Fischer 344 rats treated for 14 and 30 days contained detectable amounts of low-melting-point paraffin wax. In contrast, Sprague-Dawley rats did not accumulate detectable amounts in liver or Kupffer cells, as determined by gas chromatography and mass spectrometry (Sipes & Hoglen, 1998).

2.1.2 Biotransformation

The comparative efficiency of *n*-hydrocarbon metabolism in vitro was investigated in liver or small intestinal microsomes prepared from Wistar, Sprague-Dawley and Fischer 344 rats. Two linear saturated hydrocarbons, *n*-octadecane (C18) and *n*-heptadecane (C17), were used as representative substrates. For the assay, 2 mg of microsomal protein were incubated for 2 h with 84 μmol/l [^{14}C]heptadecane or 17, 54 or 84 μmol/l [^{14}C]octadecane. The extracts from the incubations were analysed for oxidative metabolites by high-performance liquid chromatography. No hydrocarbon metabolism was detected in the small intestinal microsomal preparations with 84 μmol/l heptadecane or 54 μmol/l octadecane. No significant hydroxylation of octadecane was seen after incubation of octadecane with liver microsomes at any of the three concentrations. Incubation of liver microsomes with 84 μmol/l heptadecane resulted in the production of three metabolites: the corresponding alcohol and acid, heptadecanol and heptadecanoic acid. There was no apparent sex-related difference in the rate of metabolism. The overall rate of hydroxylation of heptadecane was calculated to be 3.6%, 1.4% and 0.75% for Wistar, Sprague-Dawley and Fischer 344 rats, respectively (Institut National de la Recherche Agronomique, 2000).

2.2 Toxicological studies

2.2.2 Short-term studies of toxicity

The effects of low-melting point paraffin wax were compared in a 90-day study in female Fischer 344 and Sprague-Dawley rats. This mineral hydrocarbon was selected since it was reported to have more severe effects in Fischer 344 rats than the other oils and waxes examined. The study was conducted in compliance with US Food and Drug Administration Good Laboratory Practice Regulations and the OECD Principles of Good Laboratory Practice.

The study was divided into a component for determination of toxicity and one for analysis of mineral hydrocarbon concentrations. For the study of toxicity, groups of 10 female rats of each strain were fed a standard diet containing 0.0, 0.2 or 2.0% low-melting-point paraffin wax (100% pure) (equal to 0, 160 and 1600 mg/kg bw per day) for 90 days or 0.0 or 2.0% for 30 or 60 days. For the analysis of mineral hydrocarbon, groups of five female rats of each strain were fed diets containing 0.0, 0.2 or 2.0% low-melting-point paraffin wax for 90 days or 2.0% for 30 or 60 days. The food consumption and body weights of all groups were recorded weekly. The animals were killed at end of the respective feeding periods. For the study of toxicity, blood was collected from all animals before sacrifice for haematological and clinical chemical evaluations. All animals were subjected to a macroscopic post-mortem examination, and the weights of the heart, kidneys, liver, mesenteric lymph nodes, spleen and ovaries were determined. Histological examinations were performed on the heart (with mitral valve), kidneys, liver (caudal and medial lobes), cervical and mesenteric lymph nodes, ovaries and spleen from control and treated animals killed after 90 days, and on the heart (with mitral valve), liver (caudal and medial lobes) and mesenteric lymph nodes from controls and animals at the high dose killed at 30 and 60 days. In addition, sections of the duodenum, jejunum, ileum and caecum from three controls and three rats of each strain at the high dose killed at 90 days were examined histologically. For analysis of mineral hydrocarbon, heart, kidney, liver, mesenteric lymph node and spleen were collected from all animals at the 90-day sacrifice for measurement of total mineral hydrocarbon, and liver and mesenteric lymph nodes were also collected from animals at the high dose at the 30- and 60-day sacrifices. In addition, liver tissue from selected groups was collected and preserved for electron microscopic analysis, while other tissues were used to evaluate immunological and pathological end-points, as described in sections 2.2.5 and 2.2.6.

No deaths occurred during the study. Treatment did not adversely affect body-weight gain, food consumption or physical parameters in either strain. In the Fischer 344 rats, statistically significant, treatment-related effects were observed in: haematological parameters (decreased erythrocyte count, erythrocyte volume fraction, haemoglobin concentration and platelet count; increased neutrophil count; values for monocytes were not presented); clinical chemical parameters (increased serum activity of alanine and aspartate aminotransferases, alkaline phosphatase and γ-glutamyl transpeptidase; decreased serum albumin); organ weights (increased liver, mesenteric lymph node, spleen and ovary weights), macroscopic findings (discoloured liver, enlarged mesenteric lymph nodes); and histological changes in the liver (increased incidence and severity of vacuolated hepatocytes, time-related development of microgranulomas, lymphoid-cell infiltrates or aggregates and scattered foci of necrosis), mesenteric lymph nodes (increased incidence and severity of microgranulomas and reticuloendothelial-cell hyperplasia), cervical lymph node (development of microgranulomas in 3/9 animals at the high dose), spleen (increased severity of extramedullary haematopoiesis) and heart (time-related increase in incidence of lymphoid-cell infiltrates in the base of the mitral valve). In animals at the high dose, many of these effects were seen as early as 30 days into the study. With the exception of inflammation of the mitral valve (affecting 1/10, 2/10 and 5/10 rats in the control group and at the low and high doses, respectively) and microgranulomas in the cervical lymph node (affecting 0/10, 2/9 and 3/9 rats in the three groups, respectively), the histopathological changes affected all rats at the high dose (10/10). Slight reticuloendothelial cell hyperplasia was also observed in the Peyer patch of the jejunum of one of three Fischer 344 female rats at the high dose. The only treatment-related effects observed in the Sprague-Dawley animals were increased weight of the mesenteric lymph nodes accompanied by microscopic changes (microgranulomas and reticuloendothelial-cell hyperplasia).

The mineral hydrocarbon concentrations in tissues measured over the course of the study were also indicative of a clear strain difference. The Fischer 344 rats showed a significant, time- and dose-related accumulation of mineral hydrocarbon in the mesenteric lymph nodes and liver. No plateau was seen during the 90 days of the study. In Sprague-Dawley rats, slight accumulation of mineral hydrocarbon was observed only in the mesenteric lymph nodes and only after 90 days of treatment. There was no notable accumulation in the heart, kidney or spleen in either strain of rat, regardless of dose and treatment duration (Thornton et al., 2000).

A 28-day study and a 90-day study were conducted to confirm the findings of previous feeding studies with mineral oils and waxes and to extend these observations to include a synthetic wax and a wider range of tissue samples for histopathological examination and chemical analysis. Groups of 12 female Fischer 344 rats were fed diets for either 28 or 90 days containing 2% of low-melting-point wax, C80 synthetic wax or one of three white oils, N15H (mineral oil, medium and low viscosity, class

III), N70H (mineral oil, medium and low viscosity, class II) or P70H (mineral oil, medium and low viscosity, class I), providing doses of 2500, 2600, 2500, 2600 and 2600 mg/kg bw per day, respectively, for the groups treated for 28 days and 2100, 2100, 2000, 2000 and 2100 mg/kg bw per day, respectively, for those treated for 90 days. Respective control groups received standard diet containing < 0.01% mineral hydrocarbon by weight. Four animals in each group at each time were designated for chemical analysis of tissues, and the remaining eight animals per group were designated for histological examination. Body weight and food intake were determined twice a week throughout the study. On day 14 of the study and on the day before necropsy, urine and faeces were collected during 4 h from the animals in each group designated for chemical analysis; blood samples were collected from these animals at necropsy. At scheduled necropsy, the weights of the brain, heart, kidney, liver, mesenteric lymph nodes and spleen from all the animals were recorded. In addition, the kidney, liver, mesenteric and cervical lymph nodes, small intestine and spleen from all animals were evaluated histopathologically, and the mineral hydrocarbon content was measured in the kidney, liver, mesenteric and cervical lymph nodes and small intestine. Mineral hydrocarbon accumulation was assessed as part of the histopathological examination.

No deaths were reported. Consumption of 2% mineral hydrocarbon in the diet did not affect body-weight gain. Food consumption was statistically significantly increased in the groups receiving the white oils (N15H, N70H and P70H) throughout most of the treatment period of each study. While increased food consumption was noted with diets containing low-melting-point paraffin wax at various times during the first 10 weeks of the 13-week study, no difference from controls was observed during the 4-week study. Animals on the C80 wax diets had increased food intake during weeks 1–4 of both studies. Increases in the absolute and relative (to body weight) weights of liver, spleen and mesenteric lymph nodes were observed with some of the mineral hydrocarbons at both 4 and 13 weeks. The liver weights were statistically significantly increased in the group given N15H low-viscosity oil at 4 weeks and in those given N15H and low-melting-point paraffin wax at 13 weeks. The weights of the mesenteric lymph nodes were increased in the groups given low-melting-point paraffin wax, C80 wax and N15H oil at 4 and 13 weeks and in those given N70H, but not the P70H medium-viscosity oil, at 13 weeks. At 4 weeks, spleen weights were increased only with the N70H and P70H medium-viscosity oils, while at 13 weeks increases in spleen weights were observed only with the low-melting-point paraffin wax, C80 wax and N15H oil. The weights of the kidney, heart and brain were not affected by treatment. Significant histopathological alterations were found in the cervical and mesenteric lymph nodes, liver, heart and small intestine. The incidence and severity of the findings were more marked at 13 than at 4 weeks and were associated primarily with ingestion of low-melting-point paraffin wax. Histiocytosis was seen in the lymph nodes of rats in all groups except those receiving N70H medium-viscosity oil for 4 weeks or P70H medium-viscosity oil for either 4 or 13 weeks. The incidence, severity (progression to individual cell necrosis) and onset were greater in mesenteric than in cervical nodes and in proximal mesenteric than in distal mesenteric nodes. The combined incidence and severity scores indicated a rank order of low-melting-point paraffin wax > C80 wax, N15H oil > N70H oil > P70H oil. Other changes identified at 13 weeks in nearly all (6–8/8) rats receiving low-melting-point paraffin wax in the diet included: hepatic focal necrosis with inflammation, periportal vacuolation and granuloma; and focal inflammation of the mitral valve of the heart and macrophage accumulation, particularly in Peyer patches of the small intestine. Periportal vacuolation was also observed in the livers of 7/8 rats receiving dietary C80 wax; vacuolation of macrophages was found in the lamina propria of the small intestine; granuloma of lower severity was seen in the livers of 7/8 rats; and focal inflammation of the cardiac mitral valve was recorded in 3/8 rats receiving N15H low-viscosity oil. Calcification of the renal medulla was exacerbated at both 28 and 90 days by dietary treatment with the white oils and marginally by treatment with the waxes.

No hydrocarbons with carbon numbers between C16 and C56 were detected in urine samples. The hydrocarbon profiles of faecal extracts were very similar to those of the corresponding mineral hydrocarbon materials, but with small differences for some materials, suggesting selective uptake and/or metabolism. A distortion of the faecal hydrocarbon profile was observed in extracts from the faeces of rats fed the C80 wax (loss of hydrocarbons C47–C56) and N15H oil (loss of hydrocarbons below C22). A very slight change was observed in faecal extracts from rats fed low-melting-point paraffin wax (loss of hydrocarbons from both the high and low ends of the carbon range). The authors indicated that apparent loss of alkanes with high carbon numbers was due to poor solubility during extraction and analysis. No significant changes were observed in the hydrocarbon profile in faecal extracts from rats consuming the N70H or P70H medium-viscosity mineral oils. Mineral hydrocarbon was detected in samples of small intestine, heart and kidney from all treated rats and in the livers of rats given low-melting-point paraffin wax or C80 wax, the only treatments for which this tissue was analysed. In addition, mineral hydrocarbon was detected in mesenteric lymph nodes of rats from all treated groups except that given P70H oil, and in cervical lymph nodes and aortic blood of rats treated with low-melting-point paraffin wax or C80 wax. The extent of accumulation of mineral hydrocarbons in tissues ranged from 0.01% to nearly 3%; the tissues with the greatest accumulation of hydrocarbon were not those with the greatest likelihood of histopathological alterations. Qualitative analysis of the carbon number range of the accumulated mineral hydrocarbons in tissue samples from rats treated with low-melting-point paraffin wax or C80 wax suggested that material in the alkane range C20–C35 was preferentially retained, confirming observations from previous studies (Cunninghame et al., 2001).

2.2.3 Long-term studies of toxicity and carcinogenicity

Parallel studies were conducted to assess the long-term toxic and carcinogenic effects of P70(H) mineral oil (medium and low viscosity, class I) and P100(H) mineral oil (high viscosity) in male and female Fischer 344 rats after 2 years' administration in the diet. The study design also included an evaluation of the reversibility or persistence of the biological effects associated with 12 months' exposure, after a 12-month recovery period. The studies were conducted in compliance with US Food and Drug Administration Good Laboratory Practice Regulations and the OECD Principles of Good Laboratory Practice.

Five groups were used for each study: a control group and groups given concentrations in the diet corresponding to a dose of 60, 120, 240 or 1200 mg/kg bw per day. The concentrations in the diet were adjusted to achieve a constant target dose throughout the study. Groups of 50 male and 50 female animals were used in the main (24-month) study, with additional groups of 30 male and 30 female animals for the reversibility phase (treatment for 12 months followed by 12 months on control diet). Of the 30 animals of each sex per group in the reversibility phase, 10 animals of each sex per group were killed after the 12-month feeding period. Additional small groups of at least five female animals were added to evaluate tissue hydrocarbon concentrations at 3, 6, 12, 18 and 24 months in both phases of the study. The parameters investigated included body weight, food consumption, clinical observations, serum chemistry, haematology, ophthalmology, urine parameters and organ weights, including mesenteric lymph nodes. Analyses for mineral hydrocarbon were performed on liver, kidneys, mesenteric lymph nodes and spleen from female animals. Detailed histopathological examination of 48 tissues, including the liver, spleen, mesenteric and mandibular lymph nodes, Peyer patches, kidney and bone marrow, was conducted for all animals in the control group and at the highest dose in the main (2-year) study and at the 12-month sacrifice. From animals at 60, 120 or 240 mg/kg bw per day in the main study, only the lungs, liver, mesenteric lymph nodes, spleen and kidneys were examined; the mesenteric lymph nodes and livers of animals in all groups in the recovery study were also examined. Immune function was not tested, but standard end-points considered to reflect immune function (i.e. total and differential leukocyte count, albumin:globulin ratio, the weights and histological appearance of the thymus, spleen and mesenteric lymph nodes, histopathological evaluation of Peyer patches and bone-marrow cellularity) were assessed.

Administration of either oil to Fischer 344 rats in the diet for 24 months did not affect survival. The small decrease in survival rate among females at the high dose of P100H oil, starting at 15 months of the study, was within the rates of other control groups and was considered by the authors not to be biologically significant. No treatment-related effects were seen on clinical signs, body weight, food consumption, food conversion efficiency, ophthalmic, haematological, serum chemical or urinary parameters, and no treatment-related changes were seen at gross necropsy. Dietary administration of both oils was associated with increased weight of mesenteric lymph nodes and increased grade of infiltrating cell histiocytosis; increased incidence and grade of vacuolation of periportal hepatocytes; increased incidence of combined cystic degeneration or angiectasis of the liver of male rats, with no dose–response relationship; and a quantifiable, reversible accumulation of mineral hydrocarbons in the liver to a similar level regardless of dose but dependent on the type of mineral oil. The effects were more marked in rats treated with P70H oil than with P100H oil, and there were sex-related differences in response. Details of these findings are described separately for each oil. Neither oil was carcinogenic in this assay.

Treatment for 24 months with medium-viscosity white mineral oil (P70 (H)) resulted in a dose-related, statistically significant increase in mean mesenteric lymph node weight, both absolute and relative to body and brain weight, at all doses in female rats and at the highest dose, 1200 mg/kg bw per day, in males. Mesenteric lymph node weights were also statistically significantly increased in males at the highest dose after 12 months' feeding; this effect was no longer significant after the 12-month recovery period owing to an increased value for this parameter in the control group. No significant differences in mesenteric lymph node weights were observed in females after 12 months of treatment, although, after the 12-month recovery, a slight but significant increase was seen in females at the two higher doses, 240 and 1200 mg/kg bw per day. The grade, but not the incidence, of infiltrating cell histiocytosis of the mesenteric lymph nodes was increased in all treated groups after 24 months of treatment. In addition, there was a slight increase in the severity of this lesion in animals at the highest dose, the only group assessed, at the 12-month sacrifice. After the 12-month recovery period, these changes were apparent in all groups, at a severity greater than or comparable to that in the animals treated for 24 months.

A dose-related increase in the incidence and grade of vacuolation of periportal hepatocytes was observed in the livers of males in all treated groups. A smaller increase in grade only was seen in treated female rats, because the incidence and grade of portal vacuolation were already high in female, compared with male, controls. In males allowed to recover, the increased incidence and severity of vacuolation of periportal hepatocytes was still evident after 12 months without treatment. The investigators did not consider this indicative of an adverse effect but rather a marker of prolonged exposure to mineral oil. An increased incidence of combined angiectasis and cystic degeneration (focal sinusoidal dilatation) was also observed in all treated male groups compared with the control group at the 24-month sacrifice. This lesion was of minimal grade, and the incidence was similar in all treated groups. While the incidence of this lesion was not increased over that in controls in males allowed to recover for 12 months, data from the 12-month sacrifice were not available to assess the significance of this observation.

An increased incidence of mononuclear cell leukaemia was observed in treated females. This was not considered treatment-related, however, as the incidence in treated groups was not dose-related and was within the range for other control female Fischer 344 rats.

Histological analysis revealed quantifiable amounts of hydrocarbon in the livers of treated animals; the hepatic concentrations of mineral hydrocarbon in animals at the highest dose reached nearly a maximum within 3 months and increased slowly up to 24 months of treatment. The values for animals at 60, 120 and 240 mg/kg bw per day at 12 and 24 months were 1200–1500 μg/g tissue, which were similar to each other and were approximately 60% of those of animals at 1200 mg/kg bw per day (1800 and 2300 μg/g tissue, respectively). After cessation of treatment, the hepatic concentrations had dropped substantially by 6 months and had returned to background levels by 12 months. The amount of mineral hydrocarbon in the spleen and mesenteric lymph nodes of most treated animals at higher doses was below the practical limit of reliable quantification.

After 24 months of treatment with high viscosity white mineral oil (P100 (H)), statistically significant increases in mean mesenteric lymph node weights (absolute and relative to body weight and brain weight) were observed in females at 1200 mg/kg bw per day. This effect was not evident at lower doses or in male rats or after the recovery period. A marginal increase in the grade, but not the incidence, of infiltrating cell histiocytosis in the mesenteric lymph nodes was observed in all treated groups at 24 months. No clear dose–response relationship was apparent in the males, and no effects were observed at 12 months of treatment.

A slight increase in the incidence and grade of vacuolation of periportal hepatocytes was observed in the livers of treated animals of each sex in all treated groups compared with controls. An increase in the incidence of combined angiectasis and cystic degeneration (focal sinusoidal dilatation) was observed in all treated male groups, without a dose–response relationship, which was statistically significant in animals given 1200 mg/kg bw per day. This lesion was not apparent at the 12-month sacrifice. In view of the nature and severity of the response, the investigators did not consider the increased grade of vacuolation to be indicative of an adverse toxicological effect but rather a marker of prolonged administration of white oil.

Histopathological examination revealed an increased incidence of adenoma of the pars distalis of the pituitary in female rats at the highest dose (42% in controls and 60% at the highest dose). This finding was not considered treatment-related as the incidence was well within the range reported for other control female Fischer 344 rats.

Histological analysis revealed quantifiable amounts of hydrocarbon in the livers of treated animals; the hepatic concentrations of mineral hydrocarbon in animals at the highest dose reached nearly a maximum within 3 months (900 μg/g tissue) and increased slowly up to 24 months of treatment. The concentrations in the liver at 12 and 24 months were similar in all treated groups (800–900 and 1200–1400 μg/g tissue, respectively), suggesting that a steady-state level had been reached. After cessation of treatment, the hepatic levels had dropped substantially by 6 months and had returned to background levels by 12 months. Mineral hydrocarbon was found in mesenteric lymph nodes and spleen at some times. The concentrations in the spleen and mesenteric lymph nodes of most treated animals at the higher doses were below the practical limit of reliable quantification. Detectable residues of mineral hydrocarbon were not found consistently in the kidneys (Exxon Biomedical Sciences, 2001).

A medium-viscosity white mineral oil (mineral oil, medium- and low-viscosity, class I), a blend of equal quantities of eight commercially available paraffinic white mineral oils obtained from eight member companies of the Japan Liquid Paraffin Industry, was fed in the diet to Fischer 344 rats. The oils also complied with the requirements of the Japanese food additive standards and the Japanese pharmacopoeia. Five of the component white mineral oils had been derived from petroleum by acid treatment, and the other three had been derived by hydrotreatment. The physical properties of the blended mineral oil were intermediate between those of P70H and N70H.

Groups of 50 male and 50 female Fischer 344 rats were fed diets containing 2.5% or 5% of the composite medium-viscosity white mineral oil (equivalent to 1250 and 2500 mg/kg bw per day), continually for 104 weeks. Body weights and food consumption were measured throughout the study. At the end of the study, the animals were killed and blood samples were collected for haematological and clinical chemical measurements. A full necropsy was performed on all animals; the major organs were weighed, and a range of tissues, including liver, mesenteric lymph node, heart and spleen, were taken for histological examination.

The food consumption and body weights of animals of each sex given 5% mineral oil were slightly increased. The frequency of clinical signs, mortality and haematological parameters were unaffected by treatment. In the group given 5%, the absolute weights of the liver and kidney were increased in males and the absolute and relative weights of the submaxillary gland were reduced in females. The increased absolute organ weights were attributed to the slightly increased body weights of males at this concentration. The absolute and relative weights of the heart and spleen were unaffected by treatment. A variety of tumours developed in all groups, including the control group, but all the neoplastic lesions were histologically similar to those known to occur spontaneously in Fischer 344 rats, and no statistically significant increase in the incidence of any tumour type was found for either sex in the treated groups. An increased grade of small granulomatous foci of macrophages was observed in the mesenteric lymph nodes of both sexes at 2.5 and 5% in comparison with the respective control groups (Shoda et al. 1997).

2.2.5 Special studies

(a) Immunotoxicity

In a comparison of the effects of dietary administration of low melting-point paraffin wax on Kupffer-cell function and morphology in female Fischer 344 and Sprague-Dawley rats, the animals received either control diet or a diet containing 2% low melting-point paraffin wax for 60 days in two separate studies. In the first study, blood was collected for analysis of serum enzymes (alanine and aspartate aminotransferases and γ-glutamyl transpeptidase) and total leukocyte and neutrophil counts, and the liver was collected for histopathological examination and determination of its wax content. Kupffer cells were isolated from the liver and assayed for production of superoxide anion, nitrite ion, tumour necrosis factor-α (TNF-α) and leukotriene B4, phagocytic activity and morphology.

Mean body weights were not affected by consumption of diets containing low melting-point paraffin wax. In treated female Fischer 344 rats, the serum activity of the liver enzymes was increased and the total leukocyte and neutrophil counts were significantly higher than those of untreated controls. No such differences were seen in Sprague-Dawley rats. The concentration of mineral hydrocarbon in the liver of treated Fischer 344 rats was 3.6 ± 0.65 mg/g liver, but none was detected in the livers of untreated control Fischer 344 rats or treated or untreated Sprague-Dawley rats. Microgranulomas were observed in the livers of all treated Fischer 344 rats but in only one treated Sprague-Dawley rat, in addition to small areas of focal necrosis, lymphoid cell aggregates and increased frequency of cytoplasmic vacuoles.

Over half the Kupffer cells isolated from Fischer 344 rats treated with low melting-point paraffin wax contained large, irregularly shaped vacuoles, and another 10–20% had numerous smaller vacuoles. The vacuoles were membrane-associated, suggesting phagocytic uptake. Similar vacuoles were detected only rarely in Kupffer cells isolated from Sprague-Dawley rats. The indices of Kupffer cell function—production of superoxide anion and nitric oxide and phagocytic activity—were increased, and TNF-α and leukotriene B4 production were decreased in treated Fischer 344 rats compared with untreated controls. No significant changes in these functions were observed in Kupffer cells isolated from treated Sprague-Dawley rats (Hoglen et al., 1998).

Serum immunooglobin (Ig) levels were measured in samples from Fischer 344 and Sprague-Dawley rats that had been fed diets containing 0.0, 0.2 or 2.0% low melting-point paraffin wax for 90 days (see section 2.2.2), in order to determine whether the differences in response to mineral hydrocarbons in these two rat strains were due to a difference in immunological response. The study was conducted according to the Good Laboratory Practices for Nonclinical Laboratory Studies of the Food and Drug Administration in the USA. Serum immunoglobin classes (IgG, IgM, IgA) and IgG subclasses (IgG1, IgG2a, IgG2b, IgG2c) were measured in frozen serum samples and related to total serum protein in rats dosed for 30 and 60 days (control and high dose) and for 90 days (control, low and high dose). Treatment did not affect serum immunoglobin levels in Sprague-Dawley rats but increased IgA, IgM and IgG2c and decreased IgG and IgG2a levels in Fischer 344 rats. The changes were dose-related for IgG and IgG2c and both dose- and time-related for IgA, IgM and IgG2a (Burleson, 2001).

The humoral immune response to the T-dependent antigen, sheep erythrocytes, was evaluated in female Fischer 344 and Sprague-Dawley rats given P15 mineral oil (medium- and low-viscosity, class III) in the diet for 90 days. The study was conducted in compliance with the Good Laboratory Practice Regulations of the Food and Drug Administration in the USA and was divided into two phases. In phase I, groups of eight female Fischer 344 rats were fed diets containing P15 mineral oil at 0%, 0.02% or 2.0% and groups of eight female Sprague-Dawley rats were fed diets containing 0%, 1.0% or 2.0% for 90 days. In phase II of the study, groups of eight female Fischer 344 were fed diets containing P15 mineral oil at 0%, 0.02% or 2.0% and Sprague-Dawley rats were fed diets containing 0%, 1.0% or 2.0% for 120 days, followed by a 30-day recovery period. A positive control group for each strain received cyclophosphamide at a dose of 50 mg/kg bw intraperitoneally on each of 4 days after sensitization (see below). Four days before sacrifice, the rats in phase I were sensitized by intravenous injection of 2 x 10^8 sheep erythrocytes. On day 91, the day of sacrifice, spleen-cell suspensions were prepared for evaluation of the IgM response with a modified haemolytic plaque assay. The animals in phase II were sensitized with 100 μg of dinitrophenol–human serum albumin intravenously on day 85, and on day 91 a blood sample was taken for determination of the IgM antibody titres to this antigen. Depending on the results of the assay for immune function in phase I, a decision was taken whether to continue phase II past day 91. The sensitization procedure and blood collection were repeated on days 115 and 121, respectively, for determination of secondary antibody (IgG) to dinitrophenol–human serum albumin. The treated diet was discontinued at this time. Serum IgM and IgG antibody titres were determined by enzyme-linked immunosorbent assay. On day 147, the animals received 2 x 10^8 sheep erythrocytes intravenously, and on day 151 they were killed and their spleens removed for determination of the primary IgM response to sheep erythrocytes. Body weights and food consumption were monitored for 90 days during each study phase, and, at sacrifice, the weights of the brain, liver, spleen and lymph nodes were recorded. The liver and mesenteric lymph nodes with associated adipose tissue from all animals and only the spleens from Sprague-Dawley rats killed at day 91 in the phase II study were examined histopathologically. A diet containing P15 mineral oil for 90 days had no consistent effect on body weight or food consumption in female Fischer 344 or Sprague-Dawley rats in phase I or II. In the Sprague-Dawley rats, the absolute and relative liver weights were statistically significantly increased over those of controls, by approximately the same amount after treatment at 1% and 2% in both phase I and phase II. Lymph nodes (type not specified) showed a slight, non-significant, dose-related increase in weight over those of controls. At 90 days, the absolute and relative weights of the liver in Fischer 344 rats at 0.02% and 2% in the diet were statistically significantly increased in a dose-related manner; and the weights of the spleen and mesenteric lymph nodes of animals at the highest concentration were also statistically significantly increased compared with controls. After 120 days' treatment and 30 days' recovery, the livers of the Fischer 344 rats were still heavier than those of controls, but only the difference in relative weights was statistically significant; the absolute and relative spleen weights remained statistically significantly increased in rats at the highest concentration. The weights of mesenteric lymph nodes after the recovery period were not provided.

Granulomatous inflammation of the mesenteric lymph nodes was noted in nearly all control and treated Fischer 344 rats (6/8 controls, 8/8 at the low concentration and 8/8 at the high concentration); the grade of the lesion was higher at the higher dietary level than at the lower or in controls, and the granulomatous inflammation extended to the attached mesenteric adipose tissue. The incidence and severity of granulomatous inflammation of the liver were also greater in rats at the higher concentration (1/8 at the low concentration and 7/8 at the high concentration, with none in controls). After 120 days' treatment and 30 days' recovery, granulomatous inflammation of the mesenteric lymph nodes was observed in all control and treated animals, with slightly greater severity in rats at the lower concentration and no change in severity in animals at the higher concentration or in controls from that in Fischer 344 rats at the 91-day sacrifice. Granulomatous inflammation of the liver found after longer exposure and after the recovery period was also found after the 91-day sacrifice. The spleens of Fischer 344 rats were not available for examination. In the Sprague-Dawley rats killed at 91 days, granulomatous inflammation of the mesenteric lymph nodes was seen, which was generally of minimal severity, in 1/16 controls, 15/16 at the lower concentration and 15/16 at the higher; the incidence and severity of this finding were similar in the groups at 1% and 2% in the diet. No treatment-related effects were seen in the spleen or liver.

Consumption of a diet containing mineral oil for 90 days had no effect on the spleen IgM antibody response in Sprague-Dawley rats, as assessed in the plaque assay, although the number of spleen cells was increased in both treated groups, statistically significantly at 1% in the diet. The Fischer 344 rats showed a dose-related decrease (40%) in the antibody-forming cell response, expressed per 10^6 spleen cells, which was statistically significant at the higher concentration; no effect was seen when the data were expressed on the basis of total spleen activity, owing to the significant increase in spleen cell number observed at the higher concentration. Increased spleen weights were also seen in these animals. No flow cytometry data were provided to allow identification of the specific spleen cell population affected; in addition, no histopathological examination of the spleen was conducted in Fischer 344 rats to find an explanation for the increase in spleen cell number in animals at the higher concentration. After 120 days' exposure and 30 days' recovery, the responses for all end-points (spleen weight, number of spleen cells, IgM antibody-forming cells per 10^6 spleen cells, IgM antibody-forming cells per spleen x 10^3) were similar to those observed at 90 days, except that the control value for the number of antibody-forming cells per 10^6 spleen cells was lower, and consequently the decrease in this parameter with treatment was not statistically significant. The positive control, cyclophosphamide, markedly suppressed the spleen IgM antibody response. Owing to the lack of effect of mineral oil on the spleen IgM T-dependent antibody-forming cell response in Sprague-Dawley rats, the primary IgM and secondary IgG responses to dinitrophenol–human serum albumin were not assessed in this strain. Treatment had no effect on the IgM response at 90 days, and a small, dose-related but non-significant effect on the IgG response was seen at 120 days in this strain. The positive control markedly reduced both the IgM and the IgG response (ImmunoTox, 2001).

(b) Pathology

The effects of low melting-point paraffin wax on cell infiltrates in the liver, mesenteric lymph node and selected cardiac mitral valves were evaluated in Fischer 344 and Sprague-Dawley rats. Replicate sections of liver, mesenteric lymph nodes and cardiac mitral valves from rats in the 90-day study described above were stained either with haematoxylin and eosin or immunohistochemically to demonstrate the presence of cells expressing CD3 (T cells), CD45RA (B cells), CD8 (CD8a, suppressor/cytotoxic T cells) or ED2 (resident macrophages) antibodies and lysozyme (activated macrophages). In addition, sections of the liver were evaluated by electron microscopy to assess Kupffer cell activation and the nature of the accumulated lipid material.

Increased grades of hepatocellular and Kupffer cell vacuolation were apparent in both rat strains in response to a diet containing low melting-point paraffin wax. The earliest observed change in the Fischer 344 rats was small, mostly centrilobular vacuoles at day 30, which increased in size and distribution with duration of treatment. Lipid-like pseudocysts and granulomatous inflammation associated with the lipid-like material developed with increasing grade as a function of duration of treatment. Occasional single-cell necrosis was observed in hepatocytes, both in the presence and absence of inflammatory infiltrates. Distinct crystalloid structures were present within lysosomes of many Kupffer cells and hepatocytes. The lipid-like pseudocysts were considered to be consistent with accumulation of the lipid-like material in crystalline form. Many activated (lysozyme-positive) Kupffer cells were observed. The granulomas were characterized by a predominantly centrilobular–midzonal distribution and were composed of large epithelioid ED-2-negative (non-resident), lysozyme-positive macrophages. Hepatic granulomatous inflammation was accompanied by a T-cell (lymphocyte) response ($CD3^+/CD4^+/CD8^+$). Vacuolar changes were also observed in hepatocytes of Sprague-Dawley rats but to a much lesser extent than in Fischer 344 rats. In Sprague-Dawley rats, activation of Kupffer cells was much less evident and granuloma formation was observed in only one treated animal. In contrast to Fischer 344 rats, the crystalloid structures within lysosomes of hepatocytes and Kupffer cells were not observed. A higher threshold number of ED2-positive Kupffer cells was seen in control Sprague-Dawley rats than in control Fischer 344 rats.

The primary change in mesenteric lymph nodes of both strains of rat in response to a diet containing low melting-point paraffin wax was formation of granulomas composed of epithelioid macrophages. The severity and incidence of the changes was greater in the Fischer 344 rats. The granulomas were initially small and subcapsular and cortical in location. With increasing dose and duration of treatment, they increased in size and ultimately coalesced to occupy most of the nodal cortex and parts of the medulla. Macrophages of the granulomas were predominantly ED2-negative and lysozyme-positive, consistent with the non-resident nature of the cells. Resident lymphocytes of the node surrounded the granuloma; most were $CD3^+$, with equal numbers of $CD4^+$ and $CD8a^+$ cells.

The inflammatory changes in the mitral heart valve of Fischer 344 rats were slight and consisted of phenotypically mixed cell populations. They were also observed in one control animal, leading the investigators to conclude that the observation was of questionable significance. Only limited numbers of tissues were available for examination. The authors suggested that the data were consistent with a difference in pharmacokinetics between the strains, related to metabolic capacity, resulting in saturation of enzyme systems and accumulation of lipid-like crystalloids, cell disruption and inflammation (Francke et al., 2001).

3. DIETARY INTAKE

3.1 Introduction

Mineral hydrocarbons found in food include both added mineral oils and naturally occurring hydrocarbons. The former category includes white mineral oils (various viscosities), paraffin waxes (low melting-point), microcrystalline wax (high melting-point) and petroleum jelly, which are used in coating and packaging materials, during bakery and confectionery manufacture and in grain and seed dust control. These petroleum-based products are mixtures of saturated hydrocarbons with carbon numbers of 18–50; however, only hydrocarbons with a chain length - 32 are of concern because they can be absorbed in the gastrointestinal tract. Many plants and animals also synthesize long-chain hydrocarbons, and humans are thus exposed to these naturally occurring hydrocarbons.

The dietary intakes of mineral oils of various viscosities and paraffin wax (low melting-point) and microcrystalline wax (high melting-point) were assessed; the intake of petroleum jelly was not included in this assessment.

Data on mineral oils from a number of studies conducted in Europe and USA were submitted by CONCAWE, the European Wax Federation and the American Petroleum Institute and included national estimates for the countries of the European Union, the United Kingdom and the USA. Data from food basket surveys, household economic surveys or sales were provided for the European Union, and data from individual dietary records were provided for the United Kingdom and the USA. The estimated dietary intakes of various categories of mineral hydrocarbons were calculated on the basis of their concentrations (added, migrated or naturally existing) in a number of food groups and combined with food consumption data from food balance sheets for countries of the European Union and national surveys, such as the National Diet and Nutrition Survey in the United Kingdom (adults, 1986–87; preschool children, 1992–93) and the Continuing Survey of Food Intake by Individuals (1989–92) in the USA.

3.2 Sources and concentrations in foods

3.2.1 Use in foods

The main use of mineral oils in foods (both direct and indirect) is in the form of polystyrene, which represents 74% of total sales in the USA. Paraffin wax is used primarily in corrugated cartons as a saturating wax (54% of total sales), in paper cups and containers for cold drinks (22%) and in wax-coated, flexible packaging (10%). Microcrystalline wax is used primarily in wax-laminated flexible packaging (23%), curtain coating for corrugated cartons (22%) and flexible packaging with wax-coated substrates (22%).

The potential concentrations of white mineral oils, paraffin wax and microcrystalline wax in various categories of foods in the United Kingdom and the

USA, based on the amounts added to foods and studies of migration from coatings and packaging materials are presented in Tables 2–4. These concentrations were used to derive the intake estimates. The data show that: (1) the principal dietary sources of mineral oil in the United Kingdom were bread divider oils (high- and low-viscosity oils) and grain dust control (low-viscosity oils) (Food Chemical Risk Analysis, 2001), while those in the USA were fruit and vegetable coatings (low-viscosity oils), confectionery (low-viscosity oils), grain dust control (low-viscosity oils), and bakery pan release oils (high-viscosity oils) (Reich et al., 1998); and (2) the sole dietary source of paraffin wax in Europe was wax paper packaging, and the principal source of microcrystalline wax was use as a confectionery glazing agent (Food Chemical Risk Analysis, 2001), while the main sources of both waxes in the USA were fruit and vegetable coatings and flexible packaging (Reich et al., 1998).

In addition, data were available from other surveys on the concentrations in five major food groups.

Table 2. Potential concentrations (mg/kg) of mineral oils in food in the United Kingdom

Application and medium	Viscosity at 40 °C (mm^2/s)		Comment
	< 65[a]	> 65[b]	
Polystyrene			
Drink containers	–	< 1	
Milk containers	–	< 1	
Individual milk/cream containers	–	< 5–93	Occasional use
Polyethylene			
Containers	–	Not significant	
Delicatessen meats	–	7	
Bakery release agent or divider oils			
Bread	243	243	Very limited use
Confectionery			
Sugar confectionery	250	250	Now obsolete
Chocolate	50	50	Now obsolete
Jute batching oil			
Chocolate	7	–	
Temporary sausage skins			
Skinless sausages	28	–	
Grain and seed de-dusting			
Flours and cereals	9–42	–	
Vegetables	110	–	

From Food Chemical Risk Analysis (2001)

[a] Includes class III low-viscosity oils

[b] Includes high-viscosity and class I and II medium-viscosity oils

Table 3. Potential concentrations (mg/kg) of mineral waxes in food in the United Kingdom

Application and medium	Concentration (mg/kg)		Comment
	Paraffin wax	Microcrystalline wax	
Flexible wax-coated packaging			
Sliced cheese	7	7	
Sliced meats	< 1	< 1	
Biscuits	105	105	Obsolete
Cereals	105	105	
Bread	29	29	
Sugar	52	52	
Cartons with saturating wax			
Milk/cream	< 1	–	No direct contact
Cheese coating			
e.g. Gouda, Edam cheese	4.4	4.4	
Chewing-gum base	Not significant		Not significant
Fruit coating			
Tropical fruit peel	–	200	No direct contact
Poultry de-feathering			
Duck meat	9	9	
Confectionery glazing agent			
Chewing-gum	–	200	
Non-chocolate confectionery	–	200	

From Food Chemical Risk Analysis (2001)

Bakery products (bread and rolls):

Mineral oils may occur not only as residues in flour and in vegetables from de-dusted grain but may also be added to breads and rolls directly as divider oils or as lubricants to prevent bread from adhering to the baking pan. Use of mineral oils in the bakery industry in the United Kingdom is now quite limited as compared with their wide use in the 1960s (Food Chemical Risk Analysis, 2001), as the British baking industry voluntarily withdrew the use of mineral oils as pan-release agents. Furthermore, use of white mineral oils in the baking industry in several countries of the European Union is now restricted by national legislation.

Few data were available on mineral oil residues in commercially produced breads and rolls. The total concentration of mineral oils in loaves in the United Kingdom in one study ranged from 220 to 490 mg/kg. A similar study on mineral oil residues in bread in Germany showed 330 mg/kg in the bottom crust and < 1 mg/kg in the top crust and centre of the bread (Food Chemical Risk Analysis, 2001).

The combined maximum permitted level for white mineral oils and petroleum jelly in baked products in the USA is 1500 mg/kg (21 CFR 172.878 and 172.880). Although vegetable oils and sometimes animal fats are also used as divider oils, pan-release oils and trough grease, the relative market shares in these uses are

Table 4. Potential concentrations of mineral oils and waxes in food in the USA

Food group	Source	Concentration (mg/kg)
Cheese	Coating: paraffin wax	2.3
Edam and Gouda	Coating: microcrystalline wax	3.1
Barley	Dust control: mineral oil	40
Corn bran	Dust control: mineral oil	33
Corn endosperm	Dust control: mineral oil	33
Oats	Dust control: mineral oil	40
Oat bran	Dust control: mineral oil	40
Rice rough	Dust control: mineral oil	40
Rice bran	Dust control: mineral oil	40
Rice milled	Dust control: mineral oil	40
Rye rough	Dust control: mineral oil	40
Rye flour	Dust control: mineral oil	9
Wheat bran	Dust control: mineral oil	42
Wheat flour	Dust control: mineral oil	9
Wheat rough	Dust control: mineral oil	42
Soybean flour	Dust control: mineral oil	25
Sunflower seeds	Dust control: mineral oil	40
Corn oil	Dust control: mineral oil	600
Sunflower oil	Dust control: mineral oil	110
Soybean oil	Dust control: mineral oil	110
Pan breads, rolls	Pan release oils: mineral oil	210
Fruits and vegetables	Edible coating: paraffin wax	44
	Edible coating: microcrystalline wax	44
	Edible coating: mineral oil	160

From Reich et al. (1998)

unknown. It has been reported that divider oils are not absorbed by dough in measurable amounts (Reich et al., 1998).

A survey conducted by the American Bakers Association in 1998 (Reich et al., 1998) showed that the overall average total concentration of mineral hydrocarbons in bread from pan-release oils and trough grease was 533 mg/kg. Some bread may contain only mineral oil, some only petrolatum jelly, some both kinds, and some may contain no mineral oils from the baking process.

Fruits and vegetables:

Mineral oil-based coatings are used on specific types of fruits and vegetables to prevent loss of moisture, act as a carrier for fungicides, protect the commodities from bruising and add various degrees of sheen. Fruits and vegetables are coated only when intended for fresh consumption. In computing the concentrations of mineral hydrocarbons in foods resulting from use of mineral oil-based coatings, it was conservatively assumed that all the relevant fruits and vegetables are coated with a standard composite of white mineral oil, petrolatum jelly, paraffin wax and microcrystalline wax. The residual concentrations of these mineral oils in standard coatings would then be 160, 44, 44 and 44 mg/kg, respectively. This assumption therefore resulted in a greatly exaggerated estimate of intake from fruit and vegetable coatings.

Grains and oil seeds:

Grain and oilseed products can contain mineral oil residues as a result of use in grain de-dusting, which is performed to reduce the free dust in stored grain and thus prevent explosions in grain elevators.

It was assumed that mineral oil is applied to all grains and soya beans at the maximum permitted level in the USA of 200 mg/kg (21 CFR 172.878), keeping in mind that the less expensive soya bean oil is also used in some grain elevators. Reich et al. (1998) showed that the concentrations of mineral oil on wheat grain stored for 6 months after spraying with low-viscosity white mineral oil at 230 mg/kg was 180–250 mg/kg. As low-viscosity mineral oils are quite volatile, losses are expected at the high temperatures used in food processing. In the study of Reich et al., no detectable mineral oil residues were found in finished bread products from milled and baked mineral oil-treated wheat grains. When mineral oil was added at 200 or 500 mg/kg to wheat flour, which was then made into dough and baked into bread, the concentration of mineral oil was reduced by approximately 80%. As a conservative measure, an 80% reduction factor was recommended for use in estimating the residual content of mineral oil in baked products and breakfast cereals when mineral oil-treated grains were the raw material.

Except where other estimates exist, such as for maize, soya beans and wheat meals, 20% of 200 mg/kg (i.e. 40 mg/kg) was used as the mineral oil concentration in other grain products, namely rice, barley, oats and rye.

Confectionery:

White mineral oils and petrolatum jelly are approved for use as release agents and as sealing and glazing agents in the manufacture of candy, chocolate and marshmallow-type items in the USA, with a maximum permitted level of 200 mg/kg (21 CFR 172.878 and 172.880).

Foods in contact with wax packaging materials:

Flexible wax packaging includes wax paper (e.g. paper cups for cold drinks), wax-coated cereal liners and wax-coated paperboard retail packaging. Reich et al. (1998) used the consumption factor proposed by the Food and Drug Administration in the USA (the proportion of consumers in the total population) to determine dietary intake of paraffin wax and microcrystalline wax from flexible wax packaging, and these are shown in Figure 1. The overall concentration of paraffin wax in crackers, cereals, meats, cheeses, milk and beverages (alcoholic, acidic and aqueous) due to contact with wax packagings was estimated to be 17 mg/kg, and that of microcrystalline wax was 9.7 mg/kg.

3.2.2 Naturally occurring hydrocarbons

Hydrocarbons, or long-chain alkanes, are synthesized by a large number of edible plants and animals, and the background levels of these naturally occurring alkanes in food were investigated in order to estimate their contribution to overall mineral hydrocarbon intake. Like mineral hydrocarbons, biogenic alkanes have various structures, which can be divided into three categories: normal paraffins (straight-chain alkanes), isoparaffins (branched-chain alkanes) and cycloparaffins (cyclic alkanes). All these structures are saturated, i.e. without double bonds. In this monograph, the concentrations of the various naturally occurring hydrocarbons in

Figure 1. Estimated daily intake (EDI) derivation of paraffin and microcrystalline waxes from wax packaging

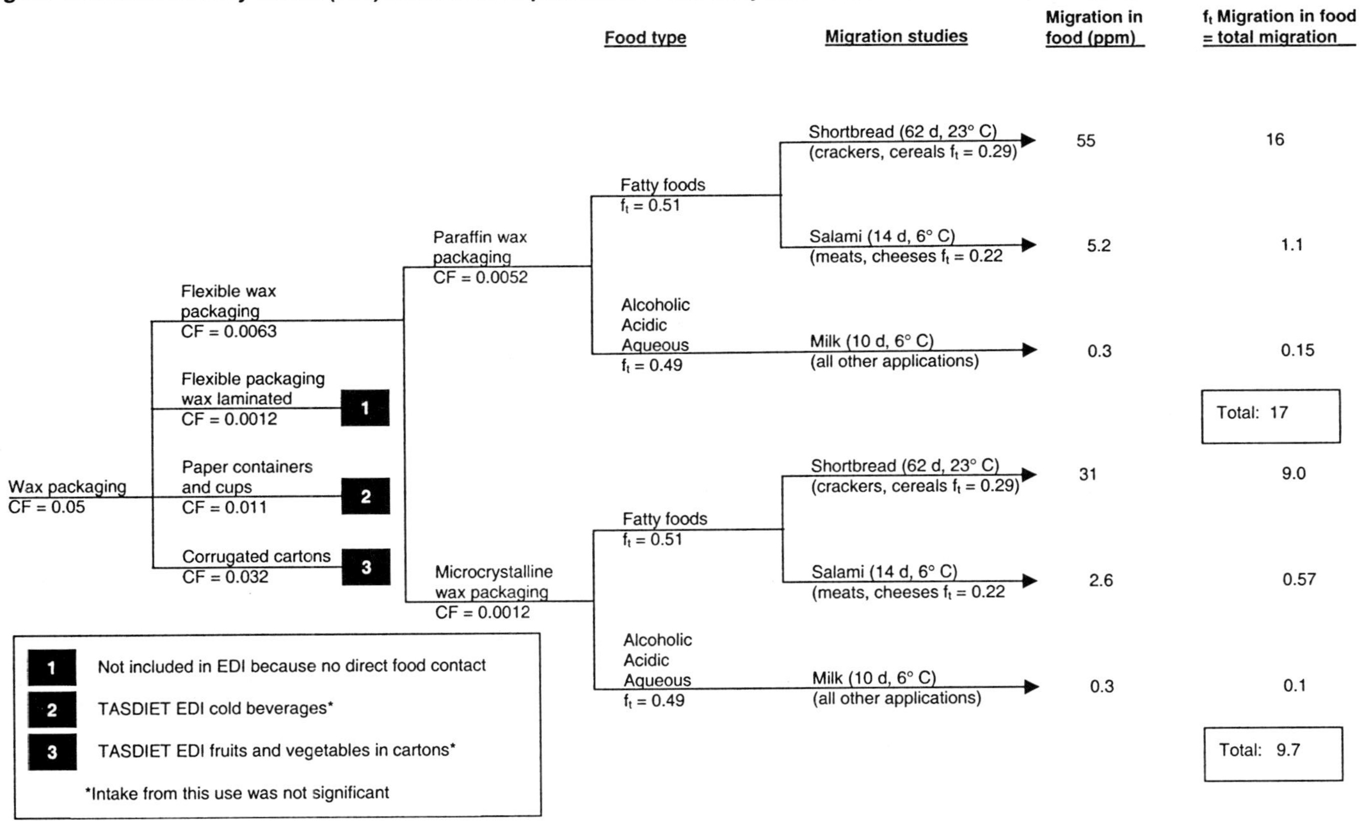

From Reich et al. (1998)

foods were combined, although the range C18–C32 was used to limit the naturally occurring hydrocarbons to those of potential concern..

Reich et al. (1997) compiled a comprehensive list of the concentrations of naturally occurring hydrocarbons in about 180 varieties of plant and animal foods, and this data set was used to estimate the dietary intakes of naturally occurring hydrocarbons in the populations of the countries of the European Union, the United Kingdom and the USA. A simplified version of this data set, based on FAO Food Balance Sheet categories, is presented in Table 5.

Table 5. Concentrations of naturally occurring hydrocarbons in plant and animal foods

Food	Concentration (mg/kg)
Wheat	15
Maize	1.4
Beans	3.1
Ground nuts	13
Sunflower seed	63
Rape, mustard, cotton seed	44
Sesame seed	24
Olives	88
Oil crops, other	44
Soya bean oil	17
Groundnut oil	34
Sunflower seed oil	120
Vegetable oil, others	59
Olive oil	60
Maize germ oil	30
Vegetables	4.6
Oranges	14
Lemons	14
Grapefruit	17
Citrus, others	14
Apples	140
Grapes	3.1
Wine	3.1
Beef	0.15
Mutton	0.15
Pork, fat	40
Pork, lean	6.8
Poultry meat	0.03
Meat, other	0.15
Animal fats	27
Butter, *ghee*	87
Cream	87
Freshwater fish	0.85
Marine fish	46
Crustaceans	0.17
Molluscs, other	0.17
Milk, skim	87
Milk fat, estimate in milk	4.4

From Food Chemical Risk Analysis (2001)

3.3 Assessment of intake

Mineral oils of various viscosities were the major contributors to the total dietary intake of all mineral hydrocarbons in both Europe (90%) and the USA (50%). Paraffin wax and microcrystalline wax contributed very little to the total intake in both regions (Table 6).

3.3.1 Use in foods

The average daily intakes of various categories of mineral hydrocarbons from different types of foods and various direct and indirect food uses in the United Kingdom and the USA are summarized in Table 6. In both countries, the average total intake of mineral carbons (excluding petroleum jelly) was estimated to be 0.47 mg/kg bw per day. Class III medium- and low-viscosity mineral oils accounted for 0.21 mg/kg bw per day in the United Kingdom and 0.25 mg/kg bw per day in the USA, whereas class I and II medium- and low-viscosity mineral oils and high-viscosity oil accounted for 0.18 and 0.19 mg/kg bw per day, respectively. The intake from paraffin wax was estimated to be 0.005 mg/kg bw per day in both countries; and the intake from microcrystalline wax was estimated to be 0.001 mg/kg bw/day in the USA (data for the United Kingdom were not available).

Data for intake in the United Kingdom were also given for consumers at a high percentile and for preschool children (Table 7). The mean, 90th and 97.5th percentile intakes, expressed per kilogram body weight, were significantly higher for children than for adults because of their lower body weight.

Table 6. Average estimated intakes of mineral hydrocarbons in the populations of the United Kingdom (UK) and the USA

Food use	Average intake (mg/kg bw per day)									
	Paraffin wax		Microcrystal-line wax		Mineral oil				Total	
					< 65 mm²/s[a]		> 65 mm²/s[b]			
	UK	USA	UK	USA	UK	USA	UK	USA	UK	USA
Direct										
Fruit and vegetable coatings	0.02	–	0.02	–	0.07	–	–	–	0.11	–
Confectionery	–	–	–	0.001	0.08	–	–	–	0.08	0.001
Bakery	–	–	–	–	–	0.13	0.17	0.19	0.17	0.32
Grain and oil, de-dusting	–	–	–	–	0.10	0.08	–	–	0.10	0.08
Others	–	0.001	–	–	–	–	–	–	–	0.001
Indirect										
Wax packaging	0.005	0.06	0.001	0.009	–	–	–	–	0.006	0.07
Plastic packaging	–	–	–	–	–	–	0.001	0.001	0.001	0.001
Total mineral hydro-carbon intake	0.03	0.06	0.02	0.01	0.25	0.21	0.17	0.19	0.47	0.47

From Reich et al. (1998); Food Chemical Risk Analysis (2001); UK, United Kingdom

[a] Viscosity at 40 °C; includes class III low-viscosity oils

[b] Viscosity at 40 °C; includes high viscosity and class I and II medium-viscosity oils

Table 7. Estimated intake of mineral oils by adults and preschool children in the United Kingdom

Mineral oil	Intake (mg/kg bw per day)		
	Mean	90th percentile	97.5th percentile
White mineral oil			
Adults	0.39	0.68	0.91
Preschool children	0.75	1.4	1.8
Paraffin and microcrystalline wax			
Adults	0.08	0.13	0.19
Preschool children	0.23	0.44	0.64
Total			
Adults	0.47	0.16	0.22
Preschool children	0.98	1.8	2.4

From Food Chemical Risk Analysis (2001)

3.3.2 *Natural sources*

(a) European Union

The average intakes of naturally occurring hydrocarbons by consumers in 15 countries of the European Union, estimated on the basis of food balance sheets, ranged from 0.33 mg/kg bw per day in the United Kingdom to 0.61 mg/kg bw per day in Germany. The European average was 0.47 mg/kg bw was day. The principal sources of intake were apples, wheat and dairy fat. Apples alone contributed about one-third of the total intake, and national intake levels were determined largely by consumption of this fruit (Food Chemical Risk Analysis, 2001).

(b) United Kingdom

The estimated average intake of mineral oils from naturally occurring hydrocarbons by adults in the United Kingdom (based on national food consumption data) was 0.25 mg/kg bw per day (Table 8), similar to the estimate of 0.33 mg/kg bw per day per capita calculated from FAO food balance sheets. Consumers at the 97.5th percentile could have an intake of up to 0.80 mg/kg bw per day. The mean intake of naturally occurring hydrocarbons by preschool children was 0.73 mg/kg bw per day, and that of consumers at the 97.5th percentile increased to 2.6 mg/kg bw per day (Table 8). The intake by children was higher than that by adults because of their lower body weight. The major source of intake was apples, followed by citrus fruits, cereals and dairy fat. The intake of naturally occurring hydrocarbons appeared to be determined largely by consumption of apples (Food Chemical Risk Analysis, 2001).

(c) USA

The total estimated per capita intake of naturally occurring hydrocarbons from all foods was 0.19 mg/kg bw per day. The highest mean intakes of naturally occurring hydrocarbons for consumers of specific foods were from kale and rutabaga tops, but their contribution to per capita intake is small as these items are not consumed

Table 8. Intakes of naturally occurring hydrocarbons by adults and preschool children in the United Kingdom

Food source	Per capita intake (mg/kg bw per day)		
	Average	90th percentile	97.5th percentile
Adults			
Brassicas	0.004	0.007	0.011
Cereals	0.034	0.056	0.071
Citrus	0.015	0.046	0.087
Dairy fat	0.031	0.060	0.085
Fish	0.018	0.043	0.066
Grapes	0.001	0.005	0.010
Meat	0.002	0.004	0.006
Nuts	0.002	0.007	0.018
Olives	0.000	Few consumers	Few consumers
Apples	0.13	0.37	0.64
Poultry	0.000	0.000	0.001
Seeds/pulses	0.002	0.004	0.006
Shellfish	0.000	0.000	0.000
Vegetable oils	0.008	0.018	0.031
Total for adults	0.25	0.50	0.80
Preschool children			
Brassicas	0.004	0.012	0.019
Cereals	0.043	0.076	0.10
Citrus	0.060	0.18	0.37
Dairy fat	0.10	0.20	0.31
Fish	0.031	0.089	0.16
Grapes	0.001	0.003	0.009
Meat	0.003	0.006	0.011
Nuts	0.005	0.017	0.050
Olives	0.000	Few consumers	Few consumers
Apples	0.43	1.2	2.2
Poultry	0.000	0.000	0.002
Seeds/pulses	0.004	0.009	0.014
Shellfish	0.000	Few consumers	Few consumers
Vegetable oils	0.014	0.031	0.048
Total for preschool children	0.73	1.5	2.6

From Food Chemical Risk Analysis (2001)

by a large proportion of the population. The next highest contributor to per capita intake was apples, which are widely consumed, so that apples contributed 34% of the total mean per capita daily intake (Reich et al., 1997).

3.4 *Conclusions and recommendations*

3.4.1 *Contribution from added mineral hydrocarbons and naturally occurring hydrocarbons*

In both the United Kingdom and the USA, the average total intake of the sum of white mineral oil, paraffin wax and microcrystalline wax from use in food (as direct and indirect food additives) was estimated to be 0.47 mg/kg bw per day, whereas the dietary intake of naturally occurring hydrocarbons from plant and animal foods was estimated to be 0.47 mg/kg bw per day for the population of the European Union, 0.25 mg/kg bw per day for that of the United Kingdom and 0.19 mg/kg bw per day for that of the USA. As naturally occurring hydrocarbons make a significant contribution to overall hydrocarbon intake in consumers in both the United Kingdom and the USA, the risk assessment of mineral oil residues from various food applications must include the dietary intake of naturally occurring hydrocarbons.

Estimates of gross intake of mineral hydrocarbons used in food tend to be overestimates because the proportion of the supply of a given food commodity that is affected is usually unknown and it is assumed that 100% of all foods contain mineral hydrocarbons. In one study in the United Kingdom (Food Chemical Risk Analysis, 2001), the gross intake level was adjusted by the actual proportion of mineral oil use derived from surveys. With this method, the gross mean intake of 0.47 mg/kg bw per day based on maximum use levels was reduced to 0.09 mg/kg bw per day (Table 9). In addition, some food uses of mineral oils are no longer common at the global level. For example, their use as pan-releasing agents is now restricted in Europe. Conversely, the intake of naturally occurring hydrocarbons may be underestimated because data on concentrations in all foods are not available.

3.4.2 Intake due to migration from coating and packaging materials

The intake of high-viscosity and classes I, II and III medium- and low-viscosity mineral oils that have migrated from coating and packaging materials was estimated to be 0.001 mg/kg bw per day in both the United Kingdom and the USA, and the intake of paraffin wax and microcrystalline wax was estimated to be 0.005 and 0.006 mg/kg bw per day, respectively (Table 6). These intakes clearly represent an insignificant proportion of the overall intake of mineral oils and waxes.

3.4.3 Comparison of intake data with ADIs

In the toxicological evaluation of mineral oils, different ADIs have been assigned to different groups of mineral oils and waxes on the basis of their physicochemical

Table 9. Intake of total added mineral hydrocarbons before and after adjustment for actual use in the United Kingdom

	Intake (mg/kg bw per day)					
	Adults			Preschool children		
	Mean	90th percentile	97.5th percentile	Mean	90th percentile	97.5th percentile
Gross total mineral hydrocarbons	0.47	0.80	1.1	0.98	1.6	2.2
Use adjusted for total mineral hydrocarbons according to actual use	0.09	0.15	0.21	0.19	0.31	0.41

From Food Chemical Risk Analysis (2001)

properties, because of differences in toxicity. The available data on intake could not, however, be matched to each group of mineral oils and waxes. Table 6 indicates that the intake of medium- and low-viscosity class III mineral oils by the populations of the United Kingdom and the USA was 0.21 and 0.25 mg/kg bw per day, respectively, which values are more than 20 times the temporary ADI for these oils, of 0–0.01 mg/kg bw. The intake of consumers at the 90th percentile in the United Kingdom, 0. 80 mg/kg bw per day, is 80 times the ADI. As separate intake estimates could not be made for high-viscosity and classes I and II medium- and low-viscosity mineral oils, these intake estimates could not be compared with the ADIs.

The intake of microcrystalline wax from food use was estimated to be 0.01 and 0.02 mg/kg bw per day in the United Kingdom and the USA, respectively, which values are well below the ADI of 0–20 mg/kg bw.

Although the intake of naturally occurring hydrocarbons was not taken into account in this evaluation, it is well known that they contribute a significant proportion of total dietary hydrocarbon intake. For example, in the United Kingdom (Food Chemical Risk Analysis, 2001), the intake of naturally occurring hydrocarbons in food (0.25 mg/kg bw per day) accounted for 35% of the total intake, 0.72 mg/kg bw per day, representing 0.47 mg/kg bw per day from total food use (including petroleum jelly) and 0.25 mg/kg bw per day from naturally occurring sources. In the USA (Reich et al., 1997, 1998), the estimated intake of naturally occurring hydrocarbons in food (0.19 mg/kg bw per day) accounted for 18% of the total intake, 1.1 mg/kg bw per day (0.88 plus 0.19 mg/kg bw per day). These estimated contributions of naturally occurring hydrocarbons are conservative, because the intake from food use was overestimated while the intake from natural sources was underestimated.

3.4.4 Estimates of intake from use in foods

The estimated average total intake of mineral hydrocarbons (excluding petroleum jelly) resulting from use in food in both the United Kingdom and the USA was 0.47 mg/kg bw per day, white mineral oil accounting for the largest proportion (0.40 and 0.43 mg/kg bw per day, respectively). As limited data were available, the intake from various types of medium- and low-viscosity mineral oils could not be estimated separately. The best estimate is that the total intake of mineral hydrocarbons from low- and medium-viscosity mineral oils is more than 0.25 mg/kg bw per day. Bakery products, fruits and vegetables, grains and oil and confectionery are the major food sources of intake. This value is likely to be an overestimate, as it was assumed that 100% of each food category contained mineral oils at the maximum permitted level of use; furthermore, some food uses described above are not currently common at the global level.

3.4.5 Estimates of intake from natural sources

Naturally occurring hydrocarbons are widely distributed in many edible plants and animals, and natural sources contribute significantly to the overall dietary intake of hydrocarbons. The dietary intake of naturally occurring hydrocarbons was estimated to be 0.47 mg/kg bw per day for the populations of the 15 countries of the European Union, 0.25 mg/kg bw per day for the population of the United Kingdom and 0.19 mg/kg bw per day for that of the USA. As data on the concentrations of naturally occurring hydrocarbons in food were limited, these values are likely to be underestimates.

3.4.6 Risk characterization

The estimated intake of medium- and low-viscosity class III mineral oils (0.21 and 0.25 mg/kg bw per day in the United Kingdom and the USA) was more than 20 times the temporary ADI for this group (0–0.01 mg/kg bw). The estimated intake of microcrystalline wax in both countries were well below its ADI (0–20 mg/kg bw); however, microcrytalline wax contributes only a small proportion to overall intake of mineral hydrocarbons. The available data on intake of other classes of white mineral oils do not allow comparisons of intake with their ADIs, as separate data were not available for class I and II mineral oils and high-viscosity mineral oils. Data on the intake of paraffin wax were available, but no ADI has been established.

3.5 Recommendations

Intake data for each category of mineral oils should be made available to allow a complete evaluation of intake.

The Committee suggested that the Codex Committee on Food Additives and Contaminants consider reviewing the standards for mineral oil, particularly for class III medium- and low-viscosity mineral oils, because mineral oil is included in the draft General Standard for Food Additives in numerous food categories, with maximum use levels ranging from 200 to 3000 mg/kg (step 6). In some food items, its use level is required to be in accordance with good manufacturing practice; however, according to the principles of the Codex Committee, substances with numerical ADIs should not be included in the category of use according to good manufacturing practice.

4. COMMENTS

Toxicological data

In the long-term studies of toxicity, Fischer 344 rats were given a diet containing P70(H) or P100(H) oil, at concentrations that varied to provide constant doses ranging from 60 to 1200 mg/kg bw per day, for up to 24 months. At 24 months, treatment with either test material was associated with mild treatment-related effects, which included: dose-related increases in the absolute and relative weights of the mesenteric lymph nodes; an increase in the severity of reticuloendothelial-cell hyperplasia (previously described as focal histiocytosis) of the mesenteric lymph nodes; increases in the incidence and/or severity of vacuolation of periportal hepatocytes; and an increased incidence of sinusoidal dilatation in the liver of male rats only, which was not dose-related. These effects were observed even at the lowest dose tested, 60 mg/kg bw per day. There were no treatment-related effects on survival with either material. Neither P70(H) nor P100(H)was carcinogenic in this study.

The effects observed at 12 months were marginal and included a slight increase in the severity of reticuloendothelial-cell hyperplasia of the mesenteric lymph nodes in male and female rats treated with P70(H) at the highest dose, the only dose for which histopathological data were available, as well as statistically significantly increased absolute and relative weights of mesenteric lymph nodes in males at this dose. Because of the marginal nature of these changes at 12 months, the reversibility of the lesions at the end of the recovery period could not be assessed; however, the results did indicate lack of progression of tissue alterations after dietary exposure to mineral hydrocarbons for 12 months.

No special indicators of immune function were measured in the long-term studies, although standard end-points of haematology, clinical chemistry, organ weight and histopathology, indicative of immune function, were evaluated. The only alteration in these end-points was exacerbation of mild cellular infiltration of the mesenteric lymph nodes, which had not been observed previously after exposure for 90 days. Aside from the earlier onset and increased severity of reticuloendothelial-cell hyperplasia of the mesenteric lymph nodes and the greater increase in the weights of mesenteric lymph nodes in response to dietary administration of P70(H), there was little difference in response to the two oils. None of the effects at 24 months was reported in 90-day studies at concentrations up to 2% of the diet (equivalent or equal to 2000 mg/kg bw per day) of either material. The pigmented macrophages observed with P70(H) oil at 2% in the diet (equivalent to 2000 mg/kg bw per day) in the 90-day study reviewed at the forty-fourth meeting, on which the ADI was based, were not observed after 24 months' intake at doses up to 1200 mg/kg bw per day.

Accumulated hydrocarbon was detected in the liver, mesenteric lymph nodes and spleen; the amounts of hydrocarbon could be quantified only in the liver. In the liver, a steady-state level of hydrocarbon accumulation appeared to have been reached in rats treated with the highest dose (1200 mg/kg bw per day) by 3 months, the only dose for which data on the complete time course were available. The concentration of P70(H) reached in the liver at any given time was about twice that of P100(H). The concentrations of mineral hydrocarbons in the liver returned to control levels within 12 months of cessation of treatment.

Medium-viscosity liquid petroleum (corresponding to class I medium- and low-viscosity mineral oils, including P70(H) oil) was not carcinogenic when fed to Fischer 344 rats at a concentration of 2.5% or 5% in the diet (equivalent to 1250 and 2500 mg/kg bw per day). Increased incidence or severity of reticuloendothelial-cell hyperplasia of the mesenteric lymph nodes was also observed in rats of each sex in this study at both dietary concentrations.

The pharmacokinetics of single doses of P15(H) oil given by gavage at a dose of 34 or 340 mg/kg bw over 96 h was studied in female Fischer 344 and Sprague-Dawley rats, with [^{14}C]1-eicosanylcyclohexane as the tracer. In one of the studies, the animals received a 2-week dietary pretreatment with low-viscosity mineral oil. The faeces were the major route of elimination, and urinary excretion was a minor route. Elimination of radiolabel by either route was more rapid in Sprague-Dawley than in Fischer 344 rats. Estimates of the area under the curve for serum concentrations of mineral hydrocarbons indicated that the bioavailability of mineral hydrocarbon was greater in Fischer 344 rats. Rats of this strain also accumulated and retained a larger proportion of the administered dose in the liver (2% vs 0.1–0.5%) and, to a much lesser extent (about 100-fold), the mesenteric lymph nodes. While a longer delay in faecal excretion of radiolabel was seen in Fischer 344 rats than in Sprague-Dawley rats in one study, suggesting a strain-related difference in intestinal absorption of mineral hydrocarbons, this difference was not observed in two similar studies. Rather, the results suggested that Sprague-Dawley rats metabolized and excreted radiolabelled mineral hydrocarbon tracer more efficiently than Fischer 344 rats, which exhibited deposition and retention in the liver. Dietary pretreatment with mineral oil did not affect the extent to which hydrocarbon accumulated in the liver of either strain of rats. In another study, the rate of metabolism of *n*-heptadecane by liver microsomes in vitro differed by rat strain, but not sex, in the order Wistar > Sprague-Dawley > Fischer 344. Metabolism of *n*-octadecane was negligible in all strains, consistent with the findings for hydrocarbons with more than 17 carbons in microsomal systems in vitro.

New short-term studies, in which the responses of female Fischer 344 and Sprague-Dawley rats to dietary administration of P15(H) oil (for 120 and 90 days, respectively) at concentrations of up to 2% were compared, indicated a more severe response in the Fischer 344 strain. The effects seen in these rats were: increased weights of the liver, mesenteric lymph nodes and spleen; granulomatous inflammation of the liver; and reticuloendothelial-cell hyperplasia of the mesenteric lymph nodes. In Sprague-Dawley rats, only statistically significant increases in liver weight and reticuloendothelial-cell hyperplasia of the lymph nodes, mostly of minimal severity, were observed. The increased liver weights in Fischer 344 rats receded during a 30-day recovery period. Reversibility was not assessed in Sprague-Dawley rats.

The results of the new 90-day study in female Fischer 344 rats with low-melting-point wax and with N15(H), N70(H) and P70(H) oils at a dietary concentration of 2% (equal to approximately 2000 mg/kg bw per day at 90 days) confirmed the previous findings in lymph nodes, liver and spleen, with an additional finding of inflammation of the cardiac mitral valve, which had not been seen previously with this material. This effect occurred at a lower incidence than with low-melting-point wax in this study.

The Committee reviewed the results of a study of T-dependent humoral immune function in response to 90 days' administration of P15(H) oil in female Fischer 344 and Sprague-Dawley rats. No effects were observed in the Sprague-Dawley rats. Interpretation of the results in the Fischer 344 rats was complicated by the increase in spleen weight that resulted from treatment. No data were provided to determine whether the increased spleen weight was associated with a change in the proportion of spleen cell types, which could alter interpretation of the results.

The responses of female Fischer 344 and Sprague-Dawley rats to mineral hydrocarbons were also compared in a 90-day study with low-melting-point wax at a dietary concentration of 0.2% or 2.0%. This substance had induced the most severe effects in similar studies in Fischer 344 rats with a range of food-grade mineral hydrocarbons, in which, in general, female rats had a stronger response to these materials. The results for Fischer 344 rats in the new study confirmed the findings of the previously evaluated 90-day studies with low-melting-point wax: altered haematological and hepatic clinical chemical parameters, increased weights of the liver, mesenteric lymph nodes and spleen; hepatic vacuolation (affecting both hepatocytes and Kupffer cells) and granulomatous inflammation; reticuloendothelial-cell hyperplasia of mesenteric and cervical lymph nodes; and focal inflammation of the cardiac mitral valve. In addition, increased severity of extramedullary haematopoiesis of the spleen, differential effects on serum immunoglobin concentrations and lipid-like material in crystalline form in hepatocytes and Kupffer cells were observed in this strain. In contrast, the only effects observed in female Sprague-Dawley rats were increased liver weights and slightly increased reticuloendothelial-cell hyperplasia of the mesenteric lymph nodes and severity of hepatocellular and Kupffer-cell vacuolation. The crystalloid structures and inflammatory response in the liver were absent in this strain.

Tissue accumulation of mineral hydrocarbon after administration of low-melting-point wax also showed a clear strain difference in terms of degree, onset and the tissues affected. Quantifiable amounts of hydrocarbon were detected in the liver and mesenteric lymph nodes of Fischer 344 rats in a time- and dose-related manner, starting at 30 days in the livers of animals at the higher dose and occurring in both tissues in animals at both doses by 90 days. However, only the mesenteric lymph nodes of Sprague-Dawley rats had quantifiable amounts of hydrocarbon, and only at 90 days in animals at the higher dose. This observation was confirmed in a qualitative investigation of the accumulation of low-melting-point wax in the liver and Kupffer cells of female Fischer 344 and Sprague-Dawley rats after administration of a diet containing 2% wax for 14, 30 or 60 days. Detectable amounts were found in liver cells of Fischer 344 rats by 14 days, but none was found at any time in Sprague-Dawley rats.

In a separate study of low-melting-point wax, feeding a diet containing 2% resulted in alterations in Kupffer-cell function and morphology after 60 days in female Fischer 344, but not Sprague-Dawley, rats. Effects were also seen on selected parameters (i.e. leukocytes, serum liver enzyme activities and hepatic histopathology) that have been associated with administration of low-melting-point wax in this strain, and these were accompanied by accumulation of mineral hydrocarbon in the liver.

Intake

At its present meeting, the Committee also assessed the intake of mineral oils. The major food uses of mineral oils are in vegetable and fruit coatings and in packaging, bakery and confectionery manufacture and grain and seed dust control, and these uses are also the major sources of dietary intake of mineral oils in some countries. Dietary intake was assessed from data on the levels of use in foods and on migration from coatings and packaging materials into foods, combined with national data on food consumption in the 1990s. In both the United Kingdom and the USA, the average total intake of mineral hydrocarbons (excluding petroleum jelly) from food use was estimated to be 0.47 mg/kg bw per day; the intake at the 90th percentile of consumption by the population of the United Kingdom was 0.80 mg/kg bw per day. Class III medium- and low-viscosity mineral oils (including P15(H) oil) accounted for 0.21 mg/kg bw per day in the United Kingdom and 0.25 mg/kg bw per day in the USA; these values are 21 and 25 times the temporary ADI of 0–0.01 mg/kg bw, respectively, whereas class I and II medium- and low-viscosity mineral oils and high-viscosity mineral oil (including P70(H) and N70(H) oils, respectively) accounted for 0.18 and 0.19 mg/kg bw per day, respectively. As these different categories of mineral oil have different ADIs, but data on intake are not available for separate categories, the intake of each category cannot be compared with the corresponding ADI. Use of solid hydrocarbons (e.g. microcrystalline wax (high-melting-point wax) and paraffin wax (low-melting-point wax)) accounted for the remainder of the total intake.

The intake of high-viscosity and class I, II and III medium- and low-viscosity mineral oils that have migrated into food from coating and packaging materials was estimated to be 0.001 mg/kg bw per day in both the United Kingdom and the USA, while the combined intake of paraffin wax and microcrystalline wax from this source was estimated to be 0.005 and 0.006 mg/kg bw per day, respectively.

Naturally occurring hydrocarbons are widely distributed in many edible plants and animals, and they contribute significantly to the overall dietary intake of hydrocarbons. For example, the dietary intake of naturally occurring hydrocarbons was estimated to be 0.47, 0.25 and 0.19 mg/kg bw per day in the populations of the European Union, the United Kingdom and the USA, respectively. It is clear, therefore, that account should be taken of intakes from naturally occurring hydrocarbons when evaluating the safety of mineral oils.

Mineral oil is included in the Codex draft GSFA in numerous food categories, with maximum use levels of 200–3000 mg/kg and, in some categories, in accordance with good manufacturing practice. The Committee suggested that the Codex Committee on Food Additives and Contaminants consider reviewing the standards for mineral oil, particularly in view of the fact that the intake of class III medium- and low-viscosity mineral oil is more than 20 times higher than its temporary ADI of 0–0.01 mg/kg bw.

5. EVALUATION

No new information was available on the compositional elements of mineral oils that affect their absorption or toxicity.

As the materials tested in the long-term studies reviewed at the present meeting, P70(H) and P100(H) oils, were not associated with induction of liver granulomas in Fischer 344 rats, the studies did not help the Committee to determine the long-term consequences or reversibility of the liver granulomas that had been seen in previous studies in response to consumption of low- and medium-viscosity mineral oils and low-melting-point waxes by Fischer 344 rats. In addition, the Committee was unable to interpret the effects in the study of humoral immune function in response to dietary administration of P15(H) oil.

The results of the studies on the effects of P15(H) oil and low-melting-point wax in Fischer 344 and Sprague-Dawley strains indicated that the more extensive response of Fischer 344 rats, in particular that of females, is associated with greatly enhanced retention of mineral hydrocarbons in the tissues, which is probably due to a reduced ability to metabolize absorbed hydrocarbons. The Committee concluded that additional studies are needed in order to determine whether the Fischer 344 rat is an appropriate model of human response to dietary intake of food-grade mineral hydrocarbons. In particular, elucidation of the metabolic differences between Fischer 344 rats and other strains and species, including humans, would be useful.

Neither P70(H) nor P100(H) oil was carcinogenic in the combined study of toxicity and carcinogenicity reviewed by the Committee at its present meeting. The effects observed even at the lowest dose, i.e. enhanced reticuloendothelial-cell hyperplasia, increased weights of mesenteric lymph nodes and increased incidence and grade of vacuolation of hepatocytes, were shown not to progress to more severe effects, and there was no indication that accumulated test material contributed to suppression or activation of an inflammatory response. Consequently, these effects were considered to be indicators of exposure to mineral hydrocarbon rather than adverse effects. At the present meeting, the NOEL for P70(H) oil was identified as the highest dose tested in the combined study of toxicity and carcinogenicity in rats, 1200 mg/kg bw per day, to which a safety factor of 100 was applied. An ADI of 0-10 mg/kg bw was allocated for class I medium- and low-viscosity mineral oils, which include P70(H) oil. An ADI of 0–20 mg/kg bw for P100(H) oil already exists (see Table 1).

No data were available that would permit allocation of a full ADI for medium- and low-viscosity mineral oils in classes II and III. The Committee noted that the new information reviewed at the present meeting indicated that the observed effects of these mineral oils, on which the temporary ADI is based, may be strain- and sex-specific. The Committee therefore extended the temporary group ADI of 0–0.01 mg/kg bw for class II and III medium- and low-viscosity mineral oils until 2006, pending information on the relevance to humans of the response of Fischer 344 and Sprague-Dawley rats to these materials. In order for the data to be applicable to as wide a range of mineral oils as possible, the Committee suggested that commercial mineral oils of the lowest viscosity be used in such studies. Further studies might be required, depending on the outcome of these studies.

6. REFERENCES

Baldwin, M.K., Berry, P.H., Esdaile, D.J., Linnett, S.L., Martin, J.G., Peristianis, G.C., Priston, R.A.J., Simpson, B.J.E. & Smith, J.D. (1992) Feeding studies in rats with mineral hydrocarbon food grade white oils. *Toxicol. Pathol.*, **20**, 426–435.

Burleson, G.R. (2001) Unpublished study conducted by Burleson Research Technologies, Raleigh, North Carolina, USA for American Petroleum Institute, Washington DC, USA. Submitted to WHO by CONCAWE, Brussels, Belgium.

Cunninghame, M., Brantom, P., Scotter, M. & Castle, L. (2001) Identification of the causative agents(s) for mineral hydrocarbon toxicity. Unpublished report FD 00/51prepared by TNO-BIBRA International, Carshalton, Surrey, United Kingdom, and Central Science Laboratory, Sand Hutton, New York, USA. Submitted to WHO by CONCAWE, Brussels, Belgium.

Exxon Biomedical Sciences (2001) Combined chronic toxicity/carcinogenicity study of white oil in Fischer 344 rats. Unpublished report numbers 105970 and 106870. Submitted to WHO by CONCAWE, Brussels, Belgium.

Firriolo, J.M., Morris, C.F., Trimmer, G.W., Twitty, L.D., Smith, J.H. & Freeman, J.J. (1995) Comparative 90-day feeding study with low-viscosity white mineral oil in Fischer-344 and Sprague-Dawley-derived CRL:CD rats. *Toxicol. Pathol.*, **23**, 26–33.

Food Chemical Risk Analysis (2001) The usage, occurrence and dietary intake of white mineral oils and waxes and naturally occurring hydrocarbons in Europe. Prepared for CONCAWE and EWF.

Francke, S., Rojko, J.L. & Hall, W.C. (2001) Unpublished report IM392 prepared by Pathology Associates International, Frederick, Maryland, USA for American Petroleum Institute, Washington DC. Submitted to WHO by CONCAWE, Brussels, Belgium.

Hoglen, N.C., Regan, S.P., Hensel, J.L., Younis, H.S., Sauer, J.-M., Steup, D.R., Miller, M.J., Waterman, S.J., Twerdok, L.E. & Sipes, I.G. (1998) Alteration of Kupffer cell function and morphology by low melt point paraffin wax in female Fischer-344 but not Sprague-Dawley rats. *Toxicol. Sci.*, **46**, 176–184.

ImmunoTox (2001) Effects of dietary mineral hydrocarbons on T-cell dependent antibody production in two strains of rats. Unpublished study ITI 1298 prepared for American Petroleum Institute, Washington DC. Submitted to WHO by CONCAWE, Brussels, Belgium.

Institut National de la Recherche Agronomique (2000) Unpublished report EP/98/HYDROCARB/01 prepared for American Petroleum Institute, Washington DC. Submitted to WHO by CONCAWE, Brussels, Belgium.

Reich, A,G., Waylett, D.K., van der Reit, E., Doyl, E., Douglass, J.S., Eickhoff, J.C. & Heimbach, J.T. (1997) Intake of naturally occurring alkanes. Prepared for the American Petroleum Institute by TAS-ENVIRON.

Reich, A,G., Douglass, J.S., van der Reit, B., Eickhoff, J.C. & Heimbach, J.T. (1998) Dietary exposure assessment for direct and indirect food use applications for mineral hydrocarbons. Prepared for the American Petroleum Institute by TAS-ENVIRON.

Shoda, T., Toyoda, K., Uneyama, C., Takada, K. & Takahashi, M. (1997) Lack of carcinogenicity of medium-viscosity liquid paraffin given in the diet to Fischer 344 rats. *Food Chem. Toxicol.*, **35**, 1181–1190.

Sipes, I.G. & Halladay, J.S. (undated) Pilot pharmacokinetic studies on mineral hydrocarbons (MHC). Unpublished study prepared for American Petroleum Institute, Washington DC. Submitted to WHO by CONCAWE, Brussels, Belgium.

Sipes, I.G. & Halladay, J.S. (2001a) Final report: 10-fold lower dose study. Unpublished study prepared for American Petroleum Institute, Washington DC. Submitted to WHO by CONCAWE, Brussels, Belgium.

Sipes, I.G. & Halladay, J.S. (2001b) Final report: 2-week pretreatment study. Unpublished study prepared for American Petroleum Institute, Washington DC. Submitted to WHO by CONCAWE, Brussels, Belgium.

Sipes, I.G. & Hoglen, N.C. (1998) Analysis of low melt point paraffin wax in liver digests and Kupffer cells by gas chromatography and gas chromatography–mass spectrometry. Unpublished study prepared for American Petroleum Institute, Washington DC. Submitted to WHO by CONCAWE, Brussels, Belgium.

Smith, J.H., Mallett, A.K., Priston, R.A.J., Brantom, P.G., Worrell, N.R., Sexsmith, C. & Simpson, B.J. (1996) Ninety-day feeding study in Fischer-344 rats of highly refined petroleum-derived food-grade white oils and waxes. *Toxicol. Pathol.*, **24**, 214–230.

Thornton, S.R., Auletta, C.S., Bolte, H.F., Gosselin, S.J., Dadgar, D. & Cojocaru, L.V. (2000) A subchronic 90-day dietary toxicity study of a low melting point paraffin wax in two rat strains. Unpublished report from Huntingdon Life Sciences, East Millstone, New Jersey, USA. Submitted to WHO by CONCAWE, Brussels, Belgium.

NITRITE
(and potential endogenous formation of N-nitroso compounds)

First draft prepared by
Dr G.J.A. Speijers[1] and Professor P.A. van den Brandt[2]
[1]Section Public Health of the Centre for Substances & Risk Assessment, National Institute of Public Health and Environmental Protection, Bilthoven, Netherlands; and [2]Department of Epidemiology, Maastricht University, Maastricht, Netherlands

1. EXPLANATION

Nitrite occurs in the environment, in food and water, and is produced endogenously. It is also used as a food additive, mainly as a preservative, antimicrobial agent and colour fixative. It is used in foods such as cheese and cheese products, raw and processed meats, edible casings, processed fish and fish products and roe.

Nitrite was reviewed by the Committee at its sixth, eighth, seventeenth, twentieth and forty-fourth meetings (Annex 1, references 6, 8, 32, 41 and 116). At its sixth meeting, the Committee allocated an ADI of 0–0.4 mg/kg bw to this substance, expressed as sodium nitrite. This ADI was based on a marginal reduction in body-weight gain at a dose of 100 mg/kg bw per day in a long-term study in rats. At its seventeenth meeting, the Committee lowered the ADI for sodium nitrite to 0–0.2 mg/kg bw and made it temporary At that time, the Committee used a safety factor higher than normal (500) because the ADI was based on a marginal effect and there is a possibility of the endogenous formation of *N*-nitroso compounds from nitrite and *N*-nitrosatable compounds present in food and the gastrointestinal tract.

At its twentieth meeting, the Committee considered the reports of a WHO task group and of a working group of the International Agency for Research on Cancer on *N*-nitroso compounds but concluded that they did not provide sufficient evidence to revise the temporary status of the ADI.

After that meeting, numerous toxicological and epidemiological data became available, which the Committee considered at its forty-fourth meeting. Nitrite both with and without nitrosatable precursors was found to be genotoxic in vitro in several tests, but, with the exception of a test in Drosophila melanogaster, the results obtained in vivo were negative. The results of carcinogenicity studies with nitrite were negative, except those in which extremely high doses of both nitrite and nitrosatable precursors were administered simultaneously or premixed in the diet. In addition, the epidemiological data provided no evidence for an association between exposure of humans to nitrite and nitrate and the risk for cancer.

At its forty-fourth meeting, the Committee noted that several controlled laboratory studies had shown that *N*-nitroso compounds are formed endogenously when both nitrite and *N*-nitrosatable compounds are present together at high concentrations; it observed, however, that quantitative data were available only on those *N*-nitroso compounds that are readily formed endogenously, such as *N*-nitrosoproline, which is not carcinogenic. As there was no quantitative evidence of the endogenous formation of carcinogenic *N*-nitroso compounds at the levels of intake of nitrite and nitrosatable precursors achievable in the diet, a quantitative risk assessment of nitrite on the basis of endogenously formed *N*-nitroso compounds was considered to be inappropriate. The Committee at its forty-fourth meeting therefore based its safety evaluation on studies of toxicity with nitrite. The NOELs in these studies were 5.4 mg/kg bw per day (expressed as nitrite ion) in a 90-day study in rats, in which hypertrophy of the zona glomerulosa was observed, and 6.7 mg/kg bw per day (expressed as nitrite ion) in a 2-year study, in rats in which effects on the heart and lungs were observed. On this basis, the Committee allocated an ADI of 0–0.06 mg/kg bw, expressed as nitrite ion, by applying a safety factor of 100.

At its present meeting, the Committee reviewed the data that had become available on nitrite since its forty-fourth meeting, including data on the relevance and adverse nature of certain effects, with new information on the mode of action of effects on the adrenals and on toxicokinetics.

2. *BIOLOGICAL DATA*

2.1 *Biochemical aspects*

2.1.1 *Absorption, distribution and excretion*

In an open, randomized, three-way cross-over study, seven women and two men (mean age, 23 ± 1.5 years; mean weight, 68 ± 3.3 kg) received a single oral dose of sodium nitrite at 0.06 or 0.12 mmol/mmol haemoglobin, corresponding to 140–190 mg or 290–380 mg, depending on the weight of the person, and 7 days later received one intravenous dose of sodium nitrite at 0.12 mmol/mmol haemoglobin, corresponding to 290–380 mg. All doses were administered after an overnight fast. Adverse effects, blood pressure, heart rate, haemoglobin concentration, per cent methaemoglobin in blood and plasma nitrite and nitrate concentration were recorded frequently for 24 h after each administration of sodium nitrite.

Headache was the most frequent complaint during each treatment session, and lowered blood pressure accompanied by an increased heart rate were seen after each treatment. The maximum per cent methaemoglobin in blood was 8.4–12% after intravenous administration and 7.7–11% and 3.4–4.5% after the two oral doses of sodium nitrite. The maximum methaemoglobin level was reached about 1.2 h after intravenous or oral administration of sodium nitrite at 0.12 mmol/mmol haemoglobin and 0.7 h after administration of the low oral dose. Gastrointestinal absorption of sodium nitrite was rapid, with maximum plasma nitrite concentrations observed 15–30 min after dosing. Nitrite disappeared rapidly from plasma, with an average elimination half-life of 30 min. The bioavailability of sodium nitrite was 73–120% after the high oral dose and 70–110% after the low oral dose. The authors concluded that, under fasting conditions, 90–95% of orally administered sodium nitrite is absorbed from the gastrointestinal tract (Kortboyer et al., 1997a).

2.1.2 Methaemoglobin formation

(a) In vitro

The methaemoglobin-inducing potential of nitrite was investigated in vitro in fresh, heparinized, human venous blood samples maintained at 37 °C. Ten doses of sodium nitrite with a molar ratio of nitrite to haemoglobin of 0.01–0.15 were tested in triplicate. The haemoglobin concentration, per cent methaemoglobin and the pH of the blood were determined at 2-h intervals after the start of the experiment. None of the blood samples showed visible haemolysis at the end of the experiment. Methaemoglobin contents up to 24% were induced. A linear relationship was established between the nitrite dose and the maximum methaemoglobin level induced in blood. Linear regression analyses showed an adjusted r^2 of > 0.99 ($p = 0.0001$). The molar ratio of the nitrite dose to the maximum per cent methaemoglobin induced at T_{max} was approximately 0.7 at all doses. A dose of sodium nitrite with a molar ratio to haemoglobin of 0.04 induced about 5% methaemoglobin (Kortboyer et al., 1997b).

(b) Humans

Contamination of food during manufacture and degradation of nitrate in vegetables appear to be important factors in the induction of methaemoglobinaemia. Stringent control of excessive use of nitrate and nitrites in food items such as refrigerated dim sum, stuffed pork and Chinese sausages, which are popular among some Asian populations, is required in order to prevent outbreaks of toxic methaemoglobinaemia. Persons with glucose-6-phosphatase deficiency, which is common in some Asian populations, may present with methaemoglobinaemia and intravascular haemolysis after exposure to oxidant drugs or chemicals. Methylene blue is inefficient and may exacerbate haemolysis in these patients; partial exchange transfusion may be required (Chan, 1996)

A case report described two incidents of illness due to intake of nitrite. In the first, 59% methaemoglobinaemia was found in 29 students and 20% methaemoglobinaemia in 14 of 49 students who ate leftover soup containing 460 mg/l nitrite. They also had blue lips and fingers, nausea, vomiting and headache. In the second incident, four of six office workers who drank leftover coffee containing 300 mg/l nitrite showed a 6–16% increase in methaemoglobin. No estimates of intake were made. In both cases, the nitrite originated from a contaminated hot-water tap (Centers for Disease Control, 1997).

In a case study, life-threatening methaemoglobinaemia associated with exposure to sodium nitrite was recorded in three previously healthy siblings: 4-year-old twin boys and their sister aged 2 years. Their father had used a solution of sodium nitrite as an insecticide. On the day of hospital admission, one of the children had mistaken a small bag of crystals for sugar, which was added to tea at concentrations of 5100 mg/l, 5000 mg/l and 4900 mg/l in the three cases, respectively (Finan et al., 1998).

In blood, nitrite is rapidly oxidized by haemoglobin to yield methaemoglobin and nitrate; therefore, the oral bioavailability of nitrite cannot be calculated from the plasma nitrite concentration. A study in healthy volunteers was conducted to investigate the oral bioavailability of a substantial dose of nitrite in an aqueous solution. As the vasodilating and methaemoglobin-inducing potential of nitrite limit the dose that can be administered safely, the study was preceded by a range-finding study in three volunteers to establish the dose that induced methaemoglobin of not more 10–15%. The results showed that a dose of 0.12 mmo/ mmol haemoglobin (290–370 mg of sodium nitrite) induced formation of 11% methaemoglobin in the blood. This was considered the maximum safe dose for adult volunteers, and doses of 0.04, 0.08 and 0.12 mmol were used. At the maximum dose, all volunteers reported effects of mild intensity. In addition to lowered blood pressure, a compensatory increase in heart rate was seen at all doses, which remained within acceptable limits. The peak plasma nitrite concentrations were considered acceptable for a bioavailability study (Kortboyer et al., 1998).

Two forms of methaemoglobin—methaemoglobin and nitrite methaemoglobin—were found in native human erythrocytes in the presence of sodium nitrite in suspension. The interaction of intercellular oxyhaemoglobin with nitrite ions in normal erythrocytes resulted in the formation of methaemoglobin, whereas it led predominantly to the formation of nitrite methaemoglobin in metabolically exhausted erythrocytes. Nitrite methaemoglobin reacts with hydrogen peroxide to form reactive intermediates (e.g. peroxynitrous acid) and the products of haemoglobin destruction. During storage of erythrocyte suspensions containing methaemoglobin and modified nitrite methaemoglobin, the forms of erythrocytes and their degree of haemolysis changed. It was assumed that formation of methaemoglobin leads to destruction of erythrocytes (Starodubtseva et al.,1999)

2.1.3 Endogenous formation of N*-nitroso compounds*

A study was conducted in the USA to determine the dose of ascorbic acid required to inhibit *N*-nitrosoproline formation when 400 mg of nitrate were given 1 h before a standard 400-calorie meal with 500 mg of L-proline. Volunteers ate their normal diets but restricted their intakes of nitrate, proline, *N*-nitrosoproline and ascorbic acid. *N*-Nitrosoproline and *N*-nitrosarcosine were determined in urine after 18 h by methylation and then gas chromatography with thermal energy analysis. The mean yields of *N*-nitrosoproline were 11, 42, 33, 22 and 23 nmol in groups of 9–25 subjects taking proline alone, proline plus nitrate or proline plus nitrate plus 120, 240 or

480 mg of ascorbic acid, respectively. There was a significant trend to lower *N*-nitrosoproline yields as the dose of ascorbic acid increased. These results correspond to inhibition by ascorbic acid of 28%, 62% and 60% , respectively. Pairwise comparisons showed that each group taking ascorbic acid formed significantly less *N*-nitrosoproline than the group given only proline plus nitrate. The mean *N*-nitrosarcosine yields were 9 nmol when proline was taken alone and 17–24 nmol with proline plus nitrate plus ascorbic acid, with no trend to increasing individual values. It was concluded that 120 mg of ascorbic acid taken with a meal containing nitrate at 360 mg/day would significantly reduce nitrosamine formation in vivo (Mirvish et al., 1998)

2.2 Toxicological studies

2.2.1 Short-term studies of toxicity

Mice

Groups of 10 male and 10 female B6C3F$_1$ mice were given drinking-water containing sodium nitrite at a concentration of 0, 375, 750, 1500, 3000 or 5000 mg/l, equal to average daily doses of approximately 90, 190, 340, 750 and 990 mg/kg bw for males and 120, 240, 440, 840 and 1200 mg/kg bw for females, for 14 weeks. The body weights of males at the highest dose were significantly lower than those of controls at week 13. The relative weights of the spleen in males and the absolute and relative weights of the heart, kidney, liver and spleen of females at the two higher doses were greater than those of the control groups. Sperm motility was decreased in males at the highest doses, and the estrous cycles of females at the three higher doses were significantly longer than those of controls. Increased incidences of squamous-cell hyperplasia of the forestomach were seen in males and females at the two higher doses, of extramedullary haematopoiesis of the spleen in males at the two higher doses and in females at the three higher doses and of degenaration of the testis in males at the two higher doses. The NOEL was 190 mg/kg bw per day (National Toxicology Program, 2001).

Rats

As hypertrophy of the adrenal zona glomerulosa was observed at all concentrations of potassium nitrite tested, including the lowest (100 mg/l), a study was performed to establish a NOEL for nitrite. Groups of 10 male and 10 female 6-week-old Wistar (TNO) rats received drinking-water containing potassium nitrite at a concentration of 12, 25, 50, 100 or 3000 mg/l or sodium nitrite at a concentration of 81 or 2400 mg/l for 13 weeks, the nitrite content of the drinking-water in the last two groups being equal to that of the groups given the two higher concentrations of potassium nitrite. The potassium and sodium concentrations in the drinking-water of the corresponding control groups were equalized with KCl and NaCl, respectively.

General health, behaviour and survival were not affected by ingestion of nitrite. Body weight and food and water intake were slightly decreased in animals of each sex given the highest dose of either nitrite. The methaemoglobin concentration was significantly increased in rats at these doses in weeks 4 and 12, while a slight increase in erythrocyte variables was seen during week 12 in females given the potassium salt at 3000 mg/l. The relative weight of the kidney was increased at the highest dose of each nitrite. In week 4, plasma aldosterone and corticosterone concentrations were slightly decreased in males at the higher dose of sodium nitrite, and plasma corticosterone was slightly decreased in females at the highest dose of potassium salt; these effects were not seen in week 13. Systolic blood pressure was not affected. Microscopic examination revealed slight hypertrophy of the adrenal zona glomerulosa in rats given potassium nitrite at 100 or 3000 mg/l and at the lower dose of the sodium salt, and plasma aldosterone and corticosterone concentrations were slightly decreased in males at the highest dose of sodium nitrite, the incidence and degree being dose-related. The results seen with potassium nitrite at 100 and 3000 mg/l in drinking-water were comparable to those found at the same concentrations in the previous 90-day study. The effects of sodium nitrite were similar to those of potassium nitrite. The NOEL for potassium nitrite was 50mg/l in drinking-water, equivalent to 5 mg/kg bw per day (Til et al., 1997).

In a study of the effect of age on the toxicity of nitrite, groups of 10 young adult and four aged male Wistar rats were allocated to four subgroups. Three subgroups received potassium nitrite in the drinking water at a concentration of 6 mmol/l (supplemented with 30 mmol KCl/l), 18 mmol/l (supplemented with 18 mmol KCL/l) or 36 mmol/l, equivalent to 83, 250 and 500 mg/kg bw, and a control group received 36 mmol/l. Blood samples were collected for measurement of methaemoglobin every 2 weeks, starting before the start of nitrite administration. In week 4, the rats were placed in metabolism cages, and 24-h urine samples were collected for qualitative (dipstick) and quantitative (volume, creatinine concentration, osmolality, *N*-acetylglutamate activity, nitrate and total protein concentrations) analysis. At the end of the urine collection period, blood was collected for clinical chemistry. In week 5, a test for phenolsulfthalein was conducted. In week 6, the rats were autopsied and examined macroscopically, and blood was collected for measurement of the concentrations of insulin, thyroxine and free thyroxine. Heart, liver, adrenals, brain, pituitary, spleen and thyroid gland, were weighed and fixed for histopathological examination. In addition, mesenteric lymph nodes and all organs showing macroscopic lesions were prepared for examination.

The known effects of nitrite, methaemoglobinaemia and hypertrophy of the zona glomerulosa of the adrenals, were observed in both young and aged rats. Hypertrophy of the zona glomerulosa was seen at the two higher doses, the NOEL thus being 6 mmol/l (equivalent to 84 mg/kg bw per day). Aged rats drank less water than controls (31% less) and the young adults (14% less at the highest dose). As nitrite is rapidly converted to nitrate, the plasma concentration of nitrate can be interpreted as a reflection of internal nitrite dose. The plasma nitrate concentrations of the aged rats were approximately three times higher than those of young adult rats. Because aged rats have more fat than young adults, the distribution volume is not proportional to body weight, leading to an increase in plasma nitrate. More efficient renal retention of water (and consequently of nitrite/nitrate) may also have contributed to the higher nitrate concentrations in the plasma of aged rats. Therefore, the internal nitrite dose of the aged rats was adequate and at least comparable to that of the young adults. No further treatment-related differences in effect were found between the groups. The NOEL in this strain of Wistar rats was 50 mg/kg bw per day, which is considerable higher than the NOEL in the strain used by Til et al. (1997), even though both NOELs were for hypertrophy of the renal zona glomerulosa. The authors concluded that the aged rats and young adult rats were equally susceptible to the toxic effects of nitrite (Boink et al., 1996, 1997).

Groups of 10 males and 10 females rats were given drinking-water containing sodium nitrite at a concentration of 0, 380, 750, 1500, 3000 or 5000 mg/l, equal to average daily doses of 30, 55, 120, 200 and 310 mg/kg bw per day for males and 40, 80, 130, 220 and 340 mg/kg bw per day for females, for 14 weeks. Further groups of 15 males and 15 females were given water containing the same concentrations for 70 and 71 days. One female at 220 mg/kg bw per day died before the end of the study.

The body weights of males at 310 mg/kg bw per day were significantly lower than those of controls. Males at 310 mg/kg bw per day and females at the two higher doses drank less water than controls in weeks 2 and 14. The clinical findings related to treatment included brown discolouration of the eyes and cyanosis of the mouth, tongue, ears and feet of males at 200 and 310 mg/kg bw per day and of females at the three higher doses. Reticulocyte counts were increased in both males and females at the two higher doses. The erythron was decreased on day 19 but increased by week 14 in males and females at the highest dose. Methaemoglobin concentrations were elevated in almost all treated groups throughout the 14-week study: the per cent of total haemoglobin found as methaemoglobin was 0.002% in control males, 4.4% in males at 200 mg/kg bw per day and 17% in males at 310 mg/kg bw per day, and 0.37% in control females, 5.8% in females at 220 mg/kg bw per day and 11% in females at 340 mg/kg bw per day. The authors concluded that a NOEL could not be identified for this effect. The relative weights of the kidney and spleen of males and females at the two higher doses were increased, and sperm motility was significantly decreased in males at 120 and 300 mg/kg bw per day. Increased erythropoietic activity in the bone marrow of males and females at the two higher doses was observed as an increased number of reticulocytes. The incidence of squamous-cell hyperplasia of the forestomach was significantly increased in males and females at the highest dose. As methaemoglobin formation of less than 3% is considered not to be adverse, the NOEL was 55 mg/kg bw per day on the basis of reduced sperm motlity in rats at 120 mg/kg bw per day (National Toxicology Program, 2001).

2.2.2 Long-term studies of toxicity and carcinogenicity

Mice

Groups of 50 male and 50 female B6C3F$_1$ mice were given drinking-water containing sodium nitrite at a concentration of 0, 750, 1500 or 3000 mg/l, equal to average daily doses of 60, 120 and 220 mg/kg bw for males and 45, 90 and 160 mg/kg bw for females, for 2 years. The survival rate of the treated groups was similar to that of the controls. The mean body weights of females at the highest dose were lower than those of controls throughout the study, and the treated groups generally consumed less water than the control groups. The incidences of squamous-cell papilloma and carcinoma (combined) in the forestomach of female mice showed a positive trend. The incidence of hyperplasia of the glandular epithelium was significantly greater in males at the highest dose than in controls. There was thus equivocal evidence of carcinogenic activity on the basis of the positive trend in the incidence of squamous-cell papilloma and carcinoma (combined) of the forestomach (National Toxicology Program, 2001).

Rats

The effects of long-term administration of powdered fish meal with sodium nitrite were studied in six groups of 50 Fischer 344 rats of each sex. The first three groups were fed diets supplemented with 8% (basal diet), 32% or 64% fish meal and given drinking-water containing 0.12% sodium nitrite. The other three groups were given the same concentrations of fish meal and tap water. At week 104, all surviving animals were killed and examined histologically.

Treatment with fish meal increased the incidence and multiplicity of atypical tubules, adenomas and renal-cell carcinomas in male rats in a dose-dependent manner. Females were less susceptible than males to renal tumour induction. The incidence and multiplicity of atypical tubules were also significantly increased in males given 64% fish meal as compared with those given 8% fish meal. Nephropathy occurred in a clearly dose-dependent manner in groups given fish meal, irrespective of the sodium nitrite concentration, and was more frequent in males than in females. *N*-Nitrosodimethylamine (NDMA) was found in the stomach contents after 4 weeks of treatment with 64% fish meal plus 0.12% sodium nitrite, at a level twice that in the group given 8% fish meal plus 0.12% sodium nitrite. Thus, concurrent administration of fish meal and sodium nitrite induced renal epithelial tumours. The high intake of protein should be noted (Furukawa et al., 2000).

Groups of 50 male and 50 female Fischer 344/N rats were given drinking-water containing sodium nitrite at a concentration of 0, 750, 1500 or 3000 mg/ l, equal to average daily doses of 35, 70 and 130 mg/kg bw for males and 40, 80 and 150 mg/kg bw for females, for 2 years. In studies of the toxicokinetics of plasma nitrite and blood haemoglobin, 10 male and 10 females were given the same concentrations for 12 months.

The survival of treated groups was similar to that of controls. The mean body weights of males and females at the highest dose were lower than those of controls throughout the study. Males and females at this dose drank less water than controls throughout the study, and the water consumption of the other treated group was generally lower only after week 14. The incidences of hyperplasia of the forestomach epithelium in males and females at the highest dose were significantly higher than in controls. The incidence of fibroadenoma of the mammary gland was significantly increased in females at the intermediate dose, and the incidences of multiple fibroadenoma were increased in females at the two lower doses; however, these neoplasms occur at a high background incidence, and no increase was seen at the highest dose. The incidences of mononuclear-cell leukaemia were significantly decreased in males and females at 80 or 150 mg/kg bw per day. Under the conditions of this study, there was no evidence of carcinogenicity (National Toxicology Program, 2001).

2.2.3 Reproductive toxicity

Mice

Sodium nitrite was tested for its effects on reproduction and fertility in Swiss CD-1 mice. Data on food and water intake, body weight and clinical signs in a dose range-finding study were used to select concentrations of 0.06%, 0.12% and 0.24% (w/v) in drinking-water for the continuous cohabitation phase. Water consumption was reduced by 10–17% in the group given the highest concentration of sodium nitrite, although the body weights of the F_0 generation during the cohabitation phase were not reduced. The data on body weight and water consumption allowed calculation of daily nitrite consumption, which was 120, 260 and 420 mg/kg at the low, intermediate and high doses, respectively. Nine animals died during this phase, comprising three at 0.06%, four at 0.12%, none at 0.24% and one control. There was no-treatment-related reduction in the mean number of litters per pair, the number of pups per litter or the viability or weight of the pups. The days to delivery of each litter were not affected. Each dam was allowed to nurse the last litter from the cohabitation phase until weaning. While mortality during nursing was not increased by treatment with nitrite, pup body-weight gain was reduced in the group at the highest dose on postnatal days 7–21 by 12–17%. This effect may have been due to reduced maternal water intake, with consequent lower milk production.

As no effects on reproduction were noted during this phase, only the controls and mice at the highest dose were retained after weaning and examined for potential reproductive toxicity. Water consumption was measured several times during this part of the study and was found to be reduced by 8%. The body weights of F_1 mice at the beginning of the week of mating were similar in all groups. The fertility and reproductive success of F_1 mice during the last phase were not affected by treatment, and there was no change in the ability of mice to mate, become pregnant or deliver live young. The number, weight and viability of the young were not reduced by treatment. After delivery and analysis of the F_2 pups, the F_1 mice were killed and necropsied. The terminal body weights of treated and exposed mice did not differ, and there were no treatment-related changes in the weight of any organ. The length and pattern of the estrous cycle were unchanged during the 12 days of evaluation, and the motility, concentration and morphology of epididymal sperm were also unchanged. Histologically, the liver and kidneys of the mice at 0.24% nitrite and the control mice were similar. Thus, sodium nitrite had no adverse effect on reproduction or reproductive performance at necropsy in mice treated at doses up to 420 mg/kg bw, which was the NOEL for reproductive toxicity (Chapin et al., 1997).

2.2.4 Genotoxicity

Sodium nitrite was mutagenic in *Salmonella typhimurium* strain TA100, with and without Aroclor 1254-induced hamster and rat liver enzymes; no mutagenicity was observed in strain TA98. Intraperitoneal injection of sodium nitrite to male mice and rats did not induce micronucleus formation in bone marrow, and a test for micronuclei in peripheral blood from mice in the 14-week study described above also gave negative results (National Toxicology Program, 2001).

2.2.5 Special studies

(a) Effects on the immune system

A pronounced decline in the percentage of $CD4^+$ T lymphocytes was found in mesenteric lymph nodes from groups of four $B6C3F_1$ mice infected with *Trichinella spiralis* and given nitrite for 10 or 25 days. Less pronounced differences were observed in the percentage of $CD8^+$ T lymphocytes in both mesenteric lymph nodes and Peyer patches. Infection with T. spiralis alone induced a pronounced decline in $CD4^+$ T lymphocytes in mesenteric lymph nodes, whereas the percentage of $CD8^+$ T lymphocytes remained at the level of controls. Treatment with sodium nitrite alone caused a clear increase in the percentage of $CD8^+$ cells as compared with untreated, uninfected controls. The latter effect may result in a depressed immune response, causing more severe T. spiralis infections in sodium nitrite-treated mice (Bany et al., 1995).

(b) Effects on the vascular system and adrenals

Studies of blood pressure and the role of the adrenals in the mode of action of nitrite in hypertrophy of the adrenal zona glomerulosa have been reviewed. Nitrite and organic nitrates have vasodilating properties (Nickerson, 1975), and dilatation and thinning of the intramuscular coronary blood vessels were seen in rats given drinking-water containing sodium nitrite at concentrations of 100–300 mg/l for 2 years (Shuval & Gruener, 1972). A fatal fall in blood pressure after nitrate poisoning was reported in cows given nitrate (Dinkla, 1976); as nitrate is efficiently converted to nitrite in ruminants, the fall in blood pressure was probably due to nitrite rather than to the ingested nitrate.

The effects of physiological amounts of nitrite on generation of nitric oxide and relaxation of rat aorta were studied in vitro with the pH reduced to that found in tissues during hypoxia or ischaemia. The relaxing effect of nitrite was greater in an acidic buffer solution (pH 6.6) than at neutral pH; the EC_{50} for nitrite was reduced from 200 to 40 μmol/l. Nitrite-evoked relaxation was effectively prevented by co-administration of an inhibitor of soluble guanylyl cyclase and was further potentiated by addition of ascorbic acid. Nitric oxide was generated from nitrite for 11 days in a pH-dependent manner, with even larger amounts seen after addition of ascorbic acid, and was correlated to the degree of relaxation of rat aorta (Modin et al., 2001)

The effect of sodium nitrite on blood pressure was investigated to test the hypothesis that nitrite induces hypertrophy of adrenal zona glomerulosa by reducing the blood pressure and stimulating the renin–angiotensin axis. Blood pressure was evaluated with a radiotelemetry system in conscious, freely moving animals. Two male Wistar rats (weighing 350 and 375 g) were maintained on a 12-h light–dark cycle for 11 days, and systolic and diastolic blood pressure and heart rate were measured every 5 min; mean arterial pressure was calculated. Potassium nitrite decreased the mean arterial pressure and increased the heart rate. Potassium chloride had no effect. A previous study in anaesthetized rats showed rapid conversion of plasma nitrite into nitrate (Vleeming et al., 1995a,b). As intravenouis administration of sodium nitrate at a concentration of 100 μmol/kg bw over 5 min did not decrease the mean arterial pressure, it was concluded that nitrite is responsible for hypotension and therefore for activation of the renin–angiotensin axis (Vleeming et al., 1996).

In order to test the hypothesis that the relatively low blood pressure of vegetarians is due partly to a high dietary load of nitrate, which is reduced to nitrite and finally to nitric oxide, spontaneously hypertensive rats received drinking-water containing nitrite at a concentration of 0, 25, 50 or 100 mmol/l for 56 days. Food was offered ad libitum or was restricted by 20% (pair-feeding) to simulate the lower energy consumption of vegetarians. Blood pressure, which was monitored at regular intervals, was lowered in a dose-dependent manner by nitrite. The effect was reversible and was not enhanced by energy restriction. In volunteers, the plasma nitrate concentration increased by a factor of 8–32 after ingestion of a nitrate-rich meal, and the mean methaemoglobin concentration increased from 1.2% to 2.4%, indicating endogenous formation of nitrite under these conditions (Beier et al., 1995).

To investigate whether long-term administration of nitrite lowers the blood pressure of spontaneously hypertensive rats and prevents secondary lesions induced in organs by hypertension, 96 of these rats received drinking-water containing sodium nitrite at concentrations of 50–75 mmol/l or equimolar amounts of sodium bicarbonate for 4, 8 or 12 months. At each time, arterial blood pressure (determined by the tail-cuff method) was significantly lower in the group receiving sodium nitrite than in the controls, indicating that no significant tolerance towards nitrite had developed. There was also a tendency to reduced cardiac hypertrophy and renal atrophy in this group, which was not statistically significant. Drinking-water containing sodium nitrite at 75 mmol/l was not well tolerated by young rats (Haas et al., 1999)

Intravenous administration of nitrite to anaesthetized rats induced an immediate, dose-dependent decrease in blood pressure, which preceded an increase in methaemoglobin concentration, suggesting that hypotension is the primary effect of nitrite. A single dose of 30 μmol/kg bw caused a 10–20% decrease in blood pressure. Rats drink about 50 ml/kg bw of water each day. If the drinking-water contains nitrite at 36 mmol/l, the total daily dose will therefore be 1800 μmol/kg bw. Thus, assuming 100% bioavailability, 1/60 of the total daily consumption was able to decrease the blood pressure by 10–20%. As the blood pressure had returned to normal within 30 min, continuous monitoring is necessary to detect the effect of nitrite on blood pressure. Recording of blood pressure and heart rate by means of a telemetry device in a freely moving rat given drinking-water containing potassium nitrite at a concentration of 36 mmol/l (equivalent to 500 mg/kg bw) revealed transient hypotension (Vleeming et al., 1997).

Morphometry and histology of the adrenals of 10 Wistar rats of each sex given drinking-water containing potassium nitrite at a concentration of 3.6, 12 or 36 mmol/l showed slight hypertrophy of the zona glomerulosa after 28 days, which persisted unchanged throughout the 90-day treatment but disappeared slowly after cessation of treatment. Slight focal and slight diffuse hypertrophy were still present after a 30-day recovery period but had disappeared by 60 days. Enlargement of the zona glomerulosa was not seen by morphometry in rats at the two lower doses, but histological examination showed that the adrenals of eight rats at 12 mmol/l and three at 3.6 mmol/l were minimally hypertrophic. This effect was dose-related; no NOEL could be identified. Nitrite is readily converted to nitrate, the plasma half-life of nitrite in various species being estimated as < 30 min, while the half-life of nitrate ranges from 4 h (sheep and horses) to 44 h (dogs), suggesting that nitrate plays a role in the causation of hypertrophy of the adrenal zona glomerulosa. However, no hypertrophy was observed by means of morphometry in rats given drinking-water containing a concentration of 36 mmol/l for 90 days, and histological examination revealed minimal hypertrophy in only two of 10 rats. Thus, the hypertrophy must be attributed to nitrite ion alone.

The adrenals regulate blood pressure via the renin–angiotensin–aldosterone axis. When blood pressure decreases, baroreceptors in the juxtaglomerular apparatus in the kidneys stimulate release of renin. Renin catalyses the conversion of angiotensinogen to angiotensin I, which in turn is converted to angiotensin II by a converting enzyme. Angiotensin II effectively contributes to restoration of normal blood pressure by its vasoconstrictive effect on blood vessels and by stimulating the release of aldosterone from the zona glomerulosa of the adrenals. Aldosterone enhances the reabsorption of sodium in the renal tubules, leading to expansion of the extracellular volume. This rise in blood volume also causes an increase in blood pressure. As described above, administration of nitrite to rats in the drinking-water caused repeated falls in blood pressure, thus repeatedly activating the renin–angiotensin–aldosterone-axis, which may have caused hypertrophy of the adrenal zona glomerulosa. To investigate this suggestion further, a limited number of rats were given ramipril, an inhibitor of angiotensin-converting enzyme, in their diet concurrently with nitrite. No hypertrophy of the zona glomerulosa occurred, strongly suggesting that slight hypertrophy of the adrenal zona glomerulosa represents a physiological adaptation to repeated episodes of hypotension caused by nitrite (Boink et al., 1998).

(c) Effects on tumour induction

A number of studies have addressed the effect of administration of 0.2–0.3% sodium nitrite on the incidence of forestomach tumours induced in rats by well-known carcinogens (*N*-methyl-*N*-nitro-*N*-nitrosoguanidine, phenolic compounds, catechol, 3-methoxycatechol and butylated hydroxyanisole). These concentrations of nitrite, equivalent to 200–300 mg/kg bw, enhanced the incidence of forestomach tumours induced by the carcinogens. Lower doses were not examined. As forestomach tumours are of limited relevance for humans, however, these findings have no impact on the safety assessment of nitrite in food (Hirose et al., 1993; Kawabe et al., 1994; Yoshida et al., 1994; Miyauchi et al., 2002). In a recent study of the effect of 0.2% sodium nitrite (equivalent to 200 mg/kg bw) on the incidence and volume of mammary tumours in rats induced by 2-amino-1-methyl-6-phenylimidazol[4,5-b]pyridine, the nitrite did not affect the incidence of mammary tumours but appeared to reduce their volume. This finding at a relatively high dose has no implications for the safety assessment of nitrite in food (Hirose et al., 2002).

2.3 Observations in humans

2.3.1 Effects on blood pressure and blood flow

To determine the relative contributions of endothelial nitric oxide and intravascular nitrogen oxide species in the regulation of blood flow, simultaneous measurements were made of forearm blood flow and arterial and venous concentrations of plasma nitrite, low-molecular-mass *S*-nitrosothiols, high-molecular-mass *S*-nitrosothiols and erythrocyte *S*-nitrosohaemoglobin. Measurements were made at rest and during regional inhibition of nitric oxide synthesis, followed by forearm exercise. Unexpectedly, significant gradients in circulating arterial–venous plasma nitrite, representing a novel delivery source for intravascular nitric oxide, were observed. Further support for the notion that circulating nitrite is bioactive was provided by the finding that use of nitrite increased significantly with exercise during inhibition of regional endothelial synthesis of nitric oxide. It was concluded that circulating nitrite is bioactive and provides a gradient for delivery of intravascular nitric oxide (Gladwin et al., 2000).

2.3.2 Epidemiological studies

Epidemiological studies can be ranked in increasing order, from ecological (or correlation) studies to cross-sectional studies, case–control studies, cohort studies and intervention trials. This classification of epidemiological study designs with respect to their potential for bias and, consequently, the strength of evidence they provide and the costs involved has been described in detail (e.g. van den Brandt et al., 2002). Intervention trials provide the strongest evidence for a causal relationship with risk and (because of the possibility to control for confounding and bias) have the least possible bias; however, they are usually the least feasible and the most expensive. Less expensive cohort studies allow assessment of exposure and selection of study participants before the health outcome of interest occurs and thus provide relatively strong evidence. Although the less expensive case–control studies generally involve assessment of exposure retrospectively in subjects with and without the health outcome, the resulting evidence is more debatable. This is particularly so in the case of dietary intake, because of the possibility of selection bias, recall bias and/or bias due to the presence of disease. Cross-sectional studies suffer from the additional problem that exposure and disease are measured at the same time, making it impossible to draw conclusions about cause and effect. Correlation studies cost the least, but they provide weak evidence and are much more susceptible to bias. In addition to the problem of extrapolating the data to the individual level (as the units of measurement are population groups), most such studies include limited data on exposure, rely on mortality (rather than incidence) rates and often do not include consideration of the induction period. Some investigators have stated that observational studies cannot, by definition, establish the causality of a relationship on the basis of a statistical association. However, when several studies of high quality, such as those in which the biases are shown to be minimal, are available and these consistently show a dose–response relationship, observational studies may well contribute to conclusions about causality. The power of observational epidemiological studies was established 50 years ago, when such studies revealed that smoking caused lung cancer.

(a) Cross-sectional studies

The concentration of nitrite in gastric juice and of several *N*-nitroso compounds and other analytes in urine were measured in nearly 600 residents of an area of Shading, China, where precancerous gastric lesions are common and the rates of stomach cancer are among the world's highest. The concentration of nitrite in gastric juice was considerably higher in people with pH values for gastric juice > 2.4 than in those with values < 2.4, and nitrite was detected more often and at higher concentrations in people with late-stage gastric lesions, especially when the gastric pH was high. Of patients with intestinal metaplasia, 18% had detectable gastric nitrite, whereas only 7.2% of people with less advanced lesions had this analyte. The odds ratio (OR) for intestinal metaplasia in people with detectable nitrite relative to those with undetectable nitrite increased from 1.5 (95% confidence interval [CI], 0.6–4.1) to 4.1 (95% CI, 1.8–9.3) for people with low and high nitrite concentrations, respectively. Urinary acetaldehyde and formaldehyde concentrations also tended to be higher among patients with more advanced disease, particularly dysplasia. However, the urinary excretion of total *N*-nitroso compounds and several nitroamino acids was similar among patients with chronic atrophic gastritis and intestinal metaplasia and dysplasia, as was reported by the authors to have been found in recent studies in Colombia, France and the United Kingdom. The results for this high-risk population suggested that elevated concentrations of gastric nitrite, especially in a high pH milieu, were associated with advanced precancerous gastric lesions, although specific *N*-nitroso compounds were not implicated (You et al., 1996). This cross-sectional study is of limited value, as cause and effect are confusingly mixed.

The possible role of *N*-nitroso compounds in the induction of upper aerodigestive tract tumours was considered in a case–control study conducted in the Valle d'Aosta, an Italian region with a high incidence of these neoplasms. Nitrate, nitrite and labile and stable *N*-nitroso compounds were analysed in the saliva of 36 patients with these cancers and 23 healthy individuals. After allowing for tobacco use, the presence of nitrate, nitrite and *N*-nitroso compounds in saliva was not associated with an increased risk for upper aerodigestive tract cancers. The OR for continuous units of total *N*-nitroso compounds was 0.99 (95% CI, 0.9–1.1). Thus, salivary nitrate, nitrite and *N*-nitroso compounds might not be suitable markers for assessing the risk for cancers at these sites, although a role for *N*-nitroso compounds could not be excluded (Airoldi et al., 1997).

The excretion of nitrate, nitrite, apparent total *N*-nitroso compounds and volatile nitrosamines was measured in 24-h urine from 61 Egyptians, divided into four groups: controls, *Schistosoma haemotobium*-infected patients and bladder cancer patients with and without a history of schistosomal infection. The urinary concentration of nitrate in *S. haemotobium*-infected patients was significantly higher than in the other three groups. The concentration of nitrite was below the limit of detection of the method (< 0.015 μg/mg creatinine) in all but one of the control samples. *S. haemotobium* infection significantly increased urinary nitrite to 0.9 ± 1.16 μg/mg creatinine (mean ± SD, $p = 0.001$), and the concentration of nitrite in both groups of patients with bladder cancer was about 20 times that in S. haemotobium-infected patients without bladder cancer. Excretion of apparent total *N*-nitroso compounds paralleled that of nitrite, and, overall, a good correlation was observed between these two variables ($r = 0.71$, $p = 0.0001$). NDMA was present in all the samples analysed, and S. haemotobium infection significantly increased the urinary concentration over that of controls (4.0 ± 1.6 and 2.0 ± 3.0 ng/mg creatinine, respectively; $p = 0.01$). Among the cancer patients, the concentration of NDMA was higher than that in controls only in those with schistomal infection. The presence of *N*-nitroso compounds, including NDMA, in the urine of S haemotobium-infected patients both before and after the development of cancer and the observation that these compounds also occur in bladder cancer patients with no history of schistosomal infection suggest that these compounds might have a role not only in initiation of the carcinogenic process but also in its progression (Mohsen et al., 1999). As endogenous nitric oxide production can also result from schistosomiosis, cause and effect can be reversed.

A study was conducted in Egypt, an area with high environmental levels of nitrosating agents, to investigate whether salivary nitrate and nitrite and the activity of nitrate reductase in saliva affect the risk for oral cancer. The concentrations of salivary nitrite (8.3 ± 1.0 μg/ml) and nitrate (44 ± 3.7 μg/ml) and the activity of nitrate reductase (74 ± 10 nmol/ml per min) were significantly ($p < 0.05$) higher in 42 oral cancer patients than in 40 healthy individuals (nitrite, 5.3 ± 0.3 μg/ml; nitrate, 27 ± 1.2 μg/ml; nitrate reductase activity, 46 ± 4 nmol/ml per min). The adjusted ORs and the 95% confidence intervals for oral cancer, categorized by the values for salivary nitrate and nitrite and nitrate reductase activity, showed a risk associated with a nitrite concentration > 7.5 μg/ml (OR, 3.0; 95% CI, 1.0–9.3), a nitrate concentration > 40 μg/ml (OR, 4.3; 95% CI, 1.4–13.3) and nitrate reductase activity > 50 nmol/ml per min (OR, 2.9; 95% CI, 1.1–7.4). The findings suggest that increased dietary intake of nitrate and nitrite is associated with elevated levels of salivary nitrite. Together with the increased activity of salivary nitrate reductase, these observations may explain, at least in part, the role of nitrate and nitrite in the development of oral cancer in individuals in areas with a high burden of *N*-nitroso precursors (Badawi et al., 1998).

(b) Case–control studies

A case–control study on gastric cancer and diet conducted in Marseille (France) included 92 patients with histologically confirmed adenocarcinoma and 128 controls undergoing functional reduction for injuries or trauma. The participants were interviewed by a trained dietician who administered a questionnaire about usual diet during the year preceding the first symptoms for cases or preceding the interview for controls. The intakes of nitrite, nitrite and pre-formed NDMA from food were estimated from a food composition table compiled ad hoc. Odds ratios were calculated after adjustment for age, sex, occupation and calorie intake. A high intake of preformed NDMA was associated with an increased risk for gastric cancer, the ORs for the second and third tertile of NDMA intake being: $OR_2 = 4.1$ (95% CI, 0.93–18) and $OR_3 = 7.0$ (95% CI, 1.8–26). Intake of nitrate and nitrite was not associated with an increased risk for stomach cancer. Consumption of vegetables was protective in general, independently of their estimated nitrate content (Pobel et al., 1995).

A study was conducted to investigate whether consumption of foods and beverages containing nitrosamines, nitrite and nitrates affects the risks for laryngeal, oesophageal and oral cancer. In a population-based case–control study in western Washington State, USA, dietary consumption of these substances was measured for 645 cases (169 laryngeal, 125 oesophageal and 351 oral cancer) and 458 controls. After adjustment for tobacco and alcohol use and other known risk factors, the risk for upper aerodigestive tract cancer was found to be 52% lower in persons who consumed larger amounts of nitrate (upper tertile) as compared with the lowest tertile ($p < 0.001$ for trend). Nitrate intake was associated with a reduction in the risk for cancers at all three sites. The reduction in risk for oesophageal cancer with increasing nitrate consumption was more evident in frequent tea drinkers than in other persons. The ORs for oesophageal cancer per category of nitrite intake were: 1.0 (reference) for < 1.1 mg/day, 1.2 for intakes between 1.1 and 1.6 mg/day and 1.6 (95% CI, 0.73–3.4) for intakes > 1.6 mg/day (p for trend = 0.20). There was no significant association between nitrite consumption and the risk for laryngeal or oral cancer; however, for individuals with a history of canker sores (an indicator of possible endogenous nitrosation), the risk for oesophageal cancer was seven times greater in those with a high nitrite intake than in those with a low intake. Consumption of foods with high concentrations of NDMA was associated with a 79% increase in the risk for upper aerodigestive tract cancer ($p = 0.037$ for trend). Daily consumption of beer and of nitrite-containing meats was associated with increased risks for oesophageal cancer (OR, 2.5 and 1.8, respectively). The OR for cancer of the oral cavity was also increased for people who drank beer daily (OR, 1.8). Persons who consumed large amounts of ascorbic acid from foods and from supplements were less likely to develop upper areodigestive tract cancer than were individuals with lower ascorbic acid intake (p for trend = 0.003) (Rogers et al., 1995).

Many cases of lower urinary tract cancer cannot be attributed to known risk factors, such as cigarette smoking and certain occupational exposures to chemicals. Data from a case–control study conducted on Oahu, Hawaii, between 1979 and 1986 were used to examine the role of other factors, such as total fluid intake and dietary intake of nitrites, nitrosamines and selected foods. A total of 195 male and 66 female patients of white and Japanese ancestry with lower urinary tract cancer were matched to two population-based controls on age, sex and race. Total fluid intake, and that of tap water in particular, showed an inverse dose–response relationship with cancer risk among women (OR for highest relative to lowest quartile of total fluid intake = 0.3; p for trend < 0.01). This association was stronger among smokers than nonsmokers. Although fluid intake showed no overall association among men, the findings for smokers suggested an effect similar to that found in women. Current intake of dietary nitrites and nitrosamines was positively associated with risk in Japanese men (for nitrites, OR for highest relative to lowest tertile = 2.0; p for trend = 0.05; for nitrosamines, OR for highest tertile relative to lowest tertile = 3.0; p for trend = 0.01). Consumption of processed meats, in particular bacon, sausage and ham, was also significantly associated with increased risk only in Japanese men, although associations with consumption of sausage in Japanese women and of bacon in white women were suggested. Unfortunately, it was not possible to determine whether these elevated risks were due to the fat, nitrite or sodium content of the processed meats or to their having been fried (Wilkens et al., 1996).

A case–control study conducted in northern Italy on 746 cases of histologically confirmed gastric cancer and 2053 controls who had been admitted to hospital for acute non-neoplastic, non-digestive-tract disorders addressed the interaction between gastric cancer and the intake of methionine, salt and nitrite. In relation to people with low methionine (< 1500 mg/day) and low salt intake (not defined), the ORs were 2.4 (95% CI, 1.5–4.0) for patients with low salt and high methionine intake (> 1900 mg/day), 1.5 (95% CI, 1.0–2.2) for patients with high salt and low methionine intake and 2.8 (95% CI, 1.9–4.2) for patients with high salt and high methionine intake. Likewise, in relation to people with low methionine and low nitrite intake (< 2.7 mg/day), the ORs were 1.9 (95% CI, 1.3–2.6) for high methionine and low nitrite, 1.5 (95% CI, 2.1) for low methionine and high nitrite (> 2.7 mg/day) and 2.5 (95% CI, 1.9–3.2) for high methionine and high nitrite intake. All these estimates were statistically significant. Thus, the findings from this large-scale case–control study conducted in a relatively high-incidence area for gastric cancer suggest that a diet rich in methionine, salt and nitrite is associated with an increased risk for gastric cancer (La Vecchia et al., 1997).

Studies of nasopharyngeal carcinoma have associated elevated risks with higher consumption of salted fish and preserved foods, particularly during childhood. Although these foods can contain high concentrations of nitrosamines, exposure to these compounds was not measured directly in most studies. In a case–control study conducted in Taiwan (China) to evaluate dietary intake and risk for nasopharyngeal carcinoma, interviews were conducted with a food-frequency questionnaire administered to 375 patients (99% response rate) and 327 controls (88% response rate) about their diet at age 10 and as an adult. The mothers of 96 cases and 120 controls were interviewed about their own diet during breastfeeding and the diets of their children during weaning, at the age of 3 and up to the age of 10. Concentrations of nitrosamines and nitrite were assigned to 66 foods on the basis of published values. Intake of nitrosamines and nitrite as an adult was not associated with an increased risk for nasopharyngeal carcinoma, but high intakes of nitrosamines and nitrite from foods other than soya bean products during childhood and weaning were associated with increased risks. The adjusted ORs for the highest quartile were 2.2 (95% CI, 0.8–5.6) for children aged 10, 2.6 (95% CI, 1.0–7.0) for children aged 3 and 3.9 (95% CI, 1.4–10) for the weaning diet. The intakes of nitrite and nitrosamines from soya bean products during childhood and weaning were inversely associated with risk, perhaps because soya beans contain known inhibitors of nitrosation. The results suggest that intake of nitrosamines and nitrite during childhood plays a role in the development of nasopharyngal carcinoma (Ward et al., 2000). The validity of the recall of childhood diets is uncertain, however.

In view of the rapidly rising incidence rates of adenocarcinoma of the oesophagus and gastric cardia in the USA, nutrient intake was studied as a risk factor for these cancers in a population-based case–control study in Connecticut, New Jersey and western Washington State. Interviews were conducted with 282 patients with histologically confirmed oesophageal adenocarcinoma, 255 with adenocarcinoma of the gastric cardia, 206 with oesophageal squamous-cell carcinoma, 352 with gastric adenocarcinoma other than in the cardia and 687 population controls. Associations between nutrient intake and risk for cancer were estimated as adjusted ORs, with a comparison of the 75th and 25th percentiles of intake. The nutrients that were significantly inversely associated with the risks for all four tumour types were fibre, β-carotene, folate and vitamins C and B6. In contrast, dietary cholesterol, animal protein and vitamin B12 were significantly positively associated with risk for all four tumour types. Dietary fat was significantly associated with the risk for oesophageal adenocarcinoma only (OR, 2.2; 95% CI, 1.3–3.8), and dietary nitrite was associated only with non-cardia gastric cancer (OR, 1.6; 95% CI, 1.3–2.2). Use of vitamin C supplements was associated with a significantly lower risk for non-cardia gastric cancer (OR, 0.60; 95% CI, 0.41–0.88). Higher intake of nutrients found primarily in plant-based foods was associated with reduced risks for adenocarcinomas of the oesophagus and gastric cardia, whereas higher intake of nutrients found primarily in foods of animal origin was associated with increased risks (Mayne et al., 2001).

A study of risk factors for childhood brain tumours was conducted to investigate the association between source of residential drinking-water during pregnancy and the occurrence of brain tumours among offspring. Dipstick measurements were made of nitrates and nitrites in tap water in the houses of a subset of women who were living in the same house in which they had lived during their pregnancy. A total of 540 patients with childhood brain tumours and 801 controls were identified in Los Angeles County and the San Francisco Bay Area of California and western Washington State, USA. Overall, no increased risk for childhood brain tumours was found in offspring of women for whom wells were the source of water; however, an increased risk (OR, 2.6; 95% CI, 1.3–5.2] was observed among offspring of women in Washington State and a decreased risk (OR, 0.2; 95% CI, 0.1–0.8) among those in Los Angeles County who relied exclusively on well water. Among the small subset of participants for whom dipstick measurements of tap water were available, the risk for childhood brain tumours associated with the presence of measurable nitrite and/or nitrate was 1.1 (95% CI, 0.7–2.0). Given the crude measurement method used and the fact that measures were often obtained years after the pregnancy, the relevance of the dipstick findings is unclear. The lack of consistency in the findings for residential water source does not support the hypothesis of an increased risk with consumption of well water; however, regional differences in the well water content of nitrite may exist, and the increased risk observed in western Washington State deserves further evaluation (Mueller et al., 2001).

Although brain tumours are the main cause of death from childhood cancers, the causes of most of these tumours remain obscure. Few chemicals cause brain tumours experimentally after systemic administration of low doses; a notable exception is one group of *N*-nitroso compounds, the nitrosamides (in particular the nitrosoureas). Thus, feeding pregnant animals nitrosamide precursors (e.g. sodium nitrite and an alkylamide such as ethylurea) causes a high incidence of nervous system tumours in the offspring. A population-based epidemiological study was therefore designed to test the hypothesis that maternal consumption during pregnancy of meats cured with sodium nitrite increases the risk for brain tumours in their offspring. Intake of vitamins C and E inhibits endogenous formation of *N*-nitroso compounds and was expected to be protective. The mothers of 540 children under the age of 20 during 1984–91 in whom a primary brain tumour was diagnosed at that time and 801 control children in the same 19 counties on the west coast of the USA were interviewed. The risk increased with increasing frequency of eating processed meats (OR, 2.1 for consumption at least twice a day compared with no consumption; 95% CI, 1.3–3.2; $p = 0.003$). The risk also increased with increasing average daily amount of cured meats or of nitrite from cured meats (p for each < 0.005) but not with nitrate from vegetables. Daily use of vitamins throughout pregnancy decreased the risk (OR, 0.54; 95% CI, 0.39–0.75). The risk of mothers who consumed nitrite from cured meat at above the median was higher when vitamins were not taken (OR, 2.4; 95% CI, 1.4–3.6) than when they were (OR, 1.3). These effects were seen for each of the three major histological types of brain tumour and across social classes, age groups and geographical areas. This is the largest study to date of maternal diet and childhood brain tumours and suggests that exposure to endogenously formed *N*-nitroso compounds during gestation is associated with tumour occurrence (Preston-Martin et al., 1996).

In a study to evaluate the roles of maternal nutrition during gestation and subsequently in the etiology of childhood brain tumours, all 300 incident cases of nervous system tumours diagnosed in children under 18 in Israel between 1984 and 1993 were identified. Two matched population controls per case were selected (n = 574). Personal interviews were conducted with a semi-quantified three-step food frequency questionnaire. Univariate analysis showed that increased consumption of vegetable fat (p for trend = 0.01; 95% CI, 1.1–3.2), carbohydrates (p for trend = 0.05; 95% CI, 1.0–5.9) and vitamin E (p for trend = 0.05; 95% CI, 1.0–3.3) during childhood was significantly associated with risk for brain tumour. No associations were found with nitrate, nitrite or vitamin C. A significant positive association with potassium consumption during gestation (p for trend = 0.01; 95% CI, 1.1–3.7) was noted. In a multivariate analysis, the only persistent associations were with vegetable fat (OR, 1.4; 95% CI, 1.1–1.7) in the diet during childhood and potassium intake during gestation (OR, 1.4; 95% CI, 1.0–2.0) (Lubin et al., 2000).

(c) Cohort studies

The association between intake of nitrate or nitrite and gastric cancer risk was investigated in a prospective cohort study begun in 1986 in the Netherlands, of 120 852 men and women aged 55–69 years. At baseline, data on dietary intake, smoking habits and other covariates were collected by means of a self-administered questionnaire. For data analysis, a case–cohort approach was used, in which the person–years at risk were estimated for a randomly selected subcohort of 1688 men and 1812 women. After 6.3 years of follow-up, 282 microscopically confirmed incident cases of stomach cancer were detected, with 219 in men and 63 in women. The rate ratios for gastric cancer with increasing quintiles of mean nitrite intake were: 1.0 (intake, 0.01 mg/day), 1.2 (0.04 mg/day), 1.2 (0.09 mg/day), 0.88 (0.16 mg/day) and 1.4 (0.35 mg/day). The 95% confidence interval for the rate ratio derived by contrasting the highest versus the lowest intake quintile was 0.95–2.2. A test for trend in rate ratios across quintiles gave a p value of 0.24. The authors concluded that the study did not support a strong positive association between nitrite intake and gastric cancer risk (van Loon et al., 1998).

A cohort study was conducted to investigate the relationship between intake of nitrates, nitrites and NDMA and the risk for cancers of the gastrointestinal tract in 9985 adult Finnish men and women who were initially free of cancer. During a follow-up period of up to 24 years, 189 gastrointestinal cancer cases were diagnosed in the cohort. The intakes of nitrate, nitrite and NDMA were estimated from data on food consumption obtained during an interview about the total diet of the participants over the previous year. The mean daily intake of nitrates was 77 mg, and that of nitrites was 5.3 mg. Nitrates were provided mainly by vegetables (92%), whereas nitrites were derived mainly from cured meats and sausages (94%); dietary NDMA was provided by smoked and salted fish (52%) and cured meats and sausages (48%). The mean daily intake of NDMA from the diet was 0.052 μg, and that from beer was 0.071 μg. A significant positive association was observed between intake of NDMA and subsequent occurrence of colorectal cancer, with a relative risk between the highest and lowest quartiles of intake of 2.1 (95% CI, 1.0–4.3). Of the various sources of *N*-nitroso compounds, smoked and salted fish were found to be significantly associated with the risk for colorectal cancer (relative risk, 2.6; 95% CI, 1.2–5.5), and intake of cured meat was nonsignificantly associated (relative risk, 1.8; 95 % CI, 0.98–3.5). No such association was observed with intake of other fish or other meat. No significant association were observed between NDMA intake and cancers of the head and neck or of the stomach or between nitrate or nitrite intake and the risk for cancers of the gastrointestinal tract (Dich et al., 1996; Knekt et al., 1999).

(d) Intervention trial

The concentration of salivary nitrite in patients with cancer is twice that in healthy individuals, indicating that endogenous nitrite is correlated with cancer occurrence but not necessarily causally. In a culture medium modified with vitamin B_1 or potassium chlorate, many bacteria found in the human body (especially the bowels) nitrified ammonium ion to nitrite ion. Nitrite ions can combine with a secondary amine to form nitrosamines, which are carcinogenic. A method for preventing cancer was tested in an experiment involving 7392 workers in a Chinese flax textile mill. Persons with a salivary concentration of nitrite ion > 10 mg/l for 3 consecutive days and with symptoms were targets for preventive treatment with antibiotics and 'nitrosamine destroyers'. The average cancer incidence rate in the control group was 55% higher than in the experimental group over 9 years. The difference between the two groups was statistically significant (Huang et al., 1996). Unfortunately, this study is difficult to interpret, because there were problems with follow-up, no standardization for age seems to have been performed, the rates in the control group changed dramatically over the study period and the incidences of cancers potentially related to nitrite intake, such as stomach cancer, were proportionally increased in the experimental group. In fact, the experiment addressed not nitrite but antibiotics, vitamin C and other 'nitrosamine destroyers'.

2.3.3 Relationship between nitrite intake and oxidant stress in human erythrocytes

Erythrocytes in vitro took up 0.8 mmol/l nitrite with a half-time of 11 min. Nitrite uptake was sensitive to temperature and to the pH and ionic solution of the medium but was not inhibited by a specific anion-exchange inhibitor. About 25% of nitrite uptake occurred on the sodium-dependent phosphatase transporter and the rest as diffusion of nitrous acid or other species across the plasma membrane. Methaemoglobin formation increased in proportion to the intracellular nitrite concentration. Nitrite reacted with erythrocyte ascorbate, but ascorbate loading of cells decreased nitrite-induced methaemoglobin formation only at high nitrite concentrations. It was concluded that nitrite rapidly enters erythrocytes and reacts with oxyhaemoglobin but does not exert a strong oxidant stress on these cells (May et al., 2000).

3. COMMENTS

The Committee considered a new study of toxicokinetics in human volunteers given sodium nitrite in their drinking-water. The maximum plasma nitrite concentrations were observed 15–30 min after dosing, and nitrite disappeared rapidly from plasma, with a half-life of approximately 30 min. The bioavailability of sodium nitrite was similar and virtually total after low and high oral doses. An intravenous dose of sodium nitrite of 290–380 mg per person induced methaemoglobinaemia, with maximum percentages of 8.4–12%. After oral administration, the maximum concentration of methaemoglobin was reached 0.70 h after dosing.

Several new studies were available on the acute and short-term toxicity of nitrite in humans, in which the severity of methaemoglobinaemia was reported after accidental, high intake of nitrite. Most were case reports of poisoning incidents. None of these studies added data that could be used for the safety evaluation.

The Committee considered studies recently completed within the National Toxicology Program in the USA. In 14-week studies of the toxicity of sodium nitrite administered in drinking-water to mice at doses equal to 0, 90, 190, 340, 750 or 990 mg/kg bw per day and to rats at doses equal to 0, 30, 55, 120, 200 or 310 mg/kg bw per day, elevated levels of methaemoglobinaemia were observed in rats in all treatment groups, but not in mice. The Committee considered a level of methaemoglobin formation of up to 5% not to be adverse. The lowest NOEL value was observed in a study in male rats and was equal to 55 mg/kg bw per day (37 mg/kg bw per day expressed as nitrite ion) on the basis of reduced sperm motility at higher doses. Reduced sperm motility was also seen in mice at higher doses. The results of the 2-year carcinogenicity studies with sodium nitrite added to drinking-water of mice and rats at a concentration of 750, 1500 or 3000 mg/l (equal to 45, 90 and 160 mg/kg bw per day in mice and to 35, 70 and 130 mg/kg bw per day in rats) led to the conclusion that nitrite is not carcinogenic in rats or male mice. The authors noted that there was equivocal evidence of carcinogenic activity of sodium nitrite in female mice, in view of the positive trend in the incidence of squamous-cell papilloma and carcinoma (combined) in the forestomach. As no evidence of genotoxicity in vivo or of cytotoxicity in the forestomach was found, the mode of action of sodium nitrite in inducing forestomach neoplasia is unclear.

A study of reproductive toxicity in mice given drinking-water containing sodium nitrite at concentrations up to 0.24% w/w (equivalent to 240 mg/kg bw per day) did not reveal adverse effects on reproductive performance or on end-points examined at necropsy.

New studies on genotoxicity were available. Sodium nitrite was mutagenic in a test for reverse mutation but did not induce micronucleus formation in the bone marrow or peripheral erythrocytes of mice treated in vivo, consistent with previous results.

A few studies on the endogenous formation of *N*-nitroso compounds after intake of nitrite or nitrate and of nitrosatable compounds (amines and amides) and the possible association with cancer were available. In addition, one long-term study of toxicity and carcinogenicity was reviewed, in which rats were fed fishmeal at various concentrations concomitantly with sodium nitrite. Dose-dependent increases in the incidences and multiplicity of atypical renal tubules, adenomas and renal-cell carcinomas were found. The concentration of *N*-nitrosodimethylamine in the stomach contents after 4 weeks of treatment with 64% fishmeal plus 0.12% sodium nitrite was twice that measured after administration of 8% fishmeal plus 0.12% sodium nitrite. However, the diets used in this study were considered to be nutritionally inappropriate for the rat: they had a high protein content, resulting in abnormally high renal nitrogen clearance. These studies do not provide additional insight for the safety evaluation of nitrite.

A number of epidemiological studies of the relationship between the intake of nitrite and cancer risk have been published since the forty-fourth meeting. At its present meeting, the Committee ranked the study designs according to their capacity to provide evidence of a relationship. In the descriptions below, relative risk estimates are given for those studies in which levels of intake of nitrite were provided. Four of the studies were cross-sectional, involving measurement of nitrite in, e.g., salivary or gastric juice in cancer patients and healthy subjects. Because cross-sectional studies do not take into account the time between exposure and disease, any observed differences in biomarkers of exposure might also be a consequence of the disease; therefore, these studies cannot contribute to a causal interpretation of the results of studies of nitrite and cancer risk.

Nine (longitudinal) case–control studies on previous nitrite intake and various cancer types were reviewed. For oral and laryngeal cancer, no association was found with nitrite intake. One study conducted in the USA reported a positive association with oesophageal cancer, with ORs of 1.0 (reference category), 1.2 and 1.6 for persons with a daily nitrite intake of < 1.1 mg, 1.1–1.6 mg and > 1.6 mg, respectively. The ORs and the trend across ORs were not statistically significant, however. The association between nitrite intake and oesophageal cancer was stronger, and it was significant for persons with a history of canker sores. Another study in the USA, however, found no association between nitrite intake and oesophageal cancer, nor with the subtypes adenocarcinoma and squamous-cell carcinoma; a positive association was found only with gastric cancer other than of the cardia. A positive association with gastric cancer was also reported in an Italian case–control study (average consumption, 2.4 mg/day), while no association was found in a French study (average consumption, 1.9 mg/day).

An association of borderline significance was found between nitrite intake and urinary bladder cancer in men but not women of Japanese descent, nor in whites of either sex, in Hawaii, USA. Although a positive association was reported from a study in the USA between brain tumours in children and their mothers' consumption of processed meat, no association was found with nitrite intake during gestation or in childhood in a recent case–control study from Israel. One study on nasopharyngeal cancer among Taiwanese reported no association with nitrite intake in adulthood (as reported by cases and controls), but a positive association was found with childhood nitrite intake as recalled by the mothers of the cases and controls. The validity of recall of remote dietary intake is questionable, however.

Two prospective cohort studies have been conducted on nitrite intake and cancer risk. A cohort study from the Netherlands, with 6 years of follow-up, on dietary nitrite and gastric cancer risk reported relative risks of 1.0 (reference category), 1.2, 1.2, 0.9 and 1.4 for increasing mean quintiles of nitrite intake of 0.01, 0.04, 0.09, 0.16 and 0.35 mg/day, respectively. Neither the relative risks nor the trend was significant. A Finnish cohort study, with 24 years of follow-up, reported no association with the incidence of stomach, colorectal, or head-and-neck tumours. The average nitrite intake by this cohort was reported to be 5.3 mg/day.

Thus, some studies indicated increased risks for oesophageal and gastric cancer; however, other studies – particularly prospective cohort studies – revealed no such association. The results for brain tumours in children and for urinary bladder cancer in adults were equivocal. Wide variation between the studied populations in the recorded intake of nitrite was noted. In none of these studies was a possible interaction between nitrite and nitrosatable amines evaluated in respect of cancer risk. The results of these studies and those of the epidemiological studies considered by the Committee at its forty-fourth meeting do not provide evidence that nitrite is carcinogenic to humans. In addition, studies on nitrate intake and cancer risk (of relevance because of the conversion of nitrate to nitrite) also did not provide evidence of a positive association.

A 6-week study was performed in young adult and old rats to determine whether older animals are more sensitive to the toxic effects of nitrite, particularly with regard to the known effects on the kidney and on the functions of some hormones (insulin, thyroxine). No age-related differences were found. The hypertrophy of the zona glomerulosa in the adrenals of rats of this strain was seen only at an intake equivalent to 50 mg/kg bw per day, expressed as nitrite ion.

Several studies were available on the mechanism of action of nitrite on the vascular system, blood pressure and the adrenals in rats. Doses of 50 mg/kg bw or higher, expressed as nitrite ion, are known to lower blood pressure by causing vasodilatation. The Committee calculated that an oral dose equivalent to 160 mg/kg bw of nitrate ion induced a reduction in blood pressure of 15–20 mm Hg.

Administration of nitrite to rats resulted in minimal hypertrophy of the zona glomerulosa of the adrenal gland and secretion of aldosterone, which was reversible after 60 days. Inhibition of angiotensin-converting enzyme indicated that the effect was produced indirectly via stimulation of the renin–angiotensin axis.

4. EVALUATION

The Committee concluded that the minimal hypertrophy reflected physiological adaptation to small fluctuations in blood pressure and should not be considered a direct toxic action on the adrenals. This conclusion implies that the safety evaluation should not be derived from the NOEL for minimal hypertrophy of the adrenal zona glomerulosa, used by the Committee at its forty-fourth meeting, but on NOELs for other end-points. The NOEL of 5.4 mg/kg bw per day (expressed as nitrite ion) was therefore considered to be no longer relevant, as it was based on the indirect effect on the adrenals described above. A NOEL of 6.7 mg/kg bw per day was identified in a 2-year study in rats, in which effects on the heart and lungs were observed at the next higher dose .

The Committee established an ADI of 0–0.07 mg/kg bw, expressed as nitrite ion, on the basis of the NOEL of 6.7 mg/kg bw per day for effects on the heart and lung in the 2-year study in rats and a safety factor of 100.

Nitrite causes methaemoglobinaemia. As this might occur after a single dose, it would be appropriate to establish an acute reference dose for nitrite. The data available for review at the present meeting related primarily to its long-term toxicity and were not relevant for establishing an acute reference dose. The Committee recommended that the acute toxicity of nitrite be reviewed at a future meeting.

5. REFERENCES

Airoldi, L., Galli, A., Vago, F., Fanelli, R., Negri, E., La Vecchia, C. & Vierucci, S. (1997) Salivary nitreate, nitrite and *N*-nitrosos compounds in patients with cancer of the upper aerodigestive tract. *Eur. J. Cancer Prev.*, **6**, 351–356.

Badawi, A.F., Hosny, G., El-Hadary, M. & Mostafa, H.M. (1998) Salivary nitrate, nitrite and nitrate reductase activity in relation to risk of oral cancer in Egypt. *Dis. Markers*, **14**, 91–97.

Bany, J., Golinska., Z & Kaczorowska, B. (1995) The effect of poisoning with sodium nitrite on subpopulations of T lymphocytes in mice infected with Trichinella spiralis. *Acta Parasitol.*, **40**, 107–109.

Beier, S., Classen, H.G., Loeffler, K, Schumacher, E. & Thöni, H. (1995) Antihypertensive effect of oral nitrite uptake in the spontaneously hypertensive rat. *Arz. Forsch. Drug Res.*, **45**, 258–261.

Boink, A.B.T.J., Beekhof, P.K., Dormans, J.A.M.A. & Speijers, G.J.A. (1996) On the etiology of nitrite-induced hypertrophy of the zona glomerulosa of rats: II. The possible role of feed. Report No. 235802004 from the National Institute of Public Health and the Environment (RIVM), Bilthoven, Netherlands.

Boink, A.B.T.J., Beekhof, P.K., Dormans, J.A.M.A., Dortant, P.M. & Speijers, G.J.A. (1997) A study on nitrite sensitivity in the rat: Comparison between young adult rats and aged rats. Report No. 650020001 from the National Institute of Public Health and the Environment (RIVM), Bilthoven, Netherlands.

Boink, A.B.T.J., Beekhof, P.K., te Biesebeek, J.D., Timmerman, A., Elvers, L.H., Dormans, J.A.M.A. & Vleeming, W. (1998) On the etiology of nitrite-induced hypertrophy of the zona glomerulosa of rats: III. Effect of combined administration of nitrite and an ACE inhibitor. Report No. 235802013 from the National Institute of Public Health and the Environment (RIVM), Bilthoven, Netherlands.

Boink, A.B.T.J., Dormans, J.A.M.A., Speijers, G.J.A. & Vleeming, W. (1999) *Effects of Nitrates and Nitrites in Experimental Animals* (Special Publication), London, Royal Chemistry Society, pp. 317–326.

van den Brandt, P.A., Voorrips, L., Hertz-Picciotto, I., Shuker, D., Boeing, H., Speijers, G., Guittard, C., Kleiner, J., Knowles, M., Wolk, A. & Goldbohm, A. (2002) The contribution of epidemiology to risk assessment of chemicals in food and diet. *Food Chem. Toxicol.*, **40**, 387–424.

Chan, T.Y.K. (1996) Food-borne nitrates and nitrites as a cause of methemoglobinemia. Minireview. *Southeast-Asian J. Trop. Med. Public Health*, **27**, 189–192.

Centers for Disease Control (1997) Methemoglobinemia attributable to nitrite contamination of potable water through boiler fluid additives. *Morbid. Mortal. Wkly Rep.*, **49**, 202–210.

Chapin, R., Gulati, D. & Barnes, L.H. (1997) Reproductive toxicology; sodium nitrite. *Environ. Health Perspectives*, **105** (Suppl. 1), 1–3.

Dich, J., Järvinen, R., Knekt, P. & Penttilä, P.L. (1996) Dietary intakes of nitrate, nitrite and NDMA in the Finnish mobile clinic health examination survey. *Food Addit. Contam.*, **13**, 541–552.

Dinkla, E.T. (1976) [Some cases of nitrate intoxicarion in cattle in the province of Groningen.] *Tijdschr. Diergeneeskunde*, **101**, 1096–1099 (in Dutch).

Finan, A., Keenan, P., O'Donovan, F., Mayne, P & Murphy, J. (1998) Methaemoglobinaemia associated with sodium nitrite in three siblings. *Br. Med. J.*, **317**, 1138–1139.

Furukawa, F., Nishikawa, A., Ishiwata, H., Takahashi, M., Hayashi, Y. & Hirose, M. (2000) Renal carcinogenicity of concurrently administered fish meal and sodium nitrite in Fischer 344 rats. *Jpn. J. Cancer Res.*, **91**, 139–147.

Gladwin, M.T., Shelhamer, J.H., Schechter, A.N., Pease-Fye, M.E., Waclawiw, M.A., Panza, J.A., Ognibene, F.P. & Cannon, R.O., III (2000) Role of circulating nitrite and S-nitrosohemoglobin in the regulation of regional blood flow in humans. *Proc. Natl Acad. Sci. USA*, **97**, 11482–11487.

Haas, M., Classen, H.G., Thöni, H., Classen, U.G & Drescher, B. (1999) Persistent antihypertensive effect of oral nitrite supplied up to one year via the drinking water in spontaneously hypertensive rats. *Arz. Forsch. Drug Res.*, **49**, 318–323.

Hirose, M., Tanaka, H., Takahashi, S., Futakuchi, M., Fukushima, S. & Ito, N. (1993) Effects of sodium nitrite and catechol, 3-methoxycatechol or butylated hydroxyanisole in combination in a rat multiorgan carcinogenesis model. *Cancer Res.*, **53**, 32–37.

Hirose, M,., Nishikawa, A., Shibutani, M., Imai, T & Shirai, T. (2002) Chemoprevention of heterocyclic amine-induced mammary carcinogenesis in rats. *Environ. Mol. Mutag.*, **39**, 271–278.

Huang, Y.G., Ji, J.D. & Hou, Q.N. (1996) A study on carcinogenesis of endogenous nitrite and nitrosamine, and preventive of cancer. *Mutat. Res.*, **358**, 7–14.

Kawabe, M., Takaba, K., Yoshida, Y. & Hirose, M. (1994) Effects of combined treatment with phenolic compound and sodium nitrite on two-stage carcinogenesis and cell proliferation in the rat stomach. *Jpn. J. Cancer Res.*, **85**, 17–25.

Knekt, P., Järvinen, R., Dich, J. & Hakulinen, T. (1999) Risk of colorectal and other gastro-intestinal cancers after exposure to nitrate, nitrite and *N*-nitroso compounds: A follow up study. *Int. J. Cancer*, **80**, 852–856.

Kortboyer, J.M., Ollling, M., Zeilmaker, M.J., Slob, W., Boink, A.B.T.J., Schothorst, R.C., Sips, A.J.A.M. & Meulenbelt, J. (1997a) The oral bioavailability of sodium nitrite investigated in healthy adult volunteers. Report No. 235802007 from the National Institute of Public Health and Environment (RIVM), Bilthoven, Netherlands.

Kortboyer, J.M., Boink, A.B.T.J., Zeilmaker, M.J. Slob, W. & Meulenbelt, J. (1997ab) Methemoglobin formation due to nitrite: Dose–effect relationship in vitro. Report No 235802006 from the National Institute of Public Health and Environment (RIVM), Bilthoven, Netherlands.

Kortboyer, J.M.,Schothorst, R.C., Zeilmaker, M.J.& Meulenbelt, J. (1998) Intravenous administration of sodium nitrite to healthy volunteers: A single ascending dose study. Report No. 235802011 from the National Institute of Public Health and Environment (RIVM), Bilthoven, Netherlands.

La Vecchia, C., Negri, E., Franceshi, S. & Decarli, A. (1997) Case–control study of methionine, nitrite, and salt on gastric carcinogenesis in northern Italy. *Nutr. Cancer*, **27**, 65–68.

van Loon, A.J.M., Botterweck, A.A.M., Goldbohm, R.A., Brants, H.A.M., van Klaveren, J.D. & van den Brandt, P.A. (1998) Intake of nitrate and nitrite and the risk of gastric cancer: A prospective cohort study. *Br. J. Cancer*, **78**, 129–136.

Lubin, F., Farbstein, H., Chetrit, A., Farbstein, M., Freedman, L., Alfrandy, E. & Modan, B. (2000) The role of nutritional habits during gestation and child life in pediatric brain tumor etiology. *Int. J. Cancer*, **86**, 139–143.

May, J.M., Qu, Z.-C., Xia, L. & Cobb, C.C. (2000) Nitrite uptake and metabolism and oxidant stress in human erythrocytes. *Am. J. Physiol. Cell Physiol.*, **279**, C1946–C1954.

Mayne, S.T., Risch, H.A., Dubrow, R. et al. (2001) Nutrient intake and risk of subtypes of esophageal and gastric cancer. *Cancer Epidemiol. Biomarkers Prev.*, **10**, 1055–1062.

Mirvish, S.S., Grandjean, A.C., Reimers, K.J., Connelly, B.J., Chen, S.C., Morris, C, R., Wang, X., Haorah, J.& Lyden, B.J. (1998) Effect of ascorbic acid dose taken with a meal on nitrosoproline excretion in subjects ingesting nitrate and proline. *Nutr. Cancer*, **31**, 106–110.

Miyauchi, M., Nakamura, H., Furukawa, F., Son, H.-Y., Nishikawa, A. & Hirose, M. (2002) Promoting effects of combined antioxidant and sodium nitrite treatment on forestomach carcinogenesis in rats after initiation with *N*-methyl-*N*'-nitro-*N*-nitrosoguanidine. *Cancer Lett.*, **178**, 19–24.

Modin, A., Björne, H., Herulf, M., Alving, K. & Weitzberg, E. (2001) Nitrite-derived nitric oxide; a possible mediator of 'acidic–metabolic' vasodilation. *Acta Physiol. Scand.*, **171**, 9–16.

Mohsen, M.A., Hassan, A.A.M., El-Sewedy, S.M., Aboul-Azm, T., Magagnotti, C., Fanelli, R. & Airoldi, L. (1999) Biomonitoring of *N*-nitroso compounds, nitrite and nitrate in the urine of Egyptian bladder cancer patients with or without Schistosoma haematobium infection. *Int. J. Cancer*, **82**, 789–794.

Mueller, B.A., Newton, K., Holly, E.A. & Preston-Martin, S. (2001) Residential water source and the risk of childhood brain tumors. *Environ. Health Perspectives*, **109**, 551–556.

Nagase, H. (1999) Health effects of nitrite-nitrogen. *Gifu Yakka Daigaku*, **48**, 1–8.

National Toxicology Program (2001) Toxicology and carcinogenesis studies of sodium nitrite (CAS No. 7632-00-0) in Fischer 344/N and B6C3F1 mice (drinking water studies), Research Triangle Park, North Carolina, Department of Health and Human Servioces, Public Health Services, National Institutes of Health.

Nickerson, M. (1975). Vasodilator drugs. In: L.S. Goodman, A., eds, *Pharmacological Basis of Therapeutics*, 5th Ed., London, MacMillan, pp. 727–743.

Pobel, D., Riboli, E., Cornée, J., Hémon, B. & Guyader, M. (1995) Nitrosamine, nitrate and nitrite in relation to gastric cancer: A case–control study in Marseille, France. *Eur. J. Epidemiol.*, **11**, 67–73.

Preston-Martin, S., Pogoda, J.M., Mueller, B.A., Holly, E.A., Lijinsky, W. & Davis, R.L. (1996) Maternal consumption of cured meats and vitamins in relation to pediatric brain tumors. *Cancer Epidemiol. Biomarkers Prev.*, **5**, 599–605.

Rogers, M.A., Vaughan, T.L., Davis, S. & Thomas, D.B. (1995) Consumption of nitrate, nitrite, and nitrosodimethylamine and the risk of upper aerodigestive tract cancer. *Cancer Epidemiol. Biomarkers Prev.*, **4**, 29–36.

Shuval, H.I. & Gruener, N. (1972) Epidemiological and toxicological aspects of nitrates and nitrites in the environment. *Am. J. Public Health*, **62**, 1045–1052.

Starodubtseva, M.N., Ignatenko, V.A. & Cherenkevich, S.N. (1999) Damage to erythrocytes initiated by the interaction of nitrite ions with hemoglobin. *Biofizika*, **44**, 1068–1972.

Til, H.P., Kuper, C.F. & Falke, H.E. (1997) Nitrite-induced adrenal effects in rats and the consequences for the no-observed-effect level. *Food Chem. Toxicol.*, **35**, 349–355.

Vleeming, W., van de Kuil, A., Boink, A.B.T.J., Speijers, G.J.A. & de Wild, D.J. (1995a) The effect of nitrite and nitrate on blood pressure in rats (Abstract). *Pharmacol. Res.*, **31**, 330.

Vleeming, W., van de Kuil, A., Boink, A.B.T.J., Speijers, G.J.A.& de Wild, D.J. (1995b) On the toxicity of nitrite and nitrate in rats (Abstract). *Naunyn Schmiedebergs Arch. Pharmacol.*, **352**, R22.

Vleeming, W., van de Kuil, A., te Biesebeek, J.D., Meulenbelt, J. & Boink, A.B.T.J. (1996) Effect of nitrite on blood pressure in free-moving rats. In: *Annual Scientific Report of RIVM 1995*, Bilthoven, pp. 18–20.

Vleeming, W., van de Kuil, A., te Biesebeek, J.D., Meulenbelt, J. & Boink, A.B.T.J. (1997) Effect of nitrite on blood pressure in anaesthetized and free-moving rats. *Food Chem. Toxicol.*, **35**, 615–619.

Ward, M.H., Pan, W.H., Cheng, Y.J., Li, F.H., Brinton, L.A., Chen, C.J., Hsu, M.M., Chen, I.H., Levine, P.H., Yang, C.S & Hildesheim, A. (2000) Dietary exposure to nitrite and nitrosamines and risk of nasopharyngeal carcinoma in Taiwan. *Int. J. Cancer*, **86**, 603–609.

Wilkens, L.R., Kadir, M.M., Kolonel, L.N., Nomura, Y. & Hankin, J.H. (1996) Risk factors for lower urinary tract cancer: The role of total fluid consumption, nitrites and nitrosamines, and selected foods. *Cancer Epidemiol. Biomarkers Prev.*, **5**, 161–166.

Yoshida, Y., Katsumi, K., Takaba, J. & Ito, N. (1994) Induction and promotion of forestomach tumors by sodium nitrite in combination with ascorbic acid or sodium ascorbate in rats with or without *N*-methyl-*N'*-nitro-*N*-*N*-nitrosoguanidine pre-treatment

You, W.C., Zhang, L., Yang, C.S., Chang, Y.S., Issaq, H., Fox, S.D., Utermahlen, W.E., Zhao, L., Keefer, L., Liu, W.D., Chow, W.H., Ma, J.L., Kneller, R., Ho, M.Y.K., Fraumeni, J.F., Xu, G.W. & Blot, W.J. (1996) Nitrite, *N*-nitroso compounds, and other analytes in physiological fluids in relation to precancerous gastric lesions. *Cancer Epidemiol. Biomarkers Prev.*, **5**, 47–52.

NITRATE
(and potential endogenous formation of N-nitroso compounds)

First draft prepared by
Dr G.J.A. Speijers[1] and Professor P.A. van den Brandt[2]
[1]Section Public Health of the Centre for Substances & Risk Assessment, National Institute of Public Health and Environmental Protection, Bilthoven, Netherlands; and [2]Department of Epidemiology, Maastricht University, Maastricht, Netherlands

1. EXPLANATION

Nitrate occurs in the environment, in air, food (particularly in vegetables and fruits) and water, and is produced endogenously. It is also used as a food additive, mainly as a preservative and anti-microbial agent. It is used in foods such as cheese and cheese products, raw and processed meats, edible casings, processed fish and fish products and spirits and liqueurs.

Nitrate was reviewed by the Committee at its sixth, eighth, seventeenth and forty-fourth meetings (Annex 1, references *6, 8, 32* and *116*). At its sixth meeting, the Committee allocated an ADI of 0–5 mg/kg bw to this substance, expressed as sodium nitrate. This ADI was based on a NOEL of 500 mg/kg bw per day for body-weight gain at a higher dose in a long-term study in rats and a short-term study of toxicity in dogs, with a safety factor of 100. This ADI was retained by the Committee at its eighth and seventeenth meetings. After the latter meeting, numerous toxicological and epidemiological data became available, which were considered by the Committee at its forty-fourth meeting.

At that meeting, the Committee concluded that nitrite should also be considered, because nitrate is readily converted in humans to nitrite. The rat was considered an unsuitable animal model for assessing the toxicity of nitrate, as it does not convert nitrate to nitrite in a quantitatively similar way to humans. Nevertheless, as the data on the toxicity of nitrate in other animal species were considered limited, the toxicity of nitrite in rats and the conversion rate of nitrate to nitrite were also evaluated.

The Committee at its forty-fourth meeting concluded that nitrate itself was not genotoxic, and the results of studies of carcinogenicity with nitrate were negative except when extremely high doses of both nitrate and nitrosatable precursors were administered. The available epidemiological data were considered to provide no evidence for an association between exposure of humans to nitrite and the risk for cancer. On the basis of this information, the NOEL of 370 mg/kg bw per day, expressed as nitrate ion, in a long-term study in rats was considered to be the most appropriate for the safety evaluation. When the proportion of nitrate converted to nitrite in humans was taken as 5% for the average individual and 20% for those with a high level of conversion, and when the NOEL for nitrite (6 mg/kg bw per day, expressed as nitrite ion) was used to calculate the 'transposed' NOEL for nitrate, expressed as nitrate ion, these values were estimated to be 160 and 40 mg/kg bw per day for average and high responders, respectively. As these figures were derived in part from data on human pharmacokinetics, use of a safety factor of less than 100 was considered to be justified.

On the basis of the NOEL of 370 mg/kg bw per day, expressed as nitrate ion, and a safety factor of 100, an ADI of 0–5 mg/kg bw expressed as sodium nitrate or 0–3.7 mg/kg bw expressed as nitrate ion was allocated. On the basis of the 'transposed' NOEL for nitrate of 160 mg/kg bw per day for normally responding persons (5% rate of conversion) and a safety factor of 50, an ADI of 0–3.2 mg/kg bw, expressed as nitrate ion, could be allocated. These two methods of deriving the ADI for nitrate thus resulted in similar figures, and the Committee at its forty-fourth meeting therefore retained the previously established ADI of 0–3.7 mg/kg bw, expressed as nitrate ion.

The Committee at its present meeting reviewed data relevant to the evaluation of nitrate which had become available since its forty-fourth meeting. Studies illustrating the relevance and severity of certain effects and other supporting data were also considered. New information on metabolism and toxicokinetics were included. Data on the putative health benefits of nitrate were not assessed, as this is not a safety issue and is therefore not within the purview of the Committee.

2. BIOLOGICAL DATA

2.1. Biochemical aspects

2.1.1 Absorption, distribution, metabolism and excretion

The following is based on a review of the literature (Walker, 1996), which gives a clear picture of the pharamacokinetics and metabolism of nitrate.

Ingested nitrate is readily absorbed from the proximal small intestine (Fritsch et al., 1979; Balish et al., 1981) and rapidly equilibrates with body fluids. In rats, about 50% of an oral dose was detected in the carcass within 1 h; in humans, peak levels were achieved in serum, saliva and urine within 1–3 h (Bartholomew & Hill, 1984). There is little absorption from the stomach in most species, although some has been reported from the rumen of cattle (Wright & Davison, 1964).

In humans and most laboratory animal species except the rat, nitrate is actively secreted in saliva in a dose-dependent manner (Cohen & Myant, 1959; Fritsch et al., 1985), but Spiegelhalder et al. (1976) were unable to detect an increase in salivary nitrate concentration in humans after ingestion of up to 54 mg of nitrate.

The active transport mechanism is common to iodide, thiocyanate and nitrate, in that order of affinity, and smokers who secrete high levels of thiocyanate have lower salivary concentrations of nitrate (Forman et al., 1985). It has been estimated that, in humans, about 25% of an orally ingested dose of nitrate is secreted in saliva (Spiegelhalder et al., 1976; Tannenbaum et al ., 1976), but these estimates are confounded by the great interindividual and diurnal variability in endogenous synthesis and secretion of nitrite. Although rats are reported not to have the necessary mechanism for active salivary secretion of nitrate (hindering extrapolation of toxicological studies in rats to human), they do secrete circulating nitrate into other gastric and intestinal fluids by an active transport system (Witter & Balish, 1979), so that enterosystemic cycling of nitrate may occur in this species also. In dogs given nitrate intravenously, extensive salivary secretion and extensive biliary secreation were seen, confirming this pathway of excretion for both nitrite and nitrate. Nitrate appears in milk by a passive diffusion mechanism, and the concentrations in human and canine milk did not exceed the plasma levels after ingestion of a nitrate-containing meal (Green et al., 1982). After absorption and equilibrium in body fluids, nitrate is rapidly excreted in urine. In humans, about 65–70% of any orally administered dose of nitrate was excreted in urine. Excretion was maximal about 5 h after dosage and essentially complete within 18 h (Bartholomew & Hill, 1984). The excretion followed first-order kinetics, and the elimination half-life was estimated to be about 5 h (Green et al., 1981).

Some metabolic conversion of nitrate clearly occurs, as, in humans, about 3% of a dose of $^{15}NO_3^-$ appeared in the urine as urea and ammonia (Wagner et al., 1983); in rats, 1% of a dose appeared in urea, and ammonia was found in urine and faeces (Schultz et al., 1985). Nitrate can be reduced to nitrite by both enteric bacteria and mammalian nitrate reductase activity. Many species of microorganisms resident in the gastrointestinal tract have nitrate reductase activity (Hegesh & Shiloah, 1982), and this enzyme has been detected in rat liver and intestinal mucosa, although with much lower activity (Schultz et al., 1985).

Comparative studies with germ-free and conventional rats indicated that about half of the 40–50% reduction of a dose of nitrate to nitrite in conventional animals was effected by mammalian nitrate reductase (Ward et al., 1986). However, Fritsch et al. (1985) were unable to detect this pathway in dogs and reported that reduction by the oral and gastrointestinal microflora was the most important mechanism in mammals. The major site of conversion of nitrate into nitrite varies by species and depends on the sites of microbial colonization and absorption of nitrate. Interestingly, the presence of nitrite in human saliva was first reported more than 55 years ago (Varady & Szanto, 1940), but saliva taken directly from the salivary ducts of humans or dogs contains only nitrate, indicating that a significant amount of reduction occurs in the oral cavity (Muramatsu et al., 1979). This is attributed to a stable population of nitrate-reducing bacteria established at the base of the tongue. Stephany & Schuller (1978) suggested that the salivary concentration of nitrite was directly related to the orally ingested dose of nitrate, and other workers reached similar conclusions (Harada et al., 1975; Spiegelhalder et al., 1976). However, Tannenbaum et al. (1976) suggested that the reduction process is saturable at high intakes. On the basis of the variable salivary levels of nitrate and nitrite after oral ingestion of nitrate by humans, it has been estimated that, of the 25% of ingested nitrate secreted in saliva, 20% is reduced to nitrite (I.e. about 5% of the oral dose). Oral reduction of nitrate is thus the most important source of nitrite for humans and most species that have an active salivary secretory mechanism.

The gastric pH, and hence bacterial populations, are low in the stomachs of rabbits, ferrets and healthy humans. Hence little further reduction of nitrate occurs at this site, and the nitrite concentrations in gastric contents are usually low. Conversely, rats and dogs have a higher gastric pH, and bacterial colonization can occur, with consequent further reduction of nitrate at this site (Mirvish, 1975). In ruminants, the dense population of rumen microflora and the relatively high pH make this a major site of reduction of orally ingested nitrate, leading to well-documented methaemoglobinaemia induced by the nitrite produced. In humans subject to achlorhydria, bacterial colonization of the stomach can occur, and the situation then more closely resembles that in the rat and the dog. A strong correlation has been reported between gastric pH, bacterial colonization and gastric nitrite concentration in humans over a pH range of 1–7 (Müller et al., 1984), and concentrations as high as 5 mg/l have been reported in patients with achlorhydria associated with pernicious anaemia or hypogammaglobulinaemia (Ruddell et al., 1976; Dolby et al., 1984). The situation in human neonates is less clear. It is commonly asserted that infants under 3 months of age are highly susceptible to gastric nitrite production, because they produce little gastric acid, but Agunod et al. (1969), who examined infants aged between 12 h and 3 months, found only one with achlorhydria (Walker, 1996).

An open, randomized, cross-over pilot study was performed to investigate the toxicokinetics of nitrate and nitrite after an oral dose of sodium nitrate at 10 mg/kg bw in eight healthy volunteers (four men and four women) with normal gastric pH or gastric pH increased artificially with omeprazole (40 mg/day) starting 15 days before nitrate administration. Blood pressure was recorded for 3.5 h, and the per cent methaemoglobin was determined for 10 h after administration of sodium nitrate. Bacterial growth and gastric pH were determined just before nitrate administration, and the gastric juice pH was measured continuously for 3 h thereafter. Serial samples of gastric juice, plasma and saliva were collected for 24 h after administration of nitrate. The volume of all urine produced during 10 h after administration was determined; part of every sample was stored, and all samples were analysed for nitrate and nitrite.

None of the volunteers reported adverse effects attributable to intake of sodium nitrate. Blood pressure and per cent methaemoglobin were not affected. Gastric pH correlated positively with omeprazole treatment, but the number of nitrate-reducing bacteria did not. Nevertheless, the nitrite concentration in the gastric juice was approximately sixfold higher after nitrate administration combined with omeprazole treatment than after only nitrate. Nitrate was absorbed rapidly, the concentration in plasma increasing within 10 min. The half-life of nitrate in plasma was about 6.5 h. The concentration of nitrite in plasma did not change after nitrate administration. The cumulative salivary nitrate excretion over 24 h, expressed as a percentage of ingested nitrate dose, was 28%. The amount of nitrite formed in the saliva, expressed as a percentage of the nitrate dose ingested, was 8%. Approximately 70% of the nitrate dose was excreted in urine within 10 h of dosing. Omeprazole treatment did not affect the nitrate concentrations in plasma, saliva or gastric juice, the nitrite concentrations in plasma or saliva, the per cent methaemoglobin or blood pressure. Thus, increased gastric pH had no significant effect on systemic nitrate or nitrite after an oral dose of sodium nitrate at twice the ADI of nitrate (Colbers et al., 1996).

A feasibility study on the oral bioavailability of nitrate from vegetables was conducted in healthy volunteers. Preparations of beetroot, lettuce and spinach were fed to groups of six volunteers, and the rate of absorption of nitrate from the gastrointestinal tract and the concentrations of nitrate in plasma were determined. After an overnight fast, volunteers received a single dose of 300 g of vegetables, and plasma nitrate and nitrite concentrations were determined frequently for 9 h after treatment. There was evidence that the vegetable was homogeneously mixed in the test diet. The beetroot portions administered to the volunteers contained an average of 570 mg of nitrate, the lettuce portions contained 740 mg and the spinach portions contained 540 mg. The nitrate in the vegetables was rapidly absorbed from the gastrointestinal tract, maximum plasma nitrate concentrations of 21–44 mg/l being observed 1–1.5 h after treatment. Small amounts of nitrite (0.1–0.2 mg/l) are detected under normal circumstances at the lower limit of quantification of nitrite in plasma, and intake of vegetables rich in nitrate did not increase the plasma nitrite concentration, which remained at < 0.3 mg/l in all volunteers (Kortboyer et al., 1998).

Another study was conducted to determine the oral bioavailability of nitrate from spinach, lettice and beetroot. In a cross-over design with four test periods, six men and six women were fasted and then given 500 mg of sodium nitrate (containing 360 mg of nitrate ion) intravenously, 300 g of spinach (containing 560 mg of nitrate ion), 300 g of fresh lettuce (containing 1000 mg of nitrate) or 300 g of cooked beetroot (containing 640 mg of nitrate ion). A wash-out period of at least 1 week was maintained between test periods. The bioavailability of nitrate from the vegetables was calculated with and without endogenous nitrate production by comparing nitrate concentrations in blood after vegetable administration and after intravenous administration of nitrate. The elimination half-life of nitrate from plasma was 6–7 h. The bioavailability of nitrate with correction for endogenous nitrate production was 98% ± 12% for spinach,

110% ± 14% for lettuce and 110% ± 15% for beetroot. The bioavailability without correction for endogenous nitrate production was 91% ± 10% for spinach, 89% ± 13% for lettuce and 93% ± 12% for beetroot. Thus, the bioavailability of nitrate from spinach, lettuce and beetroot was high and not significantly different from that of nitrate in drinking-water (Lambers et al., 2000a)

2.1.2 Physiologically based pharmacokinetic and mathematical models

The available assessment of risk to humans of exposure to nitrate was based on an evaluation of toxicity in rats. This requires reconsideration as, in humans, nitrite is formed by bacterial fermentation of nitrate secreted from blood into saliva, but this process is absent in the rat (Tannenbaum et al., 1979; Til et al., 1988). As nitrite is more toxic than nitrate, the toxicity of nitrite may supercede that of nitrate in humans. A nitrate–nitrite conversion in the order of 2% has already been seen to have this effect. The amount of nitrite formed from ingested nitrate should therefore be incorporated into the risk assessment of human exposure to nitrate.

The formation of nitrite in humans is complex, involving uptake of nitrate from food and or water, endogenous synthesis of nitrate, secretion of nitrate from blood into saliva, conversion of nitrate to nitrite by bacteria present in saliva and reconversion of nitrite to nitrate in blood. These processes were incorporated in a model of the toxicokinetics of nitrate and nitrite, which was calibrated with data on the toxicokinetics of nitrate and nitrite in volunteers (Wagner et al., 1983). According to the model, the average adult synthesizes nitrate at 120 mg/day, 32–60% of orally administered nitrate is secreted from blood into saliva, and 13–22% of the salivary nitrate is converted into nitrite. The nitrate–nitrite conversion rate in humans was determined to be 7–9%. The model was also used to estimate the formation of nitrite in an average adult after single and repeated (once every 24 h) doses of nitrate at its ADI (0–3.7 mg/kg bw). This calculation resulted in an exposure to nitrite of 0.27–0.36 mg/kg bw per day, which significantly exceeds the ADI for nitrite (0–0.06 mg/kg bw). Of this, 31–41% was estimated to have originated from endogenously synthesized nitrate. The calculated values for nitrite intake should be interpreted within the framework of the assumptions underlying the model, as the effects of food intake (stimulation of salivary nitrite formation versus detrimental effects on nitrite in the stomach) were not included. Furthermore, in calculating human nitrite intake, a bioavailability of 100% was assumed for administered nitrate. The risk assessment of human nitrate intake should thus be based on a toxicological evaluation of the nitrite formed from ingested nitrate (Zeilmaker et al., 1996).

In order to quantify the kinetics of nitrate and nitrite in humans, a model was developed which comprises absorption of nitrate from drinking-water and vegetables, secretion of nitrate from blood into saliva, conversion of nitrate to nitrite, absorption of nitrite and interaction of nitrite with haemoglobin, yielding methaemoglobin and nitrate. The model contains 16 parameters, seven of which were obtained from the literature, while the remaining nine parameters (rate of endogenous nitrate formation, rate constant for the reaction of nitrite with haemoglobin, parameters of methaemoglobin repair in the blood, rate constants for the secretion of nitrate from blood into saliva, conversion of nitrate to nitrite in saliva, a stoichometric constant for the reaction of nitrite with haemoglobin, the volume of the central nitrite compartment and the absorption rate constant of nitrite) were obtained by fitting the model to various experimental data sets. The studies addressed nitrite-induced methaemoglobin formation in human haemolysates and intact human erythrocytes in vitro and various studies in which volunteers were given nitrate and nitrite orally or intravenously. According to the model, a single dose of nitrate from vegetables at up to 44 mg/kg bw would not induce clinical methaemoglobinaemia, whereas the average long-term intake of adults is < 3.7 mg/kg bw per day. Doses of 88–270 mg/kg bw progressively induced clinical methaemoglobinaemia and clinical hypoxia. In the model, lethal toxicity was predicted to occur at doses > 440 mg/kg bw. At doses up to 620 mg/kg bw, peak methaemoglobin concentrations were efficiently nullified by repair activity, so that the methaemoglobin concentration retruned to the background level within 24 h of administration of nitrate. Clear accumulation of methaemoglobin was found after repeated administration of a slightly higher dose of nitrate, 800 mg/kg bw, indicating rapid saturation of methaemoglobin repair activity at this dose.

The model was scaled from adults to neonates by adjusting the blood volume and the volumes of the central nitrate and nitrite compartments. In order to mimic the situation in neonates, partial methaemoglobin repair deficiency was simulated by assuming the methaemoglobin reductase activity to be 10% of that in adults. Repeated intake of nitrate in drinking-water at doses > 44 mg/kg bw was predicted to induce methaemoglobin accumulation. This value can be compared with the daily intake for a 4-kg neonate of 6.2 mg/kg bw from drinking 1 l of water containing nitrate at 25 mg/l. Exposure of patients with inflammatory disease to nitrite was simulated as continuous, i.e. 24-h, intravenous exposure to nitrite. In this case, saturation of methaemoglobin repair occurred at doses > 0.23 mmol/24 h (2.7 mg/kg bw), and exposure was up to 0.47 mmol/24 h (5.5 mg/kg bw). It was concluded that the intake by healthy adults of nitrate from food and/or drinking-water has negligible effects on methaemoglobin formation but may significantly affect methaemoglobin formation in neonates with inflammatory disease (Zeilmaker et al., 2002).

2.1.3 Biotransformation

(a) Humans

The effectiveness of several mouthwash solutions and toothpastes with antibacterial constituents in reducing the conversion of nitrate into nitrite in the oral cavity was tested in 15 male and female volunteers aged 22–40 years. The mean per cent salivary nitrate that was reduced to nitrite after ingestion of 240 mg (3.8 mmol) of nitrate was 16 ± 6.2%. Use of an antiseptic mouthwash containing the active antibacterial constituent chlorhexidine resulted in a drastic decrease in the mean per cent reduced nitrate, to 0.9 ± 0.8%. Mouthwash solutions containing the antibacterial component triclosan or the antimicrobial enzymes amyloglucosidase and glucose oxidase did not affect the reduction of nitrate to nitrite. A toothpaste containing triclosan and zinc citrate with synergistic antiplaque activity also had no effect. Use of a pH-regulation chewing-gum raised the pH of the oral cavity from 6.8 to 7.3. By 30 min after nitrate ingestion, this rise was accompanied by a significant increase in the salivary nitrite concentration (van Maanen et al., 1996a).

(b) Methaemoglobin formation

Nitrite resulting from nitrate is oxidized in blood by a coupled oxidation reaction with oxyhaemoglobin (Smith & Beutler, 1966) in which methaemoglobin is produced, leading to the well-recognized acute toxicity of nitrite. The reaction rate between nitrite and haemoglobin is species-dependent: in humans, it is lower than that in ruminants but higher than that in pigs. In view of the methaemoglobinaemia induced by nitrite, many standards for nitrate in drinking-water have been drawn up on the basis of the levels (mainly in wells) associated with infantile methaemoglobinaemia and an assumed threshold below which the risk is minimal. This is apparently based on the assumption that infants are more likely than adults to have resident gastric microflora capable of reducing nitrate to nitrite. However, endogenous synthesis of nitric oxide and subsequently of nitrite can rise dramatically during infantile gastroenteritis. In one study, hospitalized infants with a low nitrate intake (2–7 mg/day) had raised blood nitrate and methaemoglobin concentrations associated with acute diarrhoea (Hegesh & Shiloah, 1982). In another case, a dyspeptic child had 72% methaemoglobinaemia associated with a nitrate concentration in drinking-water of < 50 mg/l (Thal et al., 1961), while healthy infants tolerated intakes of up to 21 mg/kg bw (Kubler, 1958). In these cases at least, the endogenously formed nitrite appeared to have caused the methaemoglobinaemia, together with increased sensitivity of fetal-type haemoglobin to oxidation and the low levels of methaemoglobin reductase in neonates (Walker, 1996).

Two infants became ill after being fed formula that was reconstituted with water from private wells. Water samples were collected and analysed chemically and microbiologically. In the first case, the well water was considered bacteriologically safe and the nitrate intake was 9 mg/kg bw as nitrogen(equivalent to 78 mg/kg bw as nitrate ion). In the second case, the water was found to contain *Escherichia coli*, and the nitrate concentration was 27 mg/kg bw as nitrogen (equivalent to 93 mg/kg bw as nitrate ion). The second infant had a methaemoglobin concentration of 91% on arrival at hospital, whereas that of the first had returned to < 2%, which is within the normal range (Knobeloch et al., 2000).

An epidemiological investigation was undertaken to assess the prevalence of methaemoglobinaemia in areas with a high nitrate concentration in drinking-water. Five areas were selected, with average nitrate concentrations of 26, 45, 95, 220 and 460 mg/l of drinking-water, as nitrate ion. A total of 178 persons (about 30 per group), matched for age and weight, who constituted 10% of the total population of each of these areas were selected and divided into five age groups. Detailed histories were taken of the selected populations, medical examinations were conducted, and blood samples taken to ascertain the per cent methaemoglobin. The data were analysed to verify any relationship between nitrate concentration and methaemoglobinaemia. A high nitrate concentration caused severe methaemoglobinaemia (7–27%) in all groups, especially in the age groups < 1 year and > 18 years. The lower level in the age group 1–18 years is probably due to better reserves of cytochrome b_5 reductase activity and its adaptation to increasing nitrate concentration in water. High levels of methaemoglobin were seen with nitrate contents as low as 26 mg/l and as high as 460 mg/l. The highest per cent methaemoglobin was observed in infants < 1 year of age (Gupta et al., 2000)

In Finland, the Ministry of Social Affairs and Health has regulated nitrate in drinking-water, with an upper limit of 25 mg/l. About 18% of people use well water for drinking, and as many as one-fiftieth of the wells in some areas do not meet the national quality standard for nitrate. Nevertheless, the correlation between per cent methaemoglobin and nitrate concentrations in drinking-water of 25–50 mg/l was poor (Suvanto & Pohjola, 1995).

Although ingestion of nitrates in drinking-water has been considered to be the primary cause of acquired methaemoglobinaemia, a mechanism whereby gastrointestinal inflammation leads to methaemoglobinaemia has been established. It is against this background that recent research and a review of cases offer a more complex picture of the causes of infantile methaemoglobinaemia. If this is indeed so, current limits on allowable levels of nitrates in drinking-water, which are based solely on the threat of infantile methaemoglobinaemia, may be unnecessarily strict (Avery, 1999, 2001).

(c) Endogenous synthesis of nitrate

In addition to dietary intake, there is considerable endogenous synthesis of nitrate in mammals. Even with rigorous exclusion of dietary nitrate, human volunteers excreted about 1 mmol of nitrate per day in urine (Wishnok et al., 1995), i.e. approximately the same amount as that provided by food. Endogenous nitrate arises from oxidation of nitric oxide, which is produced by a family of synthases, some constitutive and some inducible (Marletta, 1988). The constitutive enzyme produces nitric oxide for short periods (seconds) in response to intracellular messengers like bradykinin, while the inducible forms produce much higher levels over periods of hours in response to immunostimulants. This synthesis of nitric oxide occurs in activated macrophages but has also been demonstrated in other cell types, including endothelial cells, neurones, neutrophils and hepatocytes (Ignarro, 1987; Lancaster, 1992), and is highly variable and much increased during infection. Endogenous formation of nitrate independently of dietary sources has complicated studies of the metabolism and pharmacokinetics of nitrate and nitrite, many of which can provide only quantitative or semi-quantitative data on their interconversion in vivo (Walker, 1996).

2.1.4 Endogenous formation of N*-nitroso compounds*

A further concern relating to the metabolism of dietary nitrate and nitrite is the potential formation in vivo of carcinogenic *N*-nitroso compounds from nitrite, or the derived nitrosating species, N_2O_3 and N_2O_4 , and dietary amines. This reaction was first postulated more than 30 years ago and has since been studied extensively, by a number of approaches: incubation of precursors under simulated oral and gastric conditions in vitro, analysis of saliva and gastric contents after administration of precursors, determination of specific or total *N*-nitroso compounds in body fluids or excreta after treatment with precursors, and studies of carcinogenicity after co-administration of nitrate or nitrite and amines or amides. Many of the studies performed with the first approach involved unrealistically high concentrations of nitrite and, at best, provided no more than an indication of potential nitrosation. Low but measurable amounts of four volatile nitrosamines were found after incubation of luncheon meat (containing nitrate at 30 mg/kg), egg and milk with human gastric juice containing thiocyanate at a concentration of 1.2 mmol/l at pH 2 (Walthers et al., 1979). However, other workers who incubated a wide range of foods under similar conditions (containing nitrate at 5–7 mg/l) were unable to detect any *N*-nitrosamines (Groenen et al., 1982). Analysis of gastric contents after consumption of meals containing nitrite indicated that volatile nitrosamines may be formed under similar conditions in vivo (Walters et al., 1979; Groenen et al., 1985).

Measurement of urinary *N*-nitrosoproline (NPRO) has been widely used as a surrogate for nitrosation in vivo, as this compound is non-carcinogenic, is excreted unchanged and occurs at low levels (2–7 μg/day) in persons on a low-nitrate diet (Ohshima & Bartsch, 1981). Although NPRO itself is not carcinogenic, the NPRO test measures the potential for intragastric formation of carcinogenic nitrosamines in humans. Nitrate and L-proline are administered to volunteers, and NPRO is produced by an acid-catalysed reaction of proline with nitrate-derived nitrite in the stomach. It is then absorbed and excreted in urine, which is analysed for NPRO. A number of reservations have been raised about the validity of this approach, since, in rats, a significant amount (40–90%) of urinary NPRO was not derived from ingested [^{15}N]nitrate, and there was no correlation in humans between nitrate intake and urinary excretion of NPRO (Tannenbaum, 1987). Furthermore, basal excretion of non-dietary NPRO was unaffected by ascorbic acid or α-tocopherol, whereas both these vitamins inhibited synthesis of NPRO from orally administered proline and [^{15}N]nitrate (Wagner & Tannenbaum, 1985), suggesting that there are at least two sites of nitrosation and that gastric nitrosation occurs in an acid-catalysed mechanism that is inaccessible to ascorbic acid. The last pathway probably involves nitrosating agents (N_2O_3 and N_2O_4) derived from nitric oxide (Walker, 1996).

The endogenous formation of volatile *N*-nitrosamines was studied after intake of nitrate at the ADI in combination with an amine-rich diet consisting of fishmeal rich in amines as nitrosatable precursors. Twenty-five women (mean age, 23 years; mean weight, 60 kg), who were non-smokers and who were not using medicines or vitamin preparations, ate the fishmeal with nitrate at the ADI for 7 consecutive days. They received a diet low in nitrate for 1 week before and 1 week after the test week. They agreed to donate saliva samples, collect 24-h urine samples and answer a questionnaire on food consumption and life style. Nitrate intake at the ADI (100 ml of a solution of sodium nitrate at 2800 mg/l) resulted in a significant increase in mean salivary nitrate and nitrite concentrations. The mean urinary nitrate excretion increased from 76 mg/24 h in the first week to 190 and 160 mg/24 h on days 1–3 and 4–7 during the test week, followed by a decline to 77 mg/24 h in the second control week. When the urine samples were analysed for volatile *N*-nitrosamines, both NDMA and *N*-nitrosopiperidine were detected. The mean urinary excretion increased significantly from 290 ng/24 h in the control week to 870 and 640 ng/24 h on days 1–3 and 4–7 during the test week and declined to 380 ng/24 h in the second control week. Excretion of *N*-nitrosopiperidine was not directly related to nitrate intake or to the composition of the diet. Nitrate excretion and NDMA excretion were significantly correlated, as were salivary nitrate and nitrite concentration and NDMA excretion (Vermeer et al., 1998).

The effect of certain dietary and other factors on the concentrations of urinary NPRO was studied in healthy adults (mostly men) given a diet low in preformed

NPRO, nitrate and proline for 5 days and ascorbic acid on days 4 and 5, when tests were conducted. In the standard test, the volunteers took 400 mg of nitrate at 11:00, and at 12:00 they ate a standard 700-calorie meal containing 500 mg of proline. Urine was collected for 24 h, and samples were analysed for NPRO. This standard test yielded 26 ± 2 (mean ± SE) nmol of NPRO, whereas 5 ± 1 nmol of NPRO were found when proline was taken alone. In variations on the standard test, the NPRO yield was not significantly affected by the sex of the volunteers, the time at which the standard meal was eaten, the size of the meal or drinking extra water after the meal. Doses of 100 and 200 mg of nitrate had lesser effects on NPRO yield than did the dose of 400 mg. One gram of ascorbic acid given 5 or 2 h before, with or 1 or 2 h after the meal with proline inhibited NPRO formation by 0, 71, 71, 67 and 19%, respectively. Chewing gum or tobacco for 2–3 h after the test meal did not increase NPRO formation or the salivary nitrate concentration, but the salivary nitrite concentration was reduced, especially when gum was chewed (Mirvish et al., 1995).

2.2 Toxicological studies

2.2.1 Acute toxicity

The acute toxic effects of nitrate were studied in groups of five goats aged 1.5–2.0 years, which received potassium nitrate at a dose of 1.3 g/kg bw, equivalent to 0.66 g/kg bw as nitrate ion. Partial anorexia, mild depression, frequent muscle tremor, incoordination, dyspnoea and brown discolouration of visible mucous membranes appeared from 2 h onwards, followed terminally by recumbency, salivation and colonic convulsions. The irreversible signs of toxicity appeared at about 5.8 h, when the animals were killed and organs and tissues sampled for histopathological examination. The haematological and biochemical alterations included a decrease in haemoglobin, a marked increase in the methaemoglobin, nitrate and nitrite concentrations of plasma, urine and cerebrospinal fluid and in blood glucose, with significant increases in ammonia nitrogen in rumen liquor, cholesterol, urea nitrogen, creatinine and aspartate aminotransferase activity in serum. A diphenylamine blue test on blood, plasma, urine, cerebrospinal fluid, rumen liquor and aqueous humour revealed strong positive reactions at the peak of toxicity. The cerebrospinal fluid nitrite appeared to be a dependable diagnostic value, as it truly reflected the toxicity of nitrate. Histopathological examination revealed changes in the kidneys (tubular changes with massive degeneration of the epithelium), liver (degenerative changes with congestion, focal haemorrhage and extensive empty dilatation of the central vein), intestine (infiltration onto mononuclear cells in mucosa and degenerative villi), lungs (congestion and haemorrhages of alveoli), heart (degenerative changes in endo- and myocardium with haemorrhage), lymph nodes (depletion of lymphocytes) and urinary bladder (lamina propria infiltrated by lymphocytes). Electrocardiography during induced acute toxicity showed marked tachycardia, an increased duration of the mean QRS complex and deviation in the configuration of the ST segment. These changes were a result of significant myocardial hypoxia due to marked methaemoglobinaemia (Mondal & Pandey, 1999; Mondal et al., 1999a, 2000).

2.2.2 Short-term studies of toxicity

In a 32-day experiment, groups of five goats aged 1.5–2.0 years were given potassium nitrate orally at a dose of 0 or 4 mg/kg bw per day, equivalent to 2.4 mg/kg bw as nitrate ion. The animals remained apparently normal, except for development of dullness and partial anorexia from day 22. The haematological and biochemical alterations included a significant decrease in haemoglobin concentration and increases in methaemoglobin, nitrate and nitrite concentrations in plasma and urine, in urea nitrogen and in aspartate aminotransferase activity in serum. The morphological changes seen in two animals killed on day 32 were mild degeneration and congestion and haemorrhages in visceral organs. The authors concluded that there had been no cumulative effect of prolonged nitrate intake (Mondal et al., 1999b).

2.2.3 Reproductive toxicity

In an overview and evaluation of the available information on the health effects of nitrate and nitrite (Fan et al., 1987), an association was suggested between maternal ingestion of nitrate from drinking-water and developmental effects in their children, although no cause-and-effect relationship could be established. Studies in experimental animals have shown reproductive toxicity associated with intake of high levels of nitrate and nitrite, which are not likely to be encountered in drinking-water. No teratogenic effects were observed in rats, mice, rabbits or hamsters. Several cases of methaemoglobinaemis were reported among infants in the USA who drank water containing nitrate at concentrations higher than the current 'maximum contaminant level' of 45 mg/l. The uncertainties in the database have been discussed (Fan & Steinberg, 1996).

The effects of intake of nitrate on a number of biochemical and endocrine parameters and their impact on reproductive function were studied in groups of five feeder bulls aged 16–18 months given potassium nitrate or a control diet. The bulls were tested 30 days before administration of nitrate, for 30 days during administration and for 30 days afterwards. The initial dose of nitrate was 100 g/day (equivalent to 160 g/day as nitrate ion), which was increased at weekly intervals by 50 g up to 250 g (equivalent to 150 g/day as nitrate ion). Administration of nitrate resulted in a significant ($p < 0.01$) increase in methaemoglobin concentration, a significant ($p < 0.01$) increase in the serum concentration of bile acids and a prolonged biological half-life of progesterone, suggesting impairment of liver metabolism. Intake of nitrate also resulted in a significant ($p < 0.05$) increase in cortisol concentration during and after administration and depressed thyroid gland activity, as seen from a significant ($p = 0.05$) decrease in thyroxin concentration during administration. Non-detectable levels (< 0.001 µg/ml) of thyrotropin after administration indicated suppression of hypothalamic function. Effects of nitrate on the function of Leydig cells during and particularly after administration were apparent from weakened testicular responses to treatment with gonadotrophin. Analysis of seminal plasma revealed a significant ($p < 0.01$) increase in total acid phosphatase activity and a significant decrease in the concentration of fructose. Intake of nitrate also reduced sperm motility. While no difference was found in the frequency of primary morphological abnormalities, the number of secondary abnormalities rose by 115% in the post-administration period, suggesting damaged membrane integrity. Histological examination revealed degenerative lesions in cells of the spermiocyte and spermatid layers (Zraly et al., 1997).

The effect of nitrate on reproductive function was studied in groups of 23–24 sheep (aged 3.5–4.0 years) fed a diet containing potassium nitrate. In a preliminary experiment, a dose of 20 mg/kg bw (equivalent to 11 mg/kg bw as nitrate ion) induced a reliable per cent decrease (by 32%) in the number of sheep in estrus ($p < 0.05$). Concentrations in the diet resulting in a dose of 0.5, 1, 5 or 10 mg/kg bw per day for 48 days (equivalent to 0.27, 0.54, 2.7 and 5.4 mg/kg bw as nitrate ion) did not affect cyclic sexual behaviour; however, the fertility and gestation rates of sheep fed potassium nitrate were considerably lower (by 36% at 5 mg/kg bw per day and 33% at 10 mg/kg bw per day) than in the control group (Nestorova et al., 1997).

2.2.4 Special studies: Effects on adrenal and thyroid glands

Hypertrophy of the zona glomerulosa of the adrenals of rats was reported after administration of low doses of nitrite for 90 days; the effect was considered to be due to its conversion to nitrate. A study was therefore conducted to compare the effects of nitrate and nitrite on the zona glomerulosa. Three groups of 10 male Wistar rats were given drinking-water containing potassium chloride (control), potassium nitrite or potassium nitrate at a concentration of 36 mmol/l for 90 days. The body-weight gain of rats given nitrite or nitrate was slightly slower than that of controls, but no differences in food intake per kilogram body weight were observed between the three groups. The water intake of the group given nitrite was statistically significantly lower than that of the other two groups. The rats receiving nitrite appeared cyanotic during the first month of treatment but not thereafter, perhaps because the water intake was greater during the first month. At the end of the observation period, the concentrations of methaemoglobin and nitrite in blood were significantly increased; the nitrate concentration in plasma in the groups given nitrite or nitrate groups were similar but were higher than those of controls. Treatment with nitrite or nitrate had no consistent effect on the concentrations of thyroxin, free thyroxin, thyroid stimulating hormones, adrenocorticotrophic hormone, corticosterone or aldosterone in blood. Microscopic examination revealed slight hypertrophy of the adrenal zona glomerulosa in all rats given nitrite and minimal hypertrophy in 2/10 rats given nitrate. The results of morphometric analyses of the adrenals were in line with those of microscopic examination. In the rats given nitrite, the fraction of the surface area of the zona glomerulosa in median sections was significantly greater than that in the controls or rats given nitrate. The minimal hypertrophy of the adrenal zona glomerulosa observed occasionally in rats given nitrate was barely detectable by morphometric analysis. It was concluded that nitrate ion does not play a role in the etiology of hypertrophy of the zona glomerulosa of the adrenal glands in rats (Boink et al., 1996).

2.3 Observations in humans

2.3.1 Effects on the gastrointestinal tract

An epidemiological study was conducted to examine a possible correlation between the nitrate concentration of drinking-water, recurrent stomatitis and the activity of cytochrome b_5 reductase, which reduces methaemoglobin to haemoglobin. Five areas in the State of Rajasthan, India, were selected on the basis of a nitrate concentration (as nitrate ion) in drinking-water of 26, 45, 95, 220 and 460 mg/l; and 193 age- and weight-matched persons, representing 10% of the total population in each of these areas, were recruited. Detailed histories of recurrent stomatitis were recorded, a medical examination was conducted, and blood samples were taken to ascertain cytochrome b_5 reductase activity. The results of a multivariate regression analysis suggested a significant interdependence between the nitrate concentration of drinking-water, cytochrome b_5 reductase activity and recurrent stomatitis. It was concluded that increased cytochrome b_5 reductase activity, induced primarily by the presence of a high nitrate concentration in drinking-water, was the cause of recurrent stomatitis. The specific confounding variables that were taken into account were not described (Gupta et al., 1999).

A validated technique involving microdialysis probes was used to measure chemicals involved in nitrosation in the oesophagus, cardia and proximal and distal stomach of 15 healthy volunteers before and after ingestion of nitrate at a concentration of 2 mmol/l (equivalent to that in a portion of salad). Ingestion of nitrate ion increased the median concentration in saliva from 36 μmol/l to 250 μmol/l and that in the distal oesophagal from 29 μmol/l to 180 μmol/l (both $p < 0.01$). The concentration of nitrate in the stomach decreased progressively but increased progressively with distance from the gastro-oesophageal junction, so that the highest ratio of nitrate:nitric oxide was at the junction. The authors concluded that nitrosation in the stomach is maximal at the gastro-oesophageal junction and that dietary nitrate is involved in mutagenesis and carcinogenesis at this site (Suzuki et al., 2002).

The anatomical distribution of nitric oxide generation in the lumen of the upper gastrointestinal tract was investigated under basal conditions and after ingestion of a quantity of nitrate equivalent to that in a portion of salad. Probes containing custom-made sensors that could detect nitric oxide and pH were installed in the oesophagus of 15 *Helicobacter pylori*-negative volunteers (10 men) with a mean age of 32 years. Nitric oxide and pH were measured for 2 min at 1-cm increments throughout the length of the stomach and distal oesophagus in persons who ingested potassium nitrate at 2 mmol/l and in controls. Serum nitrate and saliva nitrite concentrations were also recorded. Ingestion of nitrate increased the mean (range) serum nitrate from 30 μmol/l (18–49) to 95 μmol/l (32–150) and the mean salivary nitrite from 4.7 μmol/l (1.4–7.8) to 23 μmol/l (2.1–50) (both $p < 0.05$). After ingestion of nitrate, the peak concentration of nitric oxide was found within 1 cm distal to the gastro-oesophageal junction in 11 of the 15 (73%) persons. The pH step-up after nitrate was 7.5 μmol/l (0.5–31) and was significantly higher than at all other sites. Nitric oxide concentrations > 50 μmol/l were observed at the location at which the neutral oesophageal pH fell to acidic gastric pH. Generation of nitric oxide from dietary nitrate via salivary nitrite was found to be maximal at the gastro-oesophageal junction and cardia (Iijima et al., 2002).

2.3.2 *Relationship between nitrate intake, subsequent endogenous formation of* N*-nitroso compounds and risk for stomach cancer*

In vitro

To mimic physiological exposure to nitrite, nitrite was added gradually to the gastric compartment of a dynamic gastrointestinal model (based on toxicokinetics) simulating salivary nitrite mass flow after intake of nitrate at concentrations 0.1, 1, 5 and 10 times the ADI. Concomitantly, 100 g of cod were added as the source of primary amines for the nitrosation reaction. With a rapid decrease in the pH of the stomach, 2.3, 16 and 420 μg of NDMA were formed at 0.1, 1 and 10 times the ADI of nitrate, respectively; and with a slow decrease in pH in the compartment, 1.8. 5.1, 21 and 43 μg were formed at 0.1, 1, 5 and 10 times the ADI of nitrate, respectively. Thus, relatively large amounts of endogenous NDMA were formed under realistic physiological conditions (Krul et al., 2001a).

The accuracy of quantification of the endogenous formation of *N*-nitroso compounds in the human body is still a matter of debate. A dynamic gastrointestinal model simulating various physiological conditions in the gastric compartment was used to study the formation of NDMA after intake of various fish species as a source of amines, in combination with nitrite. Cod, herring, pollack and plaice induced NDMA formation in the gastric compartment, the amounts after intake of 100 g of fish ranging from a mean of 28 μg (23–29 μg) with plaice to 190 μg (160–240 μg) with cod. Mackerel and salmon samples did not increase NDMA formation under these conditions. The amounts of dimethylamine, trimethylamine and trimethylamine-*N*-oxide in the samples of cod and herring did not correlate with the amount of NDMA produced, indicating that the extent of NDMA formation cannot be predicted from amine concentrations. When cod was introduced into the model with 250 mg of ascorbic acid and nitrite was added gradually, a 66% reduction in the amount of NDMA formed was observed; addition of 1000 mg of ascorbic acid did not reduce the amount of NDMA further. NDMA formation was also effectively inhibited, by > 90%, by the addition of green or black tea (2 g lyophilized tea dissolved in water). The addition of spinach (75 g) to food intake resulted in inhibition of NDMA formation with about the same efficacy as observed with ascorbic acid (Krul et al., 2001b).

Epidemiological studies

Epidemiological studies can be ranked in increasing order, from ecological (or correlation) studies to cross-sectional studies, case–control studies, cohort studies and intervention trials. This classification of epidemiological study designs with respect to their potential for bias and, consequently, the strength of evidence they provide and the costs involved has been described in detail (e.g. van den Brandt et al., 2002). Intervention trials provide the strongest evidence for a causal relationship with risk and (because of the possibility to control for confounding and bias) have the least possible bias; however, they are usually the least feasible and the most expensive. Less expensive cohort studies allow assessment of exposure and selection of study participants before the health outcome of interest occurs and thus provide relatively strong evidence. Although the less expensive case–control studies generally involve assessment of exposure retrospectively in subjects with and without the health outcome, the resulting evidence is more debatable. This is particularly so in the case of dietary intake, because of the possibility of selection bias, recall bias and/or bias due to the presence of disease. Cross-sectional studies suffer from the additional problem that exposure and disease are measured at the same time, making it impossible to draw conclusions about cause and effect. Correlation studies cost the least, but they provide weak evidence and are much more susceptible to bias. In addition to the problem of extrapolating the data to the individual level (as the units of measurement are population groups), most such studies include limited data on exposure, rely on mortality (rather than incidence) rates and often do not include consideration of the induction period. Some investigators have stated that observational studies cannot, by definition, establish the causality of a relationship on the basis of a statistical association. However, when several studies of high quality, such as those in which the biases are shown to be minimal, are available and these consistently show a dose–response relationship, observational studies may well contribute to conclusions about causality. The power of observational epidemiological studies was established 50 years ago, when such studies revealed that smoking caused lung cancer.

(i) Ecological studies

The possible conversion of nitrates to nitrites under conditions of gastric achlorhydria and their transformation to nitrosamines in the presence of nitrosatable compounds led to a number of epidemiological studies of the possible relationship between high nitrate levels in drinking-water and mortality due to various cancers. One study was performed in the Province of Valencia, Spain, which has the highest concentration of nitrate in drinking-water in Europe, to compare mortality due to cancers of the stomach, urinary bladder, prostate and colon in the populations of 258 municipalities with different levels of nitrate. The mortality rates from gastric and prostate cancer increased with increasing intake of nitrates. In populations whose drinking-water contained nitrate at a concentration > 50 mg/l, the relative risk for gastric cancer was 1.9 for men and 1.8 for women aged 55–75 years ($p < 0.05$) when compared with groups with low nitrate intake, and the relative risk for prostate cancer was 1.9 for men aged 55–75 and 1.8 for those aged > 75 (Morales Suarez Varela et al., 1995). It was not clear from the text whether correction was made for confounding factors.

The possible association between risk for gastric cancer and the concentration of nitrate and the hardness of drinking-water from municipal supplies was investigated in a matched case–control study in Taiwan (China). Data on deaths from gastric cancer among eligible residents between 1987 and 1991 (6766 cases) were obtained from the Bureau of Vital Statistics. The 6766 controls had died from other causes and were matched individually to the cases by sex, year of birth and year of death. Data on nitrate, nitrogen and hardness in drinking-water were collected from the Taiwan Water Supply Corporation. The municipality of residence was assumed to be the source of intake of nitrate for cases and controls. No difference in mortality rates from gastric cancer were found between groups with different levels of nitrate intake. The odds ratios (with 95% confidence interval [CI]) for death from gastric cancer were 0.95 (0.87–1.0) for the group with nitrate intakes of 0.23–0.44 mg/l and 1.0 (0.93–1.1) for the group with nitrate intakes > 0.45 mg/l. In contrast, a significant negative relationship was found between drinking-water hardness and mortality from gastric cancer, with odds ratios of 1.2 (1.1–1.3) and 1.6 (1.5–1.8), respectively, for intake of moderately hard water and soft water compared with hard water (Yang et al., 1997).

In a study of a population whose drinking-water had a high nitrate content (mean, 98 mg/l), nitrate was assessed by regular checks of water quality by local public health authorities. Empirical Bayes estimates of standardized mortality rates from gastric cancer were calculated and were correlated to average nitrate concentrations in drinking-water, with control for the confounding effects of smoking, ethnicity and population size. The average nitrate concentrations proved to be significant predictors of gastric cancer mortality ($p = 0.05$). The most probable limit value for this role of nitrate was 95 mg/l. With this reference value, the Mantel-Haenszel odds ratio for the relationship between nitrate and mortality from gastric cancer was 1.5 (95% CI, 1.1–2.1). The results of this ecological study support the hypothesis that high levels of nitrate in drinking-water have a causal role in gastric cancer development, but only an abstract was available (Sandor et al., 1998).

An ecological study in Yorkshire, northern England, was conducted to examine the hypothesis that intake of high levels of nitrate in drinking-water increases the risks for cancers of the stomach, oesophagus and brain in adults. Nitrate concentrations for the period 1990–95 and the numbers of incident cancers in 1975–94 were available for 148 geographically defined zones, each supplying water of homogeneous chemical composition to an average population of around 20 000. No relationship was found between nitrate concentration and the incidence of stomach or oesophageal cancer, but the incidence of cancers of the brain and central nervous system was higher in areas with higher nitrate levels, with a relative risk of 1.2 (95% CI, 1.1–1.3) in the quartile of the population with the highest mean intake (30 mg/l) compared with the quartile with the lowest intake (2.4 mg/l). The increase in risk remained statistically significant ($p < 0.01$) after allowance for other covariates and for extra-Poisson variation in a regression model. The observed relationship with brain cancer requires confirmation (Barrett et al., 1998).

In another ecological study, data were obtained on the incidences of various cancers, contamination of drinking-water with atrazine and nitrate and related agricultural practices in 40 districts in the Province of Ontario, Canada. The data were merged for geographical and statistical analyses. Weighted (by population size) least-squares regression analyses were conducted with control for confounding socioeconomic and lifestyle factors. Maximum likelihood spatial error models were estimated when least-squares regression error terms were found to be spatially auto-correlated, with Maran's I statistics. Nitrate levels ranging from 0 to 91 mg/l (maximum acceptable concentration, 10 mg/l) were negatively associated with the incidence of stomach cancer. Although the negative association appears to have no direct biological explanation, such counterintuitive outcomes may be found in complex systems in which social and biological variables interact (Van Leeuwen et al., 1999).

Randomly selected 24-h urine samples from 39 populations in 24 countries were obtained in an international study of electrolyte excretion and blood pressure and used to measure sodium (5756 people) and nitrate (3303 people). The median sodium and nitrate concentrations were standardized by age (20–49 years) and sex and averaged per country. Ecological correlation regression analyses were done in relation to national rates of mortality from stomach cancer. The Pearson correlation for the 24 countries was 0.70 in men and 0.74 in women(both $p < 0.001$) for sodium concentration and 0.63 ($p = 0.001$) in men and 0.56 ($p < 0.005$) in women for nitrate. In multiple regression analysis of stomach cancer mortality, with sodium and nitrate as independent variables, the adjusted correlation coefficient, r^2, was 0.61 in men and 0.54 in women (both $p < 0.001$). Addition of the interaction term (sodium × nitrate) to this model increased the adjusted r^2 to 0.77 in men and to 0.63 in women. The model showed that the importance of nitrate as a risk factor for death from stomach cancer increased markedly with sodium concentration; however, the relationship with sodium was always stronger than that with nitrate. Salt intake, measured as sodium excretion in 24-h urine, is probably the rate-limiting factor for stomach cancer at the population level. No attempt was made to take cancer induction time into account: the urine samples were collected in 1986–87, whereas the data on gastric cancer mortality were for 1986–88 (Joossens et al., 1996).

(ii) Cross-sectional studies

The excretion of nitrate, nitrite, apparent total *N*-nitroso compounds and volatile nitrosamines was measured in 24-h urine from 61 Egyptians, divided into four groups: controls, *Schistosoma haemotobium*-infected patients and bladder cancer patients with and without a history of schistosomal infection. The urinary concentration of nitrate in *S. haemotobium*-infected patients was significantly higher than in the other three groups. The concentration of nitrite was below the limit of detection of the method (< 0.015 μg/mg creatinine) in all but one of the control samples. *S. haemotobium* infection significantly increased urinary nitrite to 0.9 ± 1.16 μg/mg creatinine (mean ± SD, $p = 0.001$), and the concentration of nitrite in both groups of patients with bladder cancer was about 20 times that in *S. haemotobium*-infected patients without bladder cancer. Excretion of apparent total *N*-nitroso compounds paralleled that of nitrite, and, overall, a good correlation was observed between these two variables ($r^2 = 0.71$, $p = 0.0001$). NDMA was present in all the samples analysed, and *S. haemotobium* infection significantly increased the urinary concentration over that of controls (4.0 ± 1.6 and 2.0 ± 3.0 ng/mg creatinine, respectively; $p = 0.01$). Among the cancer patients, the concentration of NDMA was higher than that in controls only in those with schistomal infection. The presence of *N*-nitroso compounds, including NDMA, in the urine of *S haemotobium*-infected patients both before and after the development of cancer and the observation that these compounds also occur in bladder cancer patients with no history of schistosomal infection suggest that these compounds might have a role not only in initiation of the carcinogenic process but also in its progression (Mohsen et al., 1999). As endogenous nitric oxide production can also result from schistosomiosis, cause and effect can be reversed.

A study was conducted in Egypt, an area with high environmental levels of nitrosating agents, to investigate whether salivary nitrate and nitrite and the activity of nitrate reductase in saliva affect the risk for oral cancer. The concentrations of salivary nitrite (8.3 ± 1.0 μg/ml) and nitrate (44 ± 3.7 μg/ml) and the activity of nitrate reductase (74 ± 10 nmol/ml per min) were significantly ($p < 0.05$) higher in 42 oral cancer patients than in 40 healthy individuals (nitrite, 5.3 ± 0.3 μg/ml; nitrate, 27 ± 1.2 μg/ml; nitrate reductase activity, 46 ± 4 nmol/ml per min). The adjusted ORs and the 95% confidence intervals for oral cancer, categorized by the values for salivary nitrate and nitrite and nitrate reductase activity, showed a risk associated with a nitrite concentration > 7.5 μg/ml (OR, 3.0; 95% CI, 1.0–9.3), a nitrate concentration > 40 μg/ml (OR, 4.3; 95% CI, 1.4–13.3) and nitrate reductase activity > 50 nmol/ml per min (OR, 2.9; 95% CI, 1.1–7.4). The findings suggest that increased consumption of dietary nitrate and nitrite is associated with elevated levels of salivary nitrite. Together with the increased activity of salivary nitrate reductase, these observations may explain, at least in part, the role of nitrate and nitrite in the development of oral cancer in individuals in areas with a high burden of *N*-nitroso precursors (Badawi et al., 1998).

The frequency of variants of hypoxanthine-guanine phosphoribosyltransferase in peripheral blood lymphocytes and endogenous nitrosation in human populations with various nitrate levels in their drinking-water were investigated in four groups of women volunteers. The groups with low and medium intake of nitrate from tap water (14 and 21 women) had concentrations of 0.02 and 18 mg/l, respectively, from public water supplies, whereas those with medium and high intakes of nitrate from well water (6 and 9 women) had mean concentrations of nitrate of 25 and 140 mg/l, respectively. Higher nitrate intake resulted in a dose-dependent increase in 24-h urinary nitrate excretion and in increased salivary nitrate and nitrite levels. The mean log frequency of variants in peripheral lymphocytes was significantly higher in the women with medium intake from well water than in those with low or medium intake from tap water. An inverse correlation was found between the peripheral lymphocyte labelling index and the nitrate concentration of drinking-water. Analysis for *N*-nitrosamines in the urine of 22 women by gas chromatography–mass spectrometry revealed the presence of *N*-nitrosopyrrolidine in 18. Analysis of the mutagenicity of well water samples showed that a small number caused reverse mutation in *Salmonella typhimurium* after concentration on a XAD-2 resin column. Thus, consumption of drinking-water with high concentrations of nitrate can confer a genotoxic risk, as indicated by increased hypoxanthine-guanine phosphoribosyl transferase variant frequencies and endogenous formation of *N*-nitroso compounds from nitrate-derived nitrite (van Maanen et al., 1996b).

(iii) Case–control studies

The increasing incidence of non-Hodgkin lymphoma (NHL) in the USA is only partially explained by known risk factors. In a study of 156 cases and 527 controls in Nebraska, USA, the average intake of nitrate from community water supplies was estimated for 1947–79. Long-term consumption of community water with average nitrate levels in the highest quartile (> 4 mg/l as nitrate nitrogen, equivalent to 18 mg/l) was positively associated with risk for NHL (OR, 2.0; 95% CI, 1.1–3.6). Dietary nitrate, mainly from vegetable, was not associated with risk after adjustment for vitamin C and carotene intakes, as persons with a lower intake of vitamin C were at slightly greater risk for NHL than those whose daily intake was > 130 mg, for all levels of intake of nitrate in drinking-water. Similar findings were made for the combined effect of nitrate in drinking-water and carotene. The nitrate levels in private wells, measured at the time of interview for 51 cases and 150 controls, were not associated with the risk for NHL after adjustment for pesticide use. These findings indicate that long-term exposure to nitrate in drinking-water at levels > 18 mg/l may contribute to the risk for NHL (Ward et al., 1996).

In another study of the association between NHL and nitrate in drinking-water, a population-based case–control study was conducted among white men in Minnesota, USA. By linkage of residential history with community water records, the average exposure to nitrate in drinking-water in 1947–75 was estimated for 73 cases diagnosed between 1980 and 1982 and for 147 controls who had used community water supplies. No association was found between nitrate in community water supplies and NHL within the range of concentrations studied (median of highest exposure category, 2.4 mg/l; range, 0.1–7.2 mg/l) (Freedman et al., 2000).

A population-based case–control study in Denmark addressed the hypothesis that parental occupation in agriculture increased the risk for testicular cancer in the offspring. The factors investigated were childhood residence on a farm, in the country or in an area with a high nitrate concentration in groundwater and the parents' occupation in agriculture. The only association that emerged was with childhood residence in an area with a high nitrate concentration. The excess risk was, however, confined to men who had not grown up on a farm or in the country, making it unlikely that nitrate *per se* was responsible (Møller, 1997).

A study of risk factors for childhood brain tumours was conducted to investigate the association between source of residential drinking-water during pregnancy and the occurrence of brain tumours among offspring. Dipstick measurements were made of nitrates and nitrites in tap water in the houses of a subset of women who were living in the same house in which they had lived during their pregnancy. A total of 540 patients with childhood brain tumours and 801 controls were identified in Los Angeles County and the San Francisco Bay Area of California and western Washington State, USA. Overall, no increased risk for childhood brain tumours was found in offspring of women for whom wells were the source of water; however, an increased risk (OR, 2.6; 95% CI, 1.3–5.2] was observed among offspring of women in Washington State and a decreased risk (OR, 0.2; 95% CI, 0.1–0.8) among those in Los Angeles County who relied exclusively on well water. Among the small subset of participants for whom dipstick measurements of tap water were available, the risk for childhood brain tumours associated with the presence of measurable nitrite and/or nitrate was 1.1 (95% CI, 0.7–2.0). Given the crude measurement method used and the fact that measures were often obtained years after the pregnancy, the relevance of the dipstick findings is unclear. The lack of consistency in the findings for residential water source does not support the hypothesis of an increased risk with consumption of well water; however, regional differences in the well water content of nitrite may exist, and the increased risk observed in western Washington State deserves further evaluation (Mueller et al., 2001).

A case–control study on gastric cancer and diet conducted in Marseille (France) included 92 patients with histologically confirmed adenocarcinoma and 128 controls undergoing functional reduction for injuries or trauma. The participants were interviewed by a trained dietician who administered a questionnaire about usual diet during the year preceding the first symptoms for cases or preceding the interview for controls. The intakes of nitrite, nitrite and pre-formed NDMA from food were estimated from a food composition table compiled ad hoc. Odds ratios were calculated after adjustment for age, sex, occupation and calorie intake. A high intake of preformed NDMA was associated with an increased risk for gastric cancer, the ORs for the second and third tertile of NDMA intake being: $OR_2 = 4.1$ (95% CI, 0.93–18) and $OR_3 = 7.0$ (95% CI, 1.8–26). Intake of nitrate and nitrite was not associated with an increased risk for stomach cancer. Consumption of vegetables was protective in general, independently of their estimated nitrate content (Pobel et al., 1995).

A study was conducted to investigate whether consumption of foods and beverages containing nitrosamines, nitrite and nitrates affects the risks for laryngeal, oesophageal and oral cancer. In a population-based case–control study in western Washington State, USA, dietary consumption of these substances was measured for 645 cases (169 laryngeal, 125 oesophageal and 351 oral cancer) and 458 controls. After adjustment for tobacco and alcohol use and other known risk factors, the risk for upper aerodigestive tract cancer was found to be 52% lower in persons who consumed larger amounts of nitrate (upper tertile) as compared with the lowest tertile ($p < 0.001$ for trend). Nitrate intake was associated with a reduction in the risk for cancers at all three sites. The reduction in risk for oesophageal cancer with increasing nitrate consumption was more evident in frequent tea drinkers than in other persons. The ORs for oesophageal cancer per category of nitrite intake were: 1.0 (reference) for < 1.1 mg/day, 1.2 for intakes between 1.1 and 1.6 mg/day and 1.6 (95% CI, 0.73–3.4) for intakes > 1.6 mg/day (p for trend = 0.20). There was no significant association between nitrite consumption and the risk for laryngeal or oral cancer; however, for individuals with a history of canker sores (an indicator of possible endogenous nitrosation), the risk for oesophageal cancer was seven times greater in those with a high nitrite intake than in those with a low intake. Consumption of foods with high concentrations of NDMA was associated with a 79% increase in the risk for upper aerodigestive tract cancer ($p = 0.037$ for trend). Daily consumption of beer and of nitrite-containing meats was associated with increased risks for oesophageal cancer (OR, 2.5 and 1.8, respectively). The OR for cancer of the oral cavity was also increased for people who drank beer daily (OR, 1.8). Persons who consumed large amounts of ascorbic acid from foods and from supplements were less likely to develop upper areodigestive tract cancer than were individuals with lower ascorbic acid intake (p for trend = 0.003) (Rogers et al., 1995).

In a study to evaluate the roles of maternal nutrition during gestation and subsequently in the etiology of childhood brain tumours, all 300 incident cases of nervous system tumours diagnosed in children under 18 in Israel between 1984 and 1993 were identified. Two matched population controls per case were selected ($n = 574$). Personal interviews were conducted with a semi-quantified three-step food frequency questionnaire. Univariate analysis showed that increased consumption of vegetable fat (p for trend = 0.01; 95% CI, 1.1–3.2), carbohydrates (p for trend = 0.05; 95% CI, 1.0–5.9) and vitamin E (p for trend = 0.05; 95% CI, 1.0–3.3) during childhood was significantly associated with risk for brain tumour. No associations were found with nitrate, nitrite or vitamin C. A significant positive association with potassium consumption during gestation (p for trend = 0.01; 95% CI, 1.1–3.7) was noted. In a multivariate analysis, the only persistent associations were with vegetable fat (OR, 1.4; 95% CI, 1.1–1.7) in the diet during childhood and potassium intake during gestation (OR, 1.4; 95% CI, 1.0–2.0) (Lubin et al., 2000).

(iv) Cohort studies

The association between intake of nitrate or nitrite and gastric cancer risk was investigated in a prospective cohort study begun in 1986 in the Netherlands, of 120 852 men and women aged 55–69 years. At baseline, data on dietary intake, smoking habits and other covariates were collected by means of a self-administered questionnaire. For data analysis, a case–cohort approach was used, in which the person–years at risk were estimated for a randomly selected subcohort of 1688 men and 1812 women. After 6.3 years of follow-up, 282 microscopically confirmed incident cases of stomach cancer were detected, with 219 in men and 63 in women. The rate ratios for gastric cancer with increasing quintiles of mean nitrite intake were: 1.0 (intake, 0.01 mg/day), 1.2 (0.04 mg/day), 1.2 (0.09 mg/day), 0.88

(0.16 mg/day) and 1.4 (0.35 mg/day). The 95% confidence interval for the rate ratio derived by contrasting the highest versus the lowest intake quintile was 0.95–2.2. A test for trend in rate ratios across quintiles gave a *p* value of 0.24. The authors concluded that the study did not support a strong positive association between nitrite intake and gastric cancer risk (van Loon et al., 1998).

A cohort study was conducted to investigate the relationship between intake of nitrates, nitrites and NDMA and the risk for cancers of the gastrointestinal tract in 9985 adult Finnish men and women who were initially free of cancer. During a follow-up period of up to 24 years, 189 gastrointestinal cancer cases were diagnosed in the cohort. The intakes of nitrate, nitrite and NDMA were estimated from data on food consumption obtained during an interview about the total diet of the participants over the previous year. The mean daily intake of nitrates was 77 mg, and that of nitrites was 5.3 mg. Nitrates were provided mainly by vegetables (92%), whereas nitrites were derived mainly from cured meats and sausages (94%); dietary NDMA was provided by smoked and salted fish (52%) and cured meats and sausages (48%). The mean daily intake of NDMA from the diet was 0.052 μg, and that from beer was 0.071 μg. A significant positive association was observed between intake of NDMA and subsequent occurrence of colorectal cancer, with a relative risk between the highest and lowest quartiles of intake of 2.1 (95% CI, 1.0–4.3). Of the various sources of *N*-nitroso compounds, smoked and salted fish were found to be significantly associated with the risk for colorectal cancer (relative risk, 2.6; 95% CI, 1.2–5.5), and intake of cured meat was nonsignificantly associated (relative risk, 1.8; 95 % CI, 0.98–3.5). No such association was observed with intake of other fish or other meat. No significant association were observed between NDMA intake and cancers of the head and neck or of the stomach or between nitrate or nitrite intake and the risk for cancers of the gastrointestinal tract (Dich et al., 1996; Knekt et al., 1999).

Cancer incidence was studied in a cohort of 21 977 women in Iowa, USA, who were 55–69 years of age at baseline in 1986 and had used the same water supply for > 10 years (87% for > 20 years); 16 541 of these women were on a municipal water supply, and the remainder used private wells. Nitrate intake in 1955–88 was assessed from public databases for municipal water supplies (quartile cut-off points, 0.36, 1.0 and 2.4 mg/l for nitrate nitrogen, equivalent to 1.6, 4.5 and 11 mg/l expressed as nitrate ion). As no data on individual water consumption were available, each women was assigned an average intake calculated on a community basis; no data on intake were available for women using private wells. Cancer incidence in 1986–98 was determined by linkage to the Iowa Cancer Registry. A total of 3150 cases were identified. No consistent association was found with increasing nitrate concentration in drinking-water and NHL, leukaemia, melanoma or cancers of the colon, breast, lung, pancreas or kidney, but positive associations were found with bladder cancer (relative risks across nitrate quartiles: 1, 1.7, 1.1 and 2.8) and ovarian cancer (1, 1.5, 1.8 and 1.8) and inverse associations with uterine (1, 0.86, 0.86 and 0. 55) and rectal cancer (1, 0.72, 0.95 and 0.47) after adjustment for a variety of risk and protective factors, agents that affect nitrosation (smoking, vitamin C and vitamin E intake), dietary nitrate and water source. Similar results were obtained when the analyses were restricted to nitrate concentrations in drinking-water in 1955–64. The positive association with bladder cancer is consistent with some previous findings, but the association with ovarian cancer and the inverse associations with uterine and rectal cancer were unexpected (Weyer et al., 2001).

2.3.3 Relationship between nitrate intake and genotoxic effects

A study was conducted to investigate the potential genotoxicity of nitrates and nitrites in drinking-water. The frequencies of sister chromatid exchange and chromatid or chromosomal aberrations were studied in peripheral blood lymphocytes from 70 children aged 12–15 who were permanent residents of areas of Greece with high concentrations of nitrates in drinking-water (56– 88 mg/l). The control group comprised 20 healthy children living in areas with a low nitrate concentration in drinking-water (0.7 mg/l). No increase in the mean number of chromatid or chromosomal breaks was observed in children with nitrate concentrations > 55 mg/l, but the frequency increased with intakes > 70 mg/l ($p < 0.01$). The mean number of sister chromatid exchanges per cell was not increased significantly (Tsezou et al., 1996).

2.3.4 Putative relationship between nitrate intake and effects on the endocrine system

In a mechanistic study on gonadotrophin-stimulated steroidogenesis in mouse Leydig tumour cells *in vitro,* the effects of a number of chemicals, including nitrate, were examined. Nitrate completely abolished the effect of 1 IU/l of human chorionic gonadotrophin on androgen secretion. The inhibition could be partially overcome by increasing the gonadotrophin concentration. Inorganic nitrate probably inhibited steroidogenesis via conversion to nitric oxide without involving the guanylate cyclase-cyclic guanosine monophosphate pathway (Panesar, 1999).

It has been suggested that increased nitrate intake affects the function of the thyroid gland in humans, as was observed in a study in pigs (Annex 1, reference 117), by competitive inhibition of iodide transport leading to decreased thyroid hormone secretion, followed by an increase in thyroid-stimulating hormone. To test this hypothesis, a 4-week study was conducted in which 10 volunteers received sodium nitrate in 200 ml of distilled water at a dose of 15 mg/kg bw once a day for 28 days (three times the current ADI of 0–3.7 mg/kg bw expressed as nitrate ion), and 10 volunteers received 200 ml of distilled water and served as the control group. Both groups received an iodine-restricted, low-nitrate diet, and compliance was checked by measuring the urinary iodide and plasma nitrate concentrations. Before and after the 28-day treatment period, the per cent uptake of radioactive iodine was measured 5 h and 24 h after administration to investigate competition of nitrate in iodide transport. Before and 2, 3 and 4 weeks after treatment with nitrate, blood samples were taken for measurement of thyroid hormones and insulin-like growth factor I, as measures of thyroid gland function. Nitrate had no effect on hormone concentrations during the 4-week treatment but increased the 24-h per cent uptake of radioactive iodine by 1.5 times (20–29%) over that before treatment, whereas a decrease had been expected. The authors concluded that intake of nitrate at three times the ADI would not cause changes in thyroid gland function in a healthy population (Lambers et al., 2000b).

2.3.5 Putative relationship between nitrate intake and insulin-dependent diabetes mellitus

Although a number of studies performed before 1991 showed no association between nitrate levels in drinking-water and insulin-dependent diabetes mellitus (IDDM), an ecological study in the USA (Kostraba et al., 1992) showed a significant positive correlation, and a study in the United Kingdom (McKinney et al., 1996) showed a clear excess of childhood IDDM in rural areas, which were presumed to have higher concentrations of nitrate in drinking-water than urban areas. Further epidemiological studies were therefore performed to test the hypothesis that the incidence of childhood IDDM is increased in areas with higher levels of nitrate in domestic drinking-water.

A population-based study performed in conjunction with the Yorkshire Register of Childhood Diabetes included 1797 children aged 0–16 years in whom IDDM was diagnosed while they were resident in the geographical area of the former Yorkshire Regional Health Authority, which covered the period 1978–94 and was about 97% complete. Three categories of nitrate intake were defined on the basis of population density, with equal populations. In 9330 water samples taken between 1990 and 1995, the nitrate concentration exceeded 25 mg/l in 2775 (30%) and exceeded 50 mg/l in 31 of these (0.1%). A positive association was found between the nitrate content of domestic water and the incidence of IDDM (McKinney et al., 1999).

An ecological study addressed the possible relation between nitrate levels in drinking-water in the Netherlands and the incidence of type 1 IDDM, to determine whether the standard of the European Commission and WHO for nitrate in drinking-water (37 mg expressed as nitrate ion) is adequate to prevent this disease. During 1993–95 in the Netherlands, 1104 cases of type 1 diabetes were diagnosed among 2 829 020 children aged 0–14, and 1064 of these cases were used in the analysis. The mean concentrations of nitrate in drinking-water in 3932 postal code areas in 1991–95 were divided into two categories, one based on equal numbers of children with various nitrate levels (0.25–2.1, 2.1–6.4 and 6.4–41 mg/l) and the other based on cut-off values of 10 and 25 mg/l for nitrate. Standardized incidence ratios for type 1 diabetes were determined for subgroups of the 2 829 020 children with respect to nitrate category, sex and age and as compared in univariate analysis with a χ^2 test for trends. Incidence rate ratios were calculated by multivariate analysis in a Poisson regression model. An effect of increasing age on the incidence of type 1 diabetes was seen, but there was no effect of sex or of nitrate concentration in drinking-water for the two categories. For a nitrate concentration > 25 mg/l, both the standardized incidence ratio and the incidence rate ratio were increased (1.5); however, the increase was not statistically significant, probably because of the small number of cases (15 of 1064). The authors concluded that there is no convincing evidence that nitrate in drinking-water at current levels is a risk factor for childhood type 1 diabetes mellitus in the Netherlands, although a threshold > 25 mg/l could not be excluded as higher concentrations were not studied (van Maanen et al., 2000).

2.3.6 Putative relationship between nitrate intake and neural tube defects

A population-based case–control study was conducted in California, USA, between June 1989 and May 1991 to investigate a possible association between periconceptional exposure of mothers to nitrate in drinking-water and the diet and the risk for neural tube defects in their offspring. The mothers of 538 cases and 539 normal controls were interviewed regarding residential history, consumption of tap water at home and dietary intake during the periconceptional period. Dietary nitrate intake was not associated with an increased risk for neural tube defects; concentrations > 45 mg/l in drinking-water as nitrate ion were associated with an increased risk for anencephaly (OR, 4.0; 95% CI, 1.0–15) but not for spina bifida. Increased risks for anencephaly were also observed among the offspring of mothers who drank groundwater only with nitrate concentrations below the maximum level (OR, 2.1; 95% CI, 1.1–4.5 for 16–35 mg/l; and OR, 6.9; 95% CI, 1.9–25 for 36–67 mg/l compared with < 5 mg/l, all as nitrate ion). Adjustment for identified risk factors for anencephaly did not substantially alter these associations, nor did control for maternal dietary intake, total vitamin C intake or quantity of tap water consumed (Croen et al., 1997, 2001).

2.3.7 *Putative relationship between nitrate intake and sudden infant death syndrome*

It has been shown (Wiklund et al., 1998) that enteric bacterial urease is inhibited in victims of sudden infant death syndrome (SIDS). One possible inhibitor of this bacterial activity is nitrate. If ambient pollution by nitrate is involved in the etiology of SIDS, only a fraction of the nitrate concentration often found in drinking-water would be enough to cause such inhibition. The 636 cases of SIDS that occurred in Sweden during the period 1990–96 were analysed with regard to geographical and seasonal distribution, the nitrate concentration in drinking-water and changes in groundwater level. Both the birth rate and the incidence of SIDS decreased during the study period. One-quarter of the municipalities, constituting 11% of the population, had no cases, the maximum incidence being 6.5 per 1000 live births. The northernmost parts of the country had the highest incidence, and individual deaths were associated with recharging of groundwater, which increases its nitrate content. The local incidence of SIDS was correlated (r^2 = 0.34–0.87) to the maximal recorded concentration of nitrate in drinking-water. The seasonal distribution of SIDS was very different from the south to the north of the country and appeared to be associated with differences in groundwater level subsequent to precipitation, frost penetration and melting of snow. Consumption of drinking-water with high peak concentrations or wide variations in nitrate concentration was correlated to the incidence of SIDS. The regression plots showed that the incidence of SIDS would exceed 1.5% at nitrate concentrations of 40–60 mg/l (George et al., 2001).

2.4 *Miscellaneous aspects*

2.4.1 *Claimed beneficial effects of nitrate*

In a critical review of studies of the effect of nitrate, it was claimed that nitrate can have beneficial or protective effects on health. The implication of dietary nitrate (via conversion to nitrite) in the formation of methaemoglobin and carcinogenic nitrosamines in humans has led to restriction of nitrate and nitrite levels in food and drinking-water; however, the authors considered that there is no epidemiological evidence for increased risks for gastric or intestinal cancer in population groups with high dietary intake of vegetables or nitrate. A re-evaluation of the negative perception of dietary nitrate was conducted on the basis of studies on the metabolism and enterosalivary circulation of nitrate in mammals, which showed that nitrate is converted to nitrite in the oral cavity, which then 'fuels' mammalian resistance to infectious disease. In the acid environment of the stomach, nitrite is reduced to nitric and other nitrogen oxides and, conceivably, also contributes to the formation of systemic *S*-nitrosothiols. Nitric oxide and solutions of acidified nitrite, mimicking gastric conditions, have been shown to have antimicrobial activity against a wide range of organisms. In particular, acidified nitrite was bactericidal for a variety of gastrointestinal pathogens, such as *Yersinia* and *Salmonella*. Nitric oxide is also known to have vasodilatory properties and to modulate platelet function, as do *S*-nitrosothiols. Thus, the authors concluded that nitrate in the diet determines the production of reactive nitrogen oxide species in the stomach and acts as a host defence against gastrointestinal pathogens, as a modulator of platelet activity and is possibly involved in gastrointestinal motility and microcirculation. The authors concluded that dietary nitrate may have an important therapeutic role to play, for instance in immunocompromised persons and in persons at particular risk of contracting gastroenteritis (Duncan et al., 1995; Dykhuizen et al., 1996; McKnight et al., 1997; Vallance, 1997; McKnight et al., 1999). The authors also reported that the conversion of nitrate into oxides of nitrogen prevents the formation of carcinogenic nitrosamines (Duncan et al., 1997).

2.4.2 Other opinions on hazard and risk of nitrate

Other international committees, such as those convened by the European Commission and IARC, have assessed the safety of nitrate and nitrite, and some criticisms have been made of the safety evaluation performed by the Committee at its forty-fourth meeting. For instance, L'Hirondel & L'Hirondel (1997) noted that the drinking-water standard for nitrate, which is based mainly on methaemoglobin formation, was established by another WHO committee and not by the Expert Committee. The limit of 50 mg/l for sodium nitrate was considered to be conservative, as the enteritis that occurs in cases of gastrointestinal tract infection has a greater influence on methaemoglobin formation than nitrite derived from conversion of nitrate. The authors claimed that most cases of methaemoglobinaemia have been associated with intake of well water containing nitrate at > 100 mg/l and concomitant microbiological contamination.

The authors also criticized establishment of an ADI for both nitrate and nitrite and the use of a safety factor of 500 for nitrate. They further criticized use of the NOEL for nitrite to transpose the ADI to nitrate.

They considered that two of the studies with nitrite should be re-evaluated. They stated that the one showing hypertrophy of the zona glomerulosa in the adrenal cortex in rats (Til et al., 1988) should be reconsidered, as other studies (Vleeming et al., 1997; Boink et al., 1999) demonstrate that the minimal hypertrophy can be considered a physiological adaptation to the slight oscillation in blood pressure caused by nitrite via the renin–angiotensin system. They stated that it should therefore not be considered a toxic effect relevant for the situation in humans. They considered that the study of Shuval & Gruener (1972, 1977) should also be re-evaluated. Although differences were found in the hearts, lungs and coronary arteries of treated and control animals at concentrations of sodium nitrite ≥ 200 mg/l, the Environmental Protection Agency (1990) and L'Hirondel & L'Hirondel (1997) stated that these changes were due at least partly to an absence of the coronary thickening and narrowing that normally occurs in aged rats. Hence, these changes may not be inherently adverse.

3. COMMENTS

The few new studies on the toxicokinetics and metabolism of nitrate in animals that have become available since the forty-fourth meeting of the Committee confirm that the rat is not a good surrogate species for humans in this respect, as it does not show salivary transport of nitrate and therefore has limited conversion of nitrate to nitrite.

In a study of the conversion of nitrate to nitrite in humans, in which sodium nitrate was administered in drinking-water at a single dose of 7.3 mg/kg bw, expressed as nitrate ion, neither blood pressure nor methaemoglobin concentration was affected. The nitrite concentration of the gastric juice was approximately six times higher after administration of nitrate in combination with pretreatment with omeprazole at 40 mg/day (which increased the gastric pH) than after nitrate alone. Nitrate was absorbed rapidly, the concentration in plasma increasing within 10 min, and the half-life of nitrate in plasma was about 6.5 h; about 70% of the dose was excreted in urine within 10 h of dosing. The plasma concentration of nitrite did not change after nitrate administration. About 8% of the total nitrate administered was converted to nitrite in saliva, as found in other studies. The Committee at its forty-fourth meeting concluded that the range of nitrate conversion is 5–7% for normal individuals and 20% for individuals with a high rate of conversion.

The results of studies in humans on the potential of a high nitrate intake to cause methaemoglobinaemia were equivocal. Some of the studies showed an association between a high nitrate concentration in drinking-water and methaemoglobinaemia, and others indicated that gastrointestinal infections, inflammation and the ensuing overproduction of nitric oxide are major factors in infantile methaemoglobinaemia. No increase in methaemoglobin concentration was seen in volunteers after a single administration of sodium nitrate in drinking-water providing a dose of 7.3 mg/kg bw, expressed as nitrate ion.

A study in humans showed that nitrate in vegetable matrices and from other sources, such as drinking-water, is almost totally bioavailable.

As nitrate shares a common transport mechanism with iodide, studies were conducted to determine whether nitrate affects thyroid function. A 28-day study with volunteers given sodium nitrate in drinking-water at a concentration equivalent to 15 mg/kg bw per day (11 mg/kg bw per day expressed as nitrate ion) showed no effects on thyroid function and no increase in the per cent of methaemoglobinaemia. A 90-day study of toxicity in rats showed that sodium nitrate at a dose of 50 mg/kg bw per day did not affect the thyroid or the zona glomerulosa of the adrenals.

In studies in humans, consumption of drinking-water containing sodium nitrate at a concentration of 2800 mg/l concomitantly with volatile *N*-nitrosatable amines in the diet (in cod, salmon or shrimp) led to a two- to threefold increase in urinary excretion of *N*-nitrosodimethylamine and *N*-nitrosopiperidine.

Several studies were reviewed on the effect of administration of nitrate on the release of nitric oxide at the junction of the oesophagus and the stomach in humans, which, it was speculated, might be associated with an increased incidence of cancer at this site. However, no such association has been observed in epidemiological studies.

A number of epidemiological studies have been published since the forty-fourth meeting of the Committee on the relationship between nitrate intake and cancer risk. At its present meeting, the Committee ranked the study designs according to their capacity to provide evidence of a relationship. In the descriptions below, relative risk estimates are given for those studies in which levels of intake of nitrate were provided.

Six ecological (correlation) studies were reported on nitrate in drinking-water and mortality from or incidence of cancer. Elevated risks were found for prostate cancer and for brain tumours (each in one study), but the results of six studies on gastric cancer were conflicting. The results of ecological studies (in which populations are the units of measurement) cannot be extrapolated to the individual level. Furthermore, most of the ecological studies were based on limited data on nitrate concentrations and on cancer mortality rates (rather than incidence rates), and none took an induction period for cancer into account.

Three of the studies were cross-sectional, involving measurement of, e.g., salivary nitrate in cancer patients and healthy subjects. Because cross-sectional studies do not take into account the time between exposure and disease, any observed differences in biomarkers of exposure might also be a consequence of the disease; therefore these studies cannot contribute to a causal interpretation of the results of studies of nitrate intake and cancer risk.

Seven case–control studies on nitrate in drinking-water and/or food and cancers at various sites were reviewed. In the studies on nitrate in drinking-water, conflicting results were reported with regard to an association with non-Hodgkin lymphoma, and no association was found with brain tumours. In the studies on dietary nitrate, no association was found with oral, oesophageal, gastric or testicular cancer. No other cancer sites have been studied.

Three prospective cohort studies have been conducted on nitrate intake and cancer risk. A cohort study in the Netherlands, with 6 years of follow-up, found no significant association between the incidence of gastric cancer and intake of nitrate from food or drinking-water, with relative risks for increasing quintiles of total nitrate intake of 1.0 (reference quintile), 1.2, 0.7, 0.9 and 0.9 for mean intakes of 60, 85, 100, 120 and 180 mg/day, respectively. Neither the relative risks nor the trend across relative risks was significant. A further analysis of the effect of nitrate within tertiles of vitamin C intake also did not reveal a positive association between nitrate intake and gastric cancer. A Finnish cohort study on dietary nitrate, with 24 years of follow-up, reported no association with the risks for tumours of the stomach, colorectum or head and neck. The average nitrate intake in this cohort was reported to be 77 mg/day. A cohort study in Iowa, USA, with 11 years of follow-up, revealed no consistent association between intake of nitrate from drinking-water and the risks for cancers at many sites, and an inverse association was reported with cancers of the uterus and rectum. Positive associations with nitrate intake were observed only for cancers of the ovary and urinary bladder, although it was not possible to determine whether other factors in drinking-water were responsible for these associations. In addition, no evidence of a dose–response relationship was found for any of the cancer sites addressed in the study in Iowa. The cohort studies included control for various potential confounders, such as intake of vegetables, age and smoking.

Overall, the epidemiological studies showed no consistently increased risk for cancer with increasing consumption of nitrate. These data, combined with the results of the epidemiological studies considered by the Committee at its forty-fourth meeting, do not provide evidence that nitrate is carcinogenic to humans.

A number of studies were performed to determine whether there are associations between nitrate intake in drinking-water and insulin-dependent diabetes mellitus, neural tube defects or sudden infant death syndrome. In none of these studies was a hypothesis proposed for the mechanism of an association. Two studies were conducted on the incidence of insulin-dependent diabetes mellitus and nitrate intake via drinking-water. One study in Yorkshire, United Kingdom, suggested a positive association, but the authors considered that the finding required confirmation. A study in the Netherlands with a larger number of subjects did not show a positive association. The two studies on nitrate intake and neural tube defects also showed no association. In a recent ecological study in Sweden, a correlation was reported between the nitrate concentration in drinking-water and the occurrence of sudden infant death syndrome; however, no confounding factors were taken into account. The Committee considered that it would be premature to include these observations in its safety assessment.

4. *EVALUATION*

The Committee concluded that the pivotal observed toxic effects of nitrate are consequent on its conversion to nitrite in vivo. The Committee at its present meeting established an ADI of 0–0.07 mg/kg bw for nitrite (see earlier part of this section). As the new data on nitrite would not provide a basis for a significant change in the previous ADI for nitrate, the Committee retained the ADI of 0–5 mg/kg bw expressed as sodium nitrate, or 0–3.7 mg/kg bw, expressed as nitrate ion, established at its forty-fourth meeting.

5. *REFERENCES*

Agunod, M., Yamaguchi, N., Lopez, R., Luhby, A.L. & Glass, E.B. (1969) Correlative study of hydrochloric acid, pepsin, and intrinsic factor secretion in newborns and infants. *Am. J. Dig. Dis.*, **14**, 400–414.

Avery, A.A. (1999) Infantile methemoglobinemia: Re-examining the role of drinking water. *Environ. Health Perspectives*, **107**, 583–586.

Avery, A.A. (2001) Cause of methemoglobinemia: Illness versus nitrate exposure. *Environ. Health Perspectives*, **109**, A12–A14.

Badawi, A.F., Hosny, G., El-Hadary, M. & Mostafa, H.M. (1998) Salivary nitrate, nitrite and nitrate reductase activity in relation to risk of oral cancer in Egypt. *Dis. Markers*, **14**, 91–97.

Balish, E., Witter, J.P. &Gatley, S.J. (1981) Distribution and metabolism of nitrate and nitrite in rats. In: *Banbury Report No. 7*, Cold Spring Harbor, New York: CSH Press, pp. 305–341.

Barrett, J.H., Parslow, R.C., McKinney, P.A., Law, G.R. & Forman, D. (1998) Nitrate in drinking water and the incidence of gastric, esophageal, and brain cancer in Yorkshire, England. *Cancer Causes Control*, **9**, 153–159.

Bartholomew, B. & Hill, M.J. (1984) The pharmacology of dietary nitrate and the origin of urinary nitrate. *Food Chem. Toxicol.*, **22**, 789–795.

Boink, A.B.T.J., Beekhof, P.K, Dormans, J.A.M.A. & Speijers, G.J.A. (1996) On the etiology of nitrite-induced hypertrophy of the zona glomerulosa of rats: I. The possible role of nitrate. Report No. 235802003 of the National Institute of Public Health and the Environment (RIVM), Bilthoven, Netherlands, pp. 1–19.

Boink, A.B.T.J., Dormans, J.A.M.A., Speijers, G.J.A. & Vleeming, W. (1999) *Effects of Nitrates and Nitrites in Experimental Animals* (Special Publication), London, Royal Chemistry Society, pp. 317–326.

van den Brandt, P.A., Voorrips, L., Hertz-Picciotto, I., Shuker, D., Boeing, H., Speijers, G., Guittard, C., Kleiner, J., Knowles, M., Wolk, A. & Goldbohm, A. (2002) The contribution of epidemiology to risk assessment of chemicals in food and diet. *Food Chem. Toxicol.*, **40**, 387–424.

Cohen, B. & Myant, M.B. (1959) Concentration of salivary iodide: a comparative study. *J. Physiol.*, **145**, 595–610.

Colbers, E.P.H., Hegger, C., Kortboyer, J.M. & Meulenbelt, J. (1996) A pilot study to investigate nitrate and nitrite kinetics in healthy volunteers with both normal and artificially increased gastric pH after sodium nitrate ingestion. Report No. 235802001 of the National Institute of Public Health and the Environment (RIVM), Bilthoven, Netherlands, pp. 1–62.

Croen, L.A., Todoroff, K. & Shaw, G.M. (1997) Maternal dietary nitrate exposure and risk for neural tube defects (Abstract). *Am. J. Epidemiol.*, **145**, (Suppl. 11), 530.

Croen, L.A., Todoroff, K. & Shaw, G.M. (2001) Maternal exposure to nitrate from drinking water and diet and risk for neural tube defects. *Am. J. Epidemiol.*, **153**, 325–331.

Dolby, J.M., Webster, A.D., Borriello, S.P., Barcaly, F.E., Bartholomew, B.A. & Hill, M.J. (1984) *Scand. J. Gastroenterol.*, **19**, 105–110.

Duncan, C., Dougall, H., Johnston, P., Green, S., Brogan, R., Leifert, C., Smith, L., Golden, M. & Benjamin, N. (1995) Chemical generation of nitric oxide in the mouth from the enterosalivary circulation of dietary nitrate. *Nature Med.*, **1**, 546–551.

Duncan, C., Li, H., Dykhuizen, R.S., Frazer, R., Johnson, P., Macknight, G., Smith, L., Lamza, K., McKenzie, H., Batt, L., Kelly, D., Golden, M., Benjamin, N. & Leifert, C. (1997) Protection against oral and gastrointestinal diseases: Importance of dietary intake, oral nitrate reduction and entero salivary. *Comp. Biochem. Physiol.*, **118A**, 939–948.

Dykhuizen, R.S., Frazer, R., Duncan, C., Smith, C.C., Golden, M., Benjamin, N. & Leifert, C. (1996) Antimicrobial effect of acidified nitrite on gut pathogens: Importance of dietary nitrate in host defense. *Antimicrob. Agents Chemother.*, **40**, 1422–1425.

Environmental Protection Agency (1990) Drinking-water criteria document on nitrate/nitrite. TR-1242-6CB. Available from National Technical Information Service, Department of Commerce, Springifled, Virginia, USA.

Fan, A.M. & Steinberg, V.F. (1996) Health implications of nitrate and nitrite in drinking water: An update on methemoglobinemia occurrence and reproductive and developmental toxicity. *Regul. Toxicol. Pharmacol.*, **23**, 35–43.

Forman, D., Al-Dabbagh, S. & Doll, R. (1985) Nitrates, nitrites and gastric cancer in Great Britain. *Nature*, **313**, 620–625.

Freedman, D.M., Cantor, K.P., Ward, M.H. & Helzlsouer, K.J. (2000) A case–control study of nitrate in drinking water and non-Hodgkin's lymphoma in Minnesota. *Arch. Environ. Health*, **55**, 326–329.

Fritsch, P., de Saint Blanquat, G. & Derache, R. (1979) [Absorption kinetics of nitrate and nitrite in the intestine of rats.] *Toxicol. Eur. Res.*, **3**, 141–147 (in French).

Fritsch, P., de Saint Blanquat, G. & Klein, D. (1985) Excretion of nitrates and nitrites in saliva and bile in the dog. *Food Chem. Toxicol.*, **23**, 655–659.

George, M., Wiklund, L., Aastrup, M., Thunholm, B., Saldeen, T., Wernroth, L., Zaren, B. & Holmberg, L. (2001) Incidence and geographical distribution of sudden infant death syndrome in relation to content of nitrate in drinking water and groundwater levels. *Eur. J. Clin. Invest.*, **31**, 1–13.

Gupta, S.K., Gupta, R.C., Seth, A.K., Gupta, A.B., Tech, M., Bassin, KJ., Gupta, D.K. & Sharma, S. (1999) Epidemiological evaluation of recurrent stomatitis, nitrates in drinking water and cytochrome b_5 reductase activity. *Am. J. Gastroenterol.*, **94**, 1808–1812.

Gupta, S.K., Gupta, R.C., Seth, A.K., Gupta, A.B., Bassin, J.K. & Gupta, A. (2000) Methaemoglobinaemia in areas with high nitrate concentration in drinking water. *Natl Med. J. India*, **13**, 58–61.

Green, L.C., Ruiz-de-Luzuriaga, K., Wagner, D.A., Rand, W., Istfan, N., Young, V.R. & Tannenbaum, S.R. (1981) Nitrate biosynthesis in man. *Proc. Natl Acad. Sci. USA*, **78**, 7764–7768.

Green, L.C., Tannenbaum, S.R. & Fox, J.G. (1982) Nitrate in man and canine milk. *New Engl. J. Med.*, **306**, 1367–1368.

Harada, M., Ishiwata, H., Nakamura, Y., Tanimura, A. & Ishidate, M. (1975) Studies on *in vivo* formation of nitroso compounds. I. Changes of nitrate and nitrite concentrations in human saliva after ingestion of salted Chinese cabbage. *J. Food Hyg. Soc. Jpn*, **16**, 11.

Hegesh, E. & Shiloah, J. (1982) Blood nitrates and infantile methemoglobinemia. *Clin. Chim. Acta*, **125**, 107–115.

l'Hirondel, J. & l'Hirondel, J.-L. (1997) *Nitrate and Man: Toxic, Harmless or Beneficial?* New York: CABI Publishing,

Ignarro, L.J. (1987) Endothelium derived relaxing factor produced and released from artery and vein; nitirc oxide. *Annu. Rev. Pharamacol. Toxicol.*, **30**, 535–560.

Iijima, K., Henry, E., Moriya, A. Wirz, A., Kelman, A.W. & McColl, K.E.L. (2002) Dietary nitrate generates potentially mutagenic concentrations of nitric oxide at the gastroesophageal junction. *Gastroenterology* (in press).

Joossens, J.V., Hill, M.J., Elliott, P., Stamler, R., Stamler, J., Lesaffre, E., Dyer, A., Nichols, R. & Kesteloot, H. (1996) Dietary salt, nitrate and stomach cancer mortality in 24 countries. *Int. J. Epidemiol.*, **25**, 494–504.

Knekt, P., Järvinen, R., Dich, J. & Hakulinen, T. (1999) Risk of colorectal and other gastro-intestinal cancers after exposure to nitrate, nitrite and N-nitroso compounds: A follow up study. *Int. J. Cancer*, **80**, 852–856.

Knobeloch, L., Salna, B., Hogan, A., Postle, J. & Anderson, H. (2000) Blue babies and nitrate-contaminated well water. *Environ. Health Perspectives*, **108**, 675–678.

Kortboyer, J.M., Boink, A.B.T.J. Schothorst, R.C., Cleven, R.F.M.J. & Meulenbelt, J. (1998) Healthy volunteer study investigating the feasibility of an oral bioavailability study of nitrate from vegetables. Report No. 235802012 from National Institute of Public Health and the Environment (RIVM), Bilthoven, Netherlands, pp. 1–47.

Kostraba, J.N., Gay, E.C., Rewers, M. & Hamman, R.F. (1992) Nitrate levels in community drinking waters and risk of IDDM. An ecological analysis. *Diabetes Care.*, **15**, 1505–1508.

Krul, C., Zeilmaker, M., Schothorst, R., Tenfelde, A., van Doorn, D. & Havenaar, R. (2001a) Intragastric formation of N-nitrosodimethylamine under realistic, human physiological conditions studied with a dynamic *in vitro* gastrointestinal model. In: *Mutagenic and Antimutagenic Activity of Food Compounds. Application of a Dynamic in vitro Gastrointestinal Model.* PhD Thesis, University of Utrecht, Utrecht, Netherlands, pp. 104–124.

Krul, C., Zeilmaker, M., Schothorst, R., Kotterman, M., van Doorn, D. & Havenaar, R. (2001b) Influence of various fish species and nitrosation inhibitors on the intragastric formation of N-nitrosodimethylamine: A dynamic *in vitro* gastrointestinal model. In: *Mutagenic and Antimutagenic Activity of Food Compounds. Application of a Dynamic in vitro Gastrointestinal Model.* PhD Thesis, University of Utrecht, Utrecht, Netherlands, pp. 128–143.

Kubler, W. (1958) Methemoglobinemia in children. *Z. Kinderheilk.*, **81**, 405–416.

Lambers, A.C., Kortboyer, J.M., Schothorst, R.C., Sips, A.J.A.M., Cleven, R.F.M.J. & Meulenbelt, J. (2000a) The oral bioavailability study of nitrate from vegetables in healthy volunteers. Report No. 235802014 from National Institute of Public Health and the Environment (RIVM), Bilthoven, Netherlands, pp. 1–47.

Lambers, A.C., Koppeschaar, H.P.F., Van Isselt, J.W., Slob, W., Schothorst, R.C., Mensinga, T. & Meulenbelt, J. (2000b) The oral bioavailability study of nitrate from vegetables in healthy volunteers. Report No. 235802015 from National Institute of Public Health and the Environment (RIVM), Bilthoven, Netherlands, pp. 1–61.

Lancaster , J.R. (1992) Nitric oxide in cells. *Am. Sci.*, **80**, 248–259.

van Leeuwen, J.A., Waltner-Toews, D., Abernathy, T., Smit, B. & Shoukri, M. (1999) Association between stomach cancer incidence and drinking water contamination with atrazine and nitrate in Ontario (Canada) agroecosystems, 1987–1991. *Int. J. Epidemiol.*, **28**, 836–840.

van Loon, A.J.M., Botterweck, A.A.M., Goldbohm, R.A., Brants, H.A.M., van Klaveren, J.D. & van den Brandt, P.A. (1998) Intake of nitrate and nitrite and the risk of gastric cancer: A prospective cohort study. *Br. J. Cancer*, **78**, 129–136.

Lubin, F., Farbstein, H., Chetrit, A., Farbstein, M., Freedman, L., Alfrandy, E. & Modan, B. (2000) The role of nutritional habits during gestation and child life in pediatric brain tumor etiology. *Int. J. Cancer*, **86**, 139–143.

van Maanen, J.M., van Geel, A.A. & Kleinjans, J.C. (1996a) Modulation of nitrate–nitrite conversion in the oral cavity. *Cancer Detect. Prev.*, **20**, 590–596.

van Maanen, J.M., Welle, I.J., Hageman, G., Dallinga, J.W. & Mertens, P.L.J.M. (1996b) Nitrate contamination of drinking water: Relationship with HPRT variant frequency in lymphocyte DNA and urinary excretion of N-nitrosamines. *Environ. Health Perspectives*, **104**, 522–528.

van Maanen, J.M.S., Albering, H.J., de Kok, T.M.C.M., van Breda, S.G.J., Curfs, D.M.J., Vermeer, I.T.M., Ambergen, A.W., Wolffenbuttel, B.H.R., Kleinjans, J.C.S. & Reeser, H.M. (2000) Does the risk of childhood diabetes mellitus require revision of the guideline values for nitrate in drinking water? *Environ. Health Perspectives*, **108**, 457–461.

Marletta, M.A. 1988) Unravelling the biological significance of nitric oxide. *Chem. Res. Toxicol.*, **1**, 249–257.

McKinney, P.A., Law, G.R., Bodansky, H.J., Staines, A. & Williams, D.R. (1996) Geographical mapping of childhood diabetes in the northern English county of Yorkshire. *Diabetes Med.*, **13**, 734–740.

McKinney, P.A., Parslow, R. & Bodansky, H.J. (1999) Nitrate exposure and childhood diabetes. *J. R. Soc. Chem.*, **237**, 327–339.

McKnight, G., Smith, L.M., Drummond, R.S., Duncan, C.W., Goldem, M.N.H. & Benjamin, N. (1997) The chemical synthesis of nitric oxide in the stomach from dietary nitrate in man. *Gut*, **40**, 211–214.

McKnight, G., Duncan, C.W., Leifert, C. & Golden, M.H. (1999) Dietary nitrate in man: Friend or foe? *Br. J. Nutr.*, **81**, 349–358.

Mirvish, S.S. (1975) Formation of N-nitroso compounds: Chemistry, kinetics, and *in vivo* occurrence. *Toxicol. Appl. Pharmacol.*, **31**, 325–351.

Mirvish, S.S., Grandjean, A.C., Reimers, K.J., Connelly, B.J, Chen, S.-C., Gallagher, J., Rosissky, S., Nie, G., Tuatoo, H., Payne, S., Hinman, C. & Ruby, E.I. (1995) Dosing time with ascorbic acid and nitrate, gum and tobacco chewing, and other factors affecting N-nitrosoproline formation in healthy subjects taking proline with a standard meal. *Cancer Epidemiol. Biomarkers Prev.*, **4**, 775–782.

Møller, H. (1997) Work in agriculture, childhood residence, nitrate exposure, and testicular risk: A case–control study in Denmark. *Cancer Epidemiol. Biomarkers Prev.*, **6**, 141–144.

Mondal, D.B. & Pandey, N.N. (1999) Clinico-haematobiochemical observations of induced acute nitrate toxicity in goats. *Indian J. Anim. Sci.*, **69**, 157–159.

Mondal, D.B., Pandey, N.N. & Charan, K. (1999a) Pathomorphological changes in induced acute nitrate toxicity in goats. *Indian J. Anim. Sci.*, **69**, 118–120.

Mondal, D.B., Pandey, N.N. & Charan, K. (1999b) Effects of prolonged low nitrate intoxication in goats. *Indian Vet. J.*, **76**, 800–803.

Mondal, D.B., Pandey, N.N. & Gupta, O.P. (2000) Electrographic changes in induced acute nitrate toxicity in goats. *Indian Vet. J.*, **77**, 501–503.

Morales Suarez Varela, M.M., Llopis-Gonzalez, A. & Tejerizo-Perez, M.L. (1995) Impact of nitrates in drinking water on cancer mortality in Valencia, Spain. *Eur. J. Epidemiol.*, **11**, 15–21.

Moriya, A., Henry, E.B., Grant, J., Fletcher, J., Mowat, G., Williams, G. & McColl, K.E. (2000) High nitric oxide levels at gastro-oesophageal (GO) junction induced by dietary nitrate? Cause of mutagenesis at GO junction (Abstract). *Gastroenterology*, **118**, 1278.

Müller, R.L., Hagel, H.J., Grein, G., Ruppin, H. & Domsehke, W. (1984) [Dynamics of endogenous bacterial nitrite formation in the stomach. I. Communication: Evolution in man under natural conditions.]. *Zbl. Bakt. Hyg. I. Abt. Orig. B*, **179**, 381–396 (in German).
Mueller, B.A., Newton, K., Holly, E.A. & Preston-Martin, S. (2001) Residential water source and the risk of childhood brain tumors. *Environ. Health Perspectives*, **109**, 551–556.
Muramatsu, K., Maruyama, S. & Nishizawa, S. (1979) Nitrate-reducing bacterial flora and its ability to reduce nitrate in human saliva. *J. Food Hyg. Soc. Jpn*, **20**, 106.
Nestorova, J., Tsohev, I., Nikolov, I. & Ivanova, N. (1997) The effect of nitrates on sheep reproduction. *Zhivol. Nauk.*, **34**, 149–151.
Ohshima, H. & Bartsch, H. (1981) Quantitative estimation of endogenous nitrosation in humans by monitoring N-nitrosoproline excreted in the urine. *Cancer Res.*, **41**, 3658–3662.
Panesar, N.S. (1999) Role of chloride and inhibitory action of inorganic nitrate on gonadotropin-stimulated steroidogenesis in mouse Leydig tumor cells. *Metabolism*, **48**, 693–700.
Pobel, D., Riboli, E., Cornée, J., Hémon, B. & Guyader, M. (1995) Nitrosamine, nitrate and nitrite in relation to gastric cancer: A case–control study in Marseille, France. *Eur. J. Epidemiol.*, **11**, 67–73.
Rogers, M.A., Vaughan, T.L., Davis, S. & Thomas, D.B. (1995) Consumption of nitrate, nitrite, and nitrosodimethylamine and the risk of upper aerodigestive tract cancer. *Cancer Epidemiol. Biomarkers Prev.*, **4**, 29–36.
Ruddel, W.S., Bone, E.S., Hill, M.J., Blendis, L.M. & Walters, C.L. (1976) Gastric juice nitrite: A risk factor for cancer in the hypochlorhydric stomach? *Lancet*, **ii**, 1037–1039.
Sandor, J., Kiss, I. & Ember, I. (1998) Threshold for gastric cancer causing effect of nitrate in drinking water. *Anticancer Res.*, **18**, 4940.
Schultz, D.S., Deen, W.M., Karel, S.F., Wagner, D.A. & Tannenbaum, S.R. (1985) Pharmacokinetics of nitrate in humans: Role of gastrointestinal absorption and metabolism. *Carcinogenesis*, **6**, 847–852.
Smith, J.E. & Beutler, E. (1966) Methaemoglobin formation and reduction in men and various animals species. *Am. J. Physiol.*, **210**, 347–350.
Spiegelhalder, B., Eisenbrand, G. & Preussmann, R. (1976) Influence of dietary nitrate on nitrite content of human saliva: Possible relevance to *in vivo* formation of N-nitroso-compounds. *Food Cosmet. Toxicol.*, **14**, 545–548.
Stephany, R.W. & Schuller, P.L. (1978) The intake of nitrate, nitrite and volatile N-nitrosamines and the occurrence of volatile N-nitrosamine in human urine and veal calves. In: *IARC Scientific Publications*, **19**, 443–460.
Suvanto, F.M.E & Pohjola, M.L. (1995) Nitrates in drinking water and methemoglobin in blood. *Kemia-Kemi*, **22**, 400–401.
Suzuki, H., Lijima, K., Moriya, A., Fyfe, V. & McColl, K.E.L. (2002) Luminal nitosation potential following nitrate ingestion is maximal at the GO junction. *Gut*, **50** (Suppl. 11), A37.
Tannenbaum, S.R. (1987) Endogenous formation of N-nitroso compounds; a current perspective. In: *IARC Scientific Publications*, **84**, 292–296.
Tannenbaum, S.R., Weisman, M. & Fett, D. (1976) The effect of nitrate intake on nitrite formation in human saliva. *Food Cosmet. Toxicol.*, **14**, 549–552.
Tannenbaum, S.R., Moran, D., Rand, W., Cuello, C. & Correa, P. (1979) Gastric cancer in Colombia. IV. Nitrite and other ions in gastric contents of residents from a high-risk region. *J. Natl Cancer Inst.*, **62**, 9–12.
Thal, W., Lacchein, P.L. & Martinek, M. (1961) *Arch. Toxicol.*, **19**, 25–33.
Til, H.P., Falke, H.E., Kuper, C.F & Willems, M.I. (1988) Evaluation of the oral toxicity of potassium nitrite in a 13-week drinking-water study in rats. *Food Chem. Toxicol.*, **26**, 851–859.
Tsezou, A., Kitsiou-Tzeli, Galla, A., Gourgiotis, D., Mitrou, S., Molybdas, P. & Sinaniotis, C. (1996) High nitrate content in drinking water: Cytogenetic effects in exposed children. *Arch. Environ. Health*, **51**, 458–461.
Vallance, P. (1997) Dietary nitrate: Poison or panacea? *Gut*, **40**, 288.
Varady, J. & Szanto, G. (1940) [Research on the nitrite content of saliva, stomach and urine.] *Klin. Wochenschr.*, **19**, 200 (in German).

Vermeer, I.T.M., Pachen, D.M.F.A., Dallinga, J.W., Kleinjans, J.C.S. & van Maanen, J.M.S. (1998) Volatile N-nitrosamines formation after intake of nitrate at the ADI level in combination with an amine-rich diet. *Environ. Health Perspectives*, **106**, 459–463.

Vleeming, W., van de Kuil, A., te Biesebeek, J.D., Meulenbelt, J. & Boink, A.B.T.J. (1997) Effect of nitrite on blood pressure in anaesthetized and free-moving rats. *Food Chem. Toxicol.*, **35**, 615–619.

Wagner, D.A. & Tannenbaum, S.R. (1985) *In vivo* formation of N-nitroso compounds. *Food Technol*, **39**, 89–90.

Wagner, D.A., Schultz, D.S., Deen, W.M., Young, V.R. & Tannenbaum, S.R. (1983) Metabolic fate of an oral dose of ^{15}N-labeled nitrate in humans: effect of diet supplementation with ascorbic acid. *Cancer Res.*, **43**, 1921–1925.

Walker, R. (1996) The metabolism of dietary nitrites and nitrates. *Biochem. Soc. Trans.*, **24**, 780–785.

Walthers , C.L., Carr, F.P., Dyke, C.S., Saxby, M.J., Smith, P.L. & Walker, R. (1979) Nitrite sources and nitrosamine formation *in vitro* and *in vitro*. *Food Cosmet. Toxicol.*, **17**, 473–479.

Ward, F.W., Coates, M.E. & Walker, R. (1986) Nitrate reduction, gastro-intestinal pH and N-nitrosation in gnotobiotic and conventional rats. *Food Chem. Toxicol.*, **24**, 17–22.

Ward, M.H., Mark, S.D., Cantor, K.P., Weisenburger, D.D., Correa-Villasenor, D. & Zahm, S.H. (1996) Drinking water and the risk of non-Hodgkin's lymphoma. *Epidemiology*, **7**, 465–471.

Weyer, P.J., Cerhan, J.R., Kross, B.C., Hallberg, G.R., Kantamneni, J., Breuer, G., Jones, M.P., Zheng, W. & Lynch, C.F. (2001) Municipal drinking water nitrate level and cancer risk in older women: The Iowa women's health study. *Epidemiology*, **11**, 327–338.

Wiklund, L., George, M., Nord, C.E. & Saldeen, T. (1998) Sudden infant death syndrome and nitrogen metabolism: Further development of a hypothesis. *Eur. J. Clin. Invest.*, **28**, 958–965.

Wishnok, J.S., Tannenbaum, S.R., Tamir, S. & Rojas-Walker, T. (1995) Endogenous formation of nitrate. In: *Proceedings of the International Workshop, Bilthoven, 8–10 November 1994*, Strasbourg: Council of Europe Press, pp. 151–179.

Witter, J.P. & Balish, E. (1979) Distribution and metabolism of ingested NO_3^- and NO_2 in germfree and conventional-flora rats. *Appl. Environ. Microbiol.*, **38**, 861–869.

Wright, A.J. & Davison, K.L. (1964) Nitrate accumulation in crops and nitrate poisoning in animals. *Adv. Agron.*, **16**, 197–247.

Yang, C.Y., Chiu, H.F., Chiu, J.F., Cheng, M.F. & Kao, W.Y. (1997) Gastric cancer mortality and drinking water qualities in Taiwan. *Arch. Environ. Contam. Toxicol.*, **33**, 336–340.

Zeilmaker, M.J., Meulenbelt, J., Kortboyer, J.M. & Slob, W. (1996) Safety assessment of nitrate: Mathematical modeling of nitrite formation in man and its application in the risk assessment of nitrate. Report No. 235802002 from National Institute of Public Health and the Environment (RIVM), Bilthoven, Netherlands, pp. 1–38.

Zeilmaker, M.J., Meulenbelt, J. & Slob, W. (2002) Toxicokinetic modeling of nitrate and nitrite in man, including nitrite induced methemoglobin formation. In press.

Zraly, Z., Bendova, J., Svecova, D., Faldikova, Z., Veznik, H. & Zakjicova, A. (1997) Effects of oral intake of nitrates on reproductive functions of bulls. *Vet. Med. Czech.*, **42**, 345–354.

NITRATE and NITRITE
Intake assessment

First draft prepared by
T. Hambridge
Australia New Zealand Food Authority, Canberra, Australia

1. INTRODUCTION

Nitrate and nitrite are found naturally in the environment, and form part of the nitrogen cycle. Nitrate is formed from oxidation of organic wastes by the action of nitrogen-fixing bacteria. The nitrite ion is relatively unstable and is readily oxidized to nitrate. Nitrate and nitrite are also found in soil and water due to use of fertilizers (WHO, 1998a). Nitrate has been shown to be produced endogenously (McKnight et al., undated). It is actively secreted into saliva, where some is reduced to nitrite by oral microflora (Dich et al., 1996; WHO, 1998a). Salivary nitrate is swallowed and recirculated. Nitrate is also found in air, at concentrations of 0.1–0.4 $\mu g/m^3$, although concentrations up to 40 $\mu g/m^3$ have been reported (WHO, 1998a).

Nitrate is found naturally in foods and in high concentrations in certain vegetables. The concentrations in vegetables depend on a number of factors including season, light, temperature, method of growth, fertilizer use and storage (Dich et al., 1996). In Europe, the concentrations of nitrate are generally higher in winter (European Commission, 1998) owing to the low light intensity and fewer daylight hours.

Nitrate and nitrite are also used as food additives in processed foods, as preservatives, antimicrobial agents and colour fixatives (Tompkin, 1993; Institute of Medicine, 1996). The main processed foods in which it is used include cured meat, poultry and fish, other processed meats and cheese (Coultate, 1989; Tompkin, 1993; Maga & Tu, 1995).

Nitrate is permitted for use in the Codex draft General Standard for Food Additives (GSFA) in numerous food categories, including cheese and cheese products, raw and processed meats, edible casings, processed fish and fish products and spirits and liqueurs. The GSFA also permits use of nitrite in numerous food categories, including cheese and cheese products, meat and processed meats, edible casings, cooked and preserved fish and roe.

Ideally, all sources of nitrate and nitrite should be considered in order to determine total estimated intake and the risks to health; however, only the intake from food (non-additive sources and as a food additive) and water are considered below.

The ADI for nitrate of 0–3.7 mg/kg bw, expressed as nitrate ion, was retained by the Committee at its present meeting, and the ADI for nitrite was set at 0–0.07 mg/kg bw, expressed as nitrite ion (Annex 1, reference *116*). All values for nitrate and nitrite in the intake assessments are expressed as the ions to allow direct comparison with the respective ADIs.

Data on nitrate and nitrite for review by the Committee, including national intake assessments, were submitted by Australia, the Netherlands, New Zealand and the United Kingdom (Table 1). Both Australia and New Zealand submitted estimated dietary intakes of nitrate and nitrite based on individual dietary records. The Netherlands submitted data from monitoring and the results of a duplicate diet study (Vaessen & Schothorst, 1999). The United Kingdom submitted the results of monitoring that also included intake assessments, a survey of nitrate and nitrite in bacon and cured meats (Ministry of Agriculture, Fisheries and Food, 1998a), reports of two total diet studies (Ministry of Agriculture, Fisheries and Food, 1997a, 1998b), the results of a duplicate diet study of nitrate intake by vegetarians (Ministry of Agriculture, Fisheries and Food, 1998b) and a study of the metabolism of dietary nitrate (McKnight et al., undated).

1.1 Concentrations in foods

A regulation of the European Commission (194/97) requires Member States to monitor the concentrations of nitrate in lettuce and spinach to ensure that they remain acceptable. The Netherlands submitted the results of monitoring of nitrate concentrations in lettuce and spinach (van der Schee, 1997; van der Schee & Speek, 2000a,b), a survey of nitrate in other vegetables (van der Schee, 1998) and the results of monitoring for adherence to good agricultural practice in spinach and lettuce (de Kreij, 1998; de Kreij & van Winkel, 2000; de Kreij, 2001). The nitrate concentrations in lettuce collected over all seasons for a number of years ranged between 40 and 5500 mg/kg, while those in spinach were between 30 and 6000 mg/kg. The mean concentrations in other vegetables were 1900 mg/kg (maximum, 3400 mg/kg) in endive and 1800 mg/kg (maximum, 4700 mg/kg) in beetroot.

Table 1. Summary of submissions on intake of nitrate

Country	'Poundage' data	Food balance sheets, household economic surveys, sales data	Model diets	Individual dietary records	Monitoring
Australia				√	
Netherlands				√	√
New Zealand				√	
United Kingdom			√	√	√

The United Kingdom submitted the results of numerous monitoring surveys for nitrate in lettuce and spinach (Ministry of Agriculture, Fisheries and Food, 1997b, 1998c, 1999, 2001) and in other vegetables (Ministry of Agriculture, Fisheries and Food, 1996, 1998d). The nitrate concentrations in lettuce collected over all seasons for a number of years ranged between 50 and 5300 mg/kg, while those in spinach were between 25 and 4600 mg/kg. The concentrations in other vegetables were < 3 mg/kg (below the limit of detection) for leeks and 4200 mg/kg for radishes. The mean residue of nitrate in all cured meat products was 180 mg/kg. The mean residue of nitrite was 120 mg/kg in bacon and 65 mg/kg in other cured meats (Ministry of Agriculture, Fisheries and Food, 1998a).

Nitrate and nitrite occur naturally in water, but farming and effluents can increase the levels of these chemicals in some waters. Nitrite is rapidly oxidized to nitrate in water and is rarely found in well-oxygenated or chlorinated water (Cooperative Research Centre, 1996). The WHO guideline for nitrate in drinking-water is 50 mg/l, and the provisional guideline for nitrite is 0.2 mg/l (WHO, 1998b). The guideline levels in drinking-water in Australia and New Zealand are 50 mg/l for nitrate and 3 mg/l for nitrite (National Health and Medical Research Council and Agriculture and Resource Management Council of Australia and New Zealand, 1996; Ministry of Health, 2000).

The concentration of nitrate in surface water is usually low, between 0 and 18 mg/l, while the natural concentration in groundwater is only a few milligrans per litre. The concentration in drinking-water in most countries is < 10 mg/l (WHO, 1998a), although a level of 1500 mg/l was reported for nitrate in India. The nitrate concentration in water in Australia is usually < 0.15 mg/L but can range up to 18 mg/l The nitrite concentration in water in the USA does not exceed 0.3 mg/l, and that in drinking-water in the Netherlands is usually < 0.1 mg/l (Cooperative Research Centre, 1996).

2. ASSESSMENTS OF INTAKE

2.1 Intake from sources other than food additives

The intakes of nitrate and nitrite from food were calculated at a global level on the basis of mean food consumption in the GEMS/Food regional diets (WHO, 1998c), and the mean concentrations in foods in Europe from the submitted data. Intake from drinking-water was added, assuming a water consumption of 2 l/day in all diets except the European diet, in which it was considered to be 1 l/day. The mean concentration in water that was used in the intake calculations was 4 mg/l for nitrate and 0.3 mg/l for nitrite, which are representative of the usual concentrations found in water (WHO, 1998a). An average body weight of 60 kg was used for the global intake assessments. The calculations of intake (from consumption figures and mean concentrations) are shown in Table 2, and the concentrations and sources are summarized in Table 3. The estimated intakes are summarized in Table 4.

2.2 Intake from use as a food additive

2.2.1 Screening by the budget method

The budget method is used to estimate the theoretical maximum level of a food additive in the proportion of the food and/or beverage supply likely to contain it that

Table 2. Calculation of nitrate and nitrite intakes from sources other than food additives at the global level

Food analysed	Codex code	Concentration (mg/kg)	Diet														
			Middle Eastern			Far Eastern			African			Latin American			European		
			Consump tion (g/d)	Intake (mg/d)	% total intake	Consumption (g/d)	Intake (mg/d)	% total intake	Consumption (g/d)	Intake (mg/d)	% total intake	Consumption (g/d)	Intake (mg/d)	% total intake	Consumption (g/d)	Intake (mg/d)	% total intake
Nitrate																	
Lettuce	VL 482	1700	2.3	4.0	10	0	0	0	0	0	0	5.8	10	19	22	39	25
Spinach	VL 502	1900	0.5	0.93	2	0	0	0	0	0	0	0.3	0.56	1	2	3.7	2
Endive	VL 476	1900	0.5	0.97	2	0	0	0	0	0	0	0.3	0.58	1	2	3.9	2
Beetroot	VR 574	1500	0.5	0.74	2	0	0	0	0	0	0	0.3	0.45	1	2	3.0	2
Apple	FP 226	30	7.5	0.22	1	4.7	0.14	1	0.3	0.009	0	5.5	0.16	0	40	1.2	1
Asparagus	VS 621	30	0	0	0	0	0	0	0	0	0	0	0	0	1.5	0.045	0
Aubergine	VO 440	440	6.3	2.8	7	3	1.3	5	0.7	0.31	1	6	2.6	5	2.3	1.0	1
Broccoli	VB 400	440	0.5	0.22	1	1	0.44	2	0	0	0	1.1	0.49	1	2.7	1.2	1
Brussels sprouts	VB 402	35	0.5	0.018	0	1	0.035	0	0	0	0	1.1	0.038	0	2.7	0.094	0
Carrot	VR 577	220	2.8	0.62	2	2.5	0.55	2	0	0	0	6.3	1.4	3	22	4.8	3
Celery	VS 624	2900	0.5	1.5	4	0	0	0	0	0	0	0.3	0.88	2	2	5.9	4
Chinese cabbage	VL 466	3500	0.1	0.35	1	0.1	0.35	1	0.1	0.35	2	0.1	0.35	1	0.1	0.35	0
Garlic	VA 381	35	2	0.07	0	2.2	0.077	0	0	0	0	0.5	0.018	0	3	0.10	0
Grape	FB 269	70	16	1.1	3	1	0.07	0	0	0	0	1.3	0.091	0	14	1.0	1
Kohlrabi	VB 405	2700	0.1	0.27	1	0.1	0.27	1	0.1	0.27	1	0.1	0.27	1	0.1	0.27	0
Onion	VA 385	110	23	2.5	6	12	1.3	5	7.3	0.80	4	14	1.5	3	28	3.1	2
Pear	FP 230	20	3.3	0.066	0	2.8	0.056	0	0	0	0	1	0.02	0	11	0.23	0
Potato	VR 589	180	5.9	1.1	3	19	3.4	13	21	3.7	18	41	7.3	14	240	43	28
Radish	VR 494	2100	0.5	1.0	3	0	0	0	0	0	0	0.3	0.63	1	2	4.2	3
Rhubarb	VS 627	1600	0.5	0.8	2	0	0	0	0	0	0	0.3	0.50	1	2	3.3	2
Shallot	VA 388	65	0	0	0	2	0.13	0	1.5	0.10	0	4	0.26	0	1	0.065	0
Sweet	VO 445	120	3.3	0.40	1	2	0.23	1	5.3	0.61	3	2.3	0.26	0	10	1.2	1

Table 2 (contd)

Food analysed	Codex code	Concentration (mg/kg)	Diet: Middle Eastern			Far Eastern			African			Latin American			European		
			Consumption (g/d)	Intake (mg/d)	% total intake	Consumption (g/d)	Intake (mg/d)	% total intake	Consumption (g/d)	Intake (mg/d)	% total intake	Consumption (g/d)	Intake (mg/d)	% total intake	Consumption (g/d)	Intake (mg/d)	% total intake
Cabbage	VB 41	340	5	1.7	4	9.7	3.3	12	0	0	0	10	3.6	7	27	9.1	6
Swede	VR 497	120	0.5	0.06	0	0	0	0	0	0	0	0.3	0.036	0	2	0.24	0
Tomato	VO 448	15	82	1.2	3	7	0.10	0	16	0.25	1	26	0.38	1	66	0.99	1
Cereals	GC 80, total	10	430	4.3	11	450	4.5	17	320	3.2	15	250	2.5	5	230	2.3	1
Fish only	Fresh, frozen	5	3.3	0.016	0	8.3	0.042	0	8	0.04	0	3.1	0.016	0	7	0.035	0
Fats & oils	Total animal & vegetable	5	41	0.21	1	16	0.080	0	24	0.12	1	27	0.14	0	49	0.25	0
Eggs	PE 840	5	14	0.072	0	13	0.065	0	3.6	0.018	0	12	0.059	0	38	0.19	0
Fruit	Total fruit, minus apple and pear	25	19	0.48	1	78	1.9	7	94	2.4	11	260	6.6	12	160	4.0	3
Sugars	Total, not including honey	10	95	0.95	2	50	0.50	2	42	0.42	2	100	1.0	2	110	1.1	1
Milk	ML 106, all	5	120	0.58	1	32	0.16	1	42	0.21	1	160	0.8	1	290	1.5	1
Dairy	Cream	20	8.5	0.17	0	0	0	0	0	0	0	4.5	0.09	0	33	0.66	0
Nuts	TN 85	10	1	0.01	0	14	0.14	0	3.4	0.034	0	18	0.18	0	3.8	0.038	0
Water		4	2000	8	20	2000	8	29	2000	8	38	2000	8	15	1000	4	3
Total (mg/d)				40			27			21			53			160	
% ADI				20			10			10			25			70	

Table 2 (contd)

Food analysed	Codex code	Concentration (mg/kg)	Diet														
			Middle Eastern			Far Eastern			African			Latin American			European		
			Consumption (g/d)	Intake (mg/d)	% total intake	Consumption (g/d)	Intake (mg/d)	% total intake	Consumption (g/d)	Intake (mg/d)	% total intake	Consumption (g/d)	Intake (mg/d)	% total intake	Consumption (g/d)	Intake (mg/d)	% total intake
Nitrite																	
Potato	VR 589	0.7	5.9	0.0041	0	19	0.013	1	21	0.014	1	41	0.028	2	240	0.17	10
Vegetables	Total	0.6	230	0.14	7	180	0.11	5	77	0.046	3	150	0.090	5	370	0.22	13
Cereals	GC 80, total	2.6	430	1.1	54	450	1.2	59	320	0.83	52	250	0.66	40	230	0.59	35
Fish	Fresh, frozen only		0.5	3.3	0.0016	0	8.3	0.0042	0	8	0.004	0	3.1	0.0016	0	7	0.00350
Fats & oils	Total animal & vegetable	0.5	41	0.021	1	16	0.0080	0	24	0.012	1	27	0.014	1	49	0.025	1
Eggs	PE 840	2.6	14	0.038	2	13	0.034	2	3.6	0.0094	1	12	0.031	2	38	0.098	6
Fruit	Total fruit	0.5	200	0.10	5	85	0.043	2	95	0.047	3	270	0.14	8	210	0.10	6
Milk	ML 106, all	0.5	120	0.058	3	32	0.016	1	42	0.021	1	160	0.08	5	290	0.15	9
Dairy	Cream	0.5	0	0	0	0	0	0	0	0	0	0	0	0	4.8	0.0024	0
Nuts	TN 85	0.5	1	0.0005	0	14	0.0068	0	3.4	0.0017	0	18	0.0088	1	3.8	0.0019	0
Water		0.3	2000	0.6	29	2000	0.6	30	2000	0.6	38	2000	0.6	36	1000	0.3	18
Total (mg/d)			2.1			2.0			1.6			1.7			1.7		
% ADI			50			50			40			40			40		

Table 3. Concentrations of nitrate and nitrite (mg/kg) used to calculate mean concentrations for use in global intake calculations

Food	Source of data											
	van der Schee & Speek (2000a)	van der Schee & Speek (1999)	van der Schee (1996)	van der Schee (1998)	MAFF (1996)	MAFF (1998d)	MAFF (1998c)	MAFF (1997b)	MAFF (1999)	MAFF (2001)	MAFF (1997a)	MAFF (1998b)	Mean used in calculations (mg/kg)
Nitrate													
Lettuce	2200	2100	2000	2100	1600	1000	1400	1700	1500	1500			1700
Spinach	1800	1700	2500	2800	1700	1600	1300	1400	2000	1700			1900
Endive			2000	1900									1900
Beetroot				1800		1200							1500
Apple				29									30
Asparagus				31									30
Egg plant				440									440
Broccili				440									440
Brussels sprouts				11		59							35
Carrot				340		97							220
Celery				2900									2900
Chinese cabbage				3600									3500
Garlic				35									35
Grape				72									70
Kholrabi				2700									2700
Onion				170		48							110
Pear				20									20
Potato				280	180	160							180
Radish				2100									2100
Rhubarb				1700									1600
Shallot				64									65
Sweet pepper				120									120

Table 3 (contd)

Food	Source of data												
	van der Schee & Speek (2000a)	van der Schee & Speek (1999)	van der Schee (1996)	van der Schee (1998)	MAFF (1996)	MAFF (1998d)	MAFF (1998c)	MAFF (1997b)	MAFF (1999)	MAFF (2001)	MAFF (1997a)	MAFF (1998b)	Mean used in calculations (mg/kg)
Turnip				4800									4800
Cauliflower						86							85
Cabbage						340							340
Swede						120							120
Tomato						17							15
Cereals											9.2	11	10
Fish											8.6	11	5
Fats & oils											8.1	7.6	5
Eggs											8.1	5.4	5
Fruit											26	27	25
Sugars											8.0	8.7	10
Milk											8.8	8.3	5
Dairy											9.4	27	20
Nuts											9.5	6.1	10
Nitrite													
Potato										0.8	0.5	0.7	
Vegetables										0.6	0.6	0.6	
Cereals										3.3	1.8	2.6	
Fish										0.6	0.4	0.5	
Fats & oils										0.6	0.4	0.5	
Eggs										3.4	1.7	2.6	
Fruit										0.6	0.4	0.5	
Milk										0.6	0.4	0.5	
Dairy										0.6	0.4	0.5	
Nuts										0.6	0.4	0.5	

Table 4. Estimated intakes of nitrate and nitrite from sources other than food additives at the global level

Chemical	Regional diet	Intake (mg/day)	% ADI[a]	Major contributors to total intake (%)
Nitrate	Middle Eastern	40	20	Vegetables other than potato (60) Water (20) Cereals (10) Fruit (5)
	Far Eastwerm	28	10	Vegetables (45) Water (30) Cereals (15) Fruit (10)
	African	20	10	Water (40) Vegetables (30) Cereals (15) Fruit (10)
	Latin American	55	25	Vegetables (65) Water (15) Fruit (10) Cereals (50)
	European	155	70	Vegetables (90) Water (5) Fruit (5) Cereals (<5)
Nitrite	Middle Eastern	2.1	50	Cereal (55) Water (30)
	Far Eastern	2.0	50	Cereal (60) Water (30)
	African	1.6	40	Cereal (50) Water (40)
	Latin American	1.6	40	Cereal (40) Water (35)
	European	1.7	40	Cereal (35) Water (20)

[a] Based on 60 kg body weight

would not result in exceedance of the ADI by the population (Hansen, 1979). The factors taken into consideration are physiological energy requirements and the energy density of food. The budget method is used to assess whether use of the additive should be restricted and whether an intake assessment is required.

It is not entirely appropriate to use the budget method for nitrate and nitrite, because not all the intake of these chemicals, and therefore the contribution to the ADI, is derived from use as food additives, as is generally the case when the budget method is used. However, as the proportion of the ADI contributed by sources other than additives is not known, the full ADI was used for the budget method calculations. The budget method was used to calculate nitrate in food and beverages, as the GSFA has determined maximum permitted levels for both. As the proportion of use in solid foods and in beverages is unknown, it was assumed to be 50% for each. It was also assumed that 50% of the food supply and 25% of beverages contain the additive. The theoretical maximum level of 150 mg/kg for solid food (see Table 5) is below the maximum permitted level of the GSFA (360 mg/kg), and the theoretical maximum level of 72 mg/kg for beverages is similar to the maximum permitted level of the GSFA (73 mg/kg). An intake assessment is therefore required, and restricted use of the food additive should be considered.

The budget method was used to calculate nitrite only in food, as the GSFA provides a maximum permitted level only for solid foods. It was assumed that 50% of the solid food supply contains the additive. The theoretical maximum level for solid food of 5.6 mg/kg (see Table 5) is well below the maximum permitted level in the GSFA of 420 mg/kg. An intake assessment is therefore required, and restricted use of the additive should be considered.

Comparison of the theoretical maximum levels with the national maximum permitted levels resulted in conclusions similar to those based on the GSFA.

2.2.2 Assessments based on individual dietary records

Intakes of nitrate and nitrite as food additives estimated from individual dietary records were submitted by Australia and New Zealand (Table 6). In these assessments, intake was calculated for each individual on the basis of body weight, before population statistics were derived. The intake assessments of both countries included estimates of intake based on the maximum permitted level of each additive in foods in each country and on those of the GSFA. The Australian assessments also included dietary intake estimates based on levels of use of nitrite in industry. The national and GSFA maximum permitted levels and the data on food consumption aggregated from data on individual food consumption used in the intake assessments are shown in Tables 7 and 8, respectively.

2.3 Intake from all sources

2.3.1 Assessments based on total diet studies

Model diets are constructed from information on food consumption. They are designed to represent the typical diet of a general population or a specified population subgroup and include foods containing the food additive.

Table 5. Estimated theoretical maximum levels of nitrate and nitrite by the budget method

Additive	Distribution in food supply	% solid food or beverage supply	Theoretical maximum level (mg/kg)	GSFA maximum permitted level (mg/kg) [a]	National maximum permitted level (mg/kg) [b]
Nitrate	50% food	50	150	360	360
	50% beverages	25	72	73	73
Nitrite	100% food	50	5.6	420	83

[a] Highest level in food and beverages according to the GSFA
[b] Highest levels in food and beverages in national submissions

Table 6. Summary of estimates of intake of nitrate and nitrite as food additives based on individual dietary records

Country	Date	Survey	Assumptions	Additive group	Population bw per day)	Intake (mg/kg)	% ADI[a]
Australia (Aus)	1995	National nutrition survey; 24-h recall; ≥ 2 years; sample, 13 858	Maximum permitted level (Aus or GSFA)	Nitrate	Mean consumers Aus	0.34	10
			Nominated manufactures use level		GSFA	0.45	10
			Modified GSFA classification system		Mean consumers		
			Adjusted for individual body weight		95th %ile consumers Aus	1.3	35
					95th %ile consumers GSFA	1.3	35
			Does not include non-processed food or water	Nitrite	Mean consumers Aus		
					Maximum permitted level	0.09	130
					Mean consumers Aus nominated manufactures use level	0.07	100
					Mean consumers GSFA	0.35	500
					95th %ile consumers Aus maximum permitted level	0.29	420
					95th %ileconsumers Aus nominated manufacturer's use level	0.22	320
					95th %ile consumers GSFA	0.97	1400
New Zealand (NZ)	1997	National nutrition survey; 24-h recall; ≥ 15 years; ample, 4636	Maximum permitted level (Aus or GSFA)	Nitrate	Mean consumers NZ maximum permitted level	0.33	10
			Modified GSFA classification system				
			Adjusted for individual body weight		Mean consumers GSFA	0.41	10
			Does not include non-processed food or water		95th %ile consumers NZ	1.2	35
					95th %ile consumers GSFA	1.1	30
				Nitrite	Mean consumers NZ maximum permitted level	0.09	130
					Mean consumers GSFA	0.32	460
					95th %ile consumers NZ	0.25	360
					95th %ile consumers GSFA	0.90	1300

[a] ADI, 0–3.7 mg/kg bw for nitrate and 0–0.07 mg/kg bw for nitrite

Table 7. National and draft GSFA maximum permitted levels of use used in the intake assessments for Australia and New Zealand

Additive	Code[a]	Food category description	Maximum permitted level in Australia and New Zealand (industry use) (mg/kg)	GSFA maximum permitted level (mg/kg)
Nitrate	1.6	Cheese and cheese products	45	37
	8.1	Raw meat, poultry and game	–	150
	8.2	Processed meat, poultry and game products, whole	449	3650
	8.3	Processed comminuted meat, poultry and game products	449	360
	8.4	Edible casings	–	140
	9.3	Semi-preserved fish and fish products	–	220
	14.2.5	Spirits and liqueurs	–	73
Nitrite	1.6	Cheese and cheese products	–	17
	8.2	Processed meat, poultry and game products, whole pieces and cuts	83	420
	8.3	Processed comminuted meat, poultry and game products	83	130
	8.3.2	Sausage and sausage meat	83 (0)	
	9.2.2	Roe	–	5
	9.3	Semi-preserved fish and fish products	–	130

[a] Australia New Zealand Food Classification System

The intakes estimated from the total diet studies submitted by the United Kingdom are presented in Table 9. The results are presented for the whole population; the lower-bound estimates are for analytical concentrations below the limit of detection, assumed to be zero, and the upper-bound estimates are for analytical concentrations at the limit of detection. The results for adult consumers include estimated intake of nitrate from water and beer and of nitrite from water only, in order to estimate the 'total diet' intake. The assessments are based on measured concentrations in foods. The intakes in the submitted reports were compared with the ADIs of the Scientific Commission for Food of 3.6 mg/kg bw for nitrate and 0.06 mg/kg bw for nitrite. For the present assessment, the estimated intakes were compared with the ADIs established by the Committee.

The nitrate and nitrite concentrations found in the total diet studies are shown in Table 10. Table 11 shows consumption data for each food group used to estimate population intake. Data on food consumption by consumers only were not provided.

2.3.2 Assessments based on duplicate diet studies

The United Kingdom submitted a study with duplicate diets involving 35 vegetarians, which was intended to determine whether the nitrate intake of this

Table 8. Aggregated food consumption from individual data used in intake assessments with Australian, New Zealand and GSFA maximum permitted levels

Food group	Consumers		All respondents	Consumers only (mean; g/day)		
	No.	% of all respondents[a]		Mean	Median	95th %ile
Australia						
Cheese and cheese products	7000	50	18	36	24	99
Raw meat, poultry and game	9600	69	97	140	110	370
Processed meat, poultry and game products, whole cut	4000	29	12	42	29	120
Processed comminuted meat, poultry and game	3100	23	17	76	56	220
Edible casings	0	0	0	0	0	0
Semi-preserved fish and fish products	84	0.6	0.6	100	62	380
Canned roe	13	0.1	0.0	36	19	120
Spirits and liqueurs	660	4.7	3.7	78	48	270
New Zealand						
Cheese and cheese products	2200	46	17	36	26	110
Raw meat, poultry and game	3300	71	98	140	110	350
Processed meat, poultry and game products, whole cut	1200	27	14	51	31	160
Processed comminuted meat, poultry and game	1200	26	21	78	56	220
Edible casings	0	0	0	0	0	0
Semi-preserved fish and fish products	28	0.6	0.5	81	63	290
Roe	5	0.1	0.1	71	42	160
Spirits and liqueurs	320	7.0	5.0	71	44	240

[a] All respondents: 13 858 in Australia and 4636 in New Zealand

population group was higher than that of the general population owing to their greater consumption of vegetables. Respondents collected duplicate samples of all foods and some beverages (soya milk, soya milk in coffee or tea, herbal drinks and soup) that they had consumed over 7 days. Data were collected twice a year to account for seasonal variation in nitrate concentrations in vegetables. Foods were weighed and analysed, and the estimated nitrate intakes were calculated (Table 12). As the foods were weighed and analysed together, the measured concentrations in each food were not available.

A 24-h duplicate diet study was conducted in the Netherlands in 1994 to estimate the nitrate and nitrite intake in food and drinking-water of 123 adults aged 18–74 and living in various towns. Data were collected twice during the year to account for seasonal variation in chemical concentrations. The foods were weighed and analysed, and the intakes were calculated (Table 12). The mean food intake was 2600 g/day (2500 g for women and 2800 g for men), and the mean body weight was 75 kg (71 kg for women and 79 kg for men). As the foods were analysed as homogenized

Table 9. Estimates of intake of nitrate and nitrite from total diet studies

Country	Date	Survey	Assumptions	Additive group	Population	Intake (mg/kg) bw per day)	% ADI[a]
United Kingdom	1997	Total diet study: diets based on average of 3 previous years of data from National Food Survey. Proportion of food in diets represents foods' importance in diet. Samples < LOD = LOD (4 mg/kg nitrate, 0.4 mg/kg nitrite)	Analysed concentrations in composite samples. Population estimates include adults and children. Beer and water contribute significantly to nitrate intakes. Average body weight = 60kg	Nitrate	Mean adult consumers: food only	0.95	25
					97.5%ile adult consumers: food only	1.8	50
					Mean adult consumers: food, water, beer	1.5	40
					97.5%ile adult consumer: food, water, beer	2.3	60
					Whole population: lower bound	0.82	20
					Whole population upper bound	0.86	25
				Nitrite	Mean adult consumers: food only	0.022	30
					97.5%ile adult consumers: food only	0.037	55
					Mean adult consumers: food, water, beer	0.023	35
					97.5%ile adult consumer: food, water, beer	0.038	55
					Whole population: lower bound	0.012	15
					Whole population upper bound	0.021	30

Table 9 (contd)

Country	Date	Survey	Assumptions	Additive group	Population bw per day)	Intake (mg/kg)	% ADI[a]
United Kingdom	1994	Total diet study: diets based on average of 3 previous years of data from National Food Survey. Proportion of food in diets represents foods' importance of food in diet	Analysed concentrations in composite samples Population estimates include adults and children. Beer and water contribute significantly to nitrate intakes Samples < LOD = LOD (8 mg/kg nitrate, 0.6 mg/kg nitrite) Average body weight = 60kg	Nitrate	Mean adult consumers: food only	1.1	30
					97.5%ile adult consumers: food only	2.3	60
					Mean adult consumers: food, water, beer	1.7	45
					97.5%ile adult consumer: food, water, beer	2.8	75
					Whole population: lower bound	0.93	25
					Whole population upper bound	1.1	30
				Nitrite	Mean adult consumers: food only	0.028	40
					97.5%ile adult consumers: food only	0.048	70
					Mean adult consumers: food, water	0.03	45
					97.5%ile adult consumer: food, water	0.05	70
					Whole population: lower bound	0.01	15
					Whole population upper bound	0.028	40

[a] ADI, 0–3.7 mg/kg bw for nitrate, 0–0.07 mg/kg bw for nitrite

Table 10. Nitrate and nitrite concentrations used in total diet studies in the United Kingdom in 1994 and 1997

Food group	Mean nitrate concentration (mg/kg)				Mean nitrite concentrations (mg/kg)			
	1994		1997		1994		1997	
	Lower bound	Upper bound	Lower bound	Upper bound	Lower bound	Upper bound	Lower bound	Upper bound
Bread	10	13	7.2	8	0.1	0.6	ND	0.4
Miscellaneous cereals	3.2	9.2	10	11	3.3	3.3	1.8	1.8
Carcass meats	0.5	8.1	4.1	5.1	ND	0.6	ND	0.4
Offal	1.8	8.6	3.9	5.3	1.1	1.3	0.9	0.9
Meat products	49	49	45	45	2.1	2.1	4.1	4.1
Poultry	88	91	8.1	8.5	0.4	0.8	0.4	0.5
Fish	3.4	8.6	11	11	ND	0.6	ND	0.4
Oils and fats	0.5	8.1	7.2	7.6	ND	0.6	ND	0.4
Eggs	0.9	8.1	4.4	5.4	2.9	3.4	1.7	1.7
Sugars and preserves	ND	8.0	8.3	8.7	ND	0.6	ND	ND
Green vegetables	440	440	320	320	0.3	0.7	ND	0.5
Potatoes	140	140	140	140	0.4	0.8	ND	0.5
Other vegetables	180	180	100	100	0.1	0.6	0.4	0.6
Canned vegetables	17	17	18	18	0.5	0.8	ND	0.5
Fresh fruit	25	26	27	27	ND	0.6	ND	0.4
Fruit products	11	15	12	13	0.2	0.7	ND	0.4
Beverages (not further specified)	ND	8	2.5	4.7	ND	0.6	ND	0.5
Milk	1.6	8.8	3.9	5.3	ND	0.6	ND	0.4
Dairy produce	3.4	9.4	27	27	ND	0.6	ND	0.4
Nuts	3.1	9.5	5.5	6.1	ND	0.6	ND	0.5

ND, not detected; limit of detection, 4 mg/kg nitrate, 0.4 mg/kg nitrite in 1997; 8 mg/kg nitrate, 0.6 mg/kg nitrite in 1994

Table 11. Food consumption figures from the 1994 and 1997 total diet studies in the United Kingdom used for estimating the intake of the whole population

Food group	Consumption (g/day)	
	1994	1997
Bread	0.110	0.108
Miscellaneous cereals	0.100	0.101
Carcase meats	0.026	0.022
Offal	0.001	0.001
Meat products	0.044	0.047
Poultry	0.018	0.019
Fish	0.013	0.014
Oils and fats	0.029	0.027
Eggs	0.016	0.014
Sugars and preserves	0.067	0.063
Green vegetables	0.037	0.034
Potatoes	0.133	0.123
Other vegetables	0.073	0.076
Canned vegetables	0.035	0.033
Fresh fruit	0.065	0.069
Fruit products	0.043	0.044
Beverages	0.863	0.937
Milk	0.284	0.281
Dairy produce	0.057	0.060
Nuts	0.002	0.002

samples, the concentrations in individual foods were not available (Vaessen & Schothorst, 1999).

2.4 *Intake reported in the literature*

Many assessments of nitrate and nitrite intakes have been reported in the literature, but most did not provide details of how they were conducted and are therefore difficult to interpret or compare.

An assessment of intake of food additives in Ireland was conducted in response to a Directive of the European Commission (Directive 1989/107, Annex II, point 4), which requires Member States to monitor the levels of additives in foods. The intake of nitrite was estimated to be 0.20 mg/kg bw, or 290% of the ADI. This estimate was based on 1000 individual records from a 7-day dietary survey (age group not specified) and maximum permitted levels of the additive in Ireland. No non-processed foods or water were included in the report. It was concluded that a further, more detailed assessment of nitrite was required (Food Safety Authority of Ireland, 2001).

The intake of nitrate as a food additive in Finland was reported to represent 31% of the ADI (Branen et al., 1990). Another Finnish study reported a mean intake of 77 mg/day for nitrate (35% of the ADI for a 60-kg person) and 5.3 mg/day for nitrite (150% of the ADI for a 60-kg person). This survey included vegetables, fruits, cheese, meat and fish. More than 90% of the dietary nitrate was from vegetables, and nitrite was derived mainly from meat products (95%). The concentrations of the food

Table 12. Summary of estimates of intake of nitrate and nitrite based on duplicate diet studies

Country	Date	Survey	Assumptions	Additive	Population group	Intake (mg/kg) bw per day)	% ADI[a]
United Kingdom	1998	Duplicate diet of vegetarians	–	Nitrate	Mean, food only	0.87	25
		7 days			Maximum, food only	2.9	80
		35 respondents			Mean, food, water, beer	1.4	35
		Adults only			Maximum, food, water, beer	3.5	95
Netherlands	1994	Duplicate diet of adults	Individual body weights used	Nitrate	Mean	1.1	30
		1 day	LOQ, 2 mg/kg for nitrate, 0.4 mg/kg for nitrite		Maximum	4.4	120
		123 respondents		Nitrite	Median	< 0.003	< 5
		Age, 18–74 years			Maximum	0.23	330

LOQ, limit of quantification
[a] ADI, 0–3.7 mg/kg bw for nitrate and 0–0.06 mg/kg bw for nitrite

additives in foods were obtained from various European studies, and the dietary data were based on 1-year quantitative diet histories from approximately 10 000 people over 15 years of age. As the intake assessment did not include drinking-water, the reported estimated intakes are lower than would otherwise be calculated (Dich et al., 1996).

The intake of nitrate as a food additive in Japan was reported to represent 120% of the ADI. The nitrite intake in a total diet survey in Japan in 1982 was 0.0018 mg/person, which would be below the ADI, assuming 55 kg body weight (Branen et al., 1990).

The Ministry of Agriculture, Fisheries and Food (1998a) in the United Kingdom reported that the nitrate and nitrite intakes by consumers of cured meat products at a high percentile were 19% and 1% of the respective ADIs (age group not specified). This survey did not include other foods or beverages. Numerous surveillance studies of the Ministry of Agriculture, Fisheries and Food (1997b, 1998c, 1999, 2001) of spinach and lettuce included estimates of intake of nitrate; these results are summarized in Table 13. The results were similar in all the surveys. Another study (Ministry of Agriculture, Fisheries and Food, 1997b) indicated that the intake by consumers of lettuce or spinach (no other foods) at a mean or high percentile would not exceed the ADI. As the ADI of 3.7 mg/kg bw for nitrate is equal to an intake of 220 mg/day for a 60-kg person, none of the estimated intakes in these studies would exceed the ADI, even for consumers at high percentiles.

A study on the effects of cooking on nitrate concentrations in vegetables was also carried out by the Ministry of Agriculture, Fisheries and Food (1998d). Cooking by boiling reduced the nitrate concentration in most vegetables tested by up to 75%.

3. EVALUATION OF ESTIMATES OF INTAKE

3.1 Intake from sources other than food additives

The estimated intakes of nitrate and nitrite from sources other than food additives are below their respective ADIs. The intakes of nitrate from the various GEMS/Food

Table 13. Summary of studies of intake of nitrate by the Ministry of Agriculture, Fisheries and Food in the United Kingdom

Foods included	Consumer intake (mg/day)		Year of reference
	Mean	97.5 %ile	
Lettuce or spinach and beer and water	88	140	2001
Spinach or lettuce, with additional value added for rest of diet taken from estimates in the 1997 total diet study in the United Kingdom	78	140	1999
Various types of lettuce or spinach and beer and water	110	210	1998

regional diets represented 20% of the ADI for the Middle Eastern diet, 10% of the ADI for the Far Eastern diet, 10% of the ADI for the African diet, 25% of the ADI for the Latin American diet and 70% of the ADI for the European diet. The nitrite intakes represented 50% of the ADI for the Middle Eastern and Far Eastern diets and 40% of the ADI for the African, Latin American and European diets. Vegetables were the main contributor to nitrate intake, accounting for 30–90% of that estimated. Potatoes were included among vegetables, and in some diets these were major contributors, representing 15% of the estimated intake in the Far Eastern diet, 29% in the African diet, 15% in the Latin American diet and 30% in the European diet. Cereals were the main contributor to nitrite intake, accounting for 35–60% of the estimated intake. Drinking-water was the second highest contributor to the estimated intakes of both chemicals, at 5–40% of the estimated intake of nitrate and 20–40% of that of nitrite. These estimates are based on mean chemical concentrations and mean food consumption at the global level and provide a reasonable guide to estimates of intake and the contributions of various food groups.

3.2 Intake from use as a food additive

Screening for nitrate and nitrite by the budget method indicated that the theoretical maximum levels were below some of the maximum permitted levels in the GSFA. This indicated the need for intake assessments and suggested that use of nitrate and nitrite as food additives might have to be restricted.

The estimated intakes based on individual dietary records in Australia and New Zealand were compared with maximum permitted levels from national food standards and from the GSFA. The results indicated that, for nitrate, the ADI was not exceeded for consumers at the mean (10% of the ADI) or at a high percentile (35% of the ADI). These estimates were based on 24-h data on food consumption, which tend to overestimate the additive intake of consumers at high percentiles. Even after water was included in the national assessments (50 mg of nitrate based on consumption of 1 l of water per day and the maximum guideline level in water), the nitrate intake was still below the ADI.

The Australian and New Zealand assessments of intake of nitrate, however, did not include a contribution from sources other than additives. It is known that foods such as vegetables contribute substantially to nitrate intake, but it was difficult to estimate the impact of non-additive sources on the above estimates.

The intakes of nitrite estimated from individual dietary records in Australia and New Zealand indicated that the ADI might be exceeded by consumers at both the mean and high percentiles, on the basis of either national maximum permitted levels of nitrite or maximum permitted levels in the GSFA. The intakes of consumers at the mean in national estimates were 100–130% of the ADI, and those based on the GSFA were 460–500% of the ADI. The intakes by consumers at the 95th percentile in national estimates were 320–360% of the ADI, and those based on the GSFA were 1300–1400% of the ADI. However, these intakes are likely to be overestimates, as they are based on the maximum levels that can be added by manufacturers, rather than actual levels in foods as consumed, and on the basis of the assumption that all foods in the group contain the additive. The estimated intakes were based on 24-h data for food consumption, which tend to exaggerate the estimated intakes by consumers at high percentiles. The estimated intakes exceeded the ADI even before addition to national assessments of the contribution from water (3 mg of nitrite on the basis of consumption of 1 l of water per day and the maximum national guideline level in water).

The per cent contribution of each food group to total intake was also presented in the Australian and New Zealand assessments. For nitrate, the greatest contributor based on national permissions (not including water) was processed comminuted meat products (Australia, 56%; New Zealand, 52%), and the greatest contributor based on maximum permitted levels in the GSFA and national food consumption data was raw meat, poultry and game (Australia, 55%; New Zealand, 51%). For nitrite, the greatest contributor based on national permissions (not including water) was processed meat in whole cuts or pieces for Australia (58%) and processed comminuted meat products for New Zealand (60%), and the greatest contributor based on GSFA and national food consumption data was raw meat, poultry and game (Australia; 63%; New Zealand, 60%). As these assessments did not include intakes from sources other than additives, including vegetables and water, the estimated contributions of these food categories would be different if considered in the context of total dietary intakes from all sources.

In summary, the estimated intakes of nitrate in Australia and New Zealand were below the ADI on the basis of national and GSFA maximum permitted levels. Similar estimates of nitrite were above the ADI on the basis of national and GSFA maximum permitted levels. The results of these assessments are overestimates as they do not include the concentrations of the additives in foods as consumed; however, the results did not take into account sources other than food additives. For these reasons, it is difficult to derive meaningful conclusions or recommendations from the above results.

3.3 Intake from all sources

The results of the intake assessments from the total diet studies in the United Kingdom, based on measured concentrations in foods, indicated that the intakes of nitrate and nitrite by consumers at the mean and high percentiles in all the model diets assessed were below the respective ADIs (Table 9). The intakes represented ≤ 75% of the ADI for nitrate and ≤ 70% of the ADI for nitrite.

Foods that made a major contribution to the mean estimated population dietary intake of nitrate were green vegetables, potatoes and other vegetables (total, 69% in the 1997 total diet study and 76% in the 1994 total diet study). For nitrite, the highest contributor to the mean estimated population dietary intake was beverages (specific types not stated), accounting for 35% in the 1997 total diet study and 30% in that in 1994.

The study of nitrate in duplicate diets of vegetarians submitted by the United Kingdom indicated estimated intakes below the ADI for all consumers (Table 12). The per cent contributions were not provided for individual foods; however, food contributed the most to the mean estimated intake (62%), with water contributing 24% and beer, 13%. Measurement in a duplicate diet over 7 days is considered to be a relatively accurate method for estimating food chemical intake. A 7-day study does have the limitation that some respondents underreport food consumption or change their dietary patterns; however, the results were based only on duplicate diets that were validated as being representative samples. The sample size in this survey was relatively small (35 respondents) and may not be representative of all vegetarians. This study suggests that intake of nitrate and nitrite by vegetarians is not necessarily higher than that of the general population.

The duplicate diet study submitted by the Netherlands showed estimated intakes for consumers of nitrate at the mean and consumers of nitrite at the median that were below the ADI (30% of the ADI for nitrate, 5% of the ADI for nitrite). The nitrate intakes were slightly higher than in the duplicate diet study in the United Kingdom, perhaps because of the duration of the survey: the study in the Netherlands lasted 24 h, while that in the United Kingdom lasted 7 days and is therefore more representative of usual intake, providing a better estimate of long-term intakes. The intakes of nitrate by five consumers (4% of respondents) and of nitrite by four consumers (3% of respondents) exceeded the ADI, on the basis of 24-h intake. These estimates are likely to have been lower over a longer period. The major contributors to nitrate intake were leafy vegetables, beetroot and green beans (no percentages given). The major contributors to intake of nitrite were not given.

In summary, the results of studies of the intake of nitrate and nitrite from all dietary sources showed mean consumptions of both below the ADIs, although some consumers at high percentiles exceeded the ADI for both chemicals.

4. CONCLUSIONS AND RECOMMENDATIONS

These conclusions are based on the limited data that were submitted. The intake of nitrate was reported to be below its corresponding ADI more often than that of nitrite.

The estimated intakes of nitrate show that vegetables are the major contributor to total dietary intake, followed by water. The major contributors to nitrite intake are also sources other than additives, including cereals, beverages and water. Foods containing nitrate and nitrite as food additives are not major contributors to the total estimated intake in the submitted data.

Intake estimates based on national and GSFA maximum permitted levels showed that the intake of nitrate is not likely to exceed the ADI in any scenario or population assessed. In contrast, the intake of nitrite was shown to exceed the ADI in all scenarios and for all populations assessed. However, these estimates are overestimates as they are not based on actual concentrations in foods as consumed.

The intakes of nitrate based on more accurate methods, including total diet studies in which measured concentrations are used, are a good indication of intake from the total diet. The total diet studies submitted indicated that the intakes of both nitrate and nitrite were below the ADI. The duplicate diet study indicated that the intake of consumers at a high percentile approaches and in some cases exceeds the ADI.

Nitrite

Ideally, all sources of nitrite should be included in an intake assessment; however, only food and water were included in the assessments made at this meeting. The intake of nitrite from non-food additive sources, on the basis of the concentrations in food derived from European studies and the GEMS/Food regional diets and with an addition for water was below the ADI (40–50% of the ADI for the various diets). Major contributors to the estimated intake from sources other than food additives are cereals (35–60% of the estimated intake) and water (20–40% of the estimated intake).

The estimated intakes from use of nitrite as a food additive derived from national assessments were above the ADI, assuming maximum levels in the Codex draft GSFA. The intake of nitrite by consumers at the mean represented 460–500% of the ADI, while that of consumers at the 95th percentile represented 1300–1400% of the ADI. The intakes of nitrite from food additive use based on maximum permitted levels in national standards also exceeded the ADI (consumers at the mean, 100–130% of the ADI; consumers at the 95th percentile, 320–360% of the ADI). The estimated intakes based on the GSFA do not reflect reality, as the calculations include the GSFA maximum levels, which often are significantly higher than the typical levels of addition to foods by manufacturers. The calculations also do not take into consideration degradation of nitrite over time and do not reflect the concentrations in foods as consumed. The intakes are based on 24-h food consumption data, which tend to result in overestimates of intake by consumers at high percentiles.

More realistic estimates of the intake of nitrite from all dietary sources, submitted in total diet and duplicate diet studies, showed mean intakes representing 5–15% of the ADI, although the intakes by consumers at higher percentiles approached the ADI and exceeded it in one study, representing 55–330% of the ADI. A major contributor to the total intake estimated from these studies was beverages (specific types not stated), which accounted for 30–35% of the total estimated intake, owing more to the high volumes consumed than to high concentrations of nitrite. Analysed concentrations in foods as consumed were used in these studies. These results confirm that the estimates of intake from use of nitrite as a food additive at the listed maximum levels in the GSFA or national maximum permitted levels are overestimates.

Nitrate

Ideally, all sources of nitrate should be included in an intake assessment; however, contributions from air were not included here. Nitrate intake from sources other than food additives, based on concentrations in food derived from European studies and the GEMS/Food regional diets and with an addition for water, was below the ADI (10–70% of the ADI for the various diets). Major contributors to intake were vegetables (30–65% of the estimated intake) and water (5–45% of the estimated intake). The nitrate content of both groundwater and surface water is highly variable, and wide variations are seen between countries.

The estimated intakes from use of nitrate as a food additive derived from national assessments were below the ADI, assuming the maximum levels in the Codex draft GSFA. The intake by consumers at the mean was about 10% of the ADI, while the intake of those at the 95th percentile was 30–35% of the ADI, on the basis of national data on food consumption. The intakes of nitrate from food additive use based on maximum permitted levels in national standards were also below the ADI (consumers at the mean, 10% of the ADI; consumers at the 95th percentile, 35% of the ADI). Major contributors from its use as a food additive were processed comminuted meats in national assessments (50–55% of estimated intake) and raw meats on the basis of GSFA maximum levels (50–55% of the estimated intake) from national food consumption data. These intakes were based on 24-h food consumption data, which tend to result in overestimates of the intake of consumers at high percentiles.

Assessments of intake of nitrate from all dietary sources, including drinking-water, showed that they were below the ADI. In total diet studies in the United Kingdom, the mean intakes were between 40% and 45% of the ADI and those by consumers at the 95th percentile were 60–75% of the ADI. Duplicate diet studies showed mean intakes representing 30–35% of the ADI and intakes by consumers at the 95th percentile representing 95–125% of the ADI. These studies included one of vegetarians, who are likely to have higher intakes of nitrate owing to their greater consumption of vegetables; however, their intakes were not considerably different from those of the general population. Major contributors to total dietary intake of nitrate were vegetables, which accounted for 70–75% of the total estimated intake. The Committee noted that the food additive use of nitate is not a major source of intake.

4.1 General recommendation

The Committee recommended that assessments of the intake of nitrate and nitrite should include sources other than food additives, such as vegetables and drinking-water, as well as food additives. Studies should be conducted with an accurate method of assessment, such as total or duplicate diets and individual dietary records, as these give the best estimates of intakes. Also, foods should be analysed as 'ready to consume', thus accounting for losses of the chemicals during food storage, preparation and cooking. This applies specifically to nitrite. Studies should be conducted during different seasons to account for natural variation in the concentrations in foods such as vegetables, particularly for nitrate.

4.2 Recommendations to the Codex Committee on Food Additives and Contaminants

The Committee recommended that the Codex Committee on Food Additives and Contaminants reconsider the maximum permitted levels of nitrate and nitrite in the draft GSFA, as the estimated intakes of nitrate and nitrite might exceed the ADI.

5. REFERENCES

Branen, A.L., Davidson, P.M. & Salminen, S., eds (1990) *Food Additives*, New York: Marcel Dekker.

Cooperative Research Centre for Water Quality and Treatment (1996) *Fact Sheets—Inorganic Chemicals* (Fact Sheet Number 61), Canberra.

Coultate, T.C. (1989) *Food. The Chemistry of its Components*, London: Royal Society of Chemistry.

Dich, J., Järvinen, R., Knekt, P. & Penttilä, P.L. (1996) Dietary intakes of nitrate, nitrite and NDMA in the Finnish Mobile Clinic Health Examination Survey. *Food Addit. Contam.*, **13**, 541–552.

European Commission (1998) *Food Science and Techniques. Reports of the Scientific Committee for Food* (Thirty-eighth Series), Luxembourg.

Food Safety Authority of Ireland (2001) *Legislation, Intake and Usage of Food Additives in Ireland*, Dublin.

Hansen, S.C. (1979) Conditions for use of food additives based on a budget method for an acceptable daily intake. *J. Food Prot.*, **42**, 429–434.

Institute of Medicine (1996) *Food Chemicals Codex*, 4th Ed., Washington DC: National Academy Press.

de Kreij, C. (1998) *Good Agricultural Practice (GAP) in glasshouse lettuce in the Netherlands. Registration during 1995–1998*, Internal report 149, Naaldwijk, Profstation voor Blomisterij en Glasgroente.

de Kreij, C. (2001) *Good Agricultural Practice (GAP) in glasshouse lettuce and spinach cultivation in the Netherlands. Registration during 2000–2001*, Naaldwijk, Applied Plant Research B.V.

de Kreij, C. & van Winkel, A. (2000) *Good Agricultural Practice (GAP) in glasshouse lettuce and spinach cultivation in the Netherlands. Registration during 1998–2000*, Internal report 227 (revised version), Naaldwijk, Profstation voor Blomisterij en Glasgroente.

Maga, J.A. & Tu, A.T., eds (1995) *Food Additive Toxicology*, New York: Marcel Dekker.

McKnight, G., Smith, L., Golden, M.N.H. & Benjamin, N. (undated) *Metabolism of dietary nitrate in the gastrointestinal tract in man*, Study FS2188 (1CO64), London: Ministry of Agriculture, Fisheries and Food.

Ministry of Agriculture, Fisheries and Food (1996) *Nitrate in Vegetables*, Food Surveillance Information Sheet No. 91, July 1996, London.

Ministry of Agriculture, Fisheries and Food (1997a) *1994 Total Diet Study—Nitrate and Nitrite*, Food Surveillance Information Sheet No. 137, December 1997, London.

Ministry of Agriculture, Fisheries and Food (1997b) *1996/97 UK Monitoring Programme for Nitrate in Spinach and Lettuce*, Food Surveillance Information Sheet No. 121, August 1997, London.

Ministry of Agriculture, Fisheries and Food (1998a) *Survey of Nitrite and Nitrate in Bacon and Cured Meat Products*, Food Surveillance Information Sheet No. 142, February 1998, London.

Ministry of Agriculture, Fisheries and Food (1998b) *1997 Total Diet Study—Nitrate and Nitrite*, Food Surveillance Information Sheet No. 163, October 1998, London.

Ministry of Agriculture, Fisheries and Food (1998c) *1997/98 UK Monitoring Programme for Nitrate in Lettuce and Spinach*, Food Surveillance Information Sheet No. 154, August 1998, London.

Ministry of Agriculture, Fisheries and Food (1998d) *Nitrate in Lettuce and Spinach*, Food Surveillance Information Sheet No. 177, May 1998, London.

Ministry of Agriculture, Fisheries and Food (1999) *Nitrate in Vegetables*, Food Surveillance Information Sheet No. 158, September 1998, London.

Ministry of Agriculture, Fisheries and Food (2000) *Duplicate Diet Study of Vegetarians—Nitrate Analyses*, Food Surveillance Information Sheet No. 165, November 1998, revised January 2000, London

Ministry of Agriculture, Fisheries and Food (2001) *UK Monitoring Programme for Nitrate in Lettuce and Spinach*, Food Survey Information Sheet Number 16/01, July 2001, London.

Ministry of Health (2000) *Drinking-water Standards for New Zealand 2000*, Wellington.

National Health and Medical Research Council and Agriculture and Resource Management Council of Australia and New Zealand (1996) *Australian Water Quality Guidelines*, Canberra.

van der Schee, H.A. (1997) *Report. Nitrate Monitoring Results Concerning Regulation EU 194/97. The Netherlands 1996*, Amsterdam: Inspectorate for Health Protection .

van der Schee, H.A. (1998) *The Nitrate Content of Vegetables of the Dutch Market in 1996*, Amsterdam: Inspectorate for Health Protection.

van der Schee, H.A. & Speek, A.J. (2000a) *Report of Nitrate Monitoring Results Concerning Regulation EU 194/97. The Netherlands 2000*, Amsterdam, Inspectorate for Health Protection and Veterinary Public Health, Region Northwest.

van der Schee, H.A. & Speek, A.J. (2000b) *Report of Nitrate Monitoring Results Concerning Regulation EU 194/97. The Netherlands 1999*, Amsterdam, Food Inspection Agency Northwest.

Tompkin, R.B. (1993) Nitrite. In: Davidson, P.M. & Branen, A.L., eds, *Antimicrobials in Foods*, 2nd Ed., New York: Marcel Dekker.

Vaessen, H.A.M.G. & Schothorst, R.C. (1999) The oral intake of nitrate and nitrite in The Netherlands: evaluation of the results obtained by HPIC analysis of duplicate 24-hour diet samples collected in 1994. *Food Addit. Contam.*, **16**, 181–188.

WHO (1998a) *Guidelines for Drinking-water Quality*, 2nd Ed., *Addendum to Volume 2. Health Criteria and Other Supporting Information*, Geneva.
WHO (1998b) *Guidelines for Drinking-water Quality*, 2nd Ed., *Addendum to Volume 1. Recommendations*, Geneva.
WHO (1998c) *GEMS/FOOD Regional Diets*, Geneva.

SALATRIM
(Short- and long chain acyltriglyceride molecules)

First draft prepared by
Dr D. Benford[1], Dr J.B. Grieg[1], Dr J. R. Lupton[2] and Dr C. Leclercq[3]
[1]Food Standards Agency, London, England; [2]Faculty of Nutrition Texas A&M University, College Station, Texas, USA; and [3]National Institute for Food and Nutrition Research, Rome, Italy

1. EXPLANATION

Salatrim is the acronym for **s**hort- **a**nd **l**ong-chain **a**cyl **tri**glyceride **m**olecules. Salatrim preparations are tailored mixtures of triglycerides containing at least one long-chain fatty acid moiety (principally stearic acid) and one or two short-chain fatty acid moieties (acetic, propionic and/or butyric acid). Salatrim is synthesized by non-enzymatic inter-esterification of triacetin, tripropionin, tributyrin or their mixtures with hydrogenated canola, soya bean, cottonseed or sunflower oils. The physical properties of salatrim may be fitted to its food applications by selecting specific short-chain fatty acids and hydrogenated oils for synthesis. Salatrim is intended for use as a low-calorie fat in soft sweets, coatings (e.g., wafers and confectionery), dairy products (including spreads), margarines and bakery products.

Salatrim of various compositions was evaluated by the Committee at its forty-ninth meeting (Annex 1, reference *131*), when it concluded that the available studies did not provide an adequate basis for evaluating the safety and the nutritional effects of salatrim. At that meeting, the Committee recommended that additional, appropriately designed studies be performed to assess fully both the toxicological and nutritional consequences of ingestion of salatrim.

The Committee noted at its forty-ninth meeting that the specifications for salatrim permit compositions that include a triglyceride mixture with up to 0.87 g of stearate per gram of fat, but that the available biological data did not provide information on materials of such a composition. Since that time, the sponsor has requested that the specifications for salatrim be limited to preparations with ≤ 70% by weight of total saturated long-chain fatty acids. Studies were available on the biological effects in humans of salatrim containing up to 0.61 g of stearate per gram of fat.

The Committee previously evaluated studies on the caloric value of salatrim, taking into account the fact that short-chain fatty acids supply fewer kilocalories per gram than long-chain fatty acids. Moreover, it concluded that the claim of reduced absorption of stearic acid has not been proven for humans. In addition, as there is no specific composition for salatrim, a single caloric value could not be assigned.

In evaluating the safety of salatrim at its forty-ninth meeting, the Committee considered studies in which salatrim of various compositions was administered to rats or minipigs at concentrations of up to 10% in the diet. These studies showed no toxicologically significant effects, but the study in minipigs was considered to have been of insufficient duration. The Committee also discussed one home-based trial and four clinic-based studies with volunteers who ingested up to 60 g salatrim per person per day. The studies indicated that the consumption of diets containing salatrim, particularly at 60 g per day, was associated with an increased incidence of mild gastrointestinal symptoms and significantly elevated serum enzyme activities. Thus, the Committee concluded at its forty-ninth meeting that the available studies in experimental animals and humans did not provide an adequate basis for a toxicological evaluation. In addition, as the studies were not optimized to detect potential nutritional effects, the Committee concluded that they did not provide an adequate basis for a nutritional evaluation.

In response to the Committee's earlier request, four additional home-based trials in humans and one study of nutrition in rats were submitted for consideration at the present meeting.

Table 1 shows the nomenclature of salatrim products. The products that have been used in safety evaluations are listed in Table 2. The results of all these studies were published in *The Journal of Agricultural and Food Chemistry*, Volume 42 (2) in 1994. The papers contained numerous typographical and transcription errors, and three sets of errata were subsequently published in *The Journal of Agricultural and Food Chemistry*, Volume 42 (4, 9 and 12) in 1994, and the Committee was informed of additional corrections. Two of the newly submitted studies have been published (Nestel et al., 1998; Sanders et al., 2001), whereas the remaining newly submitted studies in humans were unpublished. The summaries given below contain information from the published accounts and also from the full, unpublished study reports.

Table 1. Typical molar ratios of short- and long-chain acid sources used to prepare the salatrim family of edible oils

Salatrim family	Short-chain source	Long-chain source	Molar ratio
4CA	Tributyrin	Hydrogenated canola oil	2.5:1
4SO	Tributyrin	Hydrogenated soya bean oil	12:1
23CA	Triacetin, tripropionin	Hydrogenated canola oil	11:1:1
23SO	Triacetin, tripropionin	Hydrogenated soya bean oil	11:1:1
32CA	Tripropionin, triacetin	Hydrogenated canola oil	11:1:1
43SO	Tributyrin, tripropionin	Hydrogenated soya bean oil	11:1:1
234CS	Triacetin, tripropionin, tributyrin	Hydrogenated cottonseed oil	4:4:4:1
234CA	Triacetin, tripropionin, tributyrin	Hydrogenated canola oil	4:4:4:1
234SO	Triacetin, tripropionin, tributyrin	Hydrogenated soya bean oil	4:4:4:1

The salatrim family name defines the sources of the short-chain and long-chain fatty acids, the numerals representing the carbon lengths of the short-chain acids in decreasing proportion in the mix, and the letters defining the oil that provides the source of the long-chain fatty acids; e.g., in salatrim 43SO, tributyrin and tripropionin are the short-chain fatty acids and hydrogenated soya bean oil is the source of long-chain fatty acids. The molar ratio of the mix that is used to prepare salatrim is 11 parts tributyrin:1 part tripropionin:1 part hydrogenated soya bean oil.

2. BIOLOGICAL DATA

2.1 Biochemical aspects

In an initial review of the metabolism of fats and long- and short-chain fatty acids, it was proposed that short-chain fatty acids in salatrim would be released after hydrolysis of the triglyceride in the stomach. A proportion of those released would be absorbed in the stomach and used as an energy source, while the remainder would be taken up by the liver. Hydrolysis of long-chain fatty acids from salatrim fats

Table 2. Materials used in studies of the metabolism and toxicity of salatrim

Study end-point	Material used
Reverse mutation in *Salmonella typhimurium*	4CA, 23CA, 23SO, 32CA, 234CA, 234CS
Mutation in mammalian cells in vitro	23CA
Assays for micronucleus formation in bone-marrow cells in vivo	234CA, 234SO
Metabolism in vitro (porcine pancreatic lipase)	4CA, 23CA, 32CA, 234CA
Metabolism in rats	23CA
90-day feeding studies in rats	4CA, 23CA, 32CA, 23SO, 234CA, 234CS[a]
28-day feeding study in minipigs	23SO
Effects on gut microflora in rats	23CA, 32CA
Studies I and II in volunteers	23CA
Studies III and IV in volunteers	23SO
Study V in volunteers	4SO, 23SO, 43SO
Studies VI and VII in volunteers	23SO
Studies VIII & IX in volunteers	4SO

[a] Plus supplementary 17-day test of effects on transaminases

would occur predominantly in the small intestine. The absorption of stearate would be limited. A proportion of absorbed stearic acid would be converted to oleic acid. Any short-chain fatty acids released in the small intestine would enter the hepatic portal vein (Hayes et al., 1994a).

Experiments in which salatrim 23CA was administered to rats were designed to test this hypothesis. Salatrim 23CA was chosen because it contains triacyl-glycerides unique to the salatrim family, whereas triglycerides containing butyrate are commonly consumed as part of the diet.

2.1.1 Biotransformation

Rats

Salatrim products are not absorbed intact, as they are hydrolysed and metabolized in an identical manner to triacylglycerides present in the diet. Monoacylglycerides containing stearate derived from the hydrolysis of salatrim products can be absorbed from the small intestine. After administration of a single oral dose of [^{14}C-stearate]salatrim (1.4 g/kg bw) to male Sprague-Dawley rats, 0.4% of the radiolabel was present in fat 72 h later, of which half was in oleate (Musick & Peterson, 1993; Hayes et al., 1994b).

A single oral dose of 1.4 g/kg bw of salatrim 23CA labelled with ^{14}C in the acetate, propionate, stearate or glycerol moiety or 1.4 g/kg bw of triolein labelled with ^{14}C in the oleate or glycerol moiety was administered to groups of five male Sprague-Dawley rats by gavage. Radiolabel elimination was followed for 72 h. Acetate and propionate from salatrim were exhaled as CO_2 (82% and 89% of the dose, respectively). Stearate was released from salatrim 23CA, and oleate was released from radiolabelled triolein, 22% of the labelled stearate and 44% of the labelled oleate being exhaled as CO_2. In experiments with salatrim or triolein labelled in the glycerol moiety, 75% of the dose was exhaled as CO_2. Faecal excretion of radiolabel after administration of salatrim labelled in the acetate, propionate or glycerol moiety and glycerol-labelled triolein approximated to 4–5% of the dose. More faecal excretion of radiolabel was seen when [^{14}C-stearate]salatrim (55%) and [^{14}C-oleate]triolein (38%) were administered to rats. Less than 4% was excreted in the urine, and this consisted of radiolabel that had presumably been incorporated into intermediary metabolites. Urinary excretion of radiolabel was greater in animals treated with salatrim radiolabelled at the short-chain fatty acid moiety than in those given salatrim or triolein radiolabelled in the long-chain fatty acid moiety. Approximately 10% of the radiolabel in experiments with stearate- or glycerol-labelled salatrim and oleate- or glycerol-labelled triolein was recovered from the carcass. Very small amounts of radiolabel were found in the liver, blood and fat in these studies (total, $< 1.8\%$). Feeding a diet containing 10% salatrim 23CA for 2 weeks before dosing with radiolabelled salatrim or triolein did not affect their metabolism. Slightly greater faecal excretion of radiolabel was reported in all the investigations with animals fed salatrim before testing, and the authors suggested that competition for metabolism by dietary salatrim and radiolabelled salatrim administered by gavage was responsible.

The authors concluded that the absorption, distribution and elimination of salatrim are identical to those of other triglycerides found in the diet and that the data support the observation that stearate is less well absorbed than oleate. No conclusions could be derived about the absorption of salatrim 23CA and its component fatty acids, since absorption was not measured directly. The value of this study with respect to the assessment of absorption of salatrim is limited, because the test solutions were prepared by mixing radiolabelled triolein (control fat) into the test fat (salatrim) matrix, and the absorption of the control fat may therefore have been affected. The radiolabelled triolein should have been incorporated into non-radiolabelled triolein. The data do, however, support the view that the metabolism of salatrim 23CA is similar to that of triolein once it has been absorbed (Musick & Peterson, 1993; Hayes et al., 1994b).

In vitro

The hydrolysis of a number of salatrim products by porcine pancreatic lipase was studied over 30 min. Chloroform solutions (100 mg/ml) of salatrim 4CA, 23CA, 32CA and 234CA were incubated at 37 °C for 2, 5, 10 or 30 min, and the predominant triacylglycerides, diacylglycerides, monoacylglycerides and long- and short-chain fatty acids were measured by gas chromatography with mass spectrometry. Each of the salatrim products tested showed a consistent pattern of hydrolysis, consisting mainly of a peak in diacyl-glyceride formation after about 2 min and a concurrent rapid rise in free stearate over 5 min. The rate of stearic acid formation slowed during the remainder of the 30-min period. In experiments with salatrim 23CA and 32CA, hydrolysis of the triglyceride containing two short-chain fatty acids (i.e. di-short triglyceride) was faster and more complete than that of the corresponding di-long triglyceride, which contained two stearate esterifications. The hydrolysis of triacylglycerides containing butyrate was faster than that of those containing acetate. The authors concluded that salatrim molecules undergo lipolysis in a predictable manner.

The authors speculated that the higher rate of release of short-chain as compared with long-chain fatty acids was due to the greater hydrophilicity of short-chain fatty acid-rich triacylglycerides within fat droplets and the faster diffusion of the released short-chain fatty acids from the active site of the enzyme into the aquatic phase surrounding fat droplets. The authors concluded that rapid release of short-chain fatty acids would occur in the stomach and upper intestine (Phillips, 1992; Sequeria & Gordon, 1993)

Caloric availability in rats and humans

Studies on this topic are summarized in Appendix 1 to this monograph.

2.1.2 Effects on enzymes and other biochemical parameters

Twenty-four Crl:CD BR VAF strain rats of each sex were fed a diet containing 10% salatrim 23SO for 17 days. An additional 24 rats of each sex were fed a diet containing 10% corn oil, and 12 rats of each sex received only the basal diet throughout the study. The rats were observed twice daily, and body weights were recorded weekly. Blood was collected, and serum concentrations of aspartate aminotransferase (AST) and alanine aminotransferase (ALT) and γ-glutamyltranspeptidase (GGT) were determined 12 and 4 days before and on days

3, 6, 9, 13 and 17 after initiation of the study. Neither salatrim nor corn oil had any effect on enzyme activities at any interval (Kiorpes, 1993a; Hayes et al., 1994c).

2.2 Toxicological studies

2.2.1 Short-term studies of toxicity

Rats

(i) Salatrim 4CA

Diets containing 0, 2, 5 or 10% salatrim 4CA or 10% corn oil were fed to groups of 20 Crl:CD BR VAF rats of each sex for 13 weeks. The salatrim-containing diets were supplemented with varying amounts of vitamins A, E, D and K according to the level of salatrim 4CA incorporation. A vitamin control group was also included, the level of supplementation being equivalent to that used in the 10% salatrim diet. A further group of 10 rats of each sex were fed salatrim-containing diets for 5 weeks and were used for interim histopathology and measurement of calcium, copper, iron, magnesium, phosphorus, sodium, strontium and zinc in defatted femur (10% salatrim or corn oil only). Blood and urine samples were collected at 4 and 13 weeks for clinical chemistry, urinary analysis and haematology. Bone minerals were also measured at 13 weeks. The animals were observed for signs of toxicity daily, and body-weight gain and food intake were measured weekly. At autopsy, the adrenals, brain, liver, kidneys and testes were weighed. The 10% dose represented the highest concentration that the authors considered would not cause excessive dilution of micronutrients.

No treatment-related deaths occurred, and no effects on body weight were observed in salatrim-fed animals, although increased weight gain was noted in most weeks of the study for animals fed corn oil. Decreased food consumption was observed for males fed 10% salatrim 4CA and males and females fed corn oil. There were no effects on haematological, clinical chemical or urinary end-points. The serum levels and urinary clearance of minerals were unaffected by treatment with salatrim or corn oil. No treatment-related changes in the levels of fat-soluble vitamins in serum or liver were documented in animals given 10% salatrim 4CA when compared with vitamin-supplemented controls that received the same amounts of vitamins in the diet (the occasional differences seen in single sexes at week 5 were not confirmed at termination). No definitive conclusions can be drawn regarding any potential effect on fat-soluble vitamin absorption, since an appropriate unsupplemented salatrim group was not included in the study. The concentrations of strontium and zinc in bone were higher in animals of each sex at 10% salatrim than in either control group, while the sodium concentration was higher only in females in comparison with the unsupplemented control group. Animals of each sex given 10% corn oil had higher concentrations of strontium in bone than either control group, while the concentration of zinc in bone was lower in males than in unsupplemented controls. No treatment-related effects on organ weights or histopathological changes were documented at the interim or final necropsy. A number of animals fed 10% corn oil showed hepatocellular vacuolation. The authors concluded that the changes in the concentrations of minerals in bone were directly related to the quantity of unsaturated fatty acids in the diet fed to animals. The NOEL was 10% in the diet, equal to 7.3 g/kg bw per day (Williams, 1992a; Hayes et al., 1994d).

(ii) Salatrim 23CA and 32CA

Diets containing 0, 2, 5 or 10% salatrim 23CA or 32CA or 10% corn oil were fed to groups of 20 Crl:CD BR VAF rats of each sex for 13 weeks, when blood and urine samples were collected from a subgroup of 10 rats of each sex per group for haematology, serum and urine chemistry and urinary analysis. Blood was obtained from the remaining 10 rats of each sex per group for measurement of fat-soluble vitamins in serum. Animals were observed for signs of toxicity daily, and body-weight gain and food intake were measured weekly. The 10% dose represented the highest concentration that was considered would not cause excessive dilution of micronutrients. Adrenals, brain, kidneys, liver and testes were weighed at autopsy. The caecum of each rat was ligated at the distal ileum, at the proximal colon and at approximately the distal one-third of the blind end. The distal portion of the caecum was collected for histological examination, and the remaining ligated portion of the caecum from each rat was used for a special study of effects on gut microflora (see section 2.2.5). Minerals were measured (as for salatrim 4CA) in samples of bone (defatted femur) from 10 animals per group at necropsy.

No treatment-related deaths occurred. The mean body-weight gain and feed consumption were similar to those of control animals. No toxicologically significant effects on haematology or clinical chemistry were reported. A dose-related trend to slightly increased urinary phosphorus clearance was noted in rats fed salatrim fats. A statistically significant increase in urinary phosphorus clearance was seen in males fed 10% salatrim 23CA and in males and females fed 10% salatrim 32CA. No changes in the levels of fat-soluble vitamins in serum or liver were documented, except for a reduced level of vitamin A in animals fed corn oil. The mean strontium concentration in bone was significantly higher in females fed 10% salatrim 23CA than in controls. A slight, statistically nonsignificant increase in the mean concentration of strontium in bone was also found in males fed 10% salatrim 23CA. The mean zinc concentration in bone was significantly higher in females fed 10% salatrim 23CA or 10% salatrim 32CA than in control females. The mean zinc concentration in bone was significantly lower in males given the 10% corn oil diet than in controls. No treatment-related effects on organ weights were noted. An increased incidence of renal mineralization was noted microscopically in females of groups receiving corn oil and 5% and 10% salatrim 23CA and salatrim 32CA when compared with controls. The incidence and severity of this lesion was similar in each of these groups of triacylglycerol-treated females. No treatment-related renal mineralization was noted in any group of corn oil-treated or salatrim-treated males. The authors concluded that the changes in levels of minerals in bone and renal mineralization were directly related to the quantity of unsaturated fatty acids in the salatrim diet fed to animals. A number of animals fed 10% corn oil had hepatocellular vacuolation. The Committee concluded that these two salatrim products did not induce any toxicologically significant effects. The concentration of 10% salatrim 23CA or 32CA was equal to 7.5 g/kg bw per day (Williams, 1992b; Hayes et al., 1994c).

(iii) Salatrim 234CS and 234CA

Diets containing 0 (control), 2, 5 or 10% salatrim 234CS or 234CA or 10% corn oil were fed to groups of 20 Crl:CD BR VAF plus rats of each sex for 13 weeks. At that time, blood and urine samples were collected from a subgroup of 10 rats of each sex per group for haematology, serum and urine chemistry and urinary analysis. Blood was obtained from the remaining 10 rats of each sex per group for measurement of fat-soluble vitamins in serum. Animals were observed for signs of toxicity daily, and body-weight gain and food intake were measured weekly. The 10% dose represented the highest concentration that it was considered would not cause excessive dilution of micronutrients. Adrenals, brain, kidneys, liver and testes were weighed at autopsy. The levels of minerals were measured (as for salatrim 4CA) in samples of defatted bone (femur) from 10 animals per group at autopsy.

No treatment-related deaths occurred. The mean body-weight gain and feed consumption of salatrim-treated groups were similar to those of untreated control animals, although controls given corn oil and females fed 10% salatrim 234CS gained slightly more weight. A significant decrease in food consumption was noted in animals of each sex fed 10% corn oil. No toxicologically significant effects on haematological or clinical chemical parameters were reported. Slight, statistically nonsignificant increases in urinary phosphorus clearance were seen in all groups given 10% salatrim. The mean serum vitamin A level was significantly higher in male rats fed 2% salatrim 234CA or 10% corn oil than in controls. The mean vitamin A concentration in liver was significantly lower in males fed 10% salatrim 234CA and males and females fed 10% salatrim 234CS or 10% corn oil than in controls. The mean serum 25-hydroxy vitamin D concentrations were significantly lower than those of controls in females fed 2 or 10% salatrim 234CA, 2 or 10% salatrim 234CS or 10% corn oil. The authors noted inconsistencies in the results for males and females with regard to the effects of the salatrim products on fat-soluble vitamin levels. They considered that corn oil induced a similar reduction in serum 25-hydroxy vitamin D levels and a larger increase in serum vitamin A levels in comparison with salatrim and concluded that the salatrim products tested did not substantially alter fat-soluble vitamin absorption. The mean concentration of sodium in bone was significantly lower in females fed 2% salatrim 234CA, and the mean concentration of zinc in bone was significantly higher in males fed 2% salatrim 234CA and in females fed 10% of either salatrim fat when compared with controls.

Reduced ratios of liver- and brain-to-body weight in females fed 10% salatrim 234CS were considered by the authors to be related to increased terminal body weights. Macroscopically, no treatment-related effects were observed in salatrim-treated rats, but an increased incidence of renal mineralization was seen in females fed 5% or 10% salatrim 234CA and a slightly higher incidence of renal mineralization in females fed 5 or 10% salatrim 234CS compared with controls. Except in the groups treated with 10% salatrim 234CA, the renal mineralization was similar in appearance in all groups. In females fed 10% salatrim 234CA, the severity of renal mineralization was slightly greater than in other groups. The authors concluded that the changes in the levels of minerals in bone and renal mineralization were directly related to the quantity of unsaturated fatty acids in the diet fed to treated animals. A number of animals fed 10% corn oil had hepatocellular vacuolation. The Committee concluded that these two salatrim products did not induce any toxicologically significant effects. The concentration of 10% salatrim 234CS or 234CA was equal to 7 g/kg bw per day (Williams, 1992c; Hayes et al., 1994g).

Minipigs

(i) Salatrim 23SO

Diets containing 0, 3, 6 or 10% salatrim 23SO or 10% corn oil were fed for 28 days to groups of four Hanford minipigs, 3.5–7 months old and weighing 17–30 kg at initiation of treatment. A control group was fed the basal diet. The group fed 10% corn oil served as a reference for those fed the high fat content salatrim diets. All diets (including the diet used for the untreated control group) were supplemented with 2% (w/w) corn oil, considered necessary by the authors to avoid possible induction of essential fatty acid deficiency due to dilution with the test fat. Each pig was given 500 g of the appropriate diet twice a day. The test diets were prepared every 2 weeks and stored frozen (–20 ± 10 °C) until use. After removal from the freezer, the diets were maintained at room temperature for 1–6 days (average, 3.3 days) before being fed to the pigs. Evidence of significant degradation of the test diet was reported in one stability trial in which samples of the 3% and 10% test diets were stored frozen for 63 days. The authors considered that the storage conditions used in the study would have resulted in minimal degradation (approximately 7%) of the test diets. The results of additional stability trials reported in the original unpublished account of the study support the view that limited degradation occurred during 2 weeks of frozen storage. However, the data suggest that degradation may have occurred during storage at room temperature and that the stability may have differed significantly between batches of diet. Thus, it is difficult from the available information to estimate the precise doses given to the minipigs.

Blood was collected from the vena cava of each pig 2 weeks and 3 days before initiation of salatrim feeding and on days 3, 7, 14, 21 and 29 after initiation of feeding. The pigs were fasted overnight before blood collection. Haematological and clinical chemical parameters were determined in these samples. After 28 days of treatment, all pigs were subjected to gross necropsy, and adrenals, brain, kidneys, liver, ovaries, spleen, testes, thymus and thyroid were weighed. The entire femur was removed and stored frozen at –20 ± 10 °C, and dry weight and percentage ash were determined. Each femur was assayed for calcium, phosphorus, strontium and zinc concentrations by inductively coupled plasma spectrometry.

No treatment-related effects were seen during daily physical examinations, and all pigs survived to the scheduled terminal sacrifice. The mean body weights, body-weight gains and feed consumption of pigs receiving salatrim 23SO and corn oil were comparable to those of untreated controls. Haematological and clinical chemical evaluation revealed no treatment-related effects. Fairly wide variations in ALT and AST activity in serum were seen in individual animals: 2 weeks before initiation of treatment, the activity of AST was significantly higher in males in the group destined to be fed 10% salatrim, and the activity of ALT was lower in males both before and 7 days after feeding 6% salatrim. This variation was unexplained but could have been random. The mean serum cholesterol concentration at day 3 in females given 6% salatrim 23SO was lower than that of untreated controls, and the serum concentrations of low-density lipoprotein at day 3 in females given 6 or 10% salatrim 23SO or 10% corn oil were lower than those of untreated controls. At day 29, the mean serum cholesterol and high-density lipoprotein cholesterol concentrations were higher in female pigs fed 10% corn oil than in untreated controls. No biologically significant findings were reported with respect to serum and liver vitamin A and E concentrations. No differences in percentage ash or bone concentrations of calcium, phosphorus or strontium were found between treated and untreated control pigs of either sex.

No differences in organ weights were found between pigs fed salatrim 23SO and untreated controls. Macroscopically, no treatment-related effects were observed in any animals treated with either salatrim or corn oil. Microscopically, a slight increase in the severity of focal vacuolation in hepatocytes was found in one male fed 10% corn oil and one male fed 10% salatrim. The authors considered this to be either a spurious finding, since it occurred in only two males, or a non-specific effect of fat, as it occurred only in pigs given 10% fat diets (corn oil or salatrim). The Committee concluded that 10% salatrim 23SO produced neither toxicologically nor nutritionally significant effects. The concentration of 10% salatrim 23SO was equal to 3.3 g/kg bw per day (Kiorpes, 1993b; Hayes et al., 1994e).

2.2.3 Genotoxicity

The results of tests on the genotoxicity of salatrim fats are given in Table 3.

Table 3. Results of tests for genotoxicity with salatrim fats

End-point	Test object	Material/Dose	Result	Reference
Reverse mutation	*S. typhimurium* TA98, TA100, TA1535, TA1537 TA1538 (preincubation assay)	Salatrim 4CA, 23CA, 32CA, 23SO, 234CA, 234CS 0–1000 μg/plate ± S9 (4%, 10%)[a]	Negative	Hayes & Riccio (1994)
Gene mutation 6-thioquanine resistance	Chinese hamster ovary cells	Salatrim 23CA 31–1000 μg/ml ± S9[a,b]	Negative	Hayes et al. (1994f)
Chromosomal aberration in vitro	Chinese hamster ovary cells, metaphase analysis	Salatrim 23CA 0–1000 μg/ml ± S9[a]	Negative	Hayes et al. (1994f)
Unscheduled DNA synthesis in vitro	Hepatocytes	5–1000 g/ml ± S9[a]	Negative	Hayes et al. (1994f)
Micronucleus formation in vivo	Rats fed diet containing 10% salatrim for 13 weeks	Salatrim 234CS and 234CA, 7–8 g/kg bw per day)	Negative	Hayes et al. (1994f)

[a] Maximum dose restricted by precipitation at 1000 μg/ml (or 1000 μg/plate).
[b] No evidence of cytotoxicity

2.2.4 Special studies: Effects on gut microflora

Diets containing 10% salatrim 23CA or 32CA or 10% corn oil were fed to groups of 20 Crl:CD BR VAF rats of each sex for 13 weeks. At necropsy after a 24-h fast, the caecum was ligated at the distal tip, proximal to the ileocecal junction and distal to the exit into the colon. The caecal tip was removed for histological examination, and the remainder of the caecum was removed and frozen at –20 °C until use. After thawing, the contents of each caecum were thoroughly mixed by kneading within the caecum and then removed, and a portion of the caecal contents from five male rats in each dietary group were examined by scanning electron microscopy for changes in the dominant bacterial morphotypes. The caecal contents from all animals were analysed for pH, bile acids, neutral sterols (cholesterol and its secondary metabolite coprostanol) and phytosterols, and measurements were made of the primary bile acids cholic and α- and β-muricholic acids; the secondary bile acids deoxycholic, lithocholic, hyodeoxycholic, ω-muricholic and unsaturated ω-muricholic acids; the primary phytosterols 24β-ethylcholesterol and 24 β -methylcholestero; and the secondary phytosterol metabolites 24α-methylcoprostanol (campestanol), 24β-methylcoprostanol, 24α-ethylcoprostanol (stigmastanol) and 24β-ethylcoprostanol (sitostanol).

The authors noted wide inter-animal variation in the concentrations of bile acids. No significant differences in caecal pH or in the level of secondary bile acids as a percentage of total bile acids were reported. Increased coprostanol concentrations were found in male rats fed salatrim 23CA, 32CA and corn oil, but with no effect on the ratio of coprostanol to cholesterol. Increased concentrations of all four secondary phytosterols were found in rats fed corn oil, whereas the level of only one (24α-methylcoprostanol) was increased in rats fed salatrim 32CA. In general, salatrim-fed rats of each sex produced slightly less of the three remaining secondary phytosterols than chow-fed rats, while rats fed corn oil produced more. No evidence of any alteration in the population of bacterial morphotypes was reported, although the authors considered that the inter-animal variation limited the sensitivity of scanning electron micrography to the detection of major changes (Scheinbach et al., 1994).

2.3 Observations in humans

2.3.1 Clinic-based studies

Four clinical studies of controlled diets that were reviewed by the Committee at its forty-ninth meeting were:

- an acute test for tolerance with a double-blind cross-over design and doses of 45 g and 60 g salatrim 23CA;
- a 7-day test with a double-blind protocol and doses of 45 g and 60 g salatrim 23CA;
- a four-way triple cross-over test with 30 g or 60 g of salatrim 23SO, hydrogenated soya bean oil as the control and hydrogenated coconut oil as the wash-out vehicle, with a Latin-square treatment sequence; and
- an acute bolus dose of salatrim 23SO to measure the effects on serum ketones.

(i) Acute tolerance

A randomized, double-blind, cross-over design was used, in which volunteers (six men and four women) aged 18–65 (mean, 38 years) received salatrim 23CA or coconut oil for 1 day, at either 60 g/day (for eight individuals consuming a 2500-kcal diet) or 45 g/day (for two, both female, consuming a 1800-kcal diet). The materials were introduced into the diet in the form of vanilla sandwich biscuits and chocolate bonbons (or bars), each containing 5 g of either salatrim or control fat. On day 4, five volunteers received the test material, while five received the control material. The substitution of test material was reversed on day 8. Thus, there were two treatment groups in this study, one on day 4 (group 1) and one on day 8 (group 2), for whom changes in clinical parameters on days 5 and 9, respectively, would suggest a treatment-related effect. A maintenance diet (either 1800 or 2500 kcal/day) including the control material was administered on all other study days. A standardized 4-day meal plan was repeated for three cycles.

The results from the full, unpublished report of this study are summarized here. Salatrim 23CA increased the mean activities of serum lactate dehydrogenase and γ-GGT in both treated groups and increased the mean activities of serum ALT and AST and alkaline phosphatase in group 2. The changes were modest, and the group means did not exceed the reference range. A slight increase in mean serum cholesterol concentration was reported at the end of the study in both treated groups. The authors of the published report considered that the small size of the group did not permit a conclusion about the palatability of the salatrim foods; however, the unpublished report stated that the volunteers rated the salatrim-containing foods lower than identical control food carriers. Miild adverse gastrointestinal symptoms such as flatulence, nausea and diarrhoea were reported by a number of volunteers. In the published report, these symptoms were not related to consumption of salatrim 23CA when the data were analysed by time of onset (Besselaar Clinical Research Unit, 1993a,b; Finley et al., 1994a).

(ii) 7-day test

In a randomized, double-blind design, 36 volunteers (19 men and 17 women) aged 18–65 (mean, 33 years) received either salatrim 23CA or coconut oil for 7 days, at 60 g/day (for those on a 2500-kcal diet) or 45 g/day (for those on a 1800-kcal diet). The materials were introduced into the diet in the form of biscuits, bonbons (or bars) and chocolate ice cream. All volunteers received a maintenance diet (1800 or 2500 kcal/day) containing the control material on days 1–7. On days 8–14, 12 men received the test material at 60 g/day and six women received it at 45 g/day; nine men and nine women continued on the maintenance diet with food carriers containing control fat. One woman on the test material withdrew for reasons unrelated to the test on day 10. On days 15–24, all the volunteers were returned to the maintenance diet. A standardized 7-day meal plan was followed for three cycles throughout the study.

Except where stated, the results summarized below were taken from the published report. Increases in mean serum ALT (19%) and AST (< 41%) activity over that seen before intake of salatrim 23CA were recorded during treatment: three persons had ALT values above the normal maximum of 35 mU/ml, and one had a value for AST above the normal maximum of 50 mU/ml. Lactate dehydrogenase activity was also increased during intake of salatrim 23CA,although the values remained within the normal range throughout the test. The authors reported that the serum AST and lactate dehydrogenase activities declined steadily to control levels; the unpublished report states that all parameters declined to near baseline levels after withdrawal of salatrim 23CA. Mean corpuscular volume, number of monocytes, serum calcium and carbonate concentrations and GGT activity were all significantly altered in the volunteers receiving salatrim, but all values remained within the normal range, and none of the changes was considered by the authors to be clinically relevant.

A significant increase in total cholesterol was found before the test, which was associated with ingestion of hydrogenated coconut oil. On days 8–14, a significant decrease in total and low-density lipoprotein cholesterol was found in the group receiving salatrim 23CA, whereas the values for the control group remained elevated. No significant changes in urinary parameters were reported. Large increases in the faecal excretion of fats and stearic acid were seen when salatrim 23CA was added to the diet.

The group fed salatrim 23CA reported more headaches and gastrointestinal symptoms during the test period, nausea, abdominal pain and headaches being the most frequent symptoms described. The authors considered that these effects were mild, none leading to refusal to continue the study or requiring clinical intervention. The unpublished report showed that 14/17 volunteers (eight men and six women) given salatrim 23CA (78%; 56% of controls) reported one or more adverse effect, regardless of study material. Flatulence and nausea were reported by 61% (22% of controls) and 67% (17% of controls) of volunteers given salatrim 23CA, respectively, and headache by 56% (11% of controls). The adverse gastrointestinal tract symptoms were moderate in 8/17 individuals given salatrim 23CA and mild in the remainder. None of the reported symptoms was considered to be severe. The unpublished report also suggested that the salatrim carrier foods were considered by the subjects to be of poorer palatability than the control carrier foods, and they may have been able to distinguish between the food products used in the trial. Hence, the study may have been biased (Besselaar Clinical Research Unit, 1993c,d; Finley et al., 1994b).

(iii) 4-day triple cross-over test

Volunteers received chocolate raisin crisp bars and hot chocolate drink prepared with salatrim 23SO at 30 g/day (plus 30 g of control fat) or 60 g/day or 60 g/day of hydrogenated soya bean material (control) for 4 days. Before and after receiving the test or control vehicle, the volunteers were given 60 g of hydrogenated coconut oil for 4 days, which was prepared from the same coconut oil used in the two studies summarized above and served as a wash-out medium between the salatrim and control treatments. A standardized 8-day meal plan was repeated three times throughout the study.

No clinically significant differences were reported between volunteers receiving salatrim 23SO at 30 g/day and controls; however, those given salatrim at 60 g/day had statistically significant increases in mean serum ALT, AST and lactate dehydrogenase activities and decreased mean serum cholesterol concentration. These changes were well within the normal ranges and were considered by the authors to be clinically unimportant. The values returned to pre-test levels when the volunteers were transferred to the coconut oil wash-out diets. The total stool weight was significantly higher in all volunteers given salatrim at 60 g/day than in those given salatrim at 30 g/day or soya bean oil at 60 g/day. The values for stool water, total fat and stearic acid were significantly higher in women and men given salatrim at 60 g/day than in those given 30 g/day or soya bean oil at 60 g/day. More frequent stool softening and abnormal stools were reported by women given the higher concentration of salatrim as compared with the lower concentration and by all volunteers given the higher concentration of salatrim as compared with either the lower concentration or the soya bean oil. A small but statistically significant increase in mean serum β-hydroxybutyrate concentration was reported in volunteers given 60 g of salatrim per day (0.2 ± 0.10 mmol/l, compared with 0.1 ± 0.06 mmol/l in controls). Adverse gastrointestinal effects (abdominal pain, diarrhoea and nausea) were reported by 10 women and 5 men given salatrim at 60 g/day. The authors considered that the lower body weights of the women might have exlained their more frequent complaints. These results support the conclusion that salatrim at < 30 g/day does not cause significant gastrointestinal symptoms. The Committee noted the limited duration of salatrim intake in this study.

The volunteers considered that the hot chocolate drink containing salatrim was acceptable but disliked the other salatrim food carriers. No evidence of carry-over in subjective assessments of food carriers was reported. The taste of the salatrim-containing food carriers may have biased the study (Besselaar Clinical Research Unit, 1993a; Finley et al., 1994b).

(iv) Effects on ketones

A randomized study was conducted with seven groups of six volunteers to determine the effect of a single dose of salatrim 23SO, hydrogenated soya bean oil or a medium-chain triglyceride on the serum concentrations of acetate, acetoacetate and hydroxybutyrate (monitored for up to 4 h). All fat samples were delivered in one cup of chocolate-flavoured beverage in the morning after a 10-h fast. The volunteers were assigned randomly (with equal distribution by age and sex) to salatrim 23SO at 7.5, 10, 12.5 or 15 g; control hydrogenated soya bean oil at 7.5 or 15 g; or medium-chain triglyceride at 15 g. A slight increase in serum acetate concentration was seen in volunteers receiving 15 g of salatrim, but no increase in serum ketone concentration was seen at any concentration of salatrim. As expected, slight increases in acetoacetate and β-hydroxybutyrate concentration were observed in volunteers receiving medium-chain triglyceride. Adverse gastrointestinal effects were reported by five individuals: one each given 10 g, 12.5 g and 15 g salatrim and one each given medium-chain triglyceride and soya bean oil. The authors considered that salatrim was not ketogenic at the concentrations used (Besselaar Clinical Research Unit, 1993b; Finley et al., 1994b).

2.3.2 Home-based studies

The Committee at its forty-ninth meeting evaluated a randomized, double-blind, multiple-dose, parallel comparison of fat replacement by salatrim 23SO, 4SO or 43SO with a control soya bean oil. Four additional clinical studies were conducted subsequently, comprising:

- a study in hypercholesterolaemic individuals given diets containing margarines enriched with either salatrim or palm oil, including an initial 2-week period on a low-fat diet and 5 weeks on each high-fat diet with a cross-over design;

- a randomized cross-over trial to compare the effect of a single experimental meal containing 30 g of salatrim 23SO, cocoa butter or high-oleic sunflower oil;
- a study of five men given 30 g of salatrim 4SO from whom blood samples were taken every 20 min for 6 h for measurement of plasma cholesterol, triglycerides, clotting parameters and butyric acid; and
- a study in which eight normolipidaemic volunteers were fed a single meal supplemented with either 30 g of salatrim 23SO or cocoa butter in a cross-over design and five men were given a meal containing salatrim 4SO.

(i) Gastrointestinal effects and liver enzymes

Fat replacement by salatrim 23SO, 4SO or 43SO oils was compared in a randomized, double-blind, multiple-dose study, with that by soya oil. Groups of at least 12 female and 12 male volunteers aged 19–63 (mean, 35 years) were given diets containing 30, 45 or 60 g of 23SO oil, 60 g of 4SO oil or 60 g 43SO oil for 6 weeks; two control groups were included to account for the anticipated diversity in clinical values for a typical population. The total fat intake of volunteers in all groups was 60 g/day. All persons received the control fat (soya oil) during weeks 1 and 6, and either control or test fat as assigned per group in weeks 2–5. The food products were changed weekly on a 2-week cycle to ensure variety. The volunteers were given food products for consumption each week, five of which were to be integrated into the daily diet: four contained 15 g of control or test oil, and one product (crackers or cornflakes) did not deliver test or control oil. The volunteers were free to consume a normal diet in addition to the food provided, the only restriction being that the amount of alcohol consumed be limited to no more than two glasses of wine or two servings of beer per day.

After screening, subject selection and initial check-up, which included investigations of drug use and pregnancy (day 0), the volunteers received products and diaries for reporting food consumption and health over the next 7 days. Body weight was also recorded, and blood was drawn for analysis. The volunteers returned after a 10-h fast every 7 days thereafter (days 8, 15, 22, 28 and 36) to receive food products for the next week, to return the diaries, to be weighed and to have blood drawn. On the final day of the study (day 43), the volunteers returned to the clinic to hand in the diaries, be weighed and have final samples of blood drawn. Daily health was assessed for a number of general categories.

The daily diaries were used to record the type and amount and the time of day on which all foods and beverages were consumed, to rate the palatability of the foods provided and to record side-effects and the quality of daily life. The food carriers used were ice cream, chocolate milk, pudding and yoghurt produced by conventional processes. Cinnamon raisin muffins, chocolate cake, lemon cake and waffles were prepared before the study and held frozen until they were dispensed to the volunteers. Chocolate milk was prepared in multiple batches every 2 weeks as needed. All other products were produced in a single lot before initiation of the study. The formulations of the products were adjusted so that each individually packaged serving delivered 15 g of control oil or salatrim in each unit. The food carriers were rotated on a 2-week basis:

- weeks 1, 3, 5: muffins, chocolate cake, ice-cream, yoghurt;
- weeks 2, 4, 6: waffles, lemon cake, chocolate milk/pudding.

Control diets were administered in weeks 1 and 6.

Of the 183 volunteers who started, 149 completed the study. The results of clinical assessments were reported only for those who completed the study. The Committee noted differences between the published and unpublished reports in the numbers of individuals in each group who dropped out. The data given here are derived from the unpublished clinical report prepared by the study authors. A total of 34 individuals dropped out of the study: 1/26 (control 1), 3/27 (control 2), 4/27 (30 g 23SO), 5/27 (45 g 23SO), 2/25 (60 g 23SO), 7/26 (60 g 4SO) and 12/25 (60 g 43SO). After exclusion of the controls who dropped out of the study, 20 of the 31 volunteers who dropped out and who had received salatrim in their diet eported that adverse effects due to the test material had been the reason for leaving the study. The authors reported that the salatrim food products were well tolerated.

Transient increases in mean serum AST and ALT activity over time were observed in controls and all treated groups; however, the magnitude of the increases in ALT and AST activity over that at day 8 was greater in the groups given 60 g of salatrim than in the control groups. The authors reported that the ALT and AST activities in all groups by the end of the 4-week treatment approached values equivalent to those recorded on day 8. There was no clear evidence of a reversal in enzyme activity, and appropriate statistical tests would be required to evaluate the data further. None of the group means exceeded the normal clinical limit for AST or ALT. A small, transient reduction in mean serum cholesterol concentration was recorded in the unpublished report among volunteers given salatrim 23SO or 43SO at 60 g/day.

A substantial number of volunteers who consumed 60 g/day reported stomach cramps and nausea. The authors calculated the percentage of days during treatment on which adverse effects were reported [number of individuals with adverse effect × number of days with effect / 28 × total number of individuals]. Among those at 60 g of salatrim per day, stomach cramps and nausea were reported during approximately 17–21% and 19–25%, respectively, of the 28-day treatment period, depending on which salatrim product was being evaluated. Among volunteers at 45 g/day and 30 g/day, stomach cramps and nausea were reported during approximately 8–9% and 10–11%, respectively, of the period. A similar analysis of other symptoms was not presented in the published report. The authors considered that subjects given 30 g/day salatrim 23SO did not report nausea that impaired daily function. Three volunteers out of 23 who completed the study and who were given salatrim 23SO at 30 g/day experienced stomach cramps or nausea for at least 10 days of the trial compared with one volunteer out of 49 who completed the study in the combined control groups (Harris Laboratories, 1993; Sourby, 1993; Finley et al., 1994c).

(ii) Effects of stearic acid-rich and palmitic acid-rich triacylglycerols on plasma lipid concentrations

In a study submitted for evaluation by the Committee at its present meeting, 20 middle-aged volunteers (12 men and 8 women) who had had a high plasma cholesterol concentration during the preceding year were invited to participate in a study to compare the effects on plasma lipids of diets containing margarines enriched with either salatrim or palm oil (rich in palmitic acid). The study included an initial 2-week period on a low-fat diet and 5 weeks on each high-fat diet, with a cross-over design. The average target intake of lipid was 30 g/day, but the actual intake was not specified. Weighed food records were maintained for 3 days in each period. Five volunteers failed to keep adequate records and were therefore excluded from the final analysis. The subjects' plasma total, high-density and low-density lipopotein cholesterol and triacylglycerol concentrations did not differ significantly with the salatrim and palm oil diets. None of the volunteers reported adverse symptoms such as bowel irritation, although procedures for assessing such symptoms were not described in the paper (Nestel et al., 1998).

Although the above study was designed to provide information on blood lipid concentrations, additional analyses of serum enzymes were performed and reported to the Committee on Toxicity of Chemicals in Food, Consumer Products and the Environment (1999) in the United Kingdom. The data were confidential and have not been published. It was concluded that consumption of high doses of salatrim had resulted in slight increases in AST and ALT activities in serum but that the increases were within the normal reference range.

(iii) Effect of a single intake of 23SO oil, cocoa butter or high-oleic sunflower oil on serum triacylglycerols

Thirty-five healthy adults (17 men and 18 women) participated in a randomized cross-over trial to compare the effects of a single intake of experimental meals containing 30 g of 23SO salatrim, cocoa butter or high-oleic sunflower oil. The volunteers were allocated randomly to six treatment sequences, the three test meals being consumed 1 week apart over 3 weeks, and they were asked to avoid foods with a high fat content the day before and to fast overnight before each test meal. Blood samples were taken immediately before and 3 and 6 h after the test meal. A smaller post-prandial increase in serum triacylglycerol was found after the salatrim meal than after intake of oleate or cocoa butter meals. Possible side-effects were not investigated (Sanders et al., 2001).

(iv) Effect of 30 g of 4SO oil on plasma cholesterol, triglycerides, clotting parameters and butyric acid

In an unpublished study, five men consumed 30 g of 4SO salatrim (equal to 9 g of butyric acid) in cupcakes, with a fat-filled milkshake to balance the fatty acid intake. Blood samples were taken every 20 min for 6 h for measurement of plasma cholesterol, triglycerides, clotting parameters and butyric acid. The authors reported that the post-prandial lipaemia was approximately 50% lower than expected from normal fat under these conditions, but there were no concurrent controls. The volunteers reported no gastrointestinal complaints or marked feelings of satiation.

The Advisory Committee on Novel Foods and Processes in the United Kingdom expressed concern that ingestion of butyrate-containing salatrim might increase the butyrate content of the blood, as certain short-chain fatty acids have teratogenic potential in vitro (Coakley et al., 1986; Auerbach & Frier, 1999). In a study conducted in response to that concern, free butyrate was not detected in the plasma of these men (detection limit, 1 μmol/l) at any time. It was concluded that ingestion of salatrim containing butyrate would have no effect on serum butyrate concentrations. Although the Committee would have preferred that a positive control had been included in which butyrate was detected in plasma, it agreed that butyrate from salatrim is highly unlikely to have teratogenic effects in humans with normal consumption (Pronczuk et al., 2001).

(v) Response to a single breakfast meal

Another unpublished paper reported two studies involving four male and four female volunteers with normolipidaemia, who were fed a single breakfast meal supplemented with either 30 g of 23SO salatrim or cocoa butter in a cross-over design. The second study addressed the response of five men given a 4SO salatrim meal alone. In the first study, the women complained of feeling 'bloated, cramped or even nauseated' 2–4 h after the meal, with recovery by 7 h. One also complained of general gastric discomfort after the cocoa butter. The men did not experience discomfort but noted that they felt satiated and averse to eating until the early evening after the salatrim breakfast. The post-prandial lipaemia was again about 50% lower after the salatrim meal. Less satiation was felt in the second study. There was no concurrent control group, but the post-prandial lipaemia was reported to be lower than that in the study with salatrim 23SO (Hayes et al., 2001).

3. INTAKE

The data on projected intake of salatrim submitted to the Food and Drug Administration of the USA for affirmation of 'generally recognized as safe' status were made available to the Committee at its forty-ninth meeting: consumption at the 90th percentile for all ages and for 3–5-year-olds was reported to be 37 and 26 g/day, respectively.

3.1 Assessments of intake of salatrim

The data submitted to the Food and Drug Administration were resubmitted with small revisions and complemented with projections of intake based on consumption data for the United Kingdom. In the USA, salatrim is used as a fat replacer on a weight-equivalent basis within good manufacturing practice in confectionery, fine bakery wares, dairy products, salty snacks, chips, nuts, margarine and fat-based spreads (Food and Drug Administration, 1994). The level of use ranges from 1–2 g/100 g in some dairy products and crackers to > 80 g/100 g in margarine. Use of salatrim to replace current fats would not introduce new fatty acids but would result in a different profile of fatty acids, with more saturated acids (mainly stearic and butyric). Data on the estimated intake of salatrim, stearic acid and total saturated fatty acids are thus presented (Table 4).

Substitution of all added fats with salatrim (worst case) would thus lead to a twofold increase in saturated fats and a threefold increase in fats with 18 carbons. In the USA, the highest intake of salatrim would be that of adolescents aged 13–17 (30 g/day at the 90th percentile) and that of children aged < 2 years on a grams per kilogram body weight basis (1.5 g/kg bw at the 90th percentile) (Finley, 1993).

The main contributors to the mean estimated intake of salatrim according to the simulations for the United Kingdom (Tennants & Douglass, 1999) and the USA (Finley, 1992) would be table margarine, followed by buns and pastries. In the USA, chocolate and confections were the individual food categories with the greatest single intake at the 90th percentile, on the basis of both grams per day and grams per kilogram

Table 4. Estimate of intake of salatrim and of fatty acids on basis of individual surveys

Country	Survey (year)	Assumptions group	Variable/population	Intake (g/kg) Salatrim	Total saturated fats	Stearic acid (1)
United Kingdom	National Dietary and Nutrition Survey, 7-day record, individual data, 16–65 years, 2000 persons (1986–87)[a]	No substitution	Mean %ile	0	36	9.2
			97.5th %ile	0	72	18
		Substitution of added fats in specified foods (2)	Mean %ile	11	42	14
			97.5th %ile	33	75	28
		Substitution of all added fats (3)	Mean %ile	46	62	30
			97.5th %ile	88	120	61
USA	National Food Consumption Survey, all ages, 3-day dietary records, all ages (1987–88)[b]	Substitution of all fats in specified foods (7)	Whole population:			
			Mean %ile	12		
			Mean %ile consumers only	13		
			90th %ile consumers only	28		
			12–19 years males:			
			Mean consumers only	18		
			90th %ile consumers only	35		
	National, MRCA frequency of use, 14 days (1989–90) and Department of Agriculture portion size (1987–88), all ages, number of subjects not reported[c]	Substitution of added fats in specified foods (4)	Mean %ile	13		11
			90th %ile	27		
			95th %ile	32		
			99th %ile	45		

Table 4 *(contd)*

Country	Survey (year)	Assumptions	Variable/population group	Intake (g/kg)		
				Salatrim	Total saturated fats	Stearic acid (1)
	American Heart Association / University of Minnesota, 24-h recall, 6198 subjects, year not reported, age not reported[d]	No substitution	Mean %ile	0		8.1
		Model A: Substitution of all fats in specified foods (5)	Mean %ile	Not reported		14
		Model B: Substitution of all fats in specified foods (5) and dairy products (6)	Mean %ile	19		16

[a] Tennants & Douglass (1999)
[b] Finley & Leveille (1993)
[c] Finley (1993)
[d] Finley (1992)

(1) For the United Kingdom, Tennants & Douglass (1999) assumed 58% of stearic acid and 21% of butyric acid in salatrim. For the USA, Finley (1992) assumed 62% and 85% of stearic acid in model A and B, respectively; Finley (1993) assumed 2 mol stearic acid per molecule of glycerol.
(2) Brownies, buns and pastries, chocolate, chocolate confectionery, other confectionery and biscuits
(3) Bakery foods, other cereals, oils and fats, sugars and preserves, beverages, dairy products and nuts
(4) Chocolate sweets, commercial ready-to-eat biscuits and crackers, margarine and spreads, salty snacks and nuts; with the highest likely fat level used in each food group
(5) Biscuits, crackers, chocolate and chocolate-coated confections, margarine and spreads and salty snacks
(6) Cheese, ice cream and related products, yoghurt
(7) Confectionery (chocolate and other), biscuits, crackers, ice cream and milk, cream cheese, cheese spreads and foods, margarine and spreads, salty snacks

body weight. Substitution of fats with salatrim in this food category would lead to intake of 37 g/day of salatrim by adolescent boys and 1.1 g/kg bw by children aged 1–6 years at the 90th percentile (Finley, 1993).

4. COMMENTS

Toxicological data

The Committee noted that the newly submitted study of nutrition in rats did not resolve the issue of the extent to which salatrim might reduce energy intake since, as stated previously, the rat is not a good model for humans with respect to stearic acid absorption.

The aim of the home-based trials was to investigate plasma lipid profiles in small groups of adults ingesting salatrim at 30 g/day for up to 5 weeks. Diets containing salatrim did not affect the blood concentration of total high-density or low-density lipoprotein cholesterol. The post-prandial increase in serum triglyceride concentration was lower after ingestion of a meal containing 30 g of salatrim than after ingestion of a meal containing 30 g of oleate or cocoa butter. In one study, it was reported that some of the volunteers complained of gastric discomfort, nausea or feeling 'bloated' or 'satiated and averse to eating until the early evening' after a breakfast containing 30 g of salatrim. However, adequate controls were not included in this study, and it was not clear if these subjective symptoms were related to ingestion of salatrim. The Committee was informed that, in one of these trials, serum liver enzyme activities were found to be within the normal range; however, these results were not included in the published manuscript.

Overall, the Committee concluded that the new studies did not respond to the questions raised at its forty-ninth meeting and the available information was inadequate for toxicological evaluation. The Committee noted that, in general, toxicological studies in experimental animals should be conducted at doses that allow an adequate margin of safety in relation to anticipated human intake. Such safety margins cannot feasibly be attained with substances, such as salatrim, that are intended for use as major components of food. Additionally the rat is not a good model for humans with respect to stearic acid absorption, and further studies in experimental animals would not contribute to the safety evaluation of salatrim. The Committee therefore paid particular attention to the results of the studies in humans. These studies indicated that adverse effects (gastrointestinal and hepatic effects) might be experienced by adults consuming diets containing salatrim at 60 g/day (equivalent to 1 g/kg bw per day). The available information was not adequate to evaluate whether effects might occur at intakes of less than 60 g/day, nor if the effects differ with the composition of salatrim.

Intake

Data on the projected intake of salatrim were assessed by the Committee at its forty-ninth meeting. These data, with small revisions, and with projections of intake based on data on consumption in the United Kingdom, were submitted to the present meeting. Manufacturers' use levels for salatrim vary from 1–2 g/100 g in some dairy products and crackers to more than 80 g/100 g in margarine.

The mean intake of salatrim after hypothetical substitution of added fats by this substance in all food categories would be 45 g/day, but this value is unrealistic. The mean intake of salatrim based on hypothetical substitution of fats in specified food categories would vary from 11 to 13 g/day. Intake by adolescent consumers at the 90th percentile would reach 35 g/day. On a gram per kilogram body-weight basis, the highest intake would be that of children aged less than 2 years, as those at the 90th percentile of consumption would have an intake of 1.5 g/kg bw per day. Children aged 1–6 years who are consumers of the category 'chocolates and confectionery' at the 90th percentile would have an intake of salatrim from this food category alone of 1.1 g/kg bw per day, which is higher than the intake reported to have effects in adults in clinical studies.

5. EVALUATION

The Committee concluded that the available data did not provide an adequate basis for evaluating the safety of salatrim and recommended that additional studies be performed in humans to assess fully the safety of salatrim covering the range of compositions intended for use. The studies should be designed appropriately for safety evaluation and should include investigation of the gastrointestinal and hepatic effects observed in previous studies.

6. REFERENCES

Auerbach, M.H. & Frier, H.I. (1999) Analysis of the teratogenic potential of propionate and butyrate vs likely levels of short chain organic acid levels in sytemic circulation arising from consumption of salatrim. Manuscript provided by Danisco Cultor USA, Inc., Ardsley, New York, USA.

Besselaar Clinical Research Unit (1993a) Randomised, 3-way crossover, double blind tolerance study of fat replacement compound TAG A9300 versus soybean oil administered to non-sedentary subjects by substituting 30 g/day or 60 g/day at 1800 or 2500 kcal/day diets. Unpublished report No. 8024 from Besselaar Clinical Research Unit. Submitted to WHO by Cultor Food Science, Ardsley, New York, USA).

Besselaar Clinical Research Unit (1993b) Randomized double blind study to determine the effects of a single administration of fat replacement salatrim 23SO at four different levels compared to three levels of control fat compounds on serum acetotacetate, acetate, β-hydroxybutyrate and glucose. Unpublished report No. 8165 from Besselaar Clinical Research Unit. Submitted to WHO by Cultor Food Science, Ardsley, New York, USA.

Besselaar Clinical Research Unit (1993c) Statistical report. Randomized-double blind safety and tolerance study of fat replacement compound TAG 8200 versus coconut oil administered at 60 g (2500 kcal) and 45 g (1800 kcal) utilising an acute (1-day) crossover phase and a chronic (7-day) non-cross-over phase. Protocol GHBA-105, 28 June 1993. Submitted to WHO by Cultor Food Science, Ardsley, New York, USA.

Besselaar Clinical Research Unit (1993d) Clinical report. Randomized-double blind safety and tolerance study of fat replacement compound TAG 8200 versus coconut oil administered at 60 g (2500 kcal) and 45 g (1800 kcal) utilising an acute (1-day) crossover phase and a chronic (7-day) non-cross-over phase. Unpublished report No. 6270-164 from GHBA/Hazelton, dated 27 October 1993. Submitted to WHO by Cultor Food Science, Ardsley, New York, USA.

Coakley, M.E., Rawlings, S.J., & Brown, N.A. (1986) Short-chain carboxylic acids, a new class of teratogens: Studies of potential biochemical mechanisms. *Environ. Health Perspectives*, **70**, 105–111.

Committee on Toxicity of Chemicals in Food, Consumer Products and the Environment (1999) Statement for the Advisory Committee on Novel Foods and Processes on toxicological aspects of a submission on short and long chain triacylglycerol molecules (salatrims)—A family of low calorie fats submitted for approval under the EC Novel Food Regulation. Department of Health, London, England.

Danisco Cultor USA, Inc. (2001a) Salatrim, 29 November 2001. Submitted to WHO by Danisco Cultor USA, Inc., Ardsley, New York, USA.

Danisco Cultor USA, Inc. (2001a) Salatrim, attachment B, 30 November 2001. Submitted to WHO by Danisco Cultor USA, Inc., Ardsley, New York, USA..

Finley, J.W. (1992) Estimation of dietary exposure (of) stearic acid from salatrim. Unpublished report from Nabisco Foods Group, East Hanover, New Jersey, USA. Submitted to WHO by Danisco Cultor USA, Ardsley, New York, USA.

Finley, J.W. (1993) The potential exposure of salatrim in the diet. Unpublished report from Nabisco Foods Group, East Hanover, New Jersey, USA. Submitted to WHO by Danisco Cultor USA, Ardsley, New York, USA.

Finley, J.W. & Leveille, G.A. (1993) Exposure assessment of various foods and food categories based on the 1987–88 USDA NFCS database. Unpublished report from Nabisco Foods Group, East Hanover, New Jersey, USA. Submitted to WHO by Danisco Cultor USA, Ardsley, New York, USA.

Finley, J.W., Leveille, G.A., Dixon, R.M., Walchak, C.G., Sourby, J.C., Smith, R.E., Francis, K.D. & Otterburn, M.S. (1994a) Clinical assessment of salatrim, a reduced-calory triacylglycerol. *J. Agric. Food Chem.*, **42**, 581–596.

Finley, J.W., Klemann, L.P., Leveille, G.A., Otterburn, M.S. & Walchak, C.G. (1994b) Caloric availability of salatrim in rats and humans. *J. Agric. Food Chem.*, **42**, 495–499.

Finley, J.W., Walchak, C.G., Sourby, J.C., & Leveille, G.A. (1994c) Clinical study of the effects of exposure of various salatrim preparations to subjects in a free-living environment. *J. Agric. Food Chem.*, **42**, 597–604.

Finley, J.W., Leveille, G.A., Klemann, L.P., Sourby, J.C., Ayres, P.H. & Appleton, S. (1994d) Growth method for estimating the caloric availability of fats and oils. *J. Agric. Food Chem.*, **42**, 489–494.

Food and Drug Administration (1994) GRAS Affirmation Petition GRP 4G0404, Washington DC, USA, Government Printing Office.

Harris Laboratories (1993) Clinical report. Randomized, crossover, double blind tolerance study of salatrim fat replacements. A 7 day pre-exposure to control soybean oil followed by a 28 day exposure to salatrim oils or control soybean oil followed by a 7 day exposure to control soybean oil (Project No. 15472 7). Unpublished report from Harris Clinical Laboratories, Lincoln, Nebraska, USA. Submitted to WHO by Cultor Food Science, Ardsley, New York, USA.

Hayes, J.R. & Riccio, E.S. (1994) Genetic toxicology studies of salatrim structured triacylglycerols. 1. Lack of mutagenicity in Salmonella/microsome reverse mutation assay. *J. Agric. Food Chem.*, **42**, 515–520.

Hayes, J.R., Pence, D.H., Scheinbach, S., D'Amelia, R.P., Klemann, L.P., Wilson, N.H. & Finley, J.W. (1994a) Review of triacylglycerol digestion, absorption, and metabolism with respect to salatrim triacylglycerols. *J. Agric. Food Chem.*, **42**, 474–483.

Hayes, J.R., Finley, J.W. & Leveille, G.A. (1994b) In vivo metabolism of salatrim fats in the rat. *J. Agric. Food Chem.*, **42**, 500–514.

Hayes, J.R., Wilson, N.H., Pence, D.H. & Williams, K.D. (1994c) Subchronic toxicity studies of salatrim structured triacylglycerols in rats: 2. Triacylglycerols composed of stearate, acetate, and propionate. *J. Agric. Food Chem.,* **42**, 539–551.

Hayes, J.R., Wilson, N.H., Pence, D.H. & Williams, K.D. (1994d) Subchronic toxicity studies of salatrim structured triacylglycerols in rats: 1. Triacylglycerols composed of stearate and butyrate. *J. Agric. Food Chem.*, **42**, 528–538.

Hayes, J.R., Wilson, N.H., Roblin, M.C., Mann, P.C. & Kiorpes, A.L. (1994e) 28-day continuous dosing study in minipigs with a salatrim structured triacylglycerol composed of stearate, acetate, and propionate. *J. Agric. Food Chem.*, **42**, 563–571.

Hayes, J.R., Rudd, C.J., Mirsalis, J.C., Bakke, J.P., Winegar, R.A. & Murli, H. (1994f) Genetic toxicology studies of salatrim structured triacylglycerols. 2. Lack of genetic damage in *in vitro* mammalian cell assays and the *in vivo* micronucleus assay. *J. Agric. Food Chem.*, **42**, 521–527.

Hayes, J.R., Wilson, N.H., Pence, D.H. & Williams, K.D. (1994g) Subchronic toxicity studies of salatrim structured triacylglycerols in rats: 3. Triacylglycerols composed of stearate, acetate, propionate, and butyrate. *J. Agric. Food* Chem., **42**, 552–562.

Hayes, K.C., Lipinski, B., Pronczuk, A. & Nestel, P. (2001) Structured trigyceride substitute for cocoa butter (salatrim) reduces postprandial lipemia and improves the acute, but not chronic, hemostatic profile in humans. Manuscript provided by Danisco Cultor Inc., Ardsley, New York, USA.

Kiorpes, A.L. (1993a) Final report: Exploratory study with A9300 in rats. Unpublished report No. HWI 6270-172 from Hazelton, Wisconsin, USA. Submitted to WHO by Cultor Food Science, Ardsley, New York, USA.

Kiorpes, A.L. (1993b) Final report: 28 day dietary toxicity study with A9300 in Hanford minipigs. Unpublished report No. HWI 6270-174 from Hazelton, Wisconsin, USA. Submitted to WHO by Cultor Food Science, Ardsley, New York, USA.

Musick, T. & Peterson, M. (1993) Tissue distribution and excretion of radiolabel following a single oral dose of ^{14}C-triacylglycerol in rats. Unpublished report No. HW 6270-166 from Hazelton, Wisconsin, USA. Submitted to WHO by Cultor Food Science, Ardsley, New York, USA.

Nestel, P.J., Pomeroy, S., Kay, S., Sasahara, T. & Yamashita, T. (1998) Effect of a stearic acid-rich, structured triacylglycerol on plasma lipid concentrations. *Am. J. Clin. Nutr.*, **68**, 1196–1201.

Phillips, K.D. (1992) In vitro hydrolysis of emulsified TAG A8200 by porcine pancreatic lipase. Unpublished report No. CR001-KMP-003: Final report from Tech Lab VPI Research Park, Blacksburg, Virginia, USA. Submitted to WHO by Cultor Food Science, Ardsley, New York, USA.

Pronczuk, A., Lipinski, B. & Hayes, K.C. (2001) Butyrate-rich salatrim induces minimal lipemia without altering fibrinogen or fibrinolysis in men. Submitted to WHO by Cultor Food Science, Ardsley, New York, USA.

Sanders, T.A.B., Oakley, F.R., Cooper, J.A. & Miller, G.J. (2001) Influence of a stearic acid-rich structured triacylglycerol on postprandial lipemia, factor VII concentrations, and fibrinolytic activity in healthy subjects. *Am. J. Clin. Nutr.*, **73**, 715–721.

Scheinbach, S., Hayes, J.R., Carman, R.J., Zhou, D., Van Tassell, R.L. & Wilkins, T.D. (1994) Effects of structured triacylglycerols containing stearic, acetic, and propionic acids on the intestinal microflora of rats. *J. Agric. Food Chem.*, **42**, 572–580.

Sequeria, A. & Gordon, B.M. (1993) In vitro pancreatic lipase hydrolysis of salatrim; cocoa butter, and triolein: Quantification by HTGC/F1D. Unpublished report from Experimental Pathology Laboratory, Inc., dated 17 March 1993. Submitted to WHO by Cultor Food Science, Ardsley, New York, USA.

Sourby, J.C. (1993) Statistical report. Randomised, crossover, double blind tolerance study of salatrim fat replacements. A 7 day pre-exposure to control soybean oil followed by a 28 day exposure to salatrim oils or control soybean oil followed by a 7 day exposure to control soybean oil. Salatrims under study are 23SO at exposure levels of 30 g, 45 g, and 60 g, salatrim 4SO at 60 g, and salatrim 43SO at 60 g compared to a 60 g exposure level of control soybean oil. Unpublished report from Nabisco Foods, East Hanover, New Jersey, USA .Submitted to WHO by Cultor Food Science, Ardsley, New York, USA.

Tennants, D.R. & Douglass J.S. (1999) Estimation of intakes of fatty acids from the UK diet. Unpublished report from Cultor Food Science. Submitted to WHO by Danisco Cultor USA, Ardsley, New York, USA.

Williams, K.D. (1992a) Final report: 13 week dietary toxicity study with A7200 in rats. Unpublished report No. HW 6270-157 from Hazelton, Wisconsin, USA. Submitted to WHO by Cultor Food Science, Ardsley, New York, USA.

Williams, K.D. (1992b) Final report: 13 week dietary toxicity study with A8200 and A8200 in rats. Unpublished report No. HWI 6270-162 from Hazelton, Wisconsin, USA. Submitted to WHO by Cultor Food Science, Ardsley, New York, USA.
Williams, K.D. (1992c) Final report: 13 week dietary toxicity study with A9100 and A7800 in rats. Unpublished report No. HW 6270-168 from Hazelton, Wisconsin, USA. Submitted to WHO by Cultor Food Science, Ardsley, New York, USA.

APPENDIX 1. CALORIC AVAILABILITY OF SALATRIM TRIGLYCERIDES AND OTHER NUTRITIONAL CONSIDERATIONS

A1. Explanation

Salatrim is a family of structured triglycerides prepared with combinations of short- and long-chain fatty acids, which can serve as a total replacement for conventional fats and oils in many foods. It has been claimed that it provides about half the energy content of the fats and oils it replaces. An evaluation was provided of the caloric availability of salatrim triglycerides (Howlett, 1997), and the supporting materials were reviewed by the Committee at its forty-ninth meeting (Annex 1, reference *131*). At that time, the Committee concluded that the available studies did not provide an adequate basis for evaluating the caloric availability of salatrim, since a major justification of the purportedly lowered caloric availability of salatrim was reduced absorption of stearic acid, and the rat is not a good model for stearic acid absorption. In addition, it considered that it was not possible to assign a single caloric value to salatrim as there are many different compositions. The Committee recommended that appropriate studies be conducted in humans. In response, one additional study of nutrition in rats was performed and is summarized below; furthermore, corrections of typographical and transcription errors in the reports and publications provided for the discussion at the forty-ninth meeting were provided to the Committee at its present meeting (Danisco Cultor USA, Inc., 2001a).

A2. Definitions and chemistry

According to the specifications, there is no specific requirement for the amounts of short-chain fatty acids and stearic acid in the final product, although a request has been made (Auerbach, 2000) to limit salatrim compositions to those with ≤ 70% total saturated long-chain fatty acids. Unless the composition is specified, it is not possible to assign a single caloric value to salatrim.

A3. Determination of caloric value

The lower energy content of salatrim triglycerides, in comparison with conventional fats and oils, is purportedly due to the lower caloric value of short-chain fatty acids and the reduced absorption of stearic acid. The following discussion evaluates the evidence for these two purported effects.

A3.1 Caloric value of short-chain fatty acids

It is claimed that acetic acid, propionic acid and butyric acid contribute fewer kilocalories per gram than longer-chain fatty acids. On the basis of their heats of combustion (Weast, 1992–93), these fatty acids contribute 3.5 kcal/g (acetate), 4.9 kcal/g (propionate) and 5.8 kcal/g (butyrate), while the standard physiological fuel value of lipids is 9 kcal/g. The Committee agreed with that statement, but noted that the fatty acid composition of salatrim has not been designated and therefore no specific caloric value can be attributed to it. Also, since short-chain fatty acids contribute less energy per gram than long-chain fatty acids, it can be predicted that the energy content of a salatrim product would decrease as the short-chain fatty acid:long-chain fatty acid in a salatrim product increased. In fact, in the studies that related the ratio of short- to long-chain fatty acids in salatrim to weight gain in rats, the opposite was seen. Thus, as the amount of short-chain fatty acid in salatrim increased, the energy contribution of that salatrim product increased. A ratio of 0.51 for short-chain:long-chain fatty acid was considered to provide 2.6 kcal/g, whereas a ratio of 1.99 short-chain:long-chain fatty acid was considered to provide 6.4 kcal/g (Klemann et al., 1994). In general, short-chain fatty acids provide fewer kilocalories per gram than long-chain fatty acids; however, the ratio of short-chain: long-chain fatty acids in a triglyceride may also influence its caloric availability, and, unless a salatrim product is specified, an overall energy content cannot be assigned.

A3.2 Caloric value of stearic acid

On the basis of its heat of combustion (Weast, 1992–93), stearic acid should provide 9.5 kcal/g. It has been contended that stearic acid in salatrim contributes much less energy because stearic acid is poorly absorbed (Klemann et al., 1994). Studies on the absorption of stearic acid are described below.

Rats

Reduced stearate absorption has been reported in studies of the absorption: excretion balance in rodents and humans.

Four groups of 10 rats were given a NIH-07 diet supplemented with 10% corn oil (controls) or with 5, 10 or 15% salatrim 23SO. The rats were acclimated to the diets for 5 days, and then faeces were collected for 5 days. Stearic acid was determined in the faeces by method Ce 1-62 of the American Oil Chemists' Society (1990), but the method for determining intake of stearic acid was not described. The rats given the 5% salatrim diet consumed 0.56 g of stearic acid per day and excreted 0.40 g in the faeces, representing absorption of 29%. The rats given the 10% salatrim diet consumed 1.0 g of stearic acid per day and excreted 0.67 g in the faeces, representing absorption of 35%. The rats given the 15% salatrim diet consumed 1.5 g of stearic acid and excreted 1.0 g, for a total absorption of 32% (Finley et al., 1994a). Stearic acid appeared to be poorly absorbed from salatrim 23SO by rats in this assay, but the actual absorption is unknown as the method for determining stearate intake was not described and no account was taken of the microbial contribution to stearate formation from other 18-carbon fatty acids. Furthermore, this study of stearic acid excretion was performed with only one salatrim formulation. Data for various salatrim products strongly suggest that stearic acid absorption depends on the ratio of short-chain:long-chain fatty acids in the product (Klemann et al., 1994).

Radiolabelled salatrim fats that mimic salatrim 23CA were synthesized and purified and designated salatrim APS. [^{14}C]Triolein was used as the reference fat.

Unlabelled salatrim 23CA was used to dilute the radiolabelled salatrim and triolein to the appropriate specific activity. Groups of five male Sprague-Dawley were given radiolabelled salatrim APS or triolein at a single dose of 1.4 g/kg bw by gavage. After dosing, they were housed individually in glass metabolism cages designed for collection of expired carbon dioxide, urine and faeces. One group of rats received a diet containing 10% salatrim 23CA by weight for 2 weeks before administration of the radiolabelled fats to see if pre-feeding the fat influenced its disposition. After a 72-h sample collection period, the rats were killed, and radiolabel was determined in duplicate samples of all tissues, urine, faeces and carbon dioxide. Faecal samples from the rats fed salatrim APS with radiolabelled stearate were analysed for recovery of stearate, and the amounts of radiolabel from triolein (control fat) and from the stearate carbonyl in salatrim in faecal material were plotted over time.

Although the amount of label from stearate was numerically higher than that from oleate in faeces at 12 h, the difference was not statistically significant. As the percentage of the radiolabelled dose recovered from oleate was 38% and that from stearate was 55%, oleate in triolein appeared to have been 62% absorbed, whereas stearate in salatrim 23CA was 45% absorbed. In other comparisons of the calculated caloric value of various salatrim products and other fats, however, an absorption coefficient of 99% was used for oleic acid (Klemann et al., 1994). The advantage of the study with radiolabelled stearate over that of stearate balance without radiolabel, described above and by Finley et al. (1994a), was that excretion of stearate was not overestimated in the first study because microbially derived stearate was not measured. The disadvantage of the study with radiolabel was that it involved only five rats per group instead of 10. However, when an additional five rats per group, differing from the others only in that they were pre-fed salatrim 23CA for 14 days before dosing, were also tested for excretion of labelled stearate, the values were similar to those for animals fed the basal diet or salatrim 23CA, which justifies pooling of the data on excretion. These results suggest that the stearic acid in salatrim 23CA is approximately 45% absorbed in male rats (Hayes et al., 1994).

Humans

A study was conducted without a cross-over design, in which two groups of 18 volunteers consumed products containing hydrogenated coconut oil for 7 days and then received salatrim 23CA at either 45 or 60 g/day, depending on their total caloric need (1800 kcal/day or 2500 kcal/day), or products made with hydrogenated coconut oil. During a final 10-day period, all volunteers received products made with hydrogenated coconut oil. Faecal samples were collected from each person during the last 3 days of each 7-day test period and pooled. Persons consuming 1800 kcal/day ingested 27 g/day of stearic acid (method for arriving at this value not specified) and excreted 7.6 g/day, for a net absorption of 72%. Those consuming 2500 kcal/day ingested 34 g/day of stearic acid and excreted 12 g/day, for a net absorption of 64%. The authors concluded that 64–72% of the stearic acid in salatrim was absorbed, resulting in an apparent caloric availability of 4.7–5.1 kcal/g (Finley et al., 1994a). A recent report (Danisco Cultor USA, Inc., 2001b) indicated slightly different values, of 65–74%, for the absorption of stearic acid from salatrim in the same study. The way in which the amount of stearic acid consumed in the diets was determined was not described. Without an accurate determination of the amount consumed, it is difficult to assess stearic acid absorption. Also, no data on the weight gain of the volunteers were presented. Thus, no conclusions can be drawn about the caloric value of salatrim from this study.

Olubajo et al. (1986) reported that the absorption of stearic acid in a study of 30 men aged 34–61 years was 82–88%. Jones et al. (1985) measured the absorption of ^{13}C-labelled stearic, oleic and linoleic acids in six healthy men and reported absorption coefficients of 78% for stearate, 97% for oleate and 100% for linoleic acid. In an investigation among in-patients in a metabolic ward, Denke & Grundy (1991) fed four different fats (butter, beef tallow, cocoa butter and olive oil) as part of liquid diets. Each diet was fed for 3 weeks. The dietary intakes and faecal excretion rates of three major fatty acids (palmitic, stearic and oleic) were determined and used to estimate their absorption. The highest absorption rate was that of oleic acid (approximately 99%); palmitic acid was absorbed at a rate of 95–97%, whereas the absorption of stearic acid was 90–94%. These authors made an important observation about the method for measuring stearic acid absorption. Their initial analyses were based on a method in which the fatty acid content of a sample was determined by extracting total lipids from a faecal sample and then determining the proportion of the lipid sample that was found to be stearic acid by gas chromatography. They stated, however, that faecal fatty acids could be overestimated by 100% when calculated as the percentage of extractable lipid weight. They then measured the actual mass of stearic acid by adding a known amount of 17:0 to the sample.

Bonanome & Grundy (1989) evaluated the absorption of stearic acid, relative to other fatty acids, in a group of 10 volunteers fed a meal with a high or a relatively low stearic acid content. Plasma chylomicrons were isolated 2,4, 6 and 8 h after ingestion of the meals. The fatty acid patterns of chylomicron lipids were determined and compared with those of the ingested lipids. The percentages of palmitic acid (16:0) and stearic acid, relative to other fatty acids, were only slightly lower in the lipids from chylomicrons than in those in the meal. The authors concluded that the absorption of stearic acid is similar to that of palmitic acid and that both are absorbed almost as well as oleic acid.

Stearic acid absorption in humans may thus be as low as 64–72%, as reported by Finley et al. (1994a), or, more realistically, as high as 90–94% (Bonanome & Grundy, 1989; Denke & Grundy, 1991). The rat is apparently not a good model for humans with respect to stearic acid absorption, since most studies on rats showed lower coefficients of absorption of stearic acid. Danisco Cultor USA, Inc. (2001b) have stated that stearic acid in salatrim is less well absorbed than stearic acid in conventional fats and oils, the proposed mechanism being the rapid hydrolysis of short-chain fatty acid from the triglyceride, leaving a higher melting-point lipid which is less likely to form micelles and more likely to form indigestible calcium or magnesium soaps. However, there are no experimental data to support this hypothesis.

A3.3 Caloric value of salatrim products in an assay for rat growth

The main method by which the caloric value of various salatrim products was determined was a rodent growth assay, the basic protocol of which was described by Finley et al. (1994b). A later modification to this protocol is described below. In the first set of experiments, a basal NIH-07 diet was given to groups of 10 rats, with daily feed consumption restricted to 50% of that of rats fed *ad libitum*. All rats consumed the same amount of basal diet containing different amounts of corn oil or test fat. A regression curve was calculated for the weight gain of the rats fed corn oil (considered to supply 9 kcal/g), and the caloric value of salatrim was calculated from a comparison of the weight gain of the animals fed salatrim and those fed corn oil.

The rationale behind use of this assay was that rats are an appropriate model for humans and that balance studies were not conducted because they are 'cumbersome and do not lend themselves to evaluating large numbers of materials'; studies with radiolabelled materials were not conducted because they are 'expensive and time-consuming and are not practical as a routine screening tool' (Finley et al., 1994b). The rodent growth assay has, however, a number of limitations for determining the caloric content of salatrim products for humans. First and most important, no evidence was provided that humans receiving equivalent amounts of salatrim substituted for traditional fats had decreased weight gain. As shown above, stearic acid appears to be less well absorbed in rats than in humans, and thus studies in rats are not appropriate for determining the caloric content for humans of triglycerides containing stearic acid. In addition, restriction of the food intake of the rats to 50% of that of rats fed *ad libitum* may not result in normal growth. Finally, the test fats were added to the basal diet, whereas it would have been more suitable to adjust all diets to maintain similar ratios therefore not the only variable in growth.

In summary, it must be questioned whether the growth assay in rats provides meaningful data for determining the caloric contribution of salatrim in human diets.

In a further unpublished study, the diets were less restrictive and allowed 60–80% normal weight gain as compared with 50% in the study of Finley et al. (1994b), and the test fats were blended with a defined fat in a 3:1 ratio, obviating any concern about essential fatty acid deficiency. Further, the ratios of protein plus carbohydrate:fat, fat:vitamin and fat:mineral were kept constant. The average caloric value of the four salatrims tested was 5.2 kcal/g during the 4-week feeding period. This study confirmed the finding of Finley et al. (1994b) that the higher the ratio of short-chain:long-chain fatty acids the higher the caloric value. As described above, this is counterintuitive, as short-chain fatty acids are known to have a lower energy content per gram than long-chain fatty acids. Also, in this study, the amount of stearic acid excreted in the faeces was not negatively correlated with the energy per gram of test fat, as would be expected. In fact, the diet with 82% dilong-18:0 yielded a low caloric value but was associated with low fecal 18:0 and normal total fat excretion. The results of this study confirm that the energy contribution of a particular salatrim fat cannot be predicted by assigning an absorption coefficient for stearic acid and a standard low value for short-chain fatty acids (Treadwell et al., 2001).

A3.4 Caloric value of salatrim products on the basis of stearic acid absorption coefficient

The 14-day rat growth assay was used to determine the absorption coefficient for stearic acid. Eleven compositions of salatrim were generated by inter-esterification of various starting molar ratios of tributyrin and hydrogenated canola oil. The molar short-chain:long-chain ratios of the compositions varied between 0.51 and 1.99. The caloric content of these 11 salatrim 4CA samples was determined in rats fed

50% of the basal diet of the control rats plus 21% salatrim by weight of the diet. Weight gain was calculated over 14 days, and the weight gain of rats on each of the 11 salatrim diets was compared with that of rats on 21% corn oil, considered to supply 9 kcal/g. A previously developed regression equation (Finley et al., 1994c) was used to estimate the number of kilocalories per gram for each salatrim product by comparing it to a standard curve generated with various amounts of corn oil added to the basal diet.

Stearic acid absorption was estimated from the energy (in kcal/g) of the various salatrim products in the rodent growth assay and the ratio of short-chain:long-chain fatty acids in the test salatrim products. A table was constructed of the assayed composition of a salatrim product with a ratio of 0.51:1 for short-chain:long-chain fatty acids. For each portion of the triglyceride, the component of the triglyceride, the mass fraction of that component, the gross energy from the study in rats and the absorption coefficient of that component were provided. All the absorption coefficients were derived from the literature, and the only 'unknown' was considered to be the absorption of stearic acid. An absorption coefficient for stearic acid was derived from this matrix, the 11 ratios of short-chain:long-chain fatty acid and the estimated caloric values for the various products. The absorption coefficient ranged from 0.15 with the lowest ratio to 0.70 with the highest ratio (Klemann et al., 1994).

These calculations of the absorption of stearic acid reinforced the Committee's view that a general caloric value cannot be applied to salatrim in the absence of knowledge of the specific fatty acid composition of the product.

A4. Other nutritional considerations: Potential cholesterol-raising effect of stearic acid

Although not considered by the Committee at its forty-ninth meeting, a review by the Committee on Toxicity of Chemicals in Food, Consumer Products and the Environment (1999) of the European Commission raised the issue that the stearic acid in salatrim could potentially increase blood cholesterol concentrations. This potential effect was addressed in four clinical trials, reviewed in section 2.3 above. Nestel et al. (1998) compared the effect of a stearic acid-rich diet (23SO) with that of a diet rich in palmitic acid in 15 volunteers in a cross-over study, each dietary period being 5 weeks. The outcome measures were total, low-density lipoprotein and high-density lipoprotein cholesterol and plasma triacylglycerols concentrations. The dietary intervention followed a 2-week period in which the volunteers were instructed to eat low-fat diets. Although the values for total cholesterol in the usual diets were higher than those in the low-fat diet, they did not differ significantly in the palmitic acid- and stearic acid-rich diets as compared with the low-fat diet. Furthermore, the values for triglycerides in the palmitic acid- and strearic acid-rich diets were similar. The Committee found these surprising, as it has been widely reported that cholesterol concentrations are increased by high-fat as compared with low-fat diets. Also, although some controversy exists about whether dietary stearic acid is hypercholesterolaemic, the cholesterol-raising effects of palmitic acid are well documented. The fact that neither high-fat diet raised cholesterol concentrations over those with low-fat diets, and in particular the finding that palmitic acid fed at high levels resulted in cholesterol values similar to those with a low-fat diet, suggests a lack of power to detect such differences in a sample of 15 volunteers. The report that stearic acid from salatrim does not raise plasma cholesterol levels is therefore considered not to be definitive.

Sanders et al. (2001) compared the same salatrim product (23SO) with meals enriched with cocoa butter or high-oleate sunflower oil in 35 volunteers in a randomized cross-over design. Each person received three experimental meals 1 week apart over 3 weeks. The increase in plasma triacylglycerol and in factor VII coagulant activity was less marked after consumption of a meal containing 23SO than after a meal enriched with cocoa butter or oleate. The Committee observed that the stearic acid-rich diet and the other two diets contained 77% long-chain fatty acids. As it is the long-chain fatty acids that are incorporated into chylomicrons and would thus be picked up as triglycerides in blood, it is not surprising that the diet with less long-chain fatty acids resulted in 79% less trigylcerides at 3 h than with the other two diets.

A5. References

American Oil Chemists' Society (1990) Fatty acid comparison by gas chromatography: Method Ce 1-62. In: *Official Methods of the American Oil Chemists' Society*, Champaign, Illinois, USA.

Auerbach, M.H. (2000) Letter submitted to Ms M. de L. Costarrica, Senior Officer, FAO Food and Nutrition Division, requesting a change to the specifications for salatrim. Submitted by Danisco Cultor USA, Inc., Ardsley, New York, USA.

Bonanome, A. & Grundy, S.M. (1989) Intestinal absorption of stearic acid after consumption of high fat meals in humans. *J. Nutr.*, **119**, 1556–1560.

Committee on Toxicity of Chemicals in Food, Consumer Products and the Environment (1999) Statement for the Advisory Committee on Novel Foods and Processes on toxicological aspects of a submission on short and long chain triacylglycerol molecules (salatrims)—A family of low calorie fats submitted for approval under the EC Novel Food Regulation. Department of Health, London, United Kingdom.

Danisco Cultor USA, Inc. (2001a) Salatrim, 29 November 2001. Submitted to WHO by Danisco Cultor USA, Inc., Ardsley, New York, USA.

Danisco Cultor USA, Inc. (2001a) Salatrim, Attachment B, 30 November 2001. Submitted to WHO by Danisco Cultor USA, Inc., Ardsley, New York, USA.

Denke, M.A. & Grundy, S.M. (1991) Effects of fats high in stearic acid on lipid and lipoprotein concentrations in men. *Am. J. Clin. Nutr.,* **54**, 1036–1040.

Finley, J.W., Klemann, L.P., Leveille, G.A., Otterburn, M.S. & Walchak, C.G. (1994a) Caloric availability of salatrim in rats and humans. *J. Agric. Food Chem.*, **42**, 495–499.

Finley, J.W., Leveille, G.A., Klemann, L.P., Sourby, J.C., Ayres, P.H. & Appleton, S. (1994b) Growth method for estimating the caloric availability of fats and oils. *J. Agric. Food Chem.*, **42**, 489–494.

Finley, J.W., Leveille, G.A., Dixon, R.M., Walchak, C.G., Sourby, J.C., Smith, R.E., Francis, K.D. & Otterburn, M.S. (1994c) Clinical assessment of salatrim, a reduced-calory triacylglycerol. *J. Agric. Food Chem.*, **42**, 581–596.

Hayes, J.R., Pence, D.H., Scheinbach, S., D'Amelia, R.P., Klemann, L.P., Wilson, N.H. & Finley, J.W. (1994) Review of triacylglycerol digestion, absorption, and metabolism with respect to SALATRIM triacylglycerols. *J. Agric. Food Chem.*, **42**, 474–483.

Howlett, J. F. (1997) Memorandum to Dr J.L. Herrman, 27 March 1997.

Jones, P.J.H., Pencharz, P.B. & Clandinin, M.T. (1985) Absorption of ^{13}C-labeled stearic, oleic, and linoleic acids in humans: Application to breath tests. *J. Lab. Clin. Med.*, **105**, 647–652.

Klemann, L.P., Finley, J.W. & Leveille, G.A. (1994) Estimation of the absorption coefficient of stearic acid in salatrim fats. *J. Agric. Food Chem.*, **42**, 484–488.

Nestel, P.J., Pomeroy, S., Kay, S., Sasahara, T. & Yamashita, T. (1998) Effect of a stearic acid-rich, structured triacylglycerol on plasma lipid concentrations. *Am. J. Clin. Nutr.*, **68**, 1196–1201.

Olubajo, O., Marshall, M.W., Judd, J.T. & Adkins, J.T. (1986) Effects of high and low fat diets on the bioavailability of selected fatty acids, including linoleic acid, in adult men. *Nutr. Res.*, **6**, 931–955.

Sanders, T.A.B., Oakley, F.R., Cooper, J.A. & Miller, G.J. (2001) Influence of a stearic acid-rich structured triacylglycerol on postprandial lipemia, factor VII concentrations, and fibrinolytic activity in healthy subjects. *Am. J. Clin. Nutr.*, **73**, 715–721.

Treadwell, R.M., Pronczuk, A. & Hayes, K.C. (2001) Dietary stearic acid content and glyceride structure affect caloric value of glycerides in growing rats. Unpublished report. Submitted to WHO by Danisco Cultor USA, Inc., Ardsley, New York, USA.

Weast, R.C., ed. (1992–93) *CRC Handbook of Chemistry and Physics*, 73rd Ed., Boca Raton, Florida, USA: CRC Press.

SAFETY EVALUATIONS OF GROUPS OF RELATED FLAVOURING AGENTS

SAFETY EVALUATIONS OF GROUPS OF RELATED FLAVOURING AGENTS

INTRODUCTION

Six groups of flavouring agents were evaluated by the Procedure for the Safety Evaluation of Flavouring Agents as outlined in Figure 1 (Annex 1, references *116*, *122*, *131*, *137*, *143*, *149* and *154*). In applying the Procedure, the chemical is first assigned to a structural class as identified by the Committee at its forty-sixth meeting (Annex 1, reference 122). The structural classes are as follows:

- Class I. Flavouring agents that have simple chemical structures and efficient modes of metabolism which would suggest a low order of toxicity by the oral route.
- Class II. Flavouring agents that have structural features that are less innocuous than those of substances in Class I but are not suggestive of toxicity. Substances in this class may contain reactive functional groups.
- Class III. Flavouring agents that have structural features that permit no strong initial presumption of safety, or may even suggest significant toxicity.

A key element of the Procedure involves determining whether a flavouring agent and the product(s) of its metabolism are innocuous and/or endogenous substances. For the purpose of the evaluations, the Committee used the following definitions, adapted from the report of its forty-sixth meeting:

Innocuous metabolic products are defined as products that are known or readily predicted to be harmless to humans at the estimated intake of the flavouring agent.

Endogenous substances are intermediary metabolites normally present in human tissues and fluids, whether free or conjugated; hormones and other substances with biochemical or physiological regulatory functions are not included. The estimated intake of a flavouring agent that is, or is metabolized to, an endogenous substance should be judged not to give rise to perturbations outside the physiological range.

Intake

Estimates of the intake of flavouring agents by populations typically involve the acquisition of data on the amounts used in food. These data were derived from surveys in Europe and the USA. In Europe, a survey was conducted in 1995 by the International Organization of the Flavour Industry, in which flavour manufacturers reported the total amount of each flavouring agent incorporated into food sold in the European Union during the previous year. Manufacturers were requested to exclude use of flavouring agents in pharmaceutical, tobacco or cosmetic products.

In the USA, a series of surveys was conducted between 1970 and 1987 by the National Academy of Sciences National Research Council (under contract to the

Figure 1. Procedure for the safety evaluation of flavouring agents

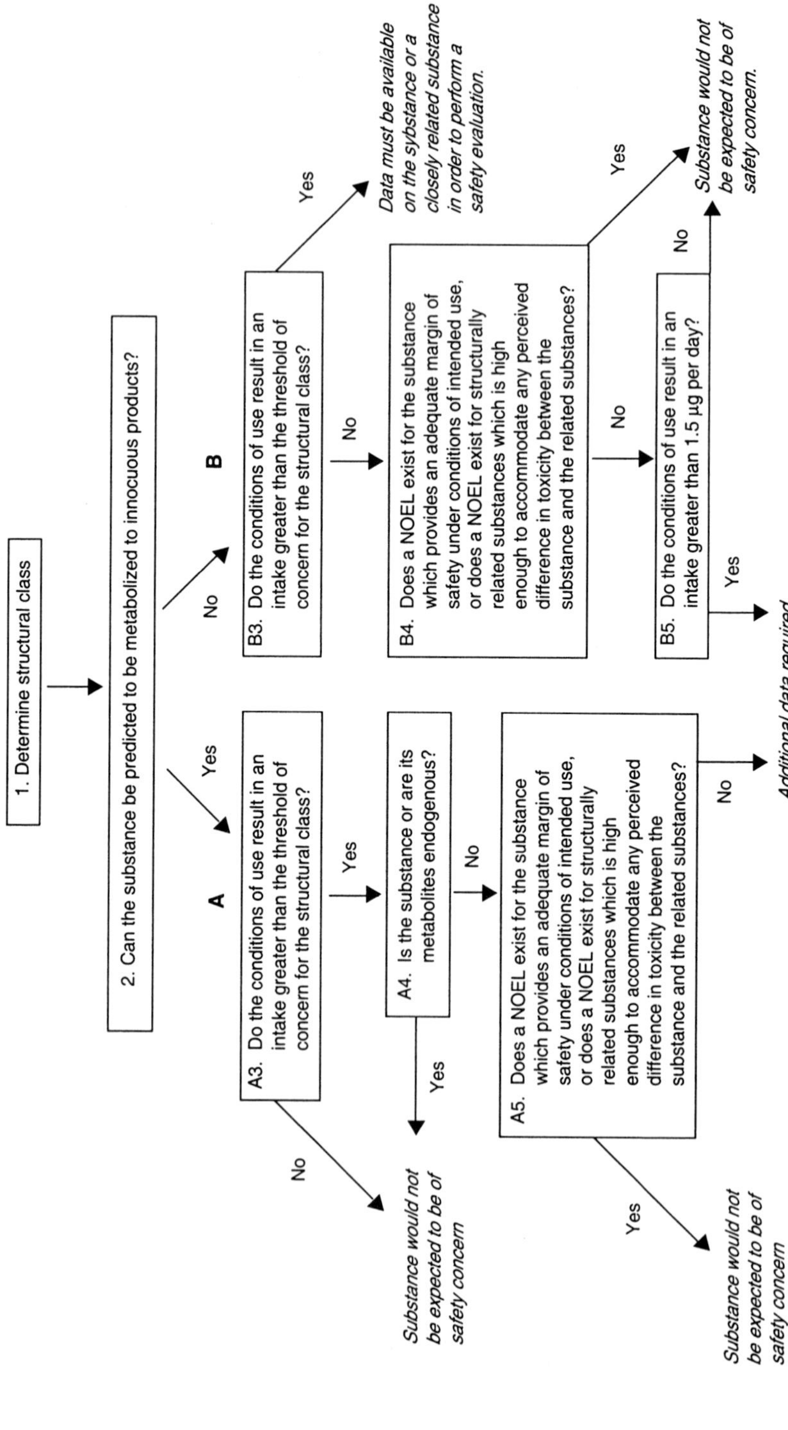

Food and Drug Administration) in which information was obtained from ingredient manufacturers and food processors on the amount of each substance destined for addition to the food supply and on the usual and maximal levels at which each substance was added in a number of broad food categories.

In using the data from these surveys to estimate intakes of flavouring agents, it was assumed that only 60% of the total amount used is reported in Europe and 80% of the amount used is reported in the USA and that the total amount used in food is consumed by only 10% of the population.

$$\text{Intake } (\mu\text{g/person per day}) = \frac{\text{annual volume of production (kg)} \times 10^{9}\ (\mu\text{g/kg})}{\text{population of consumers} \times 0.6 \text{ (or 0.8)} \times 365 \text{ days}}$$

The population of consumers was assumed to be 32×10^{6} in Europe and 26×10^{6} in the USA.

Several of the flavouring agents that were evaluated at the present meeting were not included in the above surveys or were placed on the market after the surveys were conducted. Intakes of these flavouring agents were estimated on the basis of anticipated use by the manufacturer in the USA, and the standard formula was applied.

ALICYCLIC PRIMARY ALCOHOLS, ALDEHYDES, ACIDS, AND RELATED ESTERS

First draft prepared by

Dr G.J.A. Speijers[1] and Professor A. Renwick[2]
[1]Section Public Health of the Centre for Substances & Risk Assessment, National Institute of Public Health and Environmental Protection, Bilthoven, Netherlands; and [2]Clinical Pharmacology Group, University of Southampton, Southampton, England

1. EVALUATION

The Committee evaluated a group of 26 flavouring agents that included alicyclic primary alcohols, aldehydes, acids and related esters (see Table 1), using the Procedure for the Safety Evaluation of Flavouring Agents (See Figure 1). The Committee had not evaluated any of these agents previously.

Twenty of the 26 flavouring agents in this group are terpenoid alcohols, aldehydes or carboxylic acids or their related esters. Thirteen are common components of food and have been detected in vanilla, bourbon, rum, mango, rosemary, grapefruit juice, mandarin peel oil and blackberries (Maarse et al., 1999).

1.1 Estimated daily intake

The total annual volume of the 26 flavouring agents in this group that is produced is approximately 170 kg in Europe (Insternational Organization of the Flavor Industry, 1995) and 130 kg in the USA (Lucas et al., 1999). None of the agents is produced in an annual volume greater than 55 kg. Most of the total annual volume in Europe

Table 1. Summary of results of safety evaluation of alicyclic primary alcohols, aldehydes, acids and related esters used as flavouring agents[a]

Flavouring agent	No.	CAS No. and structure	Steps A3 and B3[b] Does intake exceed the threshold for human intake?	Step B4 Adequate margin of safety for the flavouring agent or related substance?	Comments	Conclusion based on current intake
Structural class I						
Cyclohexanecarboxylic acid	961	98-89-5 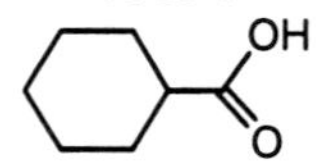	A3 – No Europe: 0.07 USA: 4	N/R	See note 1	No safety concern
Methyl cyclohexanecarboxylate	962	4630-82-4 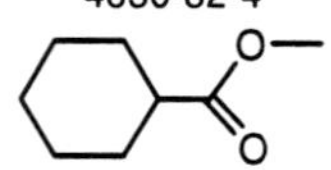	A3 – No Europe: 0.09 USA: 0.01	N/R	See notes 1,2	No safety concern
Ethyl cyclohexanecarboxylate	963	3289-28-9 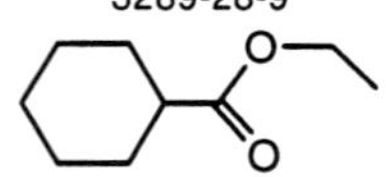	A3 – No Europe: ND USA: 0.1	N/R	See notes 1,2	No safety concern
Cyclohexaneethyl acetate	964	21722-83-8 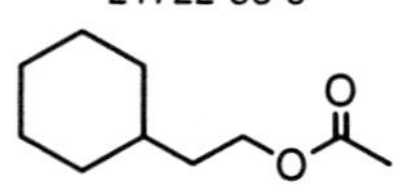	A3 – No Europe: 1 USA: ND	N/R	See notes 2,3	No safety concern
Cyclohexaneacetic acid	965	5292-21-7 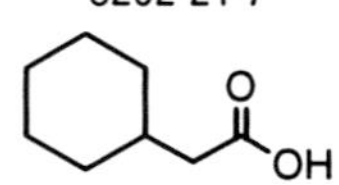	A3 – No Europe: 0.1 USA: 0.4	N/R	See note 3	No safety concern

Table 1 (continued)

Flavouring agent	No.	CAS No. and structure	Steps A3 and B3[a,b] Does intake exceed the threshold for human intake?	Step B4 Adequate margin of safety for the flavouring agent or related substance?	Comments	Conclusion based on current intake
Ethyl cyclohexanepropionate	966	10094-36-7 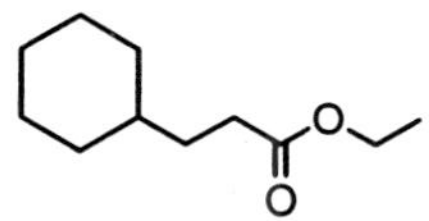	A3 – No Europe: ND USA: 0.1	N/R	See notes 1,2	No safety concern
2,2,3-Trimethylcyclopent-3-en-1-yl acetaldehyde	967	4501-58-0 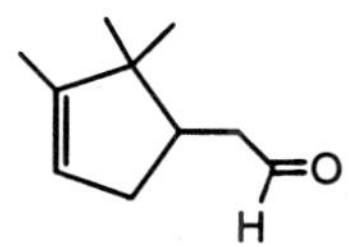	A3 – No Europe: 6 USA: ND	N/R	See note 4	No safety concern
cis-5-Isopropenyl-*cis*-2-methylcyclopentan-1-carboxaldehyde	968	55253-28-6 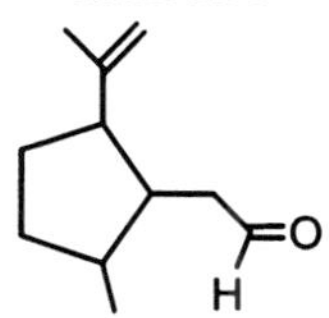	A3 – No Europe: 0.01 USA: ND	N/R	See note 4	No safety concern
Campholene acetate	969	36789-59-0 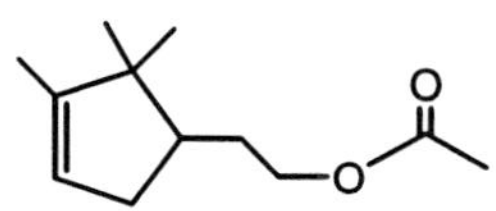	A3 – No Europe: 0.07 USA: ND	N/R	See notes 2,4	No safety concern

Table 1 (continued)

Flavouring agent	No.	CAS No. and structure	Steps A3 and B3[a,b] Does intake exceed the threshold for human intake?	Step B4 Adequate margin of safety for the flavouring agent or related substance?	Comments	Conclusion based on current intake
α-Campholenic alcohol	970	1901-38-8 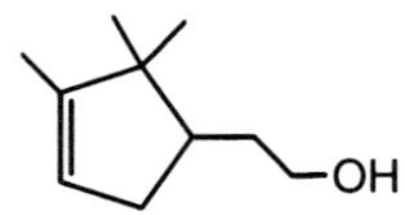	A3 – No Europe: 0.01 USA: ND	N/R	See note 4	No safety concern
para-Menth-1-ene-9-al	971	29548-14-9 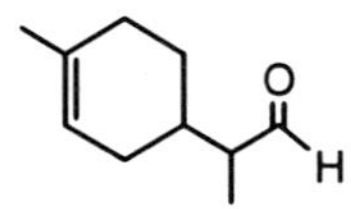	A3 – No Europe: 0.1 USA: ND	N/R	See note 4	No safety concern
1-*para*-Menthen-9-yl acetate	972	17916-91-5 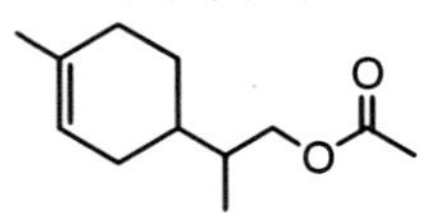	A3 – No Europe: 1 USA: ND	N/R	See notes 2,4	No safety concern
para-Mentha-1,8-dien-7-al	973	2111-75-3 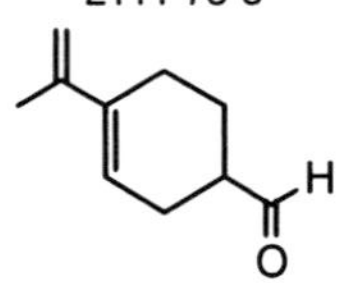	A3 – No Europe: 2 USA: 2	N/R	See note 4	No safety concern

Table 1 (continued)

Flavouring agent	No.	CAS No. and structure	Steps A3 and B3[a,b] Does intake exceed the threshold for human intake?	Step B4 Adequate margin of safety for the flavouring agent or related substance?	Comments	Conclusion based on current intake
para-Mentha-1,8-dien-7-ol	974	536-59-4 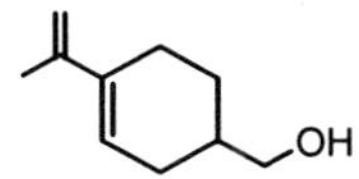	A3 – No Europe: 2 USA: 1	N/R	See note 4	No safety concern
para-Mentha-1,8-dien-7-yl acetate	975	15111-96-3 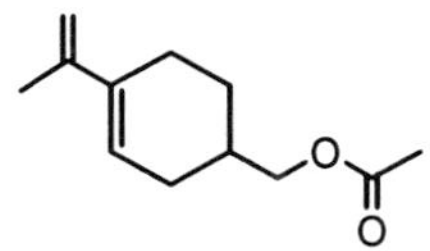	A3 – No Europe: 0.4 USA: 0.07	N/R	See notes 2,4	No safety concern
1,2,5,6-Tetrahydrocuminic acid	976	71298-42-5 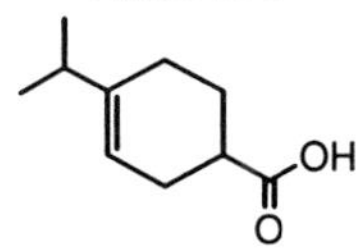	A3 – No Europe: 0.01 USA: ND	N/R	See note 1	No safety concern

Table 1 (continued)

Flavouring agent	No.	CAS No. and structure	Steps A3 and B3[a,b] Does intake exceed the threshold for human intake?	Step B4 Adequate margin of safety for the flavouring agent or related substance?	Comments	Conclusion based on current intake
2,6,6-Trimethylcyclohexa-1,3-dienyl methanal	977	116-26-7 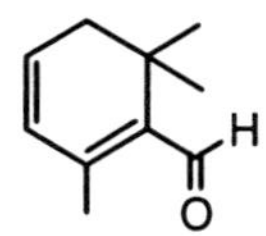	B3 – No Europe: 4 USA: 0.07	Yes. The NOELs of 120 mg/kg bw per day for perillyl alcohol in 90-day studies in rats and dogs and NOELs of 10 mg/kg bw per day for α-ionone and β-ionone in a 90-day study in rats are > 2 million and > 200 000 times, respectively, the estimated intake of 2,6,6-trimethylcyclohexa-1,3-dienyl methanal when used as a flavouring agent.		No safety concern
2,6,6-Trimethyl-1-cyclohexen-1-acetaldehyde	978	472-66-2 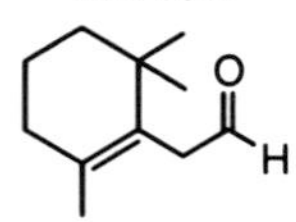	A3 – No Europe: 0.3 USA: 2	N/R	See note 3.	No safety concern

Table 1 (continued)

Flavouring agent	No.	CAS No. and structure	Steps A3 and B3[a,b] Does intake exceed the threshold for human intake?	Step B4 Adequate margin of safety for the flavouring agent or related substance?	Comments	Conclusion based on current intake
2,6,6-Trimethyl-1-cyclohexen-1-carboxaldehyde	979	432-25-7 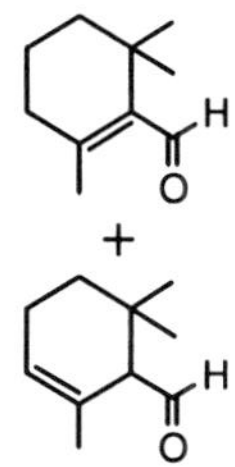	A3 – No Europe: 0.4 USA: ND	N/R	See note 1	No safety concern
2-Formyl-6,6-dimethylbicyclo[3.1.1]hept-2-ene	980	564-94-3 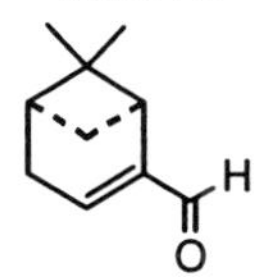	A3 – No Europe: 4 USA: 7	N/R	See note 4	No safety concern
Myrtenol	981	515-00-4 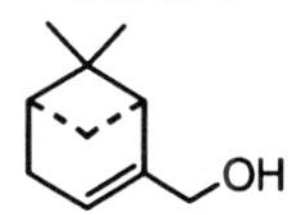	A3 – No Europe: 0.4 USA: 0.05	N/R	See note 4	No safety concern

Table 1 (continued)

Flavouring agent	No.	CAS No. and structure	Steps A3 and B3[a,b] Does intake exceed the threshold for human intake?	Step B4 Adequate margin of safety for the flavouring agent or related substance?	Comments	Conclusion based on current intake
Myrtenyl acetate	982	1079-01-2 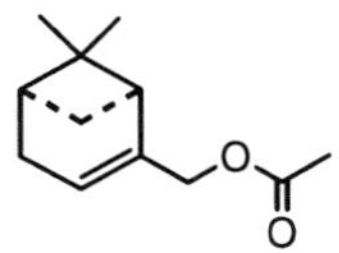	A3 – No Europe: 0.4 USA: 0.03	N/R	See note 4	No safety concern
Myrtenal formate	983	72928-52-0 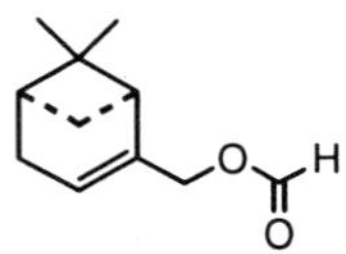	A3 – No Europe: 0.4 USA: ND	N/R	See note 4	No safety concern

Table 1 (continued)

Flavouring agent	No.	CAS No. and structure	Steps A3 and B3[a,b] Does intake exceed the threshold for human intake?	Step B4 Adequate margin of safety for the flavouring agent or related substance?	Comments	Conclusion based on current intake
Santalol (α and β)	984	115-71-9 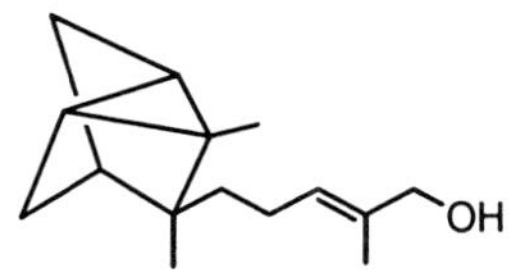 α 77-42-9 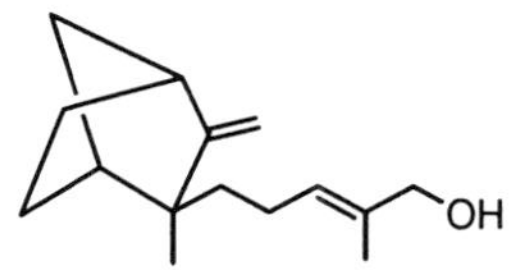 β	A3 – No Europe: 0.04 USA: ND	N/R	See note 4	No safety concern
Santalyl acetate	985	1323-00-8 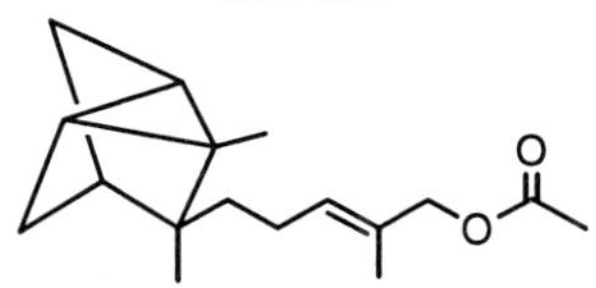	A3 – No Europe: ND USA: 0.01	N/R	See note 4	No safety concern

Table 1 (continued)

Flavouring agent	No.	CAS No. and structure	Steps A3 and B3[a,b] Does intake exceed the threshold for human intake?	Step B4 Adequate margin of safety for the flavouring agent or related substance?	Comments	Conclusion based on current intake
10-Hydroxymethylene-2-pinene	986	128-50-7 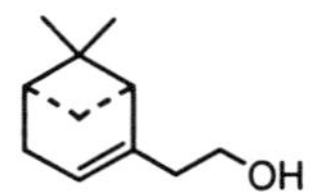	A3 – No Europe: ND USA: 0.01	N/R	See note 4	No safety concern

CAS: Chemical Abstracts Service; ND: no intake data reported; N/R: not required for evaluation

[a] Step 2: Except for No. 977, all the flavouring agents in this group were predicted to be metabolized to innocuous products.

[b] The threshold for human intake for structural class I is 1800 µg/day. All intake values are expressed in µg/day.
The combined intake of all flavouring agents in the group is 22 and 18 µg/person per day in Europe and the USA, respectively.

Notes:

1. Metabolized by β-oxidation, aromatization and conjugation with glycine and glucuronic acid
2. Metabolized by hydrolysis followed by oxidation and/or conjugation
3. Metabolized by β-oxidation and eliminated largely by metabolism via the citric acid cycle
4. Metabolized largely by oxidation of the side-chain to the corresponding carboxylic acid, which is excreted unchanged and as conjugates

is accounted for by (2,2,3-trimethylcyclopent-3-en-1-yl)acetaldehyde (No. 967; campholenic aldehyde), *para*-mentha-1,8-dien-7-al (No. 973; perillaldehyde), *para*-mentha-1,8-dien-7-ol (No. 974; perillyl alcohol), 2,6,6-trimethylcyclohexa-1,3-dienyl methanal (No. 977; safranal) and 2-formyl-6,6-dimethylbicyclo[3.1.1]hept-2-ene (No. 980; myrtenal). In the USA, approximately 91% of the total annual volume is accounted for by cyclohexanecarboxylic acid (No. 961), *para*-mentha-1,8-dien-7-al (No. 973; perillaldehyde), 2,6,6-trimethyl-1-cyclohexen-1-acetaldehyde (No. 978; β-cyclocitral) and 2-formyl-6,6-dimethylbicyclo[3.1.1]hept-2-ene (No. 980; myrtenal). The estimated daily per capita intake of 11 of the substances in this group is - 10 μg (see Table 2).

1.3 Absorption, distribution, metabolism and elimination

The number of potential pathways of metabolism of alicyclic substances increases as the number and types of functional groups and ring substituents in the molecule increase. The cyclohexane derivatives (carboxylic acids and their related esters) in this group (Nos 961–966) can undergo β-oxidation and subsequent ring cleavage and/or aromatization of the cyclohexane ring and/or can be excreted in urine as cyclohexane carboxylic acid and its glycine and glucuronide conjugates. If there is an even number of carbons in the side-chain, β-oxidative cleavage leads to ring-opened oxygenated metabolites, primarily dicarboxylic acids, which are metabolized via the citric acid cycle. If there is an odd number of carbons in the side-chain, β-oxidative cleavage leads to a cyclohexane carboxylic acid derivative that may subsequently undergo dehydrogenation to yield unsaturated derivatives. The two non-terpenoid substances containing a cyclohexene ring (Nos 978 and 979) would be metabolized and eliminated by similar pathways. The substance with two endocyclic double bonds, 2,6,6-trimethylcyclohexa-1,3-dienyl methanal (No. 977; safranal), is a structural analogue of dehydrodihydroionol (No. 397), which was evaluated by the Committee at its fifty-first meeting (Annex 1, reference *137*), and, like that agent, cannot be predicted to be metabolized to innocuous products.

The flavouring agents with a cyclopentane or cyclopentene ring (Nos 967–970) also have an oxygenated side-chain, and this is predicted to be the main site of oxidative metabolism prior to excretion of more polar acid metabolites. The processes of side-chain and alicyclic ring oxidation were considered by the Committee at its fifty-first meeting (Annex 1, reference *137*).

Monocyclic terpenoid primary alcohols (e.g., No. 974, perillyl alcohol) and aldehydes (e.g., Nos 971 and 973) would be oxidized to yield the corresponding carboxylic acid, while the esters (Nos 972 and 975) would undergo hydrolysis prior to oxidation. The acid metabolites of the above agents and 1,2,5,6-tetrahydrocuminic acid (No. 976) would be conjugated with glucuronic acid and excreted mainly in the urine. In a minor pathway, the aldehyde can be reduced to the alcohol and excreted as the glucuronic acid conjugate (Brewster et al., 1977a,b; Ishida et al., 1989; Haag & Gould, 1994). If an endocyclic double-bond is present, the flavouring agent or its metabolite could be reduced by the action of gut microflora. The acid metabolite may also undergo aromatization of the ring to yield a hippuric acid derivative (Ishida et al., 1989). Bicyclic and tricyclic compounds (e.g., Nos 980–986) would undergo similar metabolism to the monocyclic terpenoids and, in addition, may undergo ring oxidation.

Table 2. Annual volumes of production of alicyclic primary alcohols, aldehydes, acids and related esters used as flavouring agents in Europe and the USA

Substance (No.)	Most recent annual volume (kg)[a]	Intake[b]		Annual volume in naturally occurring foods (kg)[c]	Consumption ratio[d]
		μg/day	μg/kg bw per day		
Cyclohexanecarboxylic acid (961)					
Europe	0,5	0.071	0.0012	–	N/R
USA	32	4.2	0.071		
Methyl cyclohexanecarboxylate (962)					
Europe	0.6	0.086	0.0014	+	N/R
USA	0.1	0.013	0.0002		
Ethyl cyclohexanecarboxylate (963)					
Europe	ND			+	N/R
USA	0.91[e]	0.12	0.0020		
Cyclohexaneethyl acetate (964)					
Europe	8.0	1.1	0.019	–	N/R
USA	ND				
Cyclohexaneacetic acid (965)					
Europe	1.0	0.14	0.0024	–	N/R
USA	2,7	0.36	0.0059		
Ethyl cyclohexanepropionate (966)					
Europe	ND			–	N/R
USA	0.9	0.12	0.0020		
(2,2,3-Trimethylcyclopent-3-en-1-yl)acetaldehyde (967)					
Europe	41	5.8	0.098	+	N/R
USA	ND				
cis-5-Isopropenyl-*cis*-2-methylcyclopentan-1-carboxaldehyde (968)					
Europe	0.1	0.014	0.0002	–	N/R
USA	ND				
Campholene acetate (969)					
Europe	0.5	0.071	0.0012	–	N/R
USA	ND				
α-Campholenic alcohol (970)					
Europe	0.1	0.014	0.0002	+	N/R
USA	ND				
para-Menth-1-ene-9-al (971)					
Europe	1.0	0.14	0.0024	5.8	5.8
USA	ND	ND	ND		
1-*para*-Menthen-9-yl acetate (972)					
Europe	7.0	1.0	0.017	+	N/R
USA	ND				

Table 2 (contd)

Substance (No.)	Most recent annual volume (kg)[a]	Intake[b]		Annual volume in naturally occurring foods (kg)[c]	Consumption ratio[d]
		μg/day	μg/kg bw per day		
para-Mentha-1,8-dien-7-al (973)					
Europe	17	2.4	0.040	110	0.07
USA	11	1.5	0.025		
para-Mentha-1,8-dien-7-ol (974)					
Europe	13	1.8	0.031	+	N/R
USA	5.9	0.78	0.013		
para-Mentha-1,8-dien-7-yl acetate (975)					
Europe	2.9	0.41	0.069	1.2	000
USA	0.5	0.066	0.0011		
1,2,5,6-Tetrahydrocuminic acid (976)					
Europe	0.1	0.014	0.0002	–	N/R
USA	ND				
2,6,6-Trimethylcyclohexa-1,3-dienyl methanal (977)					
Europe	29	4.1	0.069	75	0.03
USA	0.5	0.066	0.0011		
2,6,6-Trimethyl-1-cyclohexen-1-acetaldehyde (978)					
Europe	2.0	0.28	0.0048	–	N/R
USA	16	2.0	0.034		
2,6,6-Trimethyl-1 and 2-cyclohexen-1-carboxaldehyde (979)					
Europe	3.0	0.43	0.0071	74	25
USA	ND				
2-Formyl-6,6-dimethylbicyclo[3.1.1]hept-2-ene (980)					
Europe	33	4.2	0.078	+	N/R
USA	54	7.2	0.12		
Myrtenol (981)					
Europe	3.0	0.43	0.0071	+	N/R
USA	0.4	0.053	0.0009		
Myrtenyl acetate (982)					
Europe	3.0	0.43	0.0071	–	N/R
USA	0.2	0.026	0.0004		
2-Hydroxymethyl-6,6-dimethylbicyclo[3.1.1]hept-2-enyl formate (983)					
Europe	2.5	0.36	0.0059	–	N/R
USA	ND				
Santalol (α and β) (984)					
Europe	0.3	0.043	0.0007	–	N/R
USA	ND				

Table 2 (contd)

Substance (No.)	Most recent annual volume (kg)[a]	Intake[b]		Annual volume in naturally occurring foods (kg)[c]	Consumption ratio[d]
		μg/day	μg/kg bw per day		
Santalyl acetate (985)					
Europe	ND			–	N/R
USA	0.1	0.013	0.0002		
10-Hydroxymethylene-2-pinene (986)					
Europe	ND			–	N/R
USA	0.1[e]	0.013	0.0002		
Total					
Europe	170				
USA	130				

ND, no data reported; N/R, not reported; +, reported to occur naturally in foods (Maarse et al., 1999) but no quantitative data available; –, not reported to occur naturally in foods

[a] From International Organization of the Flavor Industry (1995) and Lucas et al. (1999)

[b] Intake (μg/person per day) was calculated as follows: [(annual volume, kg) × (1 × 10^9 μg/kg)/(population × survey correction factor × 365 days)], where population (10%, 'eaters only') = 32 × 10^6 for Europe and 26 × 10^6 for the USA. The correction factor = 0.6 for Europe and 0.8 for the USA, representing the assumption that only 60% and 80% of the annual production volume of the flavour, resepctively, was reported in the poundage surveys. Intake (μg/kg bw per day) calculated as follows: [(μg/person per day)/body weight], where body weight = 60 kg. Slight variations may occur from rounding.

[c] Quantitative data from Stofberg & Grundschober (1987)

[d] Calculated as follows: (annual consumption in food, kg)/(most recently reported volume as a flavouring agent, kg)

[e] Anticipated annual volume in the USA as reported by the Flavor and Extract Manufacturers Association

Four agents in this group (Nos 973, 977, 979 and 980) contain an α,β-unsaturated carbonyl group, which is a structural alert for toxicity. The Committee, at previous meetings, devoted considerable attention to the safety of flavouring agents containing this reactive moiety. The Committee concluded at its fifty-seventh meeting (Annex 1, reference *154*) that the presence of protective processes in cells provides adequate detoxication capacity at the low doses associated with the use of compounds such as flavouring agents. These protective processes include conjugation with glutathione and reduction of the ketone to the corresponding alcohol (followed by conjugation of the alcohol with glucuronic acid).

1.3 Application of the Procedure for the Safety Evaluation of Flavouring Agents

Step 1 In applying the Procedure for the Safety Evaluation of Flavouring Agents, the Committee assigned all 26 agents to structural class I (Cramer et al., 1978).

Step 2 The simple cyclohexane and cyclohexene derivatives (Nos 961–966, 978 and 979), the cyclopentane and cyclopentene compounds (Nos 967–970) and the simple mono-, bi- and tricyclic terpenoid derivatives (Nos 971–976 and 980–986) were predicted to be metabolized to innocuous products. The evaluation of these substances therefore proceeded *via* the A (left-hand) side of the decision-tree (see Figure 1). The remaining substance, 2,6,6-trimethylcyclohexa-1,3-dienyl methanal (No. 977; safranal), cannot be predicted to be metabolized to innocuous products and was therefore evaluated *via* the B (right-hand) side of the decision-tree.

Step A3 The daily per capita intakes of each of the 25 flavouring agents evaluated at this step were below the daily human intake thresholds for structural class I (1800 μg per person), indicating that they present no safety concern when consumed as flavouring agents at current estimated levels.

Step B3 The daily per capita intake of the monocyclic substance with two endocyclic double-bonds evaluated at this step, 2,6,6-trimethylcyclohexa-1,3-dienyl methanal (No. 977; safranal), was below the threshold for daily human intake of compounds of structural class I , and its evaluation therefore proceeded to step B4.

Step B4 As the agent evaluated at this step, 2,6,6-trimethylcyclohexa-1,3-dienyl methanal (No. 977; safranal), is structurally related to perillyl alcohol (No. 974), data on the toxicity of perillyl alcohol were used to evaluate its safety. Perillyl alcohol given by intragastric gavage changed the weights of several organs in female rats when given at 400 mg/kg bw per day, but not at 120 mg/kg bw per day, in a 90-day study; changes in organ weights were not reported in male rats. Doses of 40, 120 and 400 mg/kg bw per day produced hyperexcitability and salivation, which the authors considered may have been due to its irritating properties (National Cancer Institute, 1996). A daily dose of 120 mg/kg bw was well tolerated by dogs in a 90-day study (National Cancer Institute, 1996). The daily intake of 2,6,6-trimethylcyclohexa-1,3-dienyl methanal (No. 977; safranal) is 4 μg/kg bw in Europe and 0.07 μg/kg bw in the USA. The margin of safety between these intakes and 120 mg/kg bw per day is > 2 000 000. The compound also shares structural similarities with α-ionone and β-ionone (Nos 388 and 389), which were evaluated by the Committee at its fifty-first meeting (Annex 1, reference *137*). The NOELs for these compounds were 10 mg/kg bw per day in a 90-day study in rats, providing a margin of safety of about 200 000. Therefore, 2,6,6-trimethylcyclohexa-1,3-dienyl methanal (No. 977; safranal) would not be a safety concern.

1.4 Consideration of combined intakes

In the unlikely event that all 26 flavouring agents were to be consumed concurrently on a daily basis, the estimated combined intake would not exceed the human intake threshold for structural class I. Therefore, the combined intake of the agents in this group would not represent a safety concern.

1.5 Conclusions

The Committee concluded that none of the 26 flavouring agents in this group of alicyclic primary alcohols, aldehydes, acids and related esters would present a safety concern at the current estimated levels of intake.

2. RELEVANT BACKGROUND INFORMATION

2.1 Explanation

The background information summarizes key data relevant to the safety evaluation of 26 alicyclic substituted primary alcohols, aldehydes, acids and related esters used as flavouring agents. Six members of the group are carboxylic acids or related esters of cyclohexane. The esters are rapidly hydrolysed to their component alcohols and cyclohexane carboxylic acids. The cyclohexane carboxylic acids are simple alicyclic substances which are expected to participate in common routes of absorption, distribution and metabolic detoxication and have similar toxicological end-points.

The remaining 20 substances are mono- or bicyclic terpenes with alkyl ring substituents and containing a primary oxygenated group (alcohol, aldehyde or carboxylic acid) or a related ester functional group. Monocyclic terpenoid primary alcohols and monocyclic and bicyclic aldehydes, with alkyl ring substituents, are generally oxidized to the corresponding carboxylic acid, conjugated with glucuronic acid and excreted as urinary metabolites. Reduction of the aldehyde to the alcohol and excretion as glucuronide is a minor pathway. When there is an endocyclic double-bond, the metabolite could be reduced by the action of gut microflora after its excretion via the bile. If the terpene is monocyclic, the resulting acid metabolite may undergo aromatization of the ring to yield a benzoic acid derivative, which may be excreted as a glycine conjugate.

2.2 Additional considerations on intake

The total annual production volume of the 26 flavouring agents in this group is approximately 170 kg in Europe (Internal Organization of the Flavor Industry, 1995) and 120 kg in the USA (Lucas et al., 1999). The annual volume of use of none of the substances in this group is greater than 55 kg. The production volumes and intakes for each substance are reported in Table 2.

Alicyclic primary alcohols, aldehydes, acids and related esters have been detected mainly in fruits and related alcoholic beverages. They have been reported to be present in mango, rosemary, grapefruit juice, mandarin peel oil, blackberries, vanilla bourbon and rum (Maarse et al., 1999). As shown in Table 2, 13 of the substances in this group have been reported to occur naturally in foods. Quantitative data on natural occurrence and consumption ratios have been reported for *para*-menth-1-ene-9-al (No. 971), *para*-mentha-1,8-dien-7-al (No. 973), *para*-mentha-1,8-dien-7-yl acetate (No. 975), 2,6,6-trimethylcyclohexa-1,3-dienyl methanal (No. 977) and 2,6,6-trimethyl-1 and 2-cyclohexen-1-carboxaldehyde (No. 979). These values indicate that their intake results mainly from traditional foods (i.e. consumption ratio > 1; see Table 2) (Stofberg & Kirschman, 1985; Stofberg & Grundschober, 1987).

2.3 Biological data

2.3.1 Biochemical data

(a) Hydrolysis

In general, aliphatic esters (Nos 962–964, 966, 969, 972, 975, 982, 983 and 985) are rapidly hydrolysed to their component alcohols and carboxylic acids. The hydrolysis is catalysed by classes of enzymes known as carboxylesterases (Heymann, 1980; Graffner-Nordberg et al., 1998), the most important of which are the β-esterases. Although these enzymes are present in most mammalian tissues, they predominate in hepatocytes (Anders, 1989; Graffner-Nordberg et al., 1998). The substrate specificity of β-carboxylesterase isoenzymes has been correlated with the structure of the alcohol and acid components (Heymann, 1980). Aliphatic esters used as flavour ingredients hydrolyse rapidly in liver homogenates, simulated pancreatic fluid, simulated gastric fluid and preparations of intestinal mucosa in vitro (Leegwater & van Straten, 1974; Butterworth et al., 1975; Grundschober, 1977; Longland & Gangolli, 1977; Junge & Heymann, 1979; Graffner-Nordberg et al., 1998).

Esters of cyclohexanecarboxylic acid (No. 961), cyclohexaneacetic acid (No. 965) and higher homologues are hydrolysed to the component alcohol and carboxylic acid (Ford & Moran, 1978). Methyl cyclohexanecarboxylate (No. 962) and ethyl cyclohexanecarboxylate (No. 963) were incubated separately with 50 ml of simulated gastric fluid at 37 °C, for 2, 4 or 6 h, and the amounts of unchanged ester and released acid were determined. The results showed approximately 20% hydrolysis of each ester in the gastric system. After a 5-h incubation in simulated intestinal fluid, 40% and 50% of methyl and ethyl cyclohexanecarboxylate were hydrolysed, respectively (Moran & Tyburcy, 1979). In a study of hydrolysis in vitro, 100% of ethyl cyclohexanepropionate (No. 966) was hydrolysed after a 2-h incubation in 5% pancreatin at 37 °C, as demonstrated by determination of the ester and the released alcohol (Grundschober, 1977; Leegwater & van Straten, 1974).

In an investigation of the hydrolysis of the structurally related aliphatic ester, *cis/trans*-1(7),8-*para*-menthadienyl acetate[1], in rat liver homogenate, menthadienyl acetate was incubated in the homogenate at 37 °C for 15, 30, 60 or 240 min. Complete (100%) hydrolysis to *cis/trans*-1(7)-8-*para*-menthadien-2-ol was observed after 60 min, with 92% hydrolysis occurring within the first 15 min and 98% within 30 min (Salzer, 1998). The ready hydrolysis of aliphatic esters in simulated gastric and intestinal fluid supports the conclusion that the ester is hydrolysed extensively in vivo before absorption.

(b) Absorption, distribution and excretion

(i) Cyclohexyl derivatives

When sodium cyclohexanecarboxylate (No. 961) with a ^{14}C-labelled ring was administered orally to male Wistar albino rats at a dose of 100 mg/kg bw, > 98% of

---

1

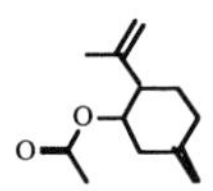

the original dose was excreted in the urine as hippuric acid metabolites and glucuronic acid conjugates (hippurate, 3,4,5,6-tetrahydrohippurate, hexahydrohippurate, benzoyl- and cyclohexylcarbonylglucuronide). Less than 1% was excreted *via* the faeces or expired air (Brewster et al., 1977a).

Similar results were obtained when [^{14}C]cyclohexanecarboxylate sodium salt was administered to male Wistar albino rats *via* a duodenal cannula at a dose of 0.5, 5.0, 35, 100 or 200 mg/kg bw. At the highest dose, 97% was recovered as five urinary (qualitatively the same as after oral dosing) and two biliary metabolites (glucuronide conjugates of cyclohexanecarboxylate and benzoate) within 7 h of dosing. As the dose increased, hippurate excretion decreased, glucuronide excretion increased, excretion of 3,4,5,6-tetrahydrohippurate and hexahydrohippurate in urine increased and biliary excretion increased. At the lowest dose (0.5 mg/kg bw), 95% of the original dose was excreted within the first 90 min (Table 3; Brewster et al., 1977a).

Dose-dependent clearance of cyclohexanecarboxylic acid and a structural analogue 1-methyl-1-cyclohexanecarboxylic acid has been reported. Female Sprague-Dawley rats were given a single intravenous dose of cyclohexanecarboxylic acid at 67, 130 or 310 mg/kg bw or 1-methyl-1-cyclohexanecarboxylic acid at 74, 150 or 340 mg/kg bw. Blood and urine were collected for 24 and 48 h after dosing, respectively. 1-Methyl-1-cyclohexanecarboxylic acid was eliminated largely via renal excretion as both the parent compound and as a base-labile conjugate (probably the glucuronide), while cyclohexanecarboxylic acid was cleared mostly via other routes, presumably the pathways identified by Brewster et al. (1977a). The plasma and renal clearances of both compounds decreased with increasing dose, suggesting that the pathways by which they are eliminated from the body are saturable (Liu & Pollack, 1993). Further studies on the pharmacokinetics of these compounds indicated that biliary excretion of parent compound and glucuronide conjugate within 5 h accounted for 0.92 and 4.3% of the dose, respectively, for cyclohexanecarboxylic acid and 1.3 and 59% of the dose, respectively, for 1-methyl-1-cyclohexanecarboxylic acid. The data for cyclohexanecarboxylic acid are the most relevant for the

Table 3. Effect of dose on metabolism and excretionof sodium cyclohexanecarboxylate in rats

Dose (mg/kg bw)	Excretion (% of dose after 7 h)					
	Bile [a]	Urine				Total
		Benzoyl and cyclohexyl-carbonyl glucuronides	Hippurate	3,4,5,6-Tetra hydrohippurate	Hexahydro-hippurate	
200	8.0	18	48	12	10	97
100	7.7	14	56	9.9	10	97
35	3.9	6.9	70	8.6	6.1	96
5	3.8	2.4	75	9.2	6.6	97
0.5	3.0	1.4	81	8.1	4.9	98

[a] Almost entirely as glucuronide conjugates of cyclohexane-carboxylate and benzoate

substances evaluated in this group. The small amount of cyclohexane carboxylic acid excreted in the bile as the glucuronide conjugate suggests that enterohepatic circulation would be minimal (Liu et al., 1992).

(ii) Terpenoid primary alcohols, aldehydes and carboxylic acids

One of the substances in this group, *para*-mentha-1,8-dien-7-ol (i.e. perillyl alcohol; No. 974), has been studied for its anticarcinogenic potential in humans. Adult patients with various advanced malignancies were treated three times a day with oral doses of 800, 1600 or 2400 mg/m^2, equivalent to approximately 20, 40 and 60 mg/kg bw. The plasma and urinary pharmacokinetics were investigated after the first dose and after 15 and 29 days of treatment. The parent drug was not detected in plasma. The peak plasma concentrations of the two main metabolites of perillyl alcohol were found at 1.5–3.5 h (perillic acid) and 3–5 h (dihydroperillic acid), and their half-lives on the basis of data for day 1 were approximately 2 h. About 9% of the two higher doses given on day 15 or 29 was recovered within 6 h of treatment, of which approximately 90% was perillic acid and 10% was dihydroperillic acid (Ripple et al., 1998).

Studies conducted in humans, dogs, and rats have shown that orally administered perillyl alcohol is rapidly absorbed and metabolized after ingestion. The compound was not detected in plasma samples taken 4 h after a single dose of 1000 mg/kg bw in sesame oil to female Wistar-Furth rats by gavage or after 3, 5 or 10 weeks' administration at 2% in the diet.. The main metabolites in plasma were perillic acid, dihydroperillic acid and their methyl esters. The ratio of perillic acid to dihydroperillic acid after a single dose by gavage was about 14, whereas this ratio was 2–4 after dietary administration (Haag & Gould, 1994).

The pharmacokinetics of perillyl alcohol in plasma was studied in 28-day studies in rats and dogs. In rats given 600 mg/kg bw per day in two divided doses, the peak plasma concentration was found at 2 h, and the elimination half-time was 4–7 h. The peak concentration of dihydroperillic acid was found at 9 h; the plasma half-life was not given. In dogs given 600 mg/kg bw per day in three divided doses, the peak concentrations of both perillic acid and dihydroperillic acid were found at 2 h, and the elimination half-life of perillic acid was 2.2 h. The peak plasma concentrations of perillic acid were two- to threefold higher in dogs than in rats, whereas the peak concentrations of dihydroperillic acid were six- to sevenfold lower in dogs (National Cancer Institute, 1996).

The peak plasma levels of oxidized metabolites of perillyl alcohol (e.g. perillic acid and dihydroperillic acid) in one male and one female beagle dog given perillyl alcohol at 250 mg/kg bw by gavage were found 1 and 5 h after administration, respectively. The half-lives of perillic acid and dihydroperillic acid appeared to be 3.2 and 3.4 h, respectively. Analysis of blood specimens collected before dosing and 10 min to 48 h after dosing indicated the presence of the oxidized metabolites 10 min after administration. The parent compound, perillyl alcohol, was not detectable in plasma (Phillips et al., 1995).

On the basis of these studies, the cyclohexyl derivatives and terpenoid primary alcohols, aldehydes and carboxylic acids are expected to be rapidly metabolized and excreted, primarily as urinary metabolites.

in plasma (Phillips et al., 1995).

On the basis of these studies, the cyclohexyl derivatives and terpenoid primary alcohols, aldehydes and carboxylic acids are expected to be rapidly metabolized and excreted, primarily as urinary metabolites.

(c) Biotransformation

(i) Cyclohexyl derivatives

Six of the 26 flavouring agents in this group have a cyclohexane ring substituent which contains a primary oxygenated functional group (Nos 961–966). The remaining 20 substances (Nos 967–986) contain a monocyclic or bicyclic terpene ring substituent. Regardless of the ring substituent, metabolism is governed mainly by the presence of a primary oxygenated functional group.

Studies on representative flavouring agents indicated that they are metabolized primarily by oxidation of the primary alcohol or aldehyde function to yield the corresponding carboxylic acid, or oxidation of the alkyl ring substituents to yield polyoxygenated polar metabolites, which are readily excreted. The metabolic options available to alicyclic substances increase as the number and types of functional groups and ring substituents in the molecule increase. If a primary alcohol, aldehyde or carboxylic acid function is present on an alkyl side-chain, the substance may undergo β-oxidation and subsequent ring cleavage. If the number of carbons in the side-chain is even, β-oxidative cleavage yields acyclic metabolites, primarily dicarboxylic acids, which presumably enter the citric acid cycle and are completely metabolized. If the number of carbons present in the side-chain is odd, β-oxidative cleavage leads to aromatization and the production of cyclic metabolites which are excreted primarily in the urine.

Figure 1. Metabolism of cyclohexanecarboxylic acid derivatives in mammals

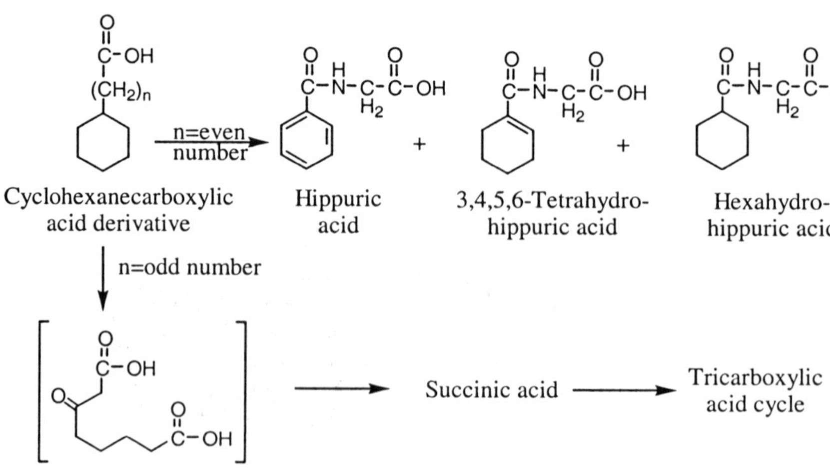

Studies with [^{14}C]cyclohexanecarboxylate sodium salt administered intraduodenally to male Wistar albino rats showed decreasing hippurate excretion with increasing dose but increasing excretion of 3,4,5,6-tetrahydrohippurate and hexahydrohippurate and glucuronide in urine. Biliary excretion of the glucuronic acid conjugates of benzoic acid and cyclohexane carboxylic acid was reported to be dose-related. The authors suggested that the glycine conjugation pathway became saturated as the cyclohexanecarboxylic acid dose increased, resulting in increased excretion of the corresponding glucuronic acid conjugates (Brewster et al., 1977a). These results also confirm the aromatization of cyclohexanecarboxylic acid to yield benzoic acid in vivo (Williams, 1959; Brewster et al., 1977a). In a parallel experiment, male Wistar albino rats given sodium cyclohexanecarboxylic acid with a ^{14}C-labelled ring orally at a dose of 100 mg/kg bw excreted > 98% of the original dose as the urinary metabolites hippurate, 3,4,5,6-tetrahydrohippurate, hexahydrohippurate and the glucuronide conjugates of benzoate and cyclohexanecarboxylate, and < 1% in faeces or expired air (Brewster et al., 1977a).

When 25 mg of [^{14}C]cyclohexanecarboxylic acid were incubated with perfused rat liver preparations for 6 h, 16% cyclohexylcarbonyl-β-D-glucuronide was excreted in the bile. The remainder was present in the perfusate as unchanged cyclohexanecarboxylic acid (10%), hippuric acid (50%), the glucuronic acid conjugate of cyclohexanecarboxylic acid (2–4%), hexahydrohippuric acid (2%), 3,4,5,6-tetrahydrohippuric acid (2%) and benzoic acid (1–2%). Unlike in the intact animal, no benzoyl glucuronide was detected. All five metabolites in the perfusate were detected by 0.5 h. After 2.5 h, the concentrations of cyclohexylcarbonyl glucuronide, hexahydrohippuric acid, tetrahydrohippuric acid and benzoic acid were at their maximum (10–12%, 4%, 4% and 2%, respectively). The amount of cyclohexylcarbonyl glucuronide diminished dramatically during the final period of perfusion owing to excretion in the bile (Brewster et al., 1977b).

Two rabbits were given deuterated cyclohexanecarboxylic acid orally at a daily dose of 1000 mg per rabbit for 3 or 4 days, resulting in cumulative intakes of 3000 and 4000 mg per rabbit. Urine collected for 4 days during and after administration contained benzoic acid, representing 27% and 53% of the two doses, respectively. In the same study, two groups of twin male dogs were maintained on a diet containing 1000 mg of ring-deuterated cyclohexanecarboxylic acid per day, resulting in approximate intakes of 77 and 72 mg/kg bw per day for 8 days or 69 and 68 mg/kg bw per day for 4 days. At the end of treatment, urinary analysis revealed benzoic acid concentrations equivalent to 20% and 24% for dogs in the first group and 33% and 25% for those in the second. Also in the same study, four men and two women were given a single dose of 4000 mg of ring-deuterated cyclohexanecarboxylic acid; urine was collected from four persons for 3 days and from the remaining two for 3 days. The concentrations of deuterium-labelled benzoic acid in urine represented 55–91% of the administered dose (Bernhard & Caflisch-Weill, 1945).

As cyclohexaneacetic acid (No. 965) contains an even-numbered carbon side-chain, it is predicted to undergo β-oxidation and subsequent ring cleavage to yield dicarboxylic acids, which can be completely oxidized via the citric acid cycle (see Figure 1; Williams, 1959). Succinic acid was identified as a urinary metabolite after subcutaneous injection of cyclohexaneacetic acid into a dog, suggesting that ring oxidation and cleavage had occurred. Unchanged cyclohexaneacetic acid was detected in the urine (Bernhard, 1937).

Cyclohexanepropionic acid undergoes oxidation to its homologue cyclohexanecarboxylic acid, which is subsequently aromatized to benzoic acid and excreted mainly as hippuric acid (Williams, 1959).

(ii) Terpenoid primary alcohols, aldehydes and carboxylic acids

In mammals, monocyclic terpenoid primary alcohols (e.g., perillyl alcohol) and monocyclic (e.g., perillaldehyde) and bicyclic (e.g., myrtenal) aldehydes which contain alkyl ring substituents are generally oxidized to the corresponding carboxylic acid, conjugated with glucuronic acid and excreted in urine. In a minor pathway, the aldehyde may be reduced to the alcohol and excreted as the glucuronide (Ishida et al., 1989; Haag & Gould, 1994). If an endocyclic double-bond is present, the metabolite can be reduced by the action of gut microflora, providing it reaches the lower intestinal tract, for example via biliary excretion. If the terpene is monocyclic (e.g., perillyl alcohol), the resulting acid metabolite can undergo aromatization of the ring to yield a benzoic acid derivative, which can be excreted as the glycine conjugate (see Figure 2; Ishida et al., 1989).

The metabolic fate of the terpenoid aldehydes (Nos 967, 968, 971 and 977–979) can be predicted on the basis of known biotransformations of the representative aldehydes *para*-mentha-1,8-dien-7-al (i.e., perillaldehyde; No. 973) and 2-formyl-6,6-dimethylbicyclo[3.1.1]hept-2-ene (i.e., myrtenal, No. 980) (Ishida et al., 1989).

Six male rabbits weighing 2.5–3.0 kg were each given 2000 mg of *para*-mentha-1,8-dien-7-al (perillaldehyde, No. 973) orally, and their urine was collected for 3

Figure 2. Proposed metabolism of perillaldehyde in mammals

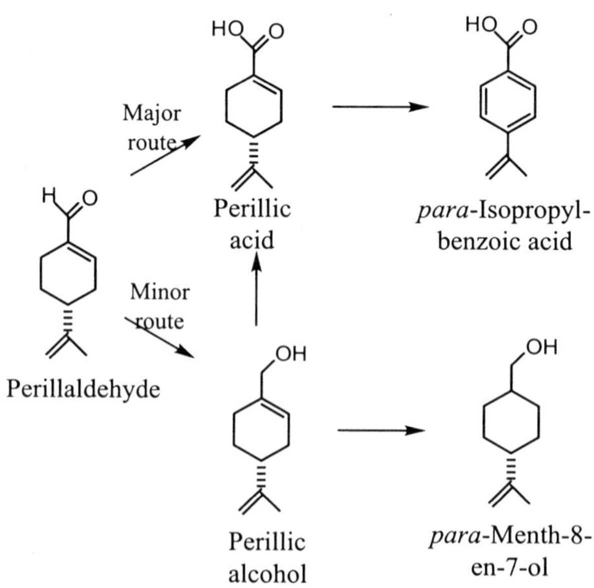

consecutive days, pooled, extracted and separated into a neutral fraction (which contained 0.83 g of metabolites) and an acidic fraction which was methylated (which contained 4.6 g of metabolites). (–)-Perillyl alcohol and (–)-*cis*-shisool (i.e., cyclohexanemethanol, 4-(1-methylethenyl)- or *para*-menth-8-en-7-ol) represented 46% and 39% of the neutral metabolites, and perillic acid represented 57% of the acidic metabolites detected in the urine. *para*-Isopropylbenzoic acid was also detected in urine. These results indicate that perillaldehyde was oxidized mainly to *para*-mentha-1,8-dien-7-carboxylic acid (perillic acid), which was converted in part to *para*-isopropylbenzoic acid by aromatization of the cyclohexene ring and reduction of the isopropenyl double bond. To a lesser extent, perillaldehyde was reduced to perillyl alcohol, which can be selectively hydrogenated to yield *para*-mentha-8-en-7-ol (see Figure 2) (Ishida et al., 1989).

A similar study was performed in which six male rabbits were given a total dose of 12 000 mg of 2-formyl-6,6-dimethylbicyclo[3.1.1]hept-2-ene (i.e., (–)-myrtenal; No. 980). The dose was excreted in the urine as myrtenol, *cis*-10-pinanol (i.e., dihydromyrtenol), myrtenic acid and perillic acid. The urines from the rabbits were pooled, extracted and separated into a neutral fraction (which contained 0.64 g of metabolites) and an acidic fraction which was methylated (which contained 2.9 g of metabolites). Myrtenol and *cis*-10-pinalol represented 55 and 44% of the neutral extract, respectively, and myrtenic acid represented 76% of the acidic metabolites detected in urine. These results indicate that myrtenal was metabolized mainly to the corresponding carboxylic acid (myrtenic acid), which is presumably excreted as the glucuronic acid conjugate. The presence of perillic acid indicates some cleavage of the strained bicyclic ring. To a lesser extent, the aldehyde can either be reduced to myrtenol (No. 981), which is conjugated with glucuronic acid and excreted, or undergoes hydrogenation of the double-bond to yield dihydromyrtenol, probably via reduction by intestinal microflora (Ishida et al., 1989; Figure 3). Workers exposed to

Figure 3. Proposed metabolism of myrtenal in rabbits

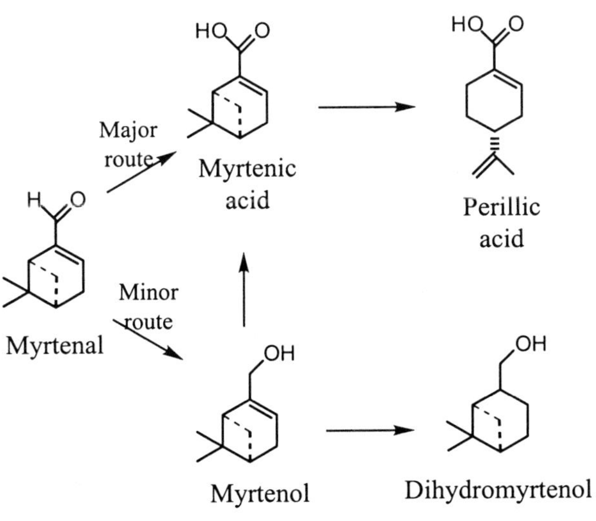

sawmill dust containing myrtenol excreted the glucuronic acid conjugate in their urine (Eriksson & Levin, 1990). Reduction of the aldehyde to the corresponding alcohol may be more important in the metabolism of sterically hindered aldehydes.

The metabolic fate of the terpenoid alcohols (Nos 970, 981 and 986) in this group is expected to be similar to that of the structurally related aldehydes, although there would probably be greater excretion of the alcohol glucuronide and less oxidation to the acid. Data on the biotransformation of *para*-mentha-1,8-dien-7-ol (i.e. perillyl alcohol, No. 974) and santalol (α and β) (No. 984) support this conclusion (reference not given). Bicyclic and tricyclic compounds (e.g., Nos 980–986) would also undergo microsomal ring oxidation (Hold et al., 2002).

In female Wistar-Furth rats fed a diet of 2% perillyl alcohol for 3, 5 or 10 weeks, perillic acid and dihydroperillic acid were major metabolites in plasma, and the methyl esters of perillic acid and dihydroperillic acid were identified as minor metabolites. The authors concluded that the methyl esters were artefacts formed during processing of the urine. Unchanged perillyl alcohol was not detected. The same plasma metabolites were identified 4 h after female Wistar-Furth rats were given perillyl alcohol at a single dose of 1000 mg/kg bw in sesame oil by gavage. No trace of perillyl alcohol was found at any time, including 15 min after gavage. These results indicate that perillyl alcohol is rapidly metabolized and absorbed from the gastrointestinal tract. The presence of dihydroperillic acid indicated that the endocyclic alkene had been hydrogenated (Haag & Gould, 1994).

A study in male Wistar rats confirmed that the oxidation of perillyl alcohol to perillic acid involves perillaldehyde as an intermediate. The rats were given perillyl alcohol, perillaldehyde or perillic acid intravenously at 80 μmol/kg bw in dimethyl sulfoxide, representing doses of 12, 12 and 13 mg/kg bw, respectively, and urine and bile were collected for 2 h after administration. In all cases, the glucuronic acid conjugate of perillic acid was the predominant metabolite detected in urine and bile. The glucuronic acid conjugate of perillyl alcohol was also a major biliary metabolite after administration of perillyl alcohol. The authors concluded that 56% of the original dose had been oxidized to perillaldehyde within 2 h, followed by conversion to perillic acid and eventually excretion as a glucuronic acid conjugate (Boon et al., 2000).

If the cyclic terpene contains an alkyl substituent of greater chain length than C-3, the primary alcohol, aldehyde, acid or related ester may undergo β-oxidation and cleavage of the side-chain to yield carboxylic acid metabolites of lower relatuve molecular mass. Elemental analysis of the urinary metabolites of rabbits given repeated oral doses of an α,β-santalol mixture (a tricyclic primary alcohol containing a C-6 alkyl ring substituent; No. 984) showed that a metabolite was formed by loss of an isoprene (C-5) unit from santalol (Hildebrandt, 1902). Structurally related triterpenoid alcohols (farnesol) and ketones (geranylgeranylacetone) undergo β-oxidation and cleavage to produce chain-shortened polar diacid fragments. For example, C-7, C-9 and C-11 diacid fragments (32%, 15% and 5.3%, respectively, of total radioactivity in urine) were identified in the urine of rats 24 h after an oral dose of [^{14}C]geranylgeranylacetone at 125 mg/kg bw (Nishizawa et al., 1987).

The studies indicate that 19 of the 26 alicyclic substances are expected to be metabolized primarily by oxidation to yield carboxylic acid derivatives that are readily conjugated and excreted primarily in the urine. There are no adequate data on the metabolism of the remaining seven polycyclic terpenoid substances and also no adequate data to demonstrate that their metabolic pathway is similar to that of the monocyclic terpenoids.

2.3.2 Toxicological studies

(a) Acute toxicity

Oral LD_{50} values have been reported for 16 of the 26 substances in this group. The values ranged from 890 to 5700 mg/kg bw for rats and > 1000 to 4000 mg/kg bw for mice, demonstrating that the oral acute toxicity of alicyclic primary alcohols, aldehydes, acids and related esters is low (Table 4).

(b) Short-term and long-term and studies of toxicity

The results of short-term and long-term studies of toxicity with representative alicyclic primary alcohols, aldehydes, carboxylic acids and related esters are summarized in Table 5 and described below.

(i) Allyl ester of cyclohexanepropionic acid

Allyl cyclohexanepropionate was evaluated by the Committee at its forty-sixth meeting in a group of allyl esters (Annex 1, reference 122). The Committee concluded that allyl cyclohexanepropionate does not present a safety concern at present levels of intake. It is expected to hydrolyse in vivo to form allyl alcohol and cyclohexanepropionic acid, and the latter would undergo β-oxidation to form cyclohexanecarboxylic acid (Williams, 1959).

Groups of five male and five female Osborne-Mendel rats were fed diets containing allyl cyclohexanepropionate at a concentration of 1000 or 2500 mg/kg, equivalent to 50 and 125 mg/kg bw per day, for 27–28 weeks or 1 year, respectively. A control group of 10 male and 10 female animals was included. The general health, food intake and weight of the rats were recorded weekly, with no significant difference found between test and control groups. Haematological examinations (haemoglobin, erythrocyte volume fraction, erythrocyte count, leukocyte count) conducted at months 3, 6 and 12 revealed normal values. At termination, the weights of the liver, kidneys, spleen, heart and testes showed no treatment-related changes. Macroscopic and microscopic examination of the liver, kidneys, spleen, heart, testes and remaining abdominal and thoracic viscera and a hind leg (for bone, bone marrow and muscle) revealed no changes that could be associated with administration of the test substance (Hagan et al., 1967).

(ii) (2,2,3-Trimethylcyclo-pent-3-en-1-yl)acetaldehyde (No. 967)

(2,2,3-Trimethylcyclo-pent-3-en-1-yl)acetaldehyde (No. 967) was administered in corn oil to groups of 16 Wistar rats of each sex by gavage for 90 days at a dose of 12 mg/kg bw per day. The vehicle control animals were given corn oil at 5 ml/kg bw per day. The rats were weighed for 6 days before treatment, on the first day of dosing and at various times throughout the study. Food intake was recorded at 3- or 4-day intervals throughout the study. Treated males showed slightly greater weight gain than controls due to greater food intake. The authors cited previous studies (Gaunt et al., 1971; Branton et al., 1972; Gaunt et al., 1974; Butterworth et al.,

Table 4. Studies of the acute toxicity of alicyclic primary alcohols, aldehydes, acids and related esters used as flavouring agents

No.	Agent	Species	Sex[1]	Route	LD_{50} (mg/kg bw)	Reference
961	Cyclohexanecarboxylic acid	Rat	M,F	Gavage	3300	Terrell & Cooper (1978)
			M	Gavage	3200	
			F	Gavage	3300	
961	Cyclohexanecarboxylic acid	Rat	M,F	Gavage	3300	Moran et al. (1980)
962	Methylcyclohexanecarboxylate	Rat	M,F	Gavage	4000	Moran et al. (1980)
963	Ethyl cyclohexanecarboxylate	Rat	M,F	Gavage	4000	Moran et al. (1980)
964	Cyclohexaneethyl acetate	Rat	NR	Oral	3200 (2 doses)	Wohl (1974)
964	Cyclohexaneethyl acetate	Rat	NR	Oral	2200	Moreno (1978)
967	2,2,3-Trimethylcyclopent-3-en-1-yl acetaldehyde	Rat	NR	Oral	4,300	British Industrial Biological Research Association (1976)
967	2,2,3-Trimethylcyclopent-3-en-1-yl acetaldehyde	Rat	NR	Oral	4,100	Moreno (1978)
969	Campholene acetate	Rat	M	Gavage	4600–5300	Piccirillo et al. (1979)
			F	Gavage	3000	
			M, F	Gavage	3900	
970	α-Campholene alcohol	Rat	NR	Gavage	1000–2000	Levenstein (1982)
973	*para*-Menth-1,8-dien-7-al	Mouse	M	Gavage	> 1000	Honda et al. (1986)
973	*para*-Menth-1,8-dien-7-al	Mouse	NR	Oral	1700	Moreno (1978)
974	*para*-Menth-1,8-dien-7-ol	Rat	NR	Oral	2,100	Moreno (1977)
976	1,2,5,6-Tetrahydrocuminic acid	Rat	NR	Gavage	> 2500– < 5000	Levenstein (1981)
977	2,6,6-Trimethylcyclohexa-1,3-dienyl methanal	Mouse	M,F	Oral	3500	Pellmont (1967)
977	2,6,6-Trimethylcyclohexa-1,3-dienyl methanal	Rat	NR	Oral	5700	Pellmont (1969)
979	2,6,6-Trimethyl-1 and 2-cyclohexen-1-carboxaldehyde	Mouse	NR	Oral	4000	Pellmont (1973)
980	2-Formyl-6,6-dimethylcicyclo-[3.1.1]hept-2-ene	Rat	NR	Oral	2300	Moreno (1982)
982	Myrtenyl acetate	Rat	M,F	Gavage	1000–2500	Kynoch & Lloyd (1977)
982	Myrtenyl acetate	Rat	NR	Oral	2600	Moreno (1982)
983	2-Hydroxy-6,6-dimethylbicyclo-[3.1.1]hept-2-enyl formate	Rat	M,F	Gavage	> 5000	Moreno (1972)
986	10-Hydroxymethylene-2-pinene	Rat	NR	Oral	890	Moreno (1977)

M, male; F, female; NR, not reported

Table 5. Results of short-term and long-term studies of toxicity and carcinogenicity on alicyclic primary alcohols, aldehydes, acids, related esters and some related compounds used as flavouring agents

No.	Substance	Species; sex	No. test groups[a]/ no. per group[b]	Route	Duration	NOEL (mg/kg bw per day)[d]	Reference
967	(2,2,3-Trimethylcyclo-pert-3-en-1-yl)acetaldehyde	Rat, MF	1/16 per sex	Gavage	90	>12	British Industrial Biological Research Association (1976)
974	*para*-Mentha-1,8-dien-7-cl	Rat, MF	3/NR	Gavage	28	200	National Cancer Institute (1996)
974	*para*-Mentha-1,8-dien-7-cl	Rat, MF	3/NR	Gavage	90	120[d] < 40[e]	National Cancer Institute (1996)
974	*para*-Mentha-1,8-dien-7-ol	Rat, F	4/20	Diet	105	1300	Haag & Gould (1994)
974	*para*-Mentha-1,8-dien-7-ol	Rat, F	1/NR	Diet	14	> 2000	Ren & Gould (1998)
974	*para*-Mentha-1,8-dien-7-ol	Rat, F	1/26	Diet	70	> 2500	Haag & Gould (1994)
974	*para*-Mentha-1,8-dien-7-ol	Rat, M	1/10	Diet	133	> 2000	Mills et al. (1995)
974	*para*-Mentha-1,8-dien-7-o	Rat, M	2/36	Diet	365	> 2000	Reddy et al. (1997)
974	*para*-Mentha-1,8-dien-7-o	Hamster, M	1/NR	Diet	42	> 1500	Stark et al. (1995)
974	*para*-Mentha-1,8-dien-7-ol	Hamster, M	1/NR	Diet	28	> 4000	Burke et al. (1997)
974	*para*-Mentha-1,8-dien-7-o	Dog, M/F	3/NR	Gelati capsule	28	300	National Cancer Institute (1996)
974	*para*-Mentha-1,8-dien-7-ol	Dog, M/F	3/NR	Gelatin capsule	90	120	National Cancer Institute (1996)

M, male; F, female; NR, not reported
[a] Total number of test groups does not include control animals.
[b] Total number per test group includes both male and female animals.
[c] > indicates highest dose tested without adverse effect.
[d] Based on systemic effects
[e] Based on hyperexcitement and o-al discharge

1975) to attribute the increased food intake to the strong flavour of the test substance. This food consumption pattern was less obvious in the female rats, but, on average, treated females consumed slightly more food than the controls. Haematological examination of eight animals of each sex per group at week 6 and of 16 of each sex per group at week 12 showed no differences in haemoglobin concentration, packed cell volume, erythrocyte or leukocyte count, differential leukocyte count or blood urea nitrogen between treated and control rats. A post-mortem examination was conducted in which the liver and kidneys were weighed and abnormalities noted, and the liver and kidneys of all rats and about 25 tissues from eight animals of each sex per group were examined microscopically. The only difference observed between test and control rats was thickened alveoli in the lungs of treated males. The authors considered that this effect was not due to the test substance (British Industrial Biological Research Association, 1976). The dose of 12 mg/kg bw per day that resulted in no adverse effects is at least 100 000 times the daily per capita intake ('eaters only') (see Table 2) of 0.073 μg/kg bw per day from use of (2,2,3-trimethyl-cyclo-pent-3-en-1-yl)acetaldehyde as a flavouring agent in Europe.

(iii) para*-Mentha-1,8-dien-7-ol (perillyl alcohol; No. 974)*

Numerous studies have shown that high dietary levels of perillyl alcohol given to experimental animals may be protective against known carcinogens. *para*-Mentha-1,8-dien-7-ol (i.e. perillyl alcohol) is a monoterpene that inhibits the growth of pancreatic, mammary and liver tumours. It has been used in animals as a chemotherapeutic agent for neuroblastoma, prostate and colon cancer and has possible chemotherapeutic applications for skin and lung cancer (Crowell, 1997; Belanger, 1998).

In preclinical studies, nude mice given perillyl alcohol by gavage at a dose of 1000 mg/kg bw in sesame oil had lost weight (3 g) by day 10 of treatment, and two of three mice died on day 13; seven of 10 nude mice bearing HT-29 xenografts that were given 1000 mg/kg bw in sesame oil by gavage died after two doses. Mice given 1800 mg/kg bw per day in two doses became lethargic after three days. Nude mice with PC-3 prostate tumor xenografts that were given 600 or 800 mg/kg bw per day showed early mortality, three of 10 mice at 600 mg/kg bw per day dying after four doses and six of 10 at 800 mg/kg bw per day dying after five doses (National Cancer Institute, 1996).

Fourteen-day range-finding studies were performed with perillyl alcohol in male Fischer 344 rats and male and female beagle dogs. The rats (number of groups and number per group not reported) received doses of 75–900 mg/kg bw per day in soya bean oil by gavage in three divided doses at an interval of 8 h. At the highest dose, the histological lesions included forestomach inflammation with hyperplasia and testicular degeneration. The dogs (number per group not given) received perillylalcohol at a dose of 60, 300, 600 or 1200 mg/kg bw per day in soya bean oil by gelatin capsule in three divided doses. Dose-related increases in the frequency of vomiting and soft or loose stools were observed. At 1200 mg/kg bw per day, leukocytosis and thrombocytosis were also seen (National Cancer Institute, 1996).

Female Wistar-Furth rats received a diet containing 2% perillyl alcohol, corresponding to about 2000 mg/kg per day (Food and Drug Administration, 1993), and pair-fed rats received the control diet of powdered laboratory chow. Two weeks after initiation of the perillyl alcohol diet, 10 μl of [^{14}C]mevalonolactone were injected into the upper four abdominal mammary glands of each rat. Analyses of the mammary epithelial cells 3 h later showed that perillyl alcohol had inhibited coenzyme Q synthesis and protein prenylation in the mammary parenchyma by approximately 20%. Cholesterol synthesis was not affected. No observable signs of toxicity were reported. The authors suggested that inhibition of coenzyme Q synthesis could be involved in the anticancer action of perillyl alcohol (Ren & Gould, 1998). The dose of 2000 mg/kg bw per day that resulted in no adverse effects is 10 000 000 times the daily per capita intake ('eaters only') of 0.023 μg/kg bw per day from use of perillyl alcohol as a flavouring agent in Europe.

Twenty-eight-day studies were performed with perillyl alcohol in male and female Fischer 344 rats and male and female beagle dogs. The rats (number per group not reported) received 200, 600 or 1000 mg/kg bw per day in soya bean oil by gavage in two divided doses because of its irritating properties. Animals at the two higher doses showed 12–13% body-weight loss on days 4–10, increased activity of alanine aminotransferase (by 32–79% in females at 600 mg/kg bw per day and in both sexes at 1000 mg/kg bw per day), dose-related forestomach hyperplasia and testicular atrophy. Animals at 1000 mg/kg bw per day showed renal tubular degeneration (with rising blood urea nitrogen) and mortality (10% of males and 50% of females) concomittant with splenic atrophy (females only) and hepatocyte cytoplasmic vacuolization.

The dogs (number per group not reported) received 300, 600 or 1200 mg/kg bw per day in soya bean oil by gelatin capsule in three divided doses. One dog at 1200 mg/kg bw died on day 13 with signs of gastrointestinal distress after consistent low food consumption; mild bone marrow atrophy and lymph node and tissue atrophy were also found. Emesis and abnormal stools were seen at the two higher doses, with dose-related renal tubular degeneration, thymic atrophy and increased blood urea nitrogen values (National Cancer Institute, 1996).

In 90-day studies, perillyl alcohol in in soya bean oil was given at a dose of 40, 120 or 400 mg/kg bw per day by gavage to groups of Fischer 344 rats (number per group not reported) and by capsule to groups of beagle dogs (number per group not reported). The animals were given food and drinking-water *ad libitum* and monitored for survival, behaviour and haematological changes. In rats, no treatment-related deaths, abnormal haematological or clinical chemical findings or gross lesions were reported at any dose. Significant weight loss (10%) was seen at 400 mg/kg bw per day. The weights of the liver, kidney and lung were increased in females at this dose, but no microscopic changes were observed. The dose-related signs (in all treated groups) included hyperexcitement and a clear oral discharge. No deaths occurred among the dogs. The highest dose was associated with emesis (more severe in females) and mild anaemia. Prothrombin time, activated partial prothrombin time and the albumin:globulin ratio were increased significantly in females at the highest dose, possibly due to severe emesis. Gross necropsy revealed testicular atrophy in one male dog at this dose. The NOEL was 120 mg/kg bw per day (National Cancer Institute, 1996).

d-Limonene, the parent compound of perillyl alcohol, has been reported in published studies to induce hyaline droplet renal nephropathyin adult male rats. The protein reponsible for this lesion has been identified and found to be unique to certain strains. Although the protein is not present in humans, members of the same superfamily have been found; thus, it is possible that this effect may occur in humans. The National Cancer Institute (1996) suggested that renal nephropathy may be induced in rats of each sex at doses greater than 400 mg/kg bw per day.

Groups of 11 male Syrian golden hamsters aged 10 weeks which had been injected with PC-1 pancreatic carcinoma cells were given diets containing 0 or 3% perillyl alcohol, equivalent to 1500 mg/kg bw per day, immediately after palpation of tumours ≥ 2 mm in diameter. After 3 weeks of treatment, perillyl alcohol had significantly reduced the tumour growth rate to less than half that of the controls. Complete regression of tumours was seen in 16% (5/31) of hamsters treated with perillyl alcohol and none of the controsl (0/25). The body-weight gain of the treated hamsters was normal, and no adverse microscopic effects were observed in the liver, kidney or pancreas (Stark et al., 1995).

In a similar study, groups pf 15 male Syrian golden hamsters received a control diet or a diet containing perillyl alcohol at a concentration of 40 000 mg/kg for 4 weeks and then 1 week later a subcutaneous injection of PC-1 hamster pancreatic ductal carcinoma cells. Tumour growth was inhibited by 50% in the group given perillyl acohol. The body weights of the animals were not affected, and the total plasma cholesterol concentrations were normal. Tumour growth was not inhibited at 20 000 mg/kg of diet (Burke et al., 1997).

To study the effect of perillyl alcohol on liver tumours, 21 male Fischer 344 rats were given drinking-water containing 50 ppm *N*-nitrosodiethylamine for 1 month. Two weeks later, 11 rats were fed a control diet of powdered food and 10 were fed powdered food containing 1% (w/w) perillyl alcohol for 1 week, followed by 2% (w/w) perillyl alcohol, equivalent to 2000 mg/kg bw per day (Food and Drug Administration, 1993), for 18 weeks. At the end of the study, the treated and untreated rats were killed and the livers removed, and total liver weight, tumour weight and number of tumours were recorded. The mean tumour mass per liver was 0.8 g in the treated and 7.0 g in the untreated animals, indicating a 10-fold decrease in the mean liver tumour mass of the perillyl alcohol-treated rats. No change in normal liver weight was observed. The authors determined that perillyl alcohol did not alter the proliferation rate of tumour cells but rather promoted apoptosis, thereby increasing the loss of tumour cells (a fivefold increase in apoptotic index in large tumours and a 10-fold increase in small tumours). The 10% decrease in body weight observed in the treated rats was attributed to a decrease in body fat (Mills et al., 1995).

In a long-term study designed to investigate the chemopreventive effects of perillyl alcohol on colon carcinogenesis, groups of 5-week-old male Fischer 344 rats were fed either a control diet (AIN-76A) or diets providing perillyl alcohol at approximately 1000 or 2000 mg/kg bw per day (Food and Drug Administration, 1993). At weeks 3 and 4 after introduction of the diet, groups of 36 rats intended for treatment with a carcinogen were given weekly subcutaneous injections of azoxymethane at 15 mg/kg bw. Control groups of 12 rats per group maintained on basal diet or a diet containing perillyl alcohol at 2000 mg/kg were injected with an equal volume of saline. The animals were maintained on the control and experimental diets for an additional 52 weeks after the last azoxymethane treatment. The general health of the animals was monitored daily, and body weights were recorded every 2 weeks for the first 10 weeks and then every 4 weeks. At the conclusion of the experiment, gross and histopathological examinations were conducted on the stomach, small intestine and large intestine.

The body weights of control animals maintained on the basal diet were similar to those fed the perillyl alcohol-supplemented diets. Gross examination of liver, kidneys, stomach, lungs and intestines at necroscopy revealed no gross changes in any of the three control groups. The body weights of the azoxymethane-treated rats on perillyl alcohol diets were comparable to those of controls throughout the experiment, and no signs of toxicity were reported at any dose. A significant decrease ($p < 0.05$) in the incidence and multiplicity of invasive adenocarcinoma of the colon was reported in the animals given 1000 mg/kg of diet, but no effect was seen on tumours in the small intestine. At 2000 mg/kg, the incidence and multiplicity of adenocarcinomas in the small intestine (specifically the invasive type) and the incidence of total (invasive plus noninvasive) adenocarcinomas in the colon were significantly decreased. Animals at both dietary concentrations showed a five- to sixfold increase in the apoptotic index ($p < 0.0001$) in colon tumours as compared with rats fed control diets. The authors suggested that perillyl alcohol-induced apoptosis inhibits tumour promotion in vivo (Reddy et al., 1997).

Studies on the related monoterpene limonene also give an indication of the potential preventive activity of perillyl alcohol against colon cancer and metastasis of established colon tumours. Limonene is metabolized to perillic acid and dihydroperillic acid, the same metabolites as perillyl alcohol. Two weeks before administration of a single dose of [^{3}H]7,12-dimethylbenz[*a*]anthracene (DMBA) by gastric intubation, groups of six female Wistar-Furth rats were placed on either a control diet of laboratory chow or a diet containing 5% limonene (w/w in basal diet). Urine was collected for 24 h after intubation of DMBA, and the rats were then killed, blood was collected and DNA was extracted from the liver, lung, kidney and spleen. About 50% less DMBA-associated radioactivity was found in the tissues of the limonene-fed animals than in controls, although the circulating levels of DMBA-associated radioactivity did not differ from those of controls. A 2.3-fold increase in DMBA and/or DMBA-derived metabolites was reported in the urine of the limonene-fed animals. These results suggest that limonene induces modifications in DMBA metabolism, resulting in the formation of more water-soluble DMBA metabolites, which are excreted (Maltzman et al., 1991).

Limonene also significantly inhibited metastasis to the liver (80%) of colonic tumour cells implanted in the spleens of mice (Broitman et al., 1996; Belanger, 1998).

A series of studies was conducted on the effect of perillyl alcohol on rats with mammary tumours induced by DMBA at 50 mg/kg bw. In the first study, groups of 26 female Wistar-Furth rats were fed a diet containing 2.5% perillyl alcohol, equivalent to 2500 mg/kg bw per day (Food and Drug Administration, 1993), for 10 weeks after palpation of the first mammary tumour $\geq$ 3 mm in diameter. After 3 weeks of treatment, 81% of the primary tumours detected in the perillyl alcohol-treated rats had regressed completely, as compared with 31% that regressed spontaneously in the pair-fed controls. The time required for complete regression was 3 and 9 weeks inthe treated and control group, respectively. Perillyl alcohol also prevented the development of secondary tumours after the initial diet assignment. No signs of toxicity were reported. Slight weight loss due to initial food aversion was reported in the rats given the diet containing 2.5% perillyl alcohol. Increased weight gain was reported as food consumption increased, but the control weights were never achieved. Several animals given 3% perillylalcohol in the diet in this study died (Haag & Gould, 1994).

In the second study, groups of 20 female Wistar-Furth rats were placed on diets containing 0, 0.5, 1, 1.5 or 2% perillyl alcohol for 15 weeks, after palpation of mammary tumours ≥ 10 mm in diameter. The maximum daily intake of perillyl alcohol on the basis of food consumption was calculated to be 0, 410, 880, 1300 and 1600 mg/kg bw, respectively. A 25–75%, dose-related regression (partial plus complete) of primary tumours was observed in the treated rats. A level of 1% in the diet was required for significant (35%) complete regression. No spontaneous regression of primary tumours was observed in the control animals. No observable signs of toxicity were reported at any dose. In all treated groups, some food aversion was seen at the beginning of feeding, which disappeared within 2 weeks. Body-weight gain was slightly depressed at the highest dietary concentration. Interestingly the concentration of perillyl alcohol in the diet that blocked formation of subsequent tumours in rats (1%) was 10 times higher than the concentration (1000 mg/kg of diet) that was chemopreventive in the colon (Haag & Gould, 1994; National Cancer Institute, 1996).

(c) Genotoxicity

No data were available for the six cyclohexane derivatives. Almost all the studies of monocyclic and polycyclic terpenoids were performed with *para*-mentha-1,8-dien-7-al (i.e., perillaldehyde), except for one assay for reverse mutation with 2,6,6-trimethylcyclohexa-1,3-dienyl methanal.

(i) In vitro

Alicyclic primary alcohols, aldehydes, acids and related esters: Two substances representative of the monocyclic and bicyclic terpenoid alcohols, aldehydes and carboxylic acids, namely 2,6,6-trimethylcyclohexa-1,3-dienyl methanal and *para*-mentha-1,8-dien-7-al (i.e., perillaldehyde), have been tested for genotoxicity. Negative results were reported in standard (Florin et al., 1980) assays for reverse mutation when *Salmonella typhimurium* strains TA98, TA100, TA1535 and TA1537 were incubated with 3 μmol of 2,6,6-trimethylcyclohexa-1,3-dienyl methanal/plate (No. 977) with and without metabolic activation. In modified assays for reverse mutation (Ishidate et al., 1984; Fujita et al., 1994), *S. typhimurium* strains TA92, TA94, TA97, TA98, TA100, TA102, TA1535, TA1537 and TA2637 were incubated with up to 1000 μg of *para*-mentha-1,8-dien-7-al (i.e., perillaldehyde) (No. 973), with and without metabolic activation. Negative results were reported (see Table 6).

para-Mentha-1,8-dien-7-al was considered to be weakly positive in an assay for DNA damage in *Bacillus subtilis* strains M45 (rec^+) and H17 (rec^-) at a concentration of 2.5 μl/disc (Yoo, 1986). Negative results were reported in a similar study at concentrations of 0.16–0.63 μl/plate, but positive results were found at higher

Table 6. Studies of genotoxicity with alicyclic primary alcohols, aldehydes, acids and related esters used as flavouring agents

No.	Agent	End-point	Test object	Maximum concentration	Results	Reference
In vitro						
973	*para*-Mentha-1,8-dien- 7-a	DNA damage	*Bacillus subtilis* M45 and H17	2.5 µl/disc	Weak positive	Yoo (1986)
973	*para*-Mentha-1,8-dien- 7-a	DNA damage	*Bacillus subtilis* M45 and H17	0.63 µl	Negative	Kuroda et al. (1984)
973	*para*-Mentha-1,8-dien- 7-al	DNA damage	*Bacillus subtilis* M45 and H17	1.25 µl	Positive	Kuroda et al. (1984)
973	*para*-Mentha-1,8-dien- 7-al	Reverse mutation[a]	*S. typhimurium* TA97, TA102	100 µg/plate	Negative[b]	Fujita et al. (1994)
973	*para*-Mentha-1,8-dien- 7-al	Reverse mutation[a]	*S. typhimurium* TA92, TA1535, TA100, TA1537, TA94, TA98, TA2637	1000 µg/plate	Negative[b]	Ishidate et al. (1984)
977	2,6,6-Trimethylcyclohexa-1,3-dienyl methanal	Reverse mutation	*S. typhimurium* TA98, TA100, TA1535, TA1537	3 µmol/plate	Negative[b]	Florin et al. (1980)
973	*para*-Mentha-1,8-dien- 7-al	DNA repair	*Escherichia coli* PQ37	Not reported	Negative	Eder et al. (1993)
973	*para*-Mentha-1,8-dien- 7-al	Sister chromatid exchange	Chinese hamster ovary cells	200 µg/ml	Positive[c,d]	Tayama et al. (1990)
973	*para*-Mentha-1,8-dien- 7-al	Sister chromatid exchange	Chinese hamster ovary cells	300 µg/ml	Positive[d,e]	Tayama et al. (1990)
973	*para*-Mentha-1,8-dien- 7-al	Chromosomal aberration	Chinese hamster fibroblasts	50 µg/ml	Positive[c]	Ishidate et al. (1984)
973	*para*-Mentha-1,8-dien- 7-al	Chromosomal aberration	Chinese hamster ovary cells	300 µg/ml	Negative[d,e,f]	Tayama et al. (1990)
973	*para*-Mentha-1,8-dien- 7-al	Chromosomal aberration	Chinese hamster ovary cells	200 µg/ml	Negative[c,d,f]	Tayama et al. (1990)
973	*para*-Mentha-1,8-dien- 7-al	Mutagenicity	Chinese hamster ovary cells	100 µg/ml	Negative[c,g]	Sasaki et al. (1990)

Table 6 (contd)

No.	Agent	End-point	Test object	Maximum concentration	Results	Reference
In vitro **(continued)**						
973	*para*-Mentha-1,8-dien- 7-al	Mutagenicity	Human fetus cells (Rsa)	0.025 μg/ml	Positive[h]	Suzuki et al. (1990)
973	*para*-Mentha-1,8-dien- 7-al	Mutagenicity	Human fetus cells (Rsa)	0.010 μg/ml	Negative[h]	Suzuki et al. (1990)
973	*para*-Mentha-1,8-dien- 7-al	Mutagenicity	Human fetus cells (Rsa)	200 ng/ml (1994)	Positive[i]	Suzuki & Suzuki
973	*para*-Mentha-1,8-dien- 7-al	Mutagenicity	*E. coli* WP2	0.4 mg/plate	Negative	Yoo (1986)
In vivo						
973	*para*-Mentha-1,8-dien- 7-al	Micronucleus formation	Mouse bone marrow erythrocytes	600 mg/kg bw	Negative[j]	Hayashi et al. (1988)

[a] Preincubation with exogenous metanolic system from rat liver
[b] Assay performed with and without metabolic activation
[c] Assay performed without metabolic activation
[d] Cytotoxic at 150 μg/ml
[e] Assay performed with metabolic activation
[f] Positive only at cytotoxic concentrations
[g] Cytotoxic at 12 μg/ml
[h] Cytotoxic at 0.025 μg/ml
[i] Cytotoxic at > 20 ng/ml
[j] Administered byintraperitoneal injection

concentrations (1.2–2.5 μl/plate) (Kuroda et al., 1984). Negative results were reported in tests in which *para*-mentha-1,8-dien-7-al was incubated with *Escherichia coli* WP2 cells at 0.05–0.4 mg/plate (Yoo, 1986).

The SOS Chromotest for DNA damage in genetically engineered *E. coli* PQ 37 gave negative results for *para*-mentha-1,8-dien-7-al (Eder et al., 1993). The maximum induction factor was calculated to be 1.0 (Positive results are considered to be significant if the maximum induction factor is at least 1.5). Standard assays for sister chromatid exchange and chromosomal aberrations with *para*-mentha-1,8-dien-7-al yielded equivocal results. An assay for sister chromatid exchange performed with and without metabolic activation in Chinese hamster ovary cells (CHO-K1), exposed for 3 h (expression time, 27 h) to concentrations of 50–300 and 50–100 μg, respectively, yielded positive results for genotoxicity; cytotoxicity was reported at 150 and 300 μg/ml without and with metabolic activation, respectively (Tayama et al., 1990). In another assay conducted in CHO-K1 cells, cytotoxicity was reported at concentrations ≥ 12 μg/ml, and negative results were seen at 10 μg/ml. *para*-Mentha-1,8-dien-7-al may have enhanced *N*-methyl-*N*'-nitro-*N*-nitrosoguanine-induced ouabain-resistant mutations at 10 μg/ml (Sasaki et al., 1990). Chromosomal aberrations were reported to be induced by *para*-mentha-1,8-dien-7-al in Chinese hamster fibroblast cells after 24 and 48 h of exposure to concentrations of 40 and 50 μg/ml without metabolic activation. Polyploidy was observed in this study (Ishidate et al., 1984). No chromosomal aberrations were induced in CHO-K1 cells after 3 h of exposure (expression time, 27 h) to non-cytotoxic concentrations of 50–100 μg/ml without metabolic activation. At the cytotoxic concentration of 150 μg/ml, induction of chromosomal aberrations was observed without metabolic activation. With metabolic activation, concentrations of 50–200 μg/ml did not induce chromosomal aberrations, whereas the cytotoxic concentration of 300 μg/ml did (Tayama et al., 1990).

Studies of mutagenicity in human fetus cells (Rsa) exposed to a concentration of 0, 0.010, 0.015, 0.02 or 0.025 μg/ml gave negative results for *para*-mentha-1,8-dien-7-al at the lowest concentration and positive results at concentrations of 0.015 and 0.02 μg/ml. At the highest concentration of 0.025 μg/ml, the substance was cytotoxic (Suzuki et al., 1990). In another study with Rsa cells, mutagenicity, determined by ouabain resistance, was reported at concentrations of *para*-mentha-1,8-dien-7-al > 10 ng/ml, with apparent cytotoxicity at ≥ 20 ng/ml. In this study, mutagenicity was detected in K-ras codons at concentrations of 2–200 ng/ml (Suzuki & Suzuki, 1994).

(ii) In vivo

In an assay for micronucleus formation, groups of six male ddY mice aged 8 weeks were given a single intraperitoneal injection of *para*-mentha-1,8-dien-7-al at a dose of 0, 75, 150, 300 or 600 mg/kg bw in olive oil. A positive control group received mitomycin C. After 24 h, the mice were killed, and femoral bone-marrow cells were collected, fixed and stained with Giemsa. No indication of micronucleus induction was reported at any dose.

One group of five mice received intraperitoneal injections of 200 mg/kg bw per day for 4 days. No induction of micronuclei was observed 24 h after the last dose, and the percentage of polychromatic erythrocytes was not affected at any dose (Hayashi et al., 1988).

(iii) Conclusion

The relevance to human health of positive results in Chinese hamster ovary cells in vitro is questionable, for the following reasons:

- The studies were performed at near-toxic concentrations, thousands of times higher than those that can be achieved in humans consuming alicyclic primary alcohols, aldehydes, acids and related esters as flavouring agents.
- In many of the assays for chromosomal aberrations, the influence of cytotoxicity (e.g., effect of lysosomal breakdown) was not studied adequately (Zajac-Kaye & Ts'o, 1984; Bradley et al., 1987) under physiological, controlled conditions (Brusick, 1986; Heck et al., 1989).
- The positive results obtained in vitro were not been confirmed in a standard assay in vivo (Hayashi et al., 1988).

2,6,6-Trimethylcyclohexa-1,3-dienyl methanal and *para*-mentha-1,8-dien-7-al did not induce reverse mutation. Mixed results were reported with *para*-mentha-1,8-dien-7-al in assays for DNA damage, sister chromatid exchange, chromosomal aberration and other geneotoxic end-points. As these substances are rapidly metabolized in vivo to compounds of lower toxicological potential, the Committee concluded that the monocyclic and bicyclic terpenes with alkyl ring substituents and containing an alcohol, aldehyde or carboxylic acid group would have little genotoxic potential in vivo.

(d) Other relevant studies

A number of effects that did not contribute to the safety evaluation have been reported in studies on substances in this group. Increased hexobarbital sleeping time, lowered rectal temperature, diminished spontaneous locomotor activity and reduced acetic acid-induced writhing were reported in male ddY mice given a single oral dose of 100 mg or more of α- or β-santalol (Okugawa et al., 1995).

A dose-related decrease in cell proliferation was reported in human and hamster pancreatic tumour cell lines treated with perillyl alcohol. A concentration-dependent increase in apoptosis was reported in rat phaeochromocytoma cells exposed to perillyl alcohol or perillaldehyde at a concentration of 500 and 200 mol/l, respectively (Boon et al., 2000).

The number of mutations induced by *N*-methyl-*N*-nitro-*N*-nitrosoguanidine in CHO-K1 cells was reduced by treatment with perillaldehyde at 15 μg/ml (Sasaki et al., 1990). In the same study, perillaldehyde at 50 μg/ml showed immunosuppressive effects in mouse spleen cells.

2.3.3 Studies in humans

A phase I clinical trial was conducted with perillyl alcohol in patients with advanced malignancies. The substance was administered orally in gelatin capsules, three times a day, at a dose of 800, 1600 or 2400 mg/m^2; two, four and three patients were treated at these doses for 3 months or more. Episodes of gastrointestinal symptoms (such as nausea, unpleasant taste and satiety) and fatigue were reported. Two patients with ovarian cancer showed reversible granulocytopenia. There was no evidence of renal, hepatic or neurological toxicity (Ripple et al., 1998).

The phase I clinical trial was repeated with smaller, more frequent doses to reduce the gastrointestinal effects. Sixteen patients with advanced refractory malignancies were treated orally with perillyl alcohol in soya bean oil at a dose of 800, 1200 or 1600 mg/m^2 four times a day. Mild gastrointestinal effects were observed (mainly at 1200 and 1600 mg/m^2), and again there was no evidence of renal, hepatic or neurological toxicity (Ripple et al., 2000).

Nausea, vomiting, loss of appetite, weight loss, fatigue, short-term memory loss, slurred speech and loss of balance were reported in further phase I trials in which patients with cancer were given perillyl alcohol orally at a dose of 2100 or 2800 mg/m^2. Similar signs of toxicity were not observed in the patients given a lower dose of 1600 mg/m^2 (Belanger, 1998).

3. REFERENCES

Anders, M.W. (1989) Biotransformation and bioactivation of xenobiotics by the kidney. In: Huston, D.H., Caldwell, J. & Paulson, G. D., eds, *Intermediary Xenobiotic Metabolism in Animals*, New York: Taylor & Francis, pp. 81–97.

Belanger, J.T. (1998) Perillyl alcohol: Applications in oncology. *Altern. Med. Rev.*, **3**, 448–457.

Bernhard, K. (1937) Metabolic hydrogenation of the cyclohexane ring. *Hoppe-Seyler's Z. Physiol. Chem.*, **248**, 256–276.

Bernhard, K. & Caflisch-Weill, H. (1945) On the dehydrogenation of hexahydro-benzoic acid in the animal body. *Helv. Chim. Acta*, **28**, 1697–1707.

Boon, P.J.M., van der Boon, D. & Mulder, G.J., (2000) Cytotoxicity and biotransformation of the anticancer drug perillyl alcohol in PC12 cells and in the rat. *Toxicol. Appl. Pharmacol.*, **167**, 55–62.

Bradley, M.O., Taylor, V.I., Armstrong, M.J. & Galloway, S.M. (1987) Relationships among cytotoxicity, lysosomal breakdown, chromosome aberrations, and DNA double-strand breaks. *Mutat. Res.*, **189**, 69–79.

Brantom, P.G., Gaunt, I.F., Grasso, P. & Lansdown, A.B.G. (1972) Short-term toxicity of tolualdehyde in rats. *Food Cosmet. Toxicol.*, **10**, 637–647.

Brewster, D., Jones, R.S. & Parke, D.V. (1977a) The metabolism of cyclohexanecarboxylic acid in the isolated perfused rat liver. *Xenobiotica*, **7**, 601–609.

Brewster, D., Jones, R.S. & Parke, D.V. (1977b) The metabolism of cyclohexanecarboxylic acid in the rat. *Biochem. J.*, **164**, 595–600.

British Industrial Biological Research Association (1976) Short-term toxicity of samples TT171, TT172, TT173, and TT174 in rats. Unpublished report. Submitted to WHO by the Flavor and Extract Manufacturers Association.

Broitman, S.A, Wilkinson, J., 4th, Cerda, S. & Branch, S.K. (1996) Effects of monoterpenes and mevinolin on murine colon tumor CT-26 *in vitro* and its hepatic 'metastases' *in vivo*. *Adv. Exp. Med. Biol.*, **401**, 111–130.

Brusick, D. (1986) Genotoxic effects in cultured mammalian cells produced by low pH treatment conditions and increased Ion concentrations. *Environ. Mutag.*, **8**, 879–886.

Burke, Y.D., Stark, M.J., Roach, S.L.., Sen, S.E. & Crowell, P.L (1997) Inhibition of pancreatic cancer growth by the dietary isopernoids farnesol and geraniol. *Lipids,* **32**, 151–156.

Butterworth, K.B., Carponini, F:M.B., Gaunt, P.F., Grasso, P. & Lloyd, A.G. (1975) *A New Approach to the Evaluation of the Safety of Flavouring Esters*, Carshalton, Surrey: British Industrial Biological Research Association.

Cramer, G.M., Ford, R.A. & Hall, R.L. (1978) Estimation of toxic hazard—A decision tree approach. *Food Cosmet. Toxicol.*, **16**, 255–276.

Crowell, P.L. (1997) Monoterpenes in breast cancer chemoprevention. *Breast Cancer Res. Treat.*, **46**, 191–197.

Eder, E., Scheckenbach, S., Deininger, C. & Hoffman, C. (1993) The possible role of α,β-unsaturated carbonyl compounds in mutagenesis and carcinogenesis. *Toxicol. Lett.*, **67**, 87–103.

Eriksson, K. & Levin, J.-O. (1990) Identification of *cis*-and *trans*-verbenol in human urine after occupational exposure to terpenes. *Int. Arch. Occup. Environ. Health*, **62**, 379–383.

Florin, I., Rutberg, L., Curvall, M. & Enzell, C.R. (1980) Screening of tobacco smoke constituents for mutagenicity using the Ames' test. *Toxicology*, **18**, 219–232.

Food and Drug Administration (1993) *Priority-based Assessment of Food Additives (PAFA) Database*, Washington DC: Center for Food Safety and Applied Nutrition, p. 58.

Ford, D.M. & Moran, E.J. (1978) Preliminary indications of *in vitro* hydrolysis of two flavor chemical esters. Private communication. Submiited to WHO by the Flavor and Extract Manufacturers Association.

Fujita, H., Aoki, N. & Sasaki, M. (1994) Mutagenicity test of food additives with Salmonella typhirurium TA97 and TA102 (IX*). *Ann. Rep. Tokyo Metr. Res. Lab. Public Health*, **45**, 191–199 (in Japanese; abstract and tables in English).

Gaunt, I.F., Agrelo, C.E., Colley, J., Lansdown, A.B.G. & Grasso, P. (1971) Short-term toxicity of isobornyl acetate in rats. *Food Cosmet. Toxicol.*, **9**, 355–366.

Gaunt, I.F., Mason, P.L., Hardy, J., Lansdown, A.B.G. & Gangolli, S.D. (1974) Short-term toxicity of methylpheylcarbinyl acetate in rats. *Food Chem. Toxicol.*, **12**, 185–194.

Graffner-Nordberg, M., Karin, S., Anders, T. & Anders, H. (1998) Synthesis and enzymie hydrolysis of esters, constituting simple models of soft drugs. *Chem. Pharm. Bull.*, **46**, 591–601.

Grundschober, F. (1977) Toxicological assessment of flavoring esters. *Toxicology*, **8**, 387–390.

Haag, J.D. & Gould, M.N. (1994) Mammary carcinoma regression induced by perillyl alcohol, a hydroxylated analog of limonene. *Cancer Chemother. Pharmacol.*, **34**, 477–483.

Hagan, E.C., Hansen, W.H., Fitzhugh, O.G., Jenner, P.M., Jones, W.I. & Taylor, J.M. (1967) Food flavourings and compounds of related structure. II. Subacute and chronic toxicity. *Food Cosmet. Toxicol.*, **5**, 141–157.

Hayashi, M., Kishi, M., Sofuni, T. & Ishidate, M., Jr (1988) Micronucleus tests in mice on 39 food additives and eight miscellaneous chemicals. *Food Chem. Toxicol.*, **26**, 487–500.

He, L., Huanbiao, M., Hadisusilo, S., Qureshi, A.A. & Elson, C.E. (1997) Isoprenoids suppress the growth of murine B16 melanomas *in vitro* and *in vivo. J. Nutr.*, **127**, 668–674

Heck, J.D., Vollmuth, T.A., Cifone, M.A. Jagannath, D.R., Myhr, B. & Curren, R.D. (1989) An evaluation of food flavoring ingredients in a genetic toxicity screening battery. *Toxicologist,* **9**, 206.

Heymann, E. (1980) Carboxlesterases and amidases. In: Jakoby, W.B., ed., *Enzymatic Basics of Detoxication*, 2nd Ed., New York: Academic Press, pp. 291–323.

Hildebrandt, H. (1902) On the behavior of carvone and santalol in the animal body. *Hoppe Seyler's Z. Physiol. Chem.*, **36**, 452–461.

Hold, K.M., Sirisoma, N.S., Sparks, S.E. & Casida, J.E. (2002) Metabolism and mode of action of *cis-* and *trans-*3-pinanones (the active ingredients of hyssop oil). *Xenobiotica*, **32**, 251–265.

Honda, G., Koeauka, Y., Kamisako, W. & Tabata, M. (1986) Isolation of sedative principles from *Perilla frutescens. Chem. Pharm. Bull.*, **34**, 1672–1677.

International Organization of the Flavor Industry (1995) European inquiry on volume of use. Unpublished report. Submitted to WHO by the Flavor and Extract Manufacturers Association

Ishida, T., Toyota, M. & Asakawa, Y.(1989) Terpenoid biotransformation in mammals. V. Metabolism of (+)-citronellal, (±)-citronellal, (±)-7-hydroxycitronellal, citral, (–)-perillaldehyde, (–)-myrtenal, cuminaldehyde, thujone, and (±)-carvone in rabbits. *Xenobiotica*, **19**, 843–855.

Ishidate, M., Jr, Sofuni, T., Yoshikawa, K., Hayashi, M., Nohmi, T., Sawada, M. & Matsouoka, A. (1984) Primary mutagenicity screening of food additives currently used in Japan. *Food Chem. Toxicol.*, **22**, 623–636.

Junge, W. & Heymann, E. (1979) Characterization of the isoenzymes of pig-liver esterase. *Eur. J. Biochem.*, **95**, 519–525.

Kynoch, S.R. & Lloyd, G.K. (1977) Acute oral toxicity to rats of myrtenyl acetate. Unpublished report. Submitted to WHO by the Flavor and Extract Manufacturers Association.

Kuroda, K., Setsuko, I., Young, S. & Takeji, Y. (1984) Rec-assay of food additives. *Nippon Kosnu Eisei Zassni*, **31**, 277–281 (in Japanese; abstract and tables in English).

Leegwater, D.C. & van Straten, S. (1974) In vitro study on the hydrolysis of twenty-six organic esters by pancreatin. Unpublished report. Submitted to WHO by the Flavor and Extract Manufacturers Association.

Levenstein, I. (1981) Oral toxicity testing of 1,2,5,6-tetrahydrocuminic acid in rats. Unpublished report. Submitted to WHO by the Flavor and Extract Manufacturers Association.

Levenstein, I. (1982a) Acute oral, LD50 of alpha-campholenic alcohol in rats. Unpublished report. Submitted to WHO by the Flavor and Extract Manufacturers Association.

Liu, M.-J. & Pollack, G.M. (1993) Pharmacokinetics and pharmacodynamics of valproate analogs in rats. II. Pharmacokinetics of octanoic acid, cyclohexanecarboxylic acid, and 1-methyl-1-cyclohexanecrboxylic acid. *Biopharm. Drug Disposition*, **14**, 325–339.

Liu, M.-J., Brouwer, K.L.R. & Pollack, G.M. (1992) Pharmacokinetics and pharmacodynamics of valproate analogs in rats. III. Pharmacokinetics of valproic acid, cyclohexanecarboxylic acid, and 1-methyl-1-cyclohexanecarboxylic acid in the bile-exteriorized rat. *Drug Metab. Disposition*, **20**, 810–815.

Longland, R.C. & Gangolli, S.D. (1977) The hydrolysis of flavouring esters by artificial gastrointestinal juices and rat tissue preparations. *Toxicology*, **8**, 197–204.

Lucas, C.D., Putnam, J.M. & Hallagan, J.B. (1999) Flavor and Extract Manufacturers Association of the United States 1995 poundage and technical effects update survey. Submitted to WHO by the Flavor and Extract Manufacturers Association.

Maarse, H., Visscher, C.A., Willemsens, L.C. & Boelens, M.H. (1999) *Volatile Components in Food—Qualitative and Quantitative Data*, Zeist: Centraal Instituut Voor Voedingsonderzioek TNO.

Maltzman, T.H., Christou, M., Gould, M.N. & Jefcoate, C.R. (1991) Effects of monoterpenoids on in vivo DMBA-DNA adduct formation and on phase I hepatic metabolising enzymes. *Carcinogenesis*, **12**, 2081–2087.

Mills, J.J., Chari, R.S., Boyer, I.J., Gould, M.N. & Jirtle, R.L. (1995) Induction of apoptosis in liver tumors by the monoterpene perillyl alcohol. *Cancer Res.*, **55**, 979–983.

Moran, E.J., Esterday, O.D. & Oser, B.L. (1980) Acute oral toxicity of selected flavor chemicals. *Drug Chem. Toxicol.*, **3**, 249–258.

Moran, E.J. & Tyburcy, K.R. (1979) In vitro hydrolysis of flavoring esters in a simulated gastric system and a simulated intestinal system. Unpublished report. Submitted to WHO by the Flavor and Extract Manufacturers Association.

Moreno, O.M. (1972) Acute toxicity studies in mice, rats and rabbits (2-hydroxymethyl-6,6-dimethylbicyclo[3.1.1]hept-2-enylformate). Unpublished report to RIFM. Submitted to WHO by the Flavor and Extract Manufacturers Association.

Moreno, O.M. (1977) Acute oral toxicity in rats and dermal toxicity in rabbits (*p*-menth-1,8-dien-7-ol and 10-hydroxymethylene-2-pinene). Unpublished report to RIFM. Submitted to WHO by the Flavor and Extract Manufacturers Association.

Moreno, O.M. (1978) Acute oral toxicity in rats and acute dermal toxicity in rabbits (cyclohexaneethyl acetate; 2,2,3-trimethylcyclo-pent-3-en-1-ylacetaldehyde; *p*-menth-1,8-dien-7-al). Unpublished report to RIFM. Submitted to WHO by the Flavor and Extract Manufacturers Association.

Moreno, O.M. (1982) Acute oral toxicity in rats and acute dermal toxicity in rabbits (myrtenal; myrtenal acetate). Unpublished report to RIFM. Submitted to WHO by the Flavor and Extract Manufacturers Association.

Naarden, Inc. (1974) In vitro study on the hydrolysis of twenty-six organic esters by pancreatin. Unpublished report. Submitted to WHO by the Flavor and Extract Manufacturers Association.

National Cancer Institute (1996) Clinical development plan: I—Perillyl alcohol. *J. Cell. Biochem.*, **265**, 137–148.

Nishizawa, Y., Abe, S., Yamada, K., Nakamura, T., Yamatsu, I. & Kinoshita, K. (1987) Identification of urinary and microsomal metabolites of geranylgeranylacetone in rats. *Xenobiotica*, **17**, 575–584.

Okugawa, H., Ueda, R., Matsumoto, K., Kawanishi, K. & Kato, A. (1995) Effects of α-santalol and β-santolol from sandalwood on the central nervous system in mice. *Phytomedicine*, **2**, 119–126.

Ohta, T., Watanabe, M., Watanabe, K. & Shirasu, Y. (1986) Inhibitory effects of flavourings on mutagenesis induced by chemicals in bacteria. *Food Chem. Toxicol.*, **24**, 51–54.

Pellmont (1967) Acute toxicity in mice with 2,6,6-trimethylcyclohexa-1,3-dienyl methanal. Unpublished report. Submitted to WHO by the Flavor and Extract Manufacturers Association.

Pellmont (1969) Acute toxicity in rats with 2,6,6-trimethylcyclohexa-1,3-dienyl methanal. Unpublished report. Submitted to WHO by the Flavor and Extract Manufacturers Association.

Pellmont (1973) Acute oral toxicity in mice with cyclocitral. Unpublished report. Submitted to WHO by the Flavor and Extract Manufacturers Association.

Phillips, L.R., Malspeis, L. & Supko, J.G. (1995) Pharmacokinetics of active drug metabolites after oral administration of perillyl alcohol, an investigational antineoplastic agent, to the dog. *Drug Metab. Disposition*, **23**, 676–680.

Piccirillo, V.J., Leatherwood, L., Swidersky, P. & Shopp, G. (1979) Acute oral LD50 determination of campholene acetate in rats. Unpublished report. Submitted to WHO by the Flavor and Extract Manufacturers Association.

Reddy, B.S., Wang, C.X., Samaha, H., Lubet, R., Steele, V.E., Kelloff, G.J. & Rao, C.V. (1997) Chemoprevention of colon carcinogenesis by dietary perillyl alcohol. *Cancer Res.*, **57**, 420–425.

Ren, Z. & Gould, M.N. (1998) Modulation of small G protein isoprenylation by anticancer monoterpenes in *in situ* mammary gland epithelial cells. *Carcinogenesis*, **19**, 827–832.

Ripple, G.H., Gould, M.N.,Arzoomanian, R.Z., Alberti, D., Feierabend, C., Simon, K., Binger, K., Tutsch, K.D., Pomplun, M., Wahamaki, A., Marnocha, R., Wilding, G. & Bailey, H.H. (1998) Phase I clinical trial of perillyl alcohol administered daily. *Clin. Cancer Res.*, **4**, 1159–1164.

Ripple, G.H., Gould, M.N., Stewart, J.A., Tutsch, K.D., Arzoomanian, R.Z., Alberti, D., Feierabend, C., Pomplun, M., Wilding, G. & Bailey, H.H. (2000) Phase I clinical and pharmacokinetic study of perillyl alcohol administered four times a day. *Clin. Cancer Res.*, **6**, 390–396.

Salzer (1998) In vitro hydrolysis test. Unpublished report. Submitted to WHO by the Flavor and Extract Manufacturers Association.

Sasaki, M., Yoshida, S., Ichikawa, H., Takano, I., Yasuda, I. & Akiyama, K. (1990) Cytological activities of perilla extracts. *Ann. Rep. Tokyo Metr. Res. Lab. Public Health*, **41**, 275–281 (in Japanese; abstract and tables in English).

Stark, M.J., Burke, Y.D., McKinzie, J.H., Ayoubi, A.S. & Crowell, P.L. (1995) Chemotherapy of pancreatic cancer with the monoterpene perillyl alcohol. *Cancer Lett.*, **96**, 15–21.

Stofberg, J. & Grundschober, F. (1987) Consumption ratio and food predominance of flavoring materials. *Perfumer Flavorist,* **12**, 27.

Stofberg, J. & Kirschman, J.C. (1985) The consumption ratio of flavoring materials: A mechanism for setting priorities for safety evaluation. *Food Chem. Toxicol.*, **23**, 857–860.

Suzuki, N. & Suzuki, H. (1994) Induction of K-ras codon 12 mutation by saccharin sodium with perillaldehyde-induced mutagenicity. *Hen'igensei Shiken*, **3**, 155–160 (in Japanese; abstract and tables in English).

Suzuki, H., Ichikawa, H. & Sasaki, M. (1990) Mutagenicity of perillaldehyde in a human cell strain. *Ann. Rep. Tokyo Metr. Res. Lab. Public Health*, **41**, 265–266 (in Japanese; abstract and tables in English).

Tayama, S., Ichikawa, H. & Sasaki, M. (1990) Effects of perillaldehyde on induction of chromosomal aberrations and sister-chromatid exchanges (SCEs) in CHO-KI-cells. *Ann. Rep. Tokyo Metr. Res. Lab. Public Health*, **41**, 267–269 (in Japanese; abstract and tables in English).

Terrell & Cooper (1978) Unpublished report. Submitted to WHO by the Flavor and Extract Manufacturers Association.

Williams, R.T. (1959) *Detoxication Mechanisms. The Metabolism and Detoxication of Drugs, Toxic Substances, and Other Organic Compounds*, 2nd Ed., London: Chapman & Hall, pp. 119–120.

Wohl, A. (1974) Acute oral and dermal toxicity studies (cyclohexyl ethyl acetate). Report to RIFM. Unpublished report. Submitted to WHO by the Flavor and Extract Manufacturers Association.

Yoo, Y.S. (1986) Mutagenic and antimutagenic activities of flavoring agents used in foodstuffs. *J. Osaka City Med. Center*, **34**, 267–288 (in Japanese; abstract and tables in English).

Zajac-Kaye, M. & Ts'O, P.O.P. (1984) DNAase I encapsulated in liposomes can induce neoplastic transformation of Syrian hamster embryo cells in culture. *Cell*, **39**, 427–437.

PHENYLETHYL ALCOHOL, ALDEHYDE, ACID AND RELATED ACETALS AND ESTERS AND RELATED SUBSTANCES

First draft prepared by

Dr A. Mattia[1] and Professor A.G. Renwick[2]
[1]Division of Petition Review, Office of Food Additive Safety, Center for Food Safety and Applied Nutrition, Food and Drug Administration, College Park, Maryland, USA; and [2]Clinical Pharmacology Group, University of Southampton, Southampton, England

1. EVALUATION

1.1 Introduction

The Committee evaluated 43 flavouring agents that are derivatives of phenethyl alcohol (No. 987) and phenoxyethyl alcohol (see Table 1). The group includes 39 phenethyl derivatives, comprising phenylacetaldehyde (No. 1002), phenylacetic acid (No. 1007) and structurally related esters and acetals. The group also includes four phenoxyethyl alcohol derivatives: phenoxyacetic acid (No. 1026), the sodium salt of a structurally related phenoxyacetic acid (No. 1029), a phenoxyethyl ester (No. 1027) and a phenoxyacetic acid ester (No. 1028). The evaluations were conducted with the Procedure for the Safety Evaluation of Flavouring Agents (See Figure1). None of these agents has previously been evaluated by the Committee.

Table 1. Summary of results of safety evaluations of phenethyl alcohol, aldehyde, acid, and related acetals and esters[a]

Flavouring agent	No.	CAS No. and structure	Step A3[b] Does intake exceed the threshold for human intake?	Step A4 Is the flavouring agent or are its metabolites endogenous?	Step A5 Adequate margin of safety for the flavouring agent or related chemical?	Comments	Conclusion based on current intake
Structural class I							
Phenethyl alcohol	987	60-12-8 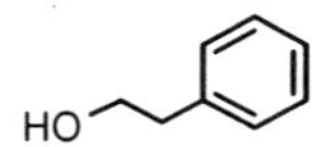	No Europe: 1400 USA: 330	N/R	N/R	See note 1.	No safety concern
Phenethyl formate	988	104-62-1 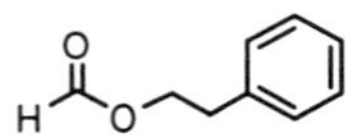	No Europe: 2 USA: 30	N/R	N/R	See note 2.	No safety concern
Phenethyl acetate	989	103-45-7 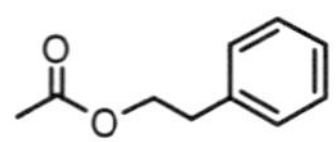	No Europe: 100 USA: 60	N/R	N/R	See note 2.	No safety concern
Phenethyl propionate	990	122-70-3 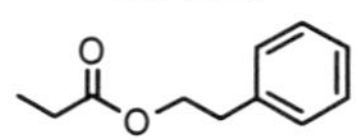	No Europe: 1 USA: 3	N/R	N/R	See note 2.	No safety concern
Phenethyl butyrate	991	103-52-6 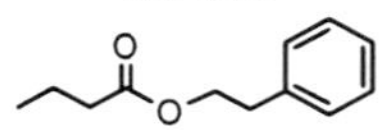	No Europe: 30 USA: 30	N/R	N/R	See note 2.	No safety concern
Phenethyl isobutyrate	992	103-48-0 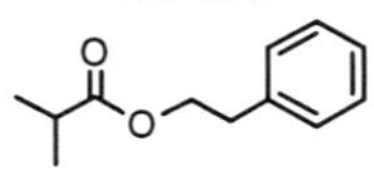	No Europe: 20 USA: 60	N/R	N/R	See note 2.	No safety concern

Table 1 (contd)

Flavouring agent	No.	CAS No. and structure	Step A3[a,b] Does intake exceed the threshold for human intake?	Step A4 Is the flavouring agent or are its metabolites endogenous?	Step A5 Adequate margin of safety for the flavouring agent or related chemical?	Comments	Conclusion based on current intake
Phenethyl 2-methylbutyrate	993	24817-51-4 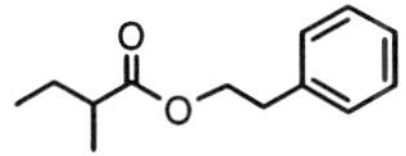	No Europe: 0.4 USA: ND	N/R	N/R	See note 2.	No safety concern
Phenethyl isovalerate	994	140-26-1 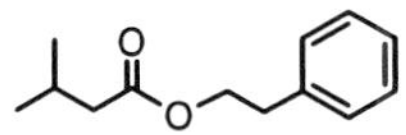	No Europe: 100 USA: 30	N/R	N/R	See note 2.	No safety concern
Phenethyl hexanoate	995	6290-37-5 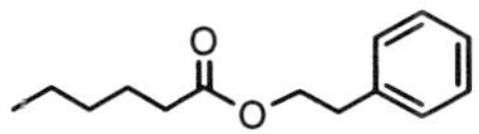	No Europe: 10 USA: 2	N/R	N/R	See note 2.	No safety concern
Phenethyl octanoate	996	5457-70-5 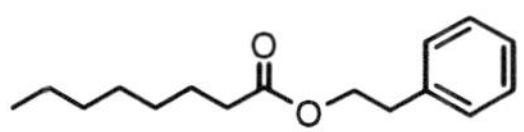	No Europe: 30 USA: 0.1	N/R	N/R	See note 2.	No safety concern
Phenethyl tiglate	997	55719-85-2 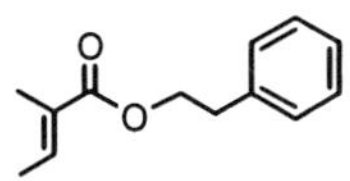	No Europe: 0.3 USA: 1	N/R	N/R	See note 2.	No safety concern
Phenethyl senecioate	998	42078-65-9 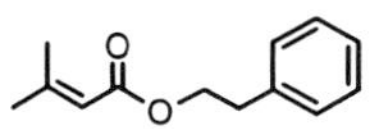	No Europe: 2 USA: ND	N/R	N/R	See note 2.	No safety concern

Table 1 (contd)

Flavouring agent	No.	CAS No. and structure	Step A3[a,b] Does intake exceed the threshold for human intake?	Step A4 Is the flavouring agent or are its metabolites endogenous?	Step A5 Adequate margin of safety for the flavouring agent or related chemical?	Comments	Conclusion based on current intake
Phenethyl phenylacetate	999	102-20-5 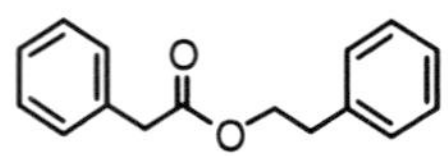	No Europe: 40 USA: 80	N/R	N/R	See note 3.	No safety concern
Acetaldehyde phenethyl propyl acetal	1000	7493-57-4 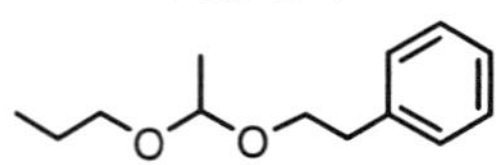	No Europe: 0.1 USA: 6	N/R	N/R	See note 4.	No safety concern
Acetaldehyde butyl phenethyl acetal	1001	64577-91-9 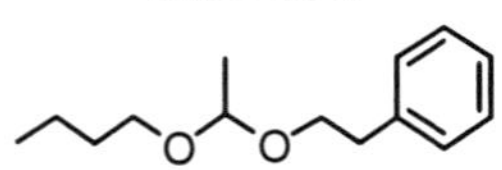	No Europe: 0.01 USA: ND	N/R	N/R	See note 4.	No safety concern
Phenylacetaldehyde	1002	122-78-1 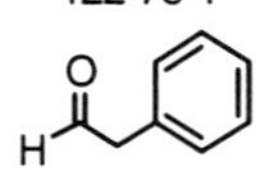	No Europe: 40 USA: 60	N/R	N/R	See note 5.	No safety concern
Phenylacetaldehyde dimethyl acetal	1003	101-48-4 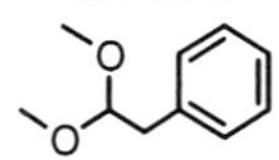	No Europe: 20 USA: 40	N/R	N/R	See note 6.	No safety concern

Table 1 (contd)

Flavouring agent	No.	CAS No. and structure	Step A3[a,b] Does intake exceed the threshold for human intake?	Step A4 Is the flavouring agent or are its metabolites endogenous?	Step A5 Adequate margin of safety for the flavouring agent or related chemical?	Comments	Conclusion based on current intake
Phenylacetaldehyde glyceryl acetal	1004	29895-73-6 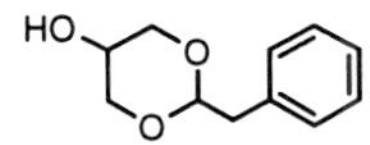 or 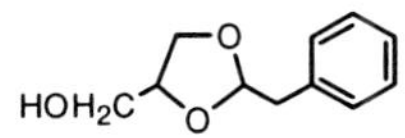	No Europe: 0.1 USA: 1	N/R	N/R	See note 6.	No safety concern
Phenylacetaldehyde 2,3-butylene glycol acetal	1005	5468-06-4 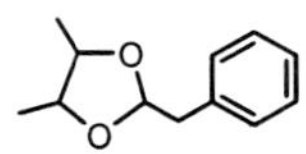	No Europe: ND USA: 1	N/R	N/R	See note 6.	No safety concern
Phenylacetaldehyde diisobutylacetal	1006	68345-22-2 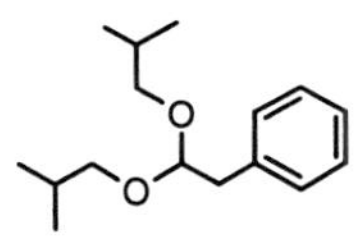	No Europe: 30 USA: 0.4	N/R	N/R	See note 6.	No safety concern
Phenylacetic acid	1007	103-82-2 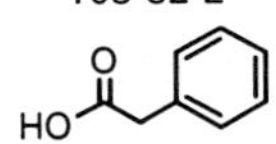	No Europe: 290 USA: 60	N/R	N/R	See note 7.	No safety concern
Methyl phenylacetate	1008	101-41-7 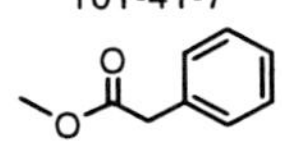	No Europe: ND USA: 20	N/R	N/R	See note 8.	No safety concern

Table 1 (contd)

Flavouring agent	No.	CAS No. and structure	Step A3[a,b] Does intake exceed the threshold for human intake?	Step A4 Is the flavouring agent or are its metabolites endogenous?	Step A5 Adequate margin of safety for the flavouring agent or related chemical?	Comments	Conclusion based on current intake
Ethyl phenylacetate	1009	101-97-3 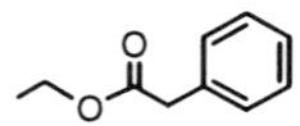	No Europe: 130 USA: 20	N/R	N/R	See note 8.	No safety concern
Propyl phenylacetate	1010	4606-15-9 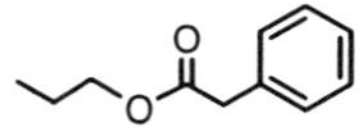	No Europe: ND USA: 0.3	N/R	N/R	See note 8.	No safety concern
Isopropyl phenylacetate	1011	4861-85-2 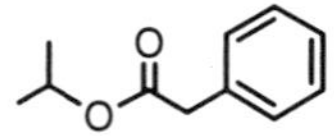	No Europe: 0.07 USA: ND	N/R	N/R	See note 8.	No safety concern
Butyl phenylacetate	1012	122-43-0 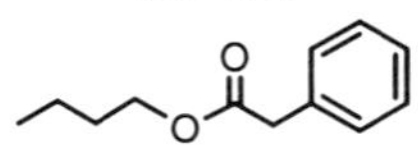	No Europe: 3 USA: 3	N/R	N/R	See note 8.	No safety concern
Isobutyl phenylacetate	1013	102-13-6 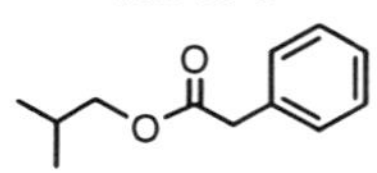	No Europe: 20 USA: 20	N/R	N/R	See note 8.	No safety concern
Isoamyl phenylacetate	1014	102-19-2 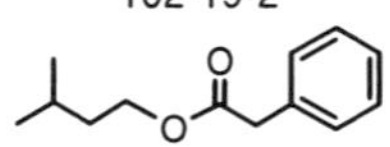	No Europe: 30 USA: 30	N/R	N/R	See note 8.	No safety concern

Table 1 (contd)

Flavouring agent	No.	CAS No. and structure	Step A3[a,b] Does intake exceed the threshold for human intake?	Step A4 Is the flavouring agent or are its metabolites endogenous?	Step A5 Adequate margin of safety for the flavouring agent or related chemical?	Comments	Conclusion based on current intake
Hexyl phenylacetate	1015	5421-17-0 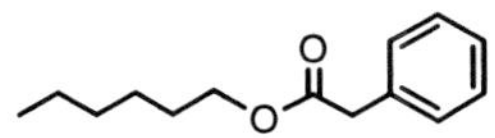	No Europe: 8 USA: ND	N/R	N/R	See note 8.	No safety concern
3-Hexenyl phenylacetate	1016	42436-07-7 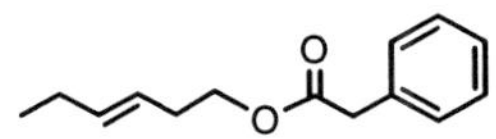	No Europe: 1 USA: 0.05	N/R	N/R	See note 8.	No safety concern
Octyl phenylacetate	1017	122-45-2 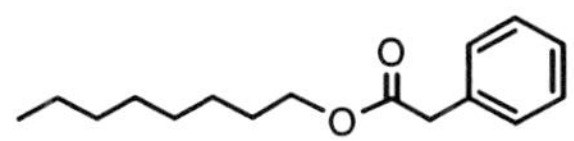	No Europe: 0.004 USA: 0.006	N/R	N/R	See note 8.	No safety concern
Rhodinyl phenylacetate	1018	10486-14-3 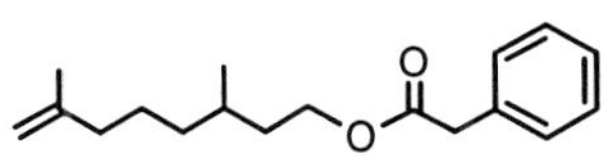	No Europe: 0.001 USA: ND	N/R	N/R	See note 9.	No safety concern
Linalyl phenylacetate	1019	7143-69-3 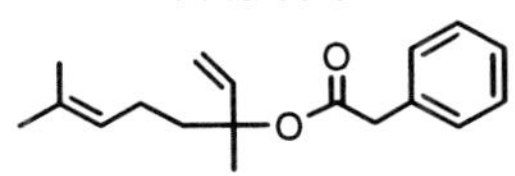	No Europe: 0.09 USA: ND	N/R	N/R	See note 9.	No safety concern
Geranyl phenylacetate	1020	102-22-7 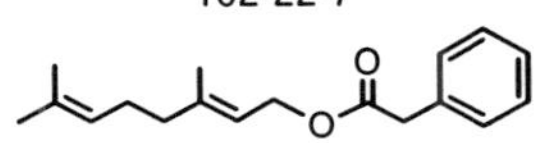	No Europe: 2 USA: 2	N/R	N/R	See note 9.	No safety concern

Table 1 (contd)

Flavouring agent	No.	CAS No. and structure	Step A3[a,b] Does intake exceed the threshold for human intake?	Step A4 Is the flavouring agent or are its metabolites endogenous?	Step A5 Adequate margin of safety for the flavouring agent or related chemical?	Comments	Conclusion based on current intake
Citronellyl phenylacetate	1021	139-70-8 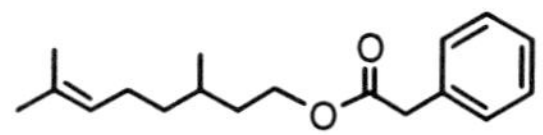	No Europe: 1 USA: 2	N/R	N/R	See note 9.	No safety concern
Santalyl phenylacetate	1022 α β	1323-75-7 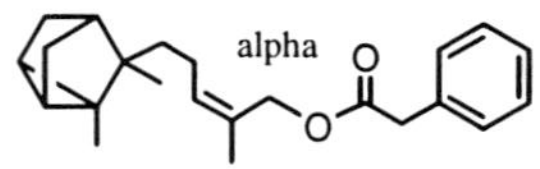 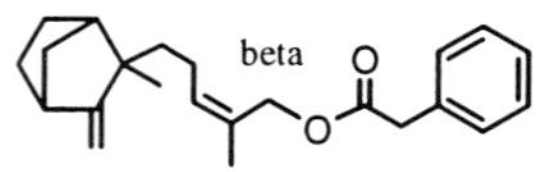	No Europe: ND USA: 1	N/R	N/R	See note 9.	No safety concern
para-Tolyacetaldehyde	1023	104-09-6 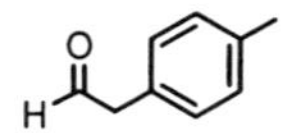	No Europe: 6 USA: 3	N/R	N/R	See note 5.	No safety concern
para-Isopropylphenyl-acetaldehyde	1024	4395-92-0 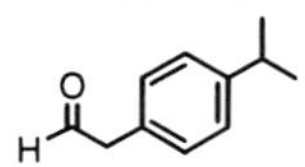	No Europe: 0.1 USA: 0.01	N/R	N/R	See note 5.	No safety concern
Methyl *para-tert*-butylphenylacetate	1025	3549-23-3 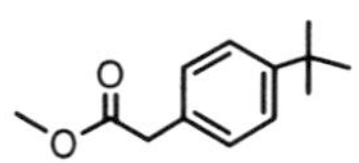	No Europe: 20 USA: 20	N/R	N/R	See note 8.	No safety concern

Table 1 (contd)

Flavouring agent	No.	CAS No. and structure	Step A3[a,b] Does intake exceed the threshold for human intake?	Step A4 Is the flavouring agent or are its metabolites endogenous?	Step A5 Adequate margin of safety for the flavouring agent or related chemical?	Comments	Conclusion based on current intake
Structural class III							
Phenoxyacetic acid	1026	122-59-8 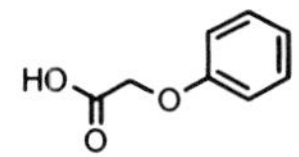	No Europe: 40 USA: 0.1	N/R	N/R	See note 10.	No safety concern
Ethyl (*para*-tolyloxy)acetate	1027	67028-40-4 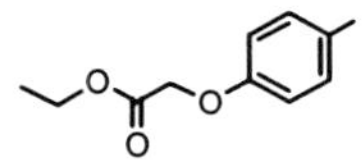	No Europe: 0.1 USA: ND	N/R	N/R	See note 11.	No safety concern
2-Phenoxyethyl isobutyrate	1028	103-60-6 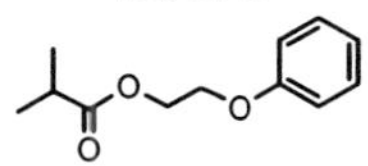	Yes Europe: 2 USA: 110	No	Yes. The NOEL of 15 mg/kg bw per day for the related chemical No. 1027 and the NOEL of 250 mg/kg bw per day for the realted chemical No. 1029 are >1000 times the estimated intake of 2-phenoxyethyl isobutyrate when used as a flavouring agent.	See note 12.	No safety concern

Table 1 (contd)

Flavouring agent	No.	CAS No. and structure	Step A3[a,b] Does intake exceed the threshold for human intake?	Step A4 Is the flavouring agent or are its metabolites endogenous?	Step A5 Adequate margin of safety for the flavouring agent or related chemical?	Comments	Conclusion based on current intake
Sodium 2-(4-methoxy-phenoxy)propanoate	1029	13794-15-5 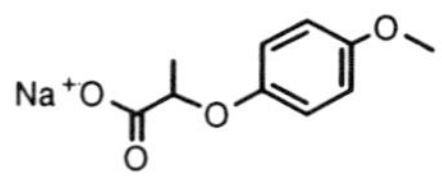	No Europe: ND USA: 6	N/R	N/R	See note 13.	No safety concern

CAS: Chemical Abstracts Service; ND: no intake data reported; N/R: not required for evaluation

[a] Step 2: All the flavouring agents in this group were predicted to be metabolized to innocuous products.

[b] The thresholds for human intake for classes I and III are 1800 µg/day and 90 µg/day, respectively. All intake values are expressed in µg/person per day. The combined intakes of flavouring agents in class I are 2300 and 920 µg/person per day in Europe and the USA, respectively. The combined intakes of flavouring agents in class III are 42 and 120 µg/person per day in Europe and the USA, respectively.

Notes:

1. Oxidized to phenylacetic acid and either completely oxidized or conjugated and excreted primarily in the urine
2. Hydrolysed to phenethyl alcohol (see note 1) and the corresponding acid, which is further oxidized to carbon dioxide and water
3. Hydrolysed to phenethyl alcohol (see note 1) and phenylacetic acid (see note 7)
4. Hydrolysed to phenethyl alcohol (see note 1) and the corresponding aldehyde, which is further oxidized to carbon dioxide and water
5. Phenylacetaldehyde derivative is oxidized to phenylacetic acid derivative (see note 7).
6. Hydrolysed to phenyacetaldehyde (see note 5) and the corresponding alcohol, which is further oxidized to carbon dioxide and water
7. Phenylacetic acid is excreted as the glutamine acid conjugate.
8. Hydrolysed to phenylacetic acid derivative (see note 7) and the corresponding alcohol, which is completely oxidized.
9. Hydrolysed to phenylacetic acid (see note 7) and terpene alcohol, which is further oxidized to a polar, excretable metabolite.
10. Excreted unchanged in the urine
11. Hydrolysed to *para*-tolyloxy acetic acid and ethanol; the acid is excreted unchanged.
12. Hydrolysed to isobutyric acid and 2-phenoxyethanol, which is oxidized to phenoxyacetic acid and excreted unchanged in the urine.
13. Excreted mainly unchanged and as the *O*-demethylated metabolite, (±)2-(4-hydroxyphenoxy)propionic acid.

Twenty of the 39 flavouring agents in this group are natural components of foods. Phenethyl alcohol has a mild rose aroma. The agents have been detected in a wide range of products; for example, the parent alcohol and its derivatives have been detected in beans, fruits, vegetables, cheeses, milk, oils and alcoholic and non-alcoholic beverages (Stofberg & Grundschober, 1987). Only one phenoxyacetic acid derivative is a natural component of food: sodium 2-(4-methoxyphenoxy)propanoate (No. 1029) has been detected in coffee (Maarse et al., 1999).

1.2 Estimated daily intake

The total annual volume of production of the 43 phenethyl alcohol and phenoxyethyl alcohol derivatives in this group is approximately 17 000 kg in Europe (International Organization of the Flavor Industry, 1995) and 7800 kg in the USA (Lucas et al., 1999) (Table 2). About 75% and 45% of the total annual volume in Europe and the USA, respectively, is accounted for by use of phenethyl alcohol (No. 987), its corresponding acetate ester (No. 989) and phenylacetic acid (No. 1007). The total annual volume of production of the four derivatives of phenoxyethyl alcohol (Nos 1026–1029) is approximately 260 kg in Europe (International Organization of the Flavor Industry, 1995) and 870 kg in the USA (Lucas et al., 1999), accounting for about 2% and 11% of the total annual volumes in Europe and the USA, respectively.

Five flavouring agents in the group have the highest estimated intakes. These are 1400 μg/day for phenethyl alcohol (No. 987), 290 μg/day for phenylacetic acid (No. 1007), 130 μg/day for ethyl phenylacetate (No. 1009) and 100 μg/day for phenethyl acetate (No. 989) and for phenethyl isovalerate (No. 994) in Europe and 330 μg/day for phenethyl alcohol (No. 987) and 110 μg/day for phenoxyethyl isobutyrate (No. 1028) in the USA. The intakes of all the other flavouring agents in the group ranged from 0.001 to 80 μg/day, the intake of about 70% of them being less than 25 μg/day.

1.3 Hydrolysis, absorption, distribution, metabolism and elimination

Phenethyl and phenylacetate esters and phenyacetaldehyde acetals are rapidly hydrolysed in vivo to yield phenethyl alcohol (No. 987), phenylacetic acid (No. 1007) and phenylacetaldehyde (No. 1002), respectively (Williams, 1959). Phenethyl alcohol and phenylacetaldehyde are both oxidized to phenylacetic acid, which is conjugated and excreted primarily in the urine (Williams, 1959). Similarly, the phenoxyethyl and phenoxyacetate esters, 2-phenoxyethyl isobutyrate (No. 1028) and ethyl(*para*-tolyloxy)acetate (No. 1027), respectively, are anticipated to be hydrolysed to their component acids and alcohols. Phenoxyethyl alcohol is rapidly oxidized to phenoxyacetic acid, which in turn is rapidly absorbed and excreted primarily unchanged in the urine (Howes, 1988). Sodium 2-(4-methoxyphenoxy)propanoate (No. 1029) is also rapidly absorbed and excreted primarily unchanged, with a small amount of the *O*-demethylated metabolite, (±)2-(4-hydroxyphenoxy) propionic acid (Brown et al., 1986; Hawkins & Mayo, 1986; Sangster & Lindley, 1986; Caldwell, 1987). Therefore, all the flavouring agents in this group were predicted to hydrolyse and/or oxidize to yield phenylacetic acid or a phenoxyacetic acid derivative that is excreted either free or in conjugated form.

Table 2. Annual volumes of usage of phenethyl alcohol, phenylacetaldehyde, phenylacetic acid and related substances used as flavouring agents in Europe and the USA

Substance (No.)	Most recent annual volume (kg)[a]	Intake ('eaters only')[b]		Annual volume in naturally occurring foods (kg)[c]	Consumption ratio[d]
		μg/day	μg/kg bw per day		
Phenethyl alcohol (987)					
Europe	10 000	1400	23		
USA	2 500	330	6	700 000	280
Phenethyl formate (988)					
Europe	17	2	0.04		
USA	200	27	0.4	14	0.1
Phenethyl acetate (989)					
Europe	730	100	2		
USA	490	65	1	30 000	61
Phenethyl propionate (990)					
Europe	8	1	0.02		
USA	24	3	0.1	220	9
Phenethyl butyrate (991)					
Europe	230	33	1		
USA	210	28	0.5	1	000
Phenethyl isobutyrate (992)					
Europe	160	22	0.4		
USA	440	57	1	+	NA
Phenethyl 2-methylbutyrate (993)					
Europe	3	0.4	0.007		
USA	N/D	NA	NA	+	NA
Phenethyl isovalerate (994)					
Europe	670	95	2		
USA	200	27	0.4	430	2
Phenethyl hexanoate (995)					
Europe	96	14	0.2		
USA	14	2	0.03	5700	410
Phenethyl octanoate (996)					
Europe	189	27	0.4		
USA	1	0.1	0.002	3200	3200
Phenethyl tiglate (997)					
Europe	2	0.3	0.005		
USA	9	1	0.02		– NA
Phenethyl senecioate (998)					
Europe	11	2	0.03		
USA	N/D	NA	NA	–	NA

Table 2 (contd)

Substance (No.)	Most recent annual volume (kg)[a]	Intake ('eaters only')[b]		Annual volume in naturally occurring foods (kg)[c]	Consumption ratio[d]
		µg/day	µg/kg bw per day		
Phenethyl phenylacetate (999)					
Europe	270	39	1		
USA	590	78	1	–	NA
Acetaldehyde phenethyl propyl acetal (1000)					
Europe	1	0.1	0.002		
USA	45	6	0.1	–	NA
Acetaldehyde butyl phenethyl acetal (1001)					
Europe	0.1	0.01	0.0002		
USA	N/D	NA	NA	–	NA
Phenylacetaldehyde (1002)					
Europe	300	43	1		
USA	430	56	1	10688	25
Phenylacetaldehyde dimethyl acetal (1003)					
Europe	143	20	0.3		
USA	345	45	1	+	NA
Phenylacetaldehyde glyceryl acetal (1004)					
Europe	1	0.1	0.002		
USA	10	1	0.02	–	NA
Phenylacetaldehyde 2,3-butylene glycol acetal (1005)					
Europe	N/D	NA	NA		
USA	11	1	0.02	–	NA
Phenylacetaldehyde disobutyl acetal (1006)					
Europe	220	31	1		
USA	3	0.4	0.01	–	NA
Phenylacetic acid (1007)					
Europe	2000	290	5		
USA	490	65	1	280	1
Methyl phenylacetate (1008)					
Europe	N/D	NA	NA		
USA	170	23	0.4	55	0.3
Ethyl phenylacetate (1009)					
Europe	900	130	2		
USA	130	17	0.3	41	0.3
Propyl phenylacetate (1010)					
Europe	N/D	NA	NA		
USA	2	0.3	0.004	–	NA

Table 2 (contd)

Substance (No.)	Most recent annual volume (kg)[a]	Intake ('eaters only')[b]		Annual volume in naturally occurring foods (kg)[c]	Consumption ratio[d]
		µg/day	µg/kg bw per day		
Isopropyl phenylacetate (1011)					
Europe	0.5	0.07	0.001		
USA	N/D	NA	NA	–	NA
Butyl phenylacetate (1012)					
Europe	20	3	0.05		
USA	25	3	0.1	+	NA
Isobutyl phenylacetate (1013)					
Europe	150	21	0.4		
USA	170	22	0.4	+	NA
Isoamyl phenylacetate (1014)					
Europe	230	33	1		
USA	220	29	0.5	+	NA
Hexyl phenylacetate (1015)					
Europe	57	8	0.1		
USA	N/D	NA	NA	32	0
3-Hexenyl phenylacetate (1016)					
Europe	6	1	0.01		
USA	0.4	0.05	0.001	+	NA
Octyl phenylacetate (1017)					
Europe	0.03	0.004	0.0001		
USA	0.045	0.006	0.0001	–	NA
Rhodinyl phenylacetate (1018)					
Europe	0.01	0.001	0.00002		
USA	N/D	NA	NA	–	NA
Linalyl phenylacetate (1019)					
Europe	0.6	0.09	0.001		
USA	N/D	NA	NA	–	NA
Geranyl phenylacetate (1020)					
Europe	14	2	0.03		
USA	14	2	0.03	–	NA
Citronellyl phenylacetate (1021)					
Europe	10	1	0.02		
USA	14	2	0.03	–	NA
Santalyl phenylacetate (1022)					
Europe	N/D	NA	NA		
USA	4	1	0.01	–	NA

Table 2 (contd)

Substance (No.)	Most recent annual volume (kg)[a]	Intake ('eaters only')[b]		Annual volume in naturally occurring foods (kg)[c]	Consumption ratio[d]
		µg/day	µg/kg bw per day		
para-Tolyacetaldehyde (1023)					
Europe	45	6	0.1		
USA	23	3	0.05	–	NA
para-Isopropylphenylacetaldehyde (1024)					
Europe	0.5	0.1	0.001		
USA	0.045	000	0.0001	–	NA
Methyl *para-tert*-butylphenylacetate (1025)					
Europe	140	19	0.3		
USA	150	20	0.3	–	NA
Phenoxyacetic acid (1026)					
Europe	250	35	1	–	NA
USA	0.9	0.1	0.002		
Ethyl (*para*-tolyloxy)acetate (1027)					
Europe	1	0.1	0.002	–	NA
USA	N/D	NA	NA		
2-Phenoxyethyl isobutyrate (1028)					
Europe	14	2	0.03	–	NA
USA	830	110	2		
Sodium 2-(4-methoxyphenoxy) propanoate (1029)					
Europe	N/D	NA	NA	+	NA
USA	43	6	0.1		
Total					
Europe	17 000				
USA	7 800				

NA, not available; N/D, no intake data reported; +, reported to occur naturally in foods (Maarse et al., 1996), but no quantitative data available; –, not reported to occur naturally in foods

[a] From International Organization of the Flavor Industry (1995) and Lucas et al. (1999)

[b] Intake (µg/person per day) was calculated as follows: [(annual volume, kg) × (1 × 10^9 µg/kg)/(population × survey correction factor × 365 days)], where population (10%, 'eaters only') = 32 × 10^6 for Europe and 26 × 10^6 for the USA. The correction factor = 0.6 for Europe and 0.8 for the USA, representing the assumption that only 60% and 80% of the annual production volume of the flavour, resepctively, was reported in the poundage surveys. Intake (µg/kg bw per day) calculated as follows: [(µg/person per day)/body weight], where body weight = 60 kg. Slight variations may occur from rounding.

[c] Quantitative data from Stofberg & Grundschober (1987)

[d] Calculated as follows: (annual consumption in food, kg)/(most recently reported volume as a flavouring agent, kg)

1.4 Application of the Procedure for the Safety Evaluation of Flavouring Agents

Step 1. In applying the Procedure, the Committee assigned phenylacetic acid (No. 1007) to structural class I (Cramer et al., 1978) because it is a normal component of human urine. The other 38 phenethyl alcohol derivatives were also assigned to structural class I because they are simple aromatic compounds with a primary oxygenated functional group.

The Committee assigned all four phenoxyacetic acid derivatives in this group to structural class III (Cramer et al., 1978). These are aromatic substances with an ether linkage between the benzene ring and an unsaturated side-chain with a primary oxygenated functional group.

Step 2. All the flavouring agents in this group were predicted to be metabolized to innocuous products. Their evaluation therefore proceeded via the A (left-hand) side of the decision tree.

Step A3. The estimated daily intakes of each of the 39 flavouring agents in structural class I are below the threshold for daily human intake for that class (1800 μg per person). According to the Procedure, the safety of these 39 flavouring agents raises no concern when they are consumed at currently estimated levels.

The estimated daily intakes of three of the four flavouring agents in structural class III are below the threshold for daily human intake for that class (90 μg per person). According to the Procedure, the safety of these three flavouring agents (Nos 1026, 1027 and 1029) raises no concern when they are consumed at their currently estimated levels. The daily intake of 2-phenoxyethyl isobutyrate (No. 1028) in the USA (110 μg/person) exceeds the threshold for daily human intake for compounds in structural class III (90 μg per person). Accordingly, the evaluation of this substance proceeded to step A4.

Step A4. 2-Phenoxyethyl isobutyrate (No. 1028) is not endogenous. Therefore, its evaluation proceeded to step A5.

Step A5. Although no NOEL was available for 2-phenoxyethyl isobutyrate (No. 1028), data were available on the toxicity of two structurally related flavouring agents in this group (Nos 1027 and 1029). The NOEL of 15 mg/kg bw per day for ethyl(*para*-tolyloxy)acetate (No. 1027) in a 90-day dietary study in rats (Posternak et al., 1969) and the NOEL of 250 mg/kg bw per day for sodium 2-(4-methoxyphenoxy)propanoate (No. 1029) in another 90-day dietary study in rats (Hill & Wood, 1986) provide adequate margins of safety (> 1000 times) in relation to the estimated daily per capita intake of 2-phenoxyethyl isobutyrate (No. 1028) in Europe (2 μg/kg bw) and the USA (110 μg/kg bw). The Committee therefore concluded that the safety of this agent would not be a concern.

Table 1 summarizes the evaluations of the 43 flavouring agents in this group.

1.5 Consideration of combined intake from use as flavouring agents

In the unlikely event that all 39 of the phenethyl alcohol derivatives (Nos 987–1025) were consumed concurrently on a daily basis, the estimated combined intake would exceed the threshold for daily human intake of compounds in structural class I (1800 µg per person). At the levels of intake associated with their use as flavouring agents, all 39 agents are expected to be metabolized efficiently, without saturating metabolic pathways. The same holds true for two of the four flavouring agents in structural class III: the estimated combined intakes of phenoxyacetic acid (No. 1026) and 2-phenoxyethyl isobutyrate (No. 1028) would exceed the threshold for daily human intake of compounds in structural class III (90 µg per person). At the levels of intake associated with their use as flavouring agents, both agents are expected to be metabolized efficiently, without saturating metabolic pathways. The other two agents, ethyl (*para*-tolyloxy)acetate (No. 1027) and sodium 2-(4-methoxyphenoxy) propanoate (No. 1029), are not converted to a common metabolite, and there is no need to consider their combined intake from use as flavouring agents. On the basis of the evaluation of all the data, combined intake would raise no safety concern.

1.6 Conclusions

The Committee concluded that none of the 43 flavouring agents in this group would present a safety concern at current estimated levels of intake. No data on toxicity were required for application of the Procedure to 42 of the 43 flavouring agents, as they were predicted to be metabolized to innocuous agents and the estimated intakes were below the human intake threshold associated with the relevant structural class. Data on the toxicity of related agents were used to evaluate 2-phenoxyethyl isobutyrate (No. 1028). Data on the toxicity and metabolism of phenethyl alcohol, phenylacetaldehyde, phenylacetic acid, phenoxyethanol and phenoxyacetic acid and related agents were consistent with the results of the safety evaluation.

2. BACKGROUND INFORMATION

2.1 Explanation

The relevant background information summarizes the key data for the safety evaluation of 39 phenethyl alcohol derivatives and four phenoxyethyl alcohol derivatives. The group of flavouring substances includes the following: phenethyl alcohol (No. 987), 28 esters formed from phenethyl alcohol or phenylacetic acid (Nos 988–999, 1008–1022 and 1025), six acetals of phenethyl alcohol or phenylacetaldehyde (Nos 1000–1001 and 1003–1006), three related phenylacetaldehydes (Nos 1002 and 1023–1024) and phenylacetic acid (No. 1007). The four phenoxyacetic acid derivatives comprise phenoxyacetic acid (No. 1026), an ester of phenoxyacetic acid (No. 1027), an ester of phenoxyethyl alcohol (No. 1028) and a homologue of phenoxyacetic acid (No. 1029) (see Table 1).

The substances in this group are structurally related because they each have a 2-phenethyl or 2-phenoxyethyl-carbon skeleton containing a primary oxygenated functional group. Studies described in more detail below indicate that the phenethyl acetals and esters in this group are readily hydrolysed to yield phenethyl alcohol, phenylacetaldehyde or phenylacetic acid. Also. phenethyl alcohol and phenylacetaldehyde are readily oxidized to phenylacetic acid. Therefore, all the substances in this group are eventually hydrolysed and oxidized to yield phenylacetic acid, which is excreted primarily in the urine in conjugated form. Similarly, the esters of phenoxyethyl alcohol or phenoxyacetic acid are hydrolysed to the parent alcohol

Figure 1. Metabolism of phenethyl alcohol and related substances in mammals

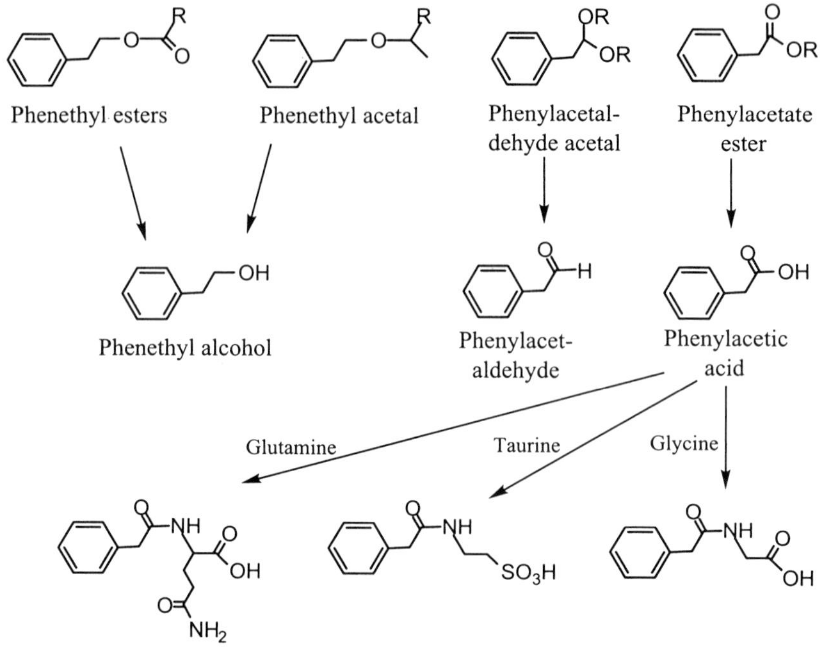

or acid. Phenoxyethyl alcohol is oxidized to phenoxyacetic acid, which is excreted primarily unchanged in the urine (Williams, 1959). Given their similar pharmacokinetics and their participation in a common metabolic pathway, the substances in this group are expected to have similar toxicological profiles (see Figure 1).

2.2 Additional considerations on intake

Production volumes and intake values for each flavouring agent in this group are reported in Table 2, which also gives the available information on natural occurrence.

Quantitative data on natural occurrence and consumption ratios have been reported for 13 phenethyl alcohol derivatives. The consumption ratios for phenethyl alcohol (No. 987), phenethyl acetate (No. 989), phenethyl propionate (No. 990), phenethyl hexanoate (No. 995), phenethyl octanoate (No. 996) and phenylacetaldehyde (No. 1002) are all > 1, indicating that they are consumed predominantly from traditional foods (Stofberg & Kirschman, 1985; Stofberg & Grundschober, 1987). The six flavouring agents that have consumption ratios > 1 represent 67% and 44% of the total annual volume of usage in Europe and the USA, respectively.

Only one phenoxyacetic acid derivative is a natural component of food. Sodium 2-(4-methoxyphenoxy) propanoate (No. 1029) has been detected in coffee (Maarse et al., 1999), but no quantitative data were available on its natural occurrence.

2.3 Biological data

2.3.1 Biochemical data

(a) Hydrolysis

If evidence can be supplied to show that an ester is readily hydrolysed in the body to constituents whose metabolic fate and biological actions are fully understood, further toxicological studies may not be necessary. Consequently, artificial gastrointestinal juices have been used to study the hydrolysis of esters (Longland et al., 1977). Before absorption in vivo, the esters and acetals in this group can be reasonably predicted to undergo hydrolysis (Williams, 1959) to yield phenethyl alcohol, phenoxyethyl alcohol, phenylacetaldehyde, phenylacetic acid and phenoxyacetic acid. Phenethyl acetate (No. 989), methyl phenylacetate (No. 1008), ethyl phenylacetate (No. 1009), isopropyl phenylacetate (No. 1011), isoamyl phenylacetate (No. 1014) and citronellyl phenylacetate (No. 1021) were rapidly hydrolysed in vitro in simulated gastric juice and pancreatic juice (Nos 989, 1008 and 1009) (Longland et al., 1977) or in a buffered solution of pancreatin (Nos 1008, 1011, 1014 and 1021) (Grundschober, 1977) to the corresponding phenethyl derivatives. Phenethyl acetate (No. 989), ethyl phenylacetate (No. 1009) and isoamyl phenylacetate (No. 1014) were not hydrolysed by partially purified human plasma arylesterase (Augustinsson & Ekedahl, 1962).

The finding of hydrolysis of aromatic acetals in simulated gastric juice and intestinal fluid leads indicates that the acetal functional group is hydrolysed before absorption in vivo. Incubation of 2-phenylpropanal dimethyl acetal (1 mmol/l) with simulated gastric juice at 37 °C resulted in 97% hydrolysis within 1 h. Under the same experimental conditions, benzaldehyde propylene glycol acetal (1 mmol/l) was 53% hydrolysed within 5 h, whereas the acetal in 0.1 N HCl under reflux was 99% hydrolysed (Morgareidge, 1962). The data on hydrolysis of acetals in vitro and in vivo indicate that aliphatic acetals undergo hydrolysis in humans before absorption.

(b) Absorption, distribution and excretion

When ingested in traditional foods, in foods to which they have been intentionally added or as hydrolysis products resulting from either condition, phenethyl and phenoxyethyl alcohols, phenylacetaldehyde and phenylacetic and phenoxyacetic acids are rapidly absorbed from the gastrointestinal tract. Once absorbed, the alcohols and aldehydes are rapidly oxidized to yield phenylacetic or phenoxyacetic acid derivatives, which are subsequently excreted in the urine, either free as in the case of phenoxyacetic acid or conjugated as in the case of phenylacetic acid (Williams, 1959; James et al., 1972; Sangster & Lindley, 1986; Hawkins & Mayo, 1986; Caldwell, 1987).

(i) Phenethyl alcohol (No. 987) and phenylacetic acid (No. 1007)

Phenylacetic acid is a normal component of human urine (250–500 mg/24 h), forming mainly from the breakdown of phenylalanine by intestinal bacteria (Seakins, 1971) or by oxidative deamination of endogenous phenethylamine (Richter, 1938; Seakins, 1971). Orally administered phenethylamine is rapidly metabolized to phenylacetylglutamine. Two volunteers, each fed a 300-mg dose of *S*-phenethylamine, excreted 60–62% of the administered dose as conjugated phenylacetic acid in the urine within 2–4.5 h (Richter, 1938; Seakins, 1971). Furthermore, > 80% of [^{14}C]-*S*-phenethylamine fed to mice was rapidly excreted from urine as conjugated [^{14}C]phenylacetic acid (Block, 1953).

One male volunteer excreted 26% of a 4000-mg oral dose of phenethyl alcohol (No. 987) in his urine as phenylacetylglutamine within 24 h (Thierfelder & Schempp, 1917). In rabbits, 42% and 5% of a single 300 mg/kg bw oral dose of phenethyl alcohol was excreted in the urine as glycine and glucuronic acid conjugates, respectively, of phenylacetic acid within 24 h. The ether-soluble acid extracted from the 24-h urine accounted for 61% of the dose (Bray et al., 1958). In an earlier study, 53% of a dose of 1300 mg bw of phenylacetic acid administered to rabbits by gavage was isolated from the 24-h urine as an ether-soluble acid. No appreciable quantity (< 0.5%) of free phenylacetic acid was recovered (Bray et al., 1946). In another study, only 0.4–3.1% of an oral dose of 740 mg/kg bw phenylacetic acid was excreted unconjugated in the urine of rabbits within 6 h (Tulane & Lewis, 1933).

More than 90% of a single oral dose of 80 mg of [carboxy-^{14}C]phenylacetic acid administered to each of three healthy volunteers and two patients with phenylketouria was excreted in the urine within 24 h as the glutamine conjugate (In a third patient with phenylketouria in this study, 12% of the dose was excreted as the glutamine conjugate, but this was attributed to incomplete collection of urine.) (James et al., 1973). Two male volunteers excreted 98% of an oral dose of 1 mg/kg bw of [carboxy-^{14}C]phenylacetic acid in their urine within 24 h (James et al., 1972), with > 90% of the dose recovered as the glutamine conjugate. The results of studies with radiolabelled phenylacetic acid indicate that it is rapidly absorbed and quantitatively excreted within 24 h.

(ii) Phenoxyethyl alcohol and phenoxyacetic acid (No. 1026) derivatives

Once hydrolysed, 2-phenoxyethanol is rapidly absorbed and oxidized to phenoxyacetic acid (No. 1026), which is then excreted almost exclusively in the urine in unconjugated form (Figure 2).

The fate of 2-phenoxyethanol in rats and humans has been investigated. More than 90% of an oral dose of 16, 27 or 160 mg/kg bw of [2-^{14}C]phenoxyethanol given to male Colworth rats by gavage was excreted in the urine within 24 h. A female rat also excreted about 90% of a dose of 27 mg/kg bw in the urine within 24 h. Approximately 2 and 1.3% of the ingested dose was recovered from expired air of female and male rats, respectively. The rate of intestinal absorption was rapid, with 60–70% of the excreted ^{14}C detected at 3 h and > 95% of the total 4-day urinary ^{14}C detected within the first 24 h. Trace amounts of radioactivity were detected in faeces.

Figure 2. Metabolism of 2-phenoxyethanol and related substances in animals

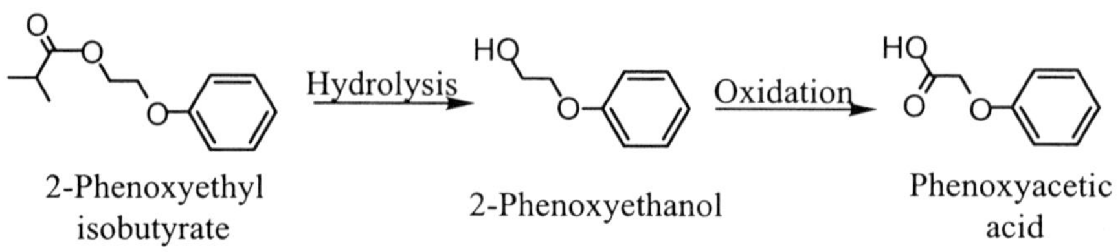

Four days after dosing, only trace amounts of radioactivity remained in the carcass, primarily in the liver (< 0.2% of the dose), fat and muscle. At 4 days, the ^{14}C concentration in blood was only 0.001% (Howes, 1988).

After ingestion of a dose of 11 mg of unlabelled 2-phenoxyethanol by a volunteer, urine samples were pooled daily for 4 days. Quantitative urinary elimination (12 mg, 104% of the dose) of phenoxyacetic acid, mainly in the unconjugated form, was reported within 24 h. Minor amounts (0.34 mg) of phenoxyacetic acid were detected at day 2, but none was detectable at day 3 or 4 (Howes, 1988).

Phenoxyacetic acid was fed to male rabbits at a dose of 100–200 mg/kg bw, and some animals also received glycine in amounts corresponding to three equivalents of the acid. In this test, 44–72% of the phenoxyacetic acid was recovered unchanged in the urine within 6 h and 82–105% within 24 h. There was no evidence of conjugation with either glucuronic acid or glycine, even when the diet was supplemented with glycine. A rabbit that received an oral dose of 500 mg of the glycine conjugate of phenoxyacetic acid excreted 30% of the dose unconjugated in the urine after 18 h (Levey & Lewis, 1947). In another study, 55% of an oral dose of an unspecified amount of phenoxyacetic acid was recovered in the urine of dogs and 61% in the urine of humans. No evidence for glycine or glucuronic acid conjugation was found (Thierfelder & Schempp, 1917).

A series of studies performed with 2-(4-methoxyphenoxy) propanoate (No. 1029) indicated that this phenoxyacetic acid derivative is also rapidly absorbed and excreted unchanged in the urine. Groups of three male and three female rats were dosed by oral gavage with [^{14}C]sodium 2-(4-methoxyphenoxy) propanoate at 50 mg/kg bw or 500 mg/kg bw (5 μCi), and urine was collected at 6, 12, 24, 48, 72, 96 and 120 h. Regardless of dose, most of the radioactivity (group mean activity, 81–95%) was eliminated in the urine within the first 24 h, most of the elimination (group mean activity, 59–76% of dose) occurring within the first 6 h. Only 0.8–4.6% was detected in the faeces at 120 h. The authors noted that contamination of faeces by urine may have been responsible for the radioactivity detected. At 120 h, the concentratuion of radioactivity in the urine of females was approximately twice that of males, suggesting that males eliminate the substance faster than females (Brown et al., 1986).

In the same study, 30 male and 27 female rats were dosed by oral gavage with [^{14}C]sodium 2-(4-methoxyphenoxy) propanoate at 50 mg/kg bw (5 μCi/ml) 18 days after mating. Groups of three male and three female animals were killed 0.5, 1, 3, 5, 7, 12, 24, 48, 72 and 96 h after dosing, and blood was sampled and tissues and organs were removed. Sodium 2-(4-methoxyphenoxy) propanoate was rapidly absorbed and eliminated by both male and female rats. Only 12% (range, 6.3–23%) and 2.6% remained in the stomach of males and females after 0.5 h, indicating that absorption was slightly faster in females than in males. The total radioactivity was greatest in the liver (2.5% for males and 3.1% for females) and kidney (2.1% for males and 2.5% for females), the concentration of radioactivity being 3.5 times greater in the kidney than in the liver. Loss of radioactivity from tissues and organs was rapid, elimination being essentially complete with 24 h (Brown et al., 1986).

In pregnant females CD rats given [^{14}C]sodium 2-(4-methoxyphenoxy) propanoate by gavage at a dose of 50 mg/kg bw (5 μCi/ml), examination of fetuses and amniotic fluid revealed that the radioactivity had crossed the placenta, although its elimination from the fetus and placenta was complete within 3 h. Elimination of radioactivity from amniotic fluid was complete within 12 h. Radioactivity was detected in uterine muscle for up to 72 h, although the activity represented < 0.05% of the dose at 7 h. The authors suggested that the acid might bind to uterine muscle protein. In a subsequent test with bovine serum albumin in vitro, sodium 2-(4-methoxyphenoxy) propanoate bound to protein, an equilibrium (83–85% in bound fraction) being reached within 2 h. The authors concluded that, although the acid binds to plasma protein, it is still rapidly eliminated from the blood (Brown et al., 1986).

Measurements in plasma confirmed that [^{14}C]sodium 2-(4-methoxyphenoxy) propanoate is rapidly absorbed and eliminated from male and pregnant female rats. The highest plasma concentration was found at the first sampling time (0.5 h); at 12 h, the concentration was only 0.3% of that achieved at 0.5 h. Repeated daily dosing resulted in 90% of the steady-state plasma concentration after 10 days. However, the steady state concentration was still < 1% of the plasma concentration at 0.5 h. Thus, repeated dosing had little effect on the plasma concentration–time profile (Brown et al., 1986).

In a subsequent study, three male and three female rats each received a single intravenous dose of [^{14}C]sodium 2-(4-methoxyphenoxy) propanoate at a level of 50 mg/kg (10 μCi). Blood samples were collected at 5, 15, 30 and 60 min and 2, 4, 6, 12 and 24 h after dosing until the concentrations of radioactivity were near the limit of accurate measurement. The highest plasma concentration, achieved at 5 min, declined rapidly over the next hour. The mean concentrations of radioactivity decreased from 190 and 210 μg/ml of plasma equivalent at 5 min to 11 and 16 μg/ml at 1 h in male and female rats, respectively. The mean elimination half-lives of radioactivity during this time were estimated to be 13 min for males and 14 min for females. The concentrations of radioactivity declined to 0.25 and 0.32 μg/ml of plasma in male and female rats, respectively, at 6 h. No radioactivity was detected in male rats after 6 h or in female rats after 24 h. The mean area under the curve of plasma concentration–time was 77 μg.h per ml for male rats and 100 μg.h per ml for females, indicating slower elimination of radioactivity by females. The authors concluded that sodium 2-(4-methoxyphenoxy) propanoate was rapidly eliminated from the vascular system of male and female rats (Hawkins & Mayo, 1986).

In a subsequent investigation of the effect of repeated intake on metabolism and excretion, five male and five female rats were fed a diet containing 5000 ppm of uniformly ring-labelled [^{14}C]sodium 2-(4-methoxyphenoxy) propanoate for 17 weeks. The study was conducted in three phases. Two unlabelled doses and one ^{14}C-labelled dose of the sodium salt were given on separate occasions with 2 weeks between each dose. In the study with the labelled dose, 99% and 104% of the radioactivity was excreted in the urine of males and females, respectively, < 1% of the dose being detected in faeces. The 0–6-h urine samples accounted for 83% of the dose and the 6–12-h samples for 17%. No radioactivity was detected in whole blood, plasma or washed red blood cells 7 days after administration. More urinary metabolites were recovered than had been administered in the studies with unlabelled substance, indicating that the test material was not completely eliminated before oral administration. However, in both studies, 90% of the recovered material was excreted within the first 6 h after dosing and most of the remainder in the 6–12-h urine sample. The author concluded that prolonged intake of sodium 2-(4-

methoxyphenoxy) propanoate does not affect the metabolism or disposition of this compound in rats (Hill, 1986).

The disposition of sodium 2-(4-methoxyphenoxy) propanoate was also investigated in three healthy male volunteers given capsules containing 114.7 mg, 112.9 mg and 114.3 mg of the test compound. About 75% of the dose was eliminated in urine within 72 h, of which 85% was recovered within 24 h (Sangster & Lindley, 1986). In another study, five healthy male volunteers were given an oral dose of 100 mg of [2-^{14}C]sodium 2-(4-methoxyphenoxy) propanoate (50 μCi). About 97% of the dose was recovered within 5 days of administration, with approximately 93% recovered within the first 24 h (Caldwell, 1987). The author considered that the data show that sodium 2-(4-methoxyphenoxy) propanoate is rapidly absorbed and rapidly eliminated from the body via the urine.

On the basis of the extensive data available on the absorption, distribution and elimination of phenethyl and phenoxyethyl alcohols and phenylacetic and phenoxyacetic acids, the Committee concluded that, after hydrolysis, the parent alcohols are converted mainly to the corresponding carboxylic acids, which are then rapidly excreted in either free or conjugated form in urine. The pharmacokinetics of phenoxyethyl alcohol derivatives in rats and humans indicate that the rat is an acceptable model for human risk assessment.

(c) Biotransformation

(i) Phenethyl alcohol derivatives

Phenethyl alcohol is successively oxidized to phenylacetaldehyde and phenylacetic acid in vivo. Phenylacetic acid undergoes species-specific conjugation with a variety of amino acids, amines or glucuronic acid, followed by excretion almost exclusively in the urine (James et al., 1972; see Figure 2). Phenethyl alcohol is readily oxidized to phenylacetaldehyde by an assortment of NAD$^+$-dependent alcohol and aldehyde dehydrogenases (Bosron & Li, 1980). The greatest activity of mammalian alcohol dehydrogenases (ALDH) occurs in the liver, where they show broad substrate specificity for the oxidation of primary aliphatic and aromatic alcohols. Human liver ALDH showed a decreased Michaelis-Menten constant (K_m, the concentration of the specific substrate at which a given enzyme yields one-half its maximum velocity with increasing lipophilicity); however, the maximum rate or velocity of an enzymatic reaction which is indicative of all the enzyme active site(s) complexed with substrate, V_{max}, remained essentially constant, suggesting that the rate-limiting step does not involve the binding or release of the alcohol or aldehyde intermediate (Pietruszko et al., 1973).

Once formed, phenylacetaldehyde is oxidized by inducible aldehyde dehydrogenases from rat liver cytosol. In rats, these isoenzymes can be induced by phenobarbital (Simpson et al., 1985). The K_m and V_{max} values of human mitochondrial ALDH (ALDH-2) and the cytosolic isoenzyme (ALDH-1) for oxidation of phenylacetaldehyde (Table 3) indicate rapid conversion to phenylacetic acid (Klyosov, 1996).

Phenylacetaldehyde and 3- and 4-chlorophenylacetaldehyde are effectively oxidized to the corresponding phenylacetic acid derivatives when incubated with rat hepatic microsomal dehydrogenase containing NAD$^+$ as a coenzyme. The rates of

Table 3. Pharmacokinetics of human mitochondrial aldehyde dehydrogenase (ALDH-2) and cytosolic isoenzyme (ALDH-1)

Pharmacokinetics	ALDH1	ALDH2
K_m (nmol/l)	5500 ± 1200	29 ± 4
$V_{max\ rel}$(%in relation to V_{max} of acetaldehyde)	380 ± 40	150 ± 20
k_{cat}[a]	3000 ± 340	1800 ± 200

From Klyosov (1996)

[a] k_{cat} values are based on the relative molecular masses of tetrameric enzymes of 230 000 (ALDH-1) and 240 000 (ALDH-2) The specific activities of purified ALDH-1 and ALDH-2 from five different batches were 3.4 ± 0.6 and 4.9 ± 0.8 μmol/min (per mg of protein), respectively, at pH 9.5. This corresponds to k_{cat} values of 782 ± 138 per min for ALDH-1 and 1200 ± 190 per min for ALDH-2.

oxidation for the 3- and 4-chloro derivatives were markedly slower than that of the parent phenylacetaldehyde (Martini & Murray, 1996). In dogs, 32% of a dose of 1900 mg/kg bw phenylacetaldehyde (No. 1002) was rapidly oxidized and excreted as the glycine conjugate within 48 h (Kay & Raper, 1922).

(ii) Conjugation of phenylacetic acid (No. 1007)

Although phenylacetic acid has been studied extensively, investigations conducted before 1950 on human metabolism (Shiple & Sherwin, 1922; Power & Sherwin, 1927; Ambrose et al., 1933; Wagreich et al., 1940) failed to account for the endogenous level of 250–500 mg/kg per day of phenylacetic acid conjugated with glutamine (Stein et al., 1954) present in human urine and did not adequately characterize the array of urinary conjugates that were formed from phenylacetic acid. More recent work demonstrates that conjugation is both dose-dependent and species-specific. The major metabolic options available to phenyl acetic acid are conjugation with glucuronic acid, glycine, taurine or glutamine and elimination as the free acid.

Two men excreted an average of 91% and 7% of an oral dose of 1 mg/kg bw of [carboxy-^{14}C]phenylacetic acid within 24 h as glutamine and taurine[1] conjugates, respectively. In contrast to most animals, humans have only traces of the glycine conjugate (James et al., 1972). The distribution and type of conjugation are virtually unaffected by continued ingestion of phenylacetic acid. One person fed 34 doses of 1000–10 000 mg of the acid over 97 days excreted > 90% of the administered dose as the phenylacetylglutamine conjugate (Ambrose et al., 1933). Like humans, Old and New World monkeys conjugate phenylacetic acid with glutamine and, to a lesser extent, taurine; however, significant quantities of acid (1–44%) are excreted free. In carnivores (e.g., dogs, cats and ferrets), glycine conjugation predominates, with no

1

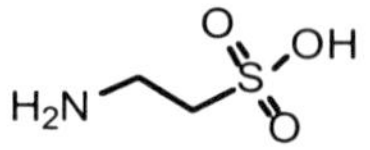

detectable glutamine conjugation. Likewise, in rodents and lagomorphs (rabbits), phenylacetic acid is excreted primarily as the glycine conjugate. Unconjugated phenylacetic acid and minute amounts of taurine conjugates are also excreted. In rats, > 94% of a dose of 80 mg/kg bw of phenylacetic acid given by intraperitoneal injection was excreted as the glycine conjugate (James et al., 1972).

Clearly, the nature of the amino acid used for conjugation is a function of species. The sources and amounts of available amino acids alter the conjugating ability of different species. In humans, endogenous sources of glutamine include those from waste urea nitrogen. Ingestion of 5000 mg/day of phenylacetic acid for 3 consecutive days resulted in a 25–78% decrease in urinary urea nitrogen. Glutamine may be supplied by blood plasma glutathione (reduced tripeptide, Glu-Cys-Gly). An 18–23% reduction in the plasma tripeptide concentration was observed within a few hours of ingestion of a 4000-mg dose of phenylacetic acid in a volunteer(Shiple & Sherwin, 1922).

The capacity for glutamine conjugation was studied in three control subjects and three patients with phenylketouria, each of whom was given a single 80-mg dose of [carboxy-^{14}C]phenylacetic acid. The average excretion of phenylacetylglutamine (measured as mmol/g creatinine) by the patients with phenylketouria was approximately five times that of two of the control subjects, indicating that the glutamine conjugation mechanism can cope with large amounts of phenylacetic acid (James & Smith, 1973). The mechanism for conjugation of glutamine with phenylacetic acid probably involves formation of a phenylacetic acid–coenzyme A intermediate. Perfusion of human kidney for 1 h or incubation of human liver homogenate with [^{14}C]glutamine and phenylacetyl–coenzyme A resulted in yields of the 5 and 13% of the respective radioactive conjugates (Moldave & Meister, 1957).

In rodents, endogenous unconjugated phenylacetic acid can be found at concentrations at which glycine conjugation is capacity-limited, presumably by the supply of endogenous glycine (Gregus et al., 1993). Only small amounts of the glycine conjugate enter the bile: < 10% of a dose of phenylacetic acid of 100 mg/kg bw delivered to the small bowel via an indwelling catheter was collected from the bile of rats over 4 h (Koss & Lamprecht, 1968). Significant concentrations of free phenylacetic acid have been found after high doses (Teuchy et al., 1971; James et al., 1972). Prolonged high concentrations of free phenylacetic acid might be associated with toxic effects similar to those observed with high doses of other carboxylic acids with which conjugation is glycine-limited. Conversely, the high levels of glutamine available for conjugation in humans might allow metabolic pathways to cope with high levels of endogenously formed phenylacetic acid.

(iii) Phenoxyethyl alcohol derivatives

In rats, > 90% of a dose of of [2-^{14}C]-2-phenoxyethanol of 16, 27 or 160 mg/kg bw in males and 27 mg/kg bw in females was excreted in the urine as phenoxyacetic acid within 24 h. An entire oral dose of 11 mg of unlabelled 2-phenoxyethanol was accounted for in the urine of one healthy male volunteer as 2-phenoxyacetic acid. Most of the acid was excreted unconjugated (Howes, 1988).

Urine of male rats collected for 6 h after an oral dose of 500 mg/kg (5 μCi) of [^{14}C]sodium 2-(4-methoxyphenoxy)propanoate contained two excretion products, identified as 2-(4-methoxyphenoxy)propanoic acid (90%) and the *O*-demethylated metabolite (±)-2-(4-hydroxyphenoxy)propionic acid (7–10%) (Brown et al., 1986).

Groups of five rats of each sex were fed a diet cointaining [^{14}C]sodium 2-(4-methoxyphenoxy)propanoate at a concentration of 5000 ppm, corresponding to 250 mg/kg bw per day, for 17 weeks. The majority (92–96%) of the dose was excreted as unchanged 2-(4-methoxyphenoxy)propanoic acid, while 6–7% of the dose in male rats and 8–9% of the dose in female rats was excreted as the *O*-demethylated metabolite (±)-2-(4-hydroxyphenoxy)propionic acid (Hill, 1986).

The metabolites in the urine of three men who received a single dose of 113–114 mg of 2-(4-methoxyphenoxy)propanoic acid included 2-(4-methoxyphenoxy) propanoic acid (65%) and 2-(4-hydroxyphenoxy)propanoic acid (10%) (Sangster & Lindley, 1986). In another study, approximately 97% an oral dose of 100 mg of [2-^{14}C]sodium 2-(4-methoxyphenoxy)propanoate (50 μCi) given to five healthy male volunteers was excreted in the urine as 2-(4-methoxyphenoxy)propanoic acid; > 93% was recovered from the urine within the first 24 h (Caldwell, 1987). Thus, sodium 2-(4-methoxyphenoxy)propanoate is rapidly absorbed and rapidly eliminated in the urine, mainly as the unchanged acid. Minor amounts undergo *O*-demethylation to yield the corresponding metabolite 2-(4-hydroxyphenoxy)propanoic acid.

In summary, phenethyl and phenoxyethyl alcohol derivatives are rapidly oxidized to the corresponding carboxylic acids. In humans, phenylacetic acid is subsequently conjugated primarily with glutamine and excreted in the urine, while phenoxyacetic acid is excreted unconjugated almost exclusively in the urine.

(d) Other biochemical studies

Other studies have been conducted in rodents and humans on the effects of phenylacetic acid on various biochemical parameters, mostly associated with phenylketouria. Phenylacetic acid administered orally to fasted guinea-pigs at a dose of 380, 500 or 750 mg/kg bw resulted in a reduction in blood sugar to 13%, 44% and 60% of the initial value, respectively, within 2–4 h. Rabbits were as sensitive as guinea-pigs to the hypoglycaemic action of phenylacetic acid. Phenylacetic acid administered intravenously at a dose of 62 mg/kg bw also produced a marked reduction in blood sugar in a cat, which was sustained for at least 5 h (Stewart, 1962).

Phenylacetic acid given orally at a single dose of 20 mg/kg bw to each of five male volunteers resulted in a two- to fourfold increase in urinary indole-3-acetic acid concentration (Tashian, 1960). Presumably, the phenylacetic acid deactivated tryptophan decarboxylase, resulting in increased conversion of tryptophan to indole-3-acetic acid.

Incubation of a suspension of 0.1 ml or 0.1 g/100 ml of phenylacetic acid with human plasma caused prolonged clotting and increased thrombin activity (Nour-Eldin, 1968).

No changes in total nitrogen or ammonia content were seen in the urine of a rabbit fed a diet containing 0.5 g on the first day and 1 g on the next. An increase in amino acid and an insignificant decrease in urea output were noted in urine. In two other rabbits, one in nitrogen balance and one fasted, fed a dose of 1.2 g or 1.0 g, respectively, for 2 days by stomach tube, increased output of amino acids, total nitrogen, urea and ammonia was observed (Hijikata, 1922).

In a study of the effects of various phenyl-substituted acids on the activity of glycosyl transferase (used in cerebral glycoprotein synthesis) in human cerebrospinal fluid, a 1-h incubation with phenylacetic acid (2–10 mmol/l) produced up to 91% inhibition (Ko et al., 1973).

Incubation of 0.25% (2500 μg/ml) phenethyl alcohol with *Escherichia coli* strains DG75 and H500T– resulted in reversible inhibition of DNA synthesis and cell division, as measured by labelled thymine uptake and cell counts, respectively. The cells remained viable and RNA and protein synthesis capable. Increased sedimentation seen during treatment was interpreted as due to either increased formation of nucleoid bodies containing the replicating mechanism of the cell or increased cell lysis due to the detergent action of phenethyl alcohol (Brunner & Treick, 1982).

Numerous studies have been conducted to evaluate the effect of phenethyl alcohol on excision repair of DNA damage induced by ultraviolet radiation (UV) in non-standard *E. coli* strains. When solutions (0.2 and 0.4%) of phenethyl alcohol were incubated with UV-damaged *E. coli* strains H/r 30 and B (wild-type), excision repair of membrane-associated DNA was inhibited, as measured by a decrease in liquid holding recovery and a decrease in the removal of thymine dimers. The authors hypothesized that excision repair may require that DNA is bound to the cell membrane, and the test article may have inhibited that binding (Tachibana & Yonei, 1985). Similar results were obtained in an earlier study, in which it was shown that 0.05–0.5% solutions of phenethyl alcohol caused detachment of DNA from the DNA–membrane complex in *E. coli* B cells. The effect was concentration-dependent. Similar results were obtained with *E. coli* NG30 (rec A mutant) (Tachibana et al., 1982). In a related study, phenethyl alcohol inhibited the removal of thymine dimers from DNA in what was considered to be the incision step of excision repair in *E. coli* (Tomiyama et al., 1986). Phenethyl alcohol thus inhibited excision repair of UV-damaged DNA in *E. coli* (Yonei, 1980; Todo & Yonei, 1983; Tachibana & Yonei, 1985).

Wingard et al. (1955) classified phenylacetaldehyde (No. 1002) as a depressor on the basis of its effects on the blood pressure of dogs given doses of 3.4–40 mg/kg bw; they typically observed an initial large, rapid fall in blood pressure followed by a smaller, longer rise. Phenylacetaldehyde (No. 1002) was reported to have decreased blood pressure by 8 mm for 40 s in one dog (Romano et al., 1954).

Phenylacetaldehyde inhibited both Na^+/K^+-activated ATPase and Mg^{2+}-ATPase. It also inhibited K^+-dependent *para*-nitrophenylphosphatase activity. The Na^+/K^+ ATPase was much more sensitive than Mg^{2+} ATPase or K^+-activated phosphatase to inhibition by aldehydes. The inhibition of Na^+/K^+ ATPase by aldehyde was reversible and was not competitive with ATP or K^+ as the variable substrate or activator, respectively. Addition of cysteine or mercaptoethanol protected the enzymes from inhibition by aldehydes. All aldehydes, including acetaldehyde, were more potent inhibitors of Na^+/K^+ ATPase activity than was ethanol (Erwin et al., 1975). Glutathione peroxidase was inactivated by phenylacetaldehyde in vitro (Tabatabaie & Floyd, 1996).

2.3.2 Toxicological studies

(a) Acute toxicity

Oral LD_{50} values have been reported for 26 of the 39 phenethyl alcohol derivatives and are summarized in Table 4. The LD_{50} values in rats ranged from 1500 mg/kg bw for phenethyl alcohol to > 15 000 mg/kg bw for phenethyl phenylacetate, showing that their oral acute toxicity is low. In mice, the oral LD_{50} values ranged from 800 mg/kg for phenylacetic acid to > 15 000 mg/kg bw for linalyl phenylacetate. The oral LD_{50} values in guinea-pigs ranged from 400 mg/kg bw for phenylacetic acid to 3900 mg/kg for phenethylacetaldehyde.

Oral LD_{50} values have been reported for three of the four phenoxyethyl alcohol derivatives (Table 4). In rats, the values ranged from 1500 mg/kg bw for phenoxyacetic acid to > 5000 mg/kg bw for phenoxyacetic acid, 2-phenoxyethyl isobutyrate and sodium 2-(4-methoxyphenoxy)propanoate. These results demonstrate that the acute toxicity of phenoxyacetic acid and related substances is low in rats.

(b) Short-term studies of toxicity

The results of short-term studies are summarized in Table 5 and described below. These studies show that phenethyl and phenoxyethyl alcohol derivatives have little toxic potential.

(i) Phenethyl alcohol (No. 987)

Groups of 20 male and 20 female Wistar albino rats were given a mixture of compounds dissolved in tap water as their only drinking-water for 56 weeks. The mixture comprised ethyl alcohol at 6000 mg/kg bw (6%), ethyl acetate at 4 mg/kg bw (0.004%), isoamyl alcohol at 120 mg/kg bw (0.12%), phenethyl alcohol at 120 mg/kg bw (0.12%), isobutyl alcohol at 200 mg/kg bw (0.2%) and acetic acid at 200 mg/kg bw (0.2%). A control group of 20 rats of each sex was maintained on tap water only. All groups of rats were fed a stock ration *ad libitum*, and their weights were recorded weekly. The activities of alcohol dehydrogenase and alanine and aspartate aminotransferases and the protein content of the liver were determined at 2–4-week intervals. At study termination, the liver, kidneys, heart, spleen and lungs were examined histologically. There was a slight, statistically nonsignificant decrease in the mean body weight of both groups at 28–29 weeks compared with that at 53–56 weeks. The absolute and relative liver weights were similar in test and control groups. A slight increase in aspartate aminotransferase activity was observed between 28 and 56 weeks in both test groups. No significant abnormalities were observed in any of the organs examined. Six animals contracted pneumonia and were discarded, and pneumonia was common in rats in all groups at termination. The authors concluded that the mixture of chemicals tested had no effects on the parameters tested (Johannsen & Purchase, 1969).

(ii) Phenethyl seneciote (No. 998), isopropylphenylacetaldehyde (No. 1024) and ethyl (para-tolyloxy)acetate (No. 1027)

Groups of 10–16 male and female Charles River CD rats were given diets adjusted during the study to provide constant concentrations of 1.5 mg/kg bw per day for males and 1.8 mg/kg bw per day for females of phenethyl senecioate,

Table 4. Studies of the acute toxicity of phenethyl alcohol, aldehyde, acid, and related acetals and esters used as flavouring agents

No.	Agent	Species	Sex[1]	Route	LD_{50} (mg/kg bw)	Reference
987	Phenethyl alcohol	Mouse	NR	Oral	800	Fassett (1963)
987	Phenethyl alcohol	Guinea-pig	NR	Oral	400	Fassett (1963)
987	Phenethyl alcohol	Rat	NR	Oral	1800	Rumyantsev et al. (1987)
987	Phenethyl alcohol	Rat	NR	Oral	1500	Moreno (1982)
987	Phenethyl alcohol	Rat	M,F	Gavage	1800	Jenner et al. (1964)
987	Phenethyl alcohol	Rat	M,F	Gavage	2500	Zaitsev & Rakhamanina (1974)
987	Phenethyl alcohol	Mouse	M,F	Gavage	2500	Zaitsev & Rakhamanina (1974)
987	Phenethyl alcohol	Guinea-pig	M,F	Gavage	2500	Zaitsev & Rakhamanina (1974)
987	Phenethyl alcohol	Rat	M	Gavage	1700	Mallory et al. (1982
988	Phenethyl formate	Rat	M,F	Gavage	3.2 ml/kg	Levenstein (1973
989	Phenethyl acetate	Rat	NR	Oral	> 5000	Moreno (1973
989	Phenethyl acetate	Rat	NR	Oral	5200	Rumyantsev et al. (1987)
989	Phenethyl acetate	Rat	M,F	Gavage	3700	Zaitsev & Rakhamanina (1974)
989	Phenethyl acetate	Mouse	M,F	Gavage	3700	Zaitsev & Rakhamanina (1974)
989	Phenethyl acetate	Guinea-pig	M,F	Gavage	3700	Zaitsev & Rakhamanina (1974)
990	Phenethyl propionate	Rat	M,F	Gavage	4000	Beroza et al. (1975)
990	Phenethyl propionate	Rat	NR	Oral	4000	Moreno (1973)
991	Phenethyl butyrate	Rat	NR	Oral	4.6 ml/kg	Levenstein (1974)
992	Phenethyl isobutyrate	Rat	M,F	Gavage	> 5000	Shelanski & Moldovan (1971)
993	Phenethyl 2-methyl-butyrate	Rat	M	Oral	> 5000	Moreno (1982)
997	Phenethyl tiglate	Rat	NR	Oral	> 5000	Levenstein (1974)
999	Phenethyl phenyl-acetate	Rat	M.F	Gavage	15 000	Jenner et al. (1964)
999	Phenethyl phenyl-acetate	Rat	M,F	Gavage	3200	Zaitsev & Rakhamanina (1974)
999	Phenethyl phenyl-acetate	Mouse	M,F	Gavage	3200	Zaitsev & Rakhamanina (1974)
999	Phenethyl phenyl-acetate	Guinea-pig	NR	Gavage	3200	Zaitsev & Rakhamanina (1974)
1001	Acetaldehyde phen-ethyl propyl acetal	Rat	NR	Oral	> 5000	Moreno (1979)
1002	Phenylacetaldehyde	Rat	NR	Oral	1600	Moreno (1977)
1002	Phenylacetaldehyde	Rat	M,F	Gavage	3900	Zaitsev & Rakhamanina (1974)
1002	Phenylacetaldehyde	Mouse	M,F	Gavage	3900	Zaitsev & Rakhamanina (1974)
1002	Phenylacetaldehyde	Guinea-pig	NR	Gavage	3900	Zaitsev & Rakhamanina (1974)

Table 4 (contd)

No.	Agent	Species	Sex[1]	Route	LD_{50} (mg/kg bw)	Reference
1003	Phenylacetaldehyde dimethyl acetal	Rat	M,F	Gavage	< 5000	Shelanski & Moldovan (1971)
1004	Phenylacetaldehyde glyceryl acetal	Rat	M	Oral	1.7 ml/kg	Moreno (1972)
1004	Phenylacetaldehyde glyceryl acetal	Rat	M	Oral	< 5000	Moreno (1972)
1007	Phenylacetic acid	Rat	NR	Oral	> 5000	Keating (1972)
1007	Phenylacetic acid	Rat	M,F	Gavage	2200	Zaitsev & Rakhamanina (1974)
1007	Phenylacetic acid	Mouse	M,F	Gavage	2200	Zaitsev & Rakhamanina (1974)
1007	Phenylacetic acid	Guinea-pig	NR	Gavage	2250	Zaitsev & Rakhamanina (1974)
1008	Methyl phenylacetate	Rat	NR	Oral	2600	Moreno (1974)
1009	Ethyl phenylacetate	Rat	NR	Oral	3300	Moreno (1973)
1012	Butyl phenylacetate	Rat	NR	Oral	> 5000	Moreno (1980)
1013	Isobutyl phenylacetate	Rat	NR	Oral	> 5000	Moreno (1973)
1014	Isoamyl phenylacetate	Rat	NR	Oral	> 5000	Moreno (1976)
1016	3-Hexenyl phenyl acetate	Rat	NR	Oral	> 5000	Moreno (1976)
1018	Rhodinyl phenyl-acetate	Rat	NR	Oral	> 5000	Moreno (1977)
1019	Linalyl phenylacetate	Rat	NR	Oral	> 5000	Moreno (1974)
1019	Linalyl phenylacetate	Mouse	M,F	Oral	15 000	Colaianni (1967)
1020	Geranyl phenylacetate	Rat	NR	Oral	> 5000	Russell (1973)
1021	Citronellyl phenyl-acetate	Rat	NR	Oral	> 5000	Moreno (1977)
1023	*para*-Tolylacet-aldehyde	Mouse	NR	Oral	> 5000	Levenstein (1975)
1024	*para*-Isopropylphenyl acetaldehyde	Rat	NR	Oral	4100	Moreno (1977)
1026	Phenoxyacetic acid	Rat	M,F	Gavage	1800	Burdock & Ford (1990)
1026	Phenoxyacetic acid	Rat	M,F	Gavage	1800	Piccirillo et al. (1983)
1026	Phenoxyacetic acid	Rat	NR	Oral	> 5000	Moreno (1976)
1028	Phenoxyethyl iso-butyrate	Rat	NR	Oral	> 5000	Moreno (1973)
1029	Sodium 2-(4-methoxy-phenoxy) propanoate	Rat	M,F	Oral	> 5000	Cummins (1985)

M, male; F, female; NR, not reported

15 mg/kg bw per day for males and 16 mg/kg bw per day for females of ethyl (*para*-tolyloxy) acetate or 17 mg/kg bw per day for males and 19 mg/kg bw per day for females of isopropylphenylacetaldehyde. The animals were housed in pairs of the same sex and given access to water and food *ad libitum*. The doses were chosen by methods described by Oser & Hall (1977). Clinical observations were recorded

Table 5. Results of short-term and long-term studies of toxicity and carcinogenicity on phenethyl alcohol, aldehyde, acid and related acetals and estersused as flavouring agents

No.	Substance	Species; sex	No. test groups[a]/ no. per group[b]	Route	Duration	NOEL (mg/kg bw per day)	Reference
987	Phenethyl alcohol	Rat; M,F	1/40	Oral	56 weeks	120[d]	Johannsen & Purchase (1969)
998	Phenethyl senecioate	Rat; M,F	1/24–28	Oral	90 days	< 1.5 (M) < 1.8 (F)	Rabinowitz (1969)
998	Phenethyl senecioate	Rat; M,F	1/20–32	Oral	90 days	1.5 (M)[c] 1.8 (F)[c]	Posternak et al. (1969)
999	Phenethyl phenylacetate	Rat; M,F	3/20	Oral	17 weeks	500[c]	Hagan et al. (1967)
1024	*para*-Isopropyl-phenylacet-aldehyde	Rat; M,F	1/20–32	Oral	90 days	17 (M)[c] 19 (F)[c]	Posternak et al. (1969)
1027	Ethyl (*para*-tolyloxy) acetate	Rat; M,F	2/20–32	Oral	90 days	15 (M) 16 (F)	Posternak et al. (1969)
1029	Sodium 2-(4-methoxyphenoxy) propanoate	Rat; M,F	5/20	Oral	14 days	200	Hill & Wood (1985)
1029	Sodium 2-(4- methoxyphenoxy) propanoate	Rat; M,F	4/20	Oral	90 days	250	Hill & Wood (1986)

M, male; F, female; NR, not reported

[a] Total number of test groups does not include control animals.

[b] Total number per test group includes both male and female animals.

[c] Study performed at either a single or multiple doses that had no adverse effect. Therefore, this dose is not a true NOEL but is the highest dose tested that had no adverse effect. The actual NOEL would be higher.

daily, and food consumption and body weights were determined weekly. During weeks 7 and 13, haematological and clinical chemical (blood urea) parameters were measured. After 90 days, all the animals were killed and subjected to detailed necropsy, and the liver and kidneys were weighed. A wide range of tissues and organs from each animal was preserved, and major organs and tissues were examined histologically.

No significant difference in body-weight gain was found between test and control animals. The kidneys of male rats given phenethyl senecioate (No. 998) were slightly enlarged, but the slight increase in relative weight became less marked when the values were compared with those for control groups in the same laboratory. The authors concluded that the changes in kidney weight were not toxicologically significant. No effects were observed in either male or female rats given isopropylphenylacetaldehyde (No. 1024) or ethyl (*para*-tolyloxy) acetate (No. 1027) (Posternak et al., 1969).

(iii) Phenethyl seneciote (No. 998)

Groups of 24 male and 28 female albino Charles River rats were fed a diet containing phenethyl seneociate in a gum arabic solution, at a concentration of 13 ppm for weeks 1–4, 22 ppm for weeks 5–10 and 26 ppm for weeks 11–13, corresponding to average concentrations ingested during the 13 weeks of the study of 1.5 (range, 1.1–1.9) mg/kg bw per day for males and 1.8 (range, 1.3–2.2) mg/kg bw per day for females. Losses in both body weight and food efficiency were noted in treated groups ($p < 0.1$), and the weight of the kidneys was significantly increased. At week 7 only, the per cent haemoglobin was significantly increased, accompanied by a significant increase in mean corpuscular haemoglobin concentration. Histological examination revealed the presence of renal lesions in both males and females, which were less severe in females (Rabinowitz, 1969).

In a study reported only as a memo, groups of 10 male and 10 female Osborne-Mendel rats were maintained on diets containing 0, 20, 1000, 2500 or 10 000 ppm of phenethyl senecioate, calculated to provide an average daily intake of 0, 1, 50, 125 or 500 mg/kg bw, for 4 months. Slight degenerative changes were seen in the kidneys of saome animals, which were not unexpected, as Osborne-Mendel rats of this age commonly show all gradations of chronic interstitial nephritis. Gross and histological examination of a wide variety of tissues revealed no alterations that could be related to administration of the test material (Bierbower, 1970).

(iv) Phenethyl phenylacetate (No. 999)

Groups of 10 male and 10 female Osborne-Mendel rats were given diets containing phenethyl phenylacetate at a concentration of 0, 1000, 2500 or 10 000 ppm, corresponding to an average daily intake of 0, 50, 250 or 500 mg/kg bw, for 17 weeks. Weekly measurement of body weight and food intake showed no significant difference between test and control animals. At termination, haematological examination revealed no effects due to treatment. At necropsy, no difference between test and control animals in the weights of major organs was reported. Gross examination of tissues from all animals revealed no remarkable changes, and histological examination of three to four animals of each sex at the high dose and from the control group revealed no treatment-related lesions (Hagan et al., 1967).

(v) Sodium 2-(4-methoxyphenoxy) propanoate (No. 1029)

Groups of 10 male and 10 female Sprague-Dawley-derived CD rats were maintained on diets containing sodium 2-(4-methoxyphenoxy)propanoate at a concentration of 0, 4000, 12 000, 25 000 or 50 000 ppm, calculated to provide an estimated daily intake of 0, 200, 600, 1250 or 2500 mg/kg bw, for 14 days. Reduced body-weight gain associated with reduced food consumption was found at the two higher doses, and water consumption was increased in animals of each sex at these doses. Increased activity of plasma alkaline phosphatase and aspartate aminotransferase was reported in half the animals at the highest dose, but the increases in females at the two higher doses were within the normal range. Male and female rats at 2500 mg/kg bw per day had lowered blood reticulocyte counts but no other haematological changes associated with treatment. No histopathological changes were found that were associated with treatment. The authors concluded that a dietary dose of 1000–1250 mg/kg bw per day would be suitable for a subsequent 90-day study (Hill & Wood, 1985).

In the 90-day study, groups of 20 male and 20 female CD rats received sodium 2-(4-methoxyphenoxy)propanoate in the diet at a concentration of 0, 5000, 10 000 or 20 000 ppm, corresponding to 0, 250, 500 or 1000 mg/kg bw per day. Body-weight gain was reduced in males (17%) and females (20%) at the highest dose, and males at 500 mg/kg bw per day showed a 7% reduction in body-weight gain. The authors did not consider these figures to be statistically significant. Haematological examination and blood chemistry revealed normal values. No treatment-related changes were found in urine cellularity, volume, specific gravity, pH or protein, bilirubin or ketone content or in blood glucose or sodium concentration, however, a reduction in urine potassium concentration was seen in males at the highest dose during weeks 6 and 13 and in females at 500 and 1000 mg/kg bw per day during week 13 of the treatment period. No microscopic changes were found that could be related to administration of the test substance (Hill & Wood, 1986).

(c) Genotoxicity

Tests for genotoxicity have been performed on seven representative phenethyl alcohol derivatives and three phenoxyethyl alcohol derivatives. The results are summarized in Table 6 and described below.

(i) In vitro

Phenethyl alcohol (No. 987), phenylacetaldehyde (No. 1002), phenylacetic acid (No. 1007), ethyl phenylacetate (No. 1009), isobutyl phenylacetate (No. 1013), isoamyl phenylacetate (No. 1014) and *para*-tolylacetaldehyde (No. 1023) have been tested for their ability to induce reverse mutation in various strains of *Salmonella typhimurium* (e.g., TA98, TA100, TA1535, TA1537 and TA1538) in the presence or absence of an exogenous metabolic activation system. None of the compounds was mutagenic when tested at concentrations up to 5000 μg/ml or 50 mg/plate (Oda et al., 1979; Florin et al., 1980; Rapson et al., 1980; Ishidate et al., 1984; Heck et al., 1989; Kato et al., 1989; Fujita et al., 1994). No reverse mutation was seen when various strains of *S. typhimurium* (TA98, TA100, TA1535, TA1537 and TA1538) were incubated with ethyl (*para*-tolyloxy)acetate (No. 1027) at up to 3600 μg per plate (Wild et al., 1983), 2-phenoxyethyl isobutyrate (No. 1028) at 3600 μg per plate

Table 6. Studies of genotoxicity with phenethyl alcohol, aldehyde, acid and related acetals and esters used as flavouring agents

No.	Agent	End-point	Test object	Maximum concentration	Result	Reference
In vitro						
987	Phenethyl alcohol	Reverse mutation TA1535, TA1537	*S. typhimurium* TA98, TA100,	3 mmol/plate	Negative[a]	Florin et al. (1980)
987	Phenethyl alcohol exchange	Sister chromatid	Human lymphocytes	Not specified	Negative (1983)	Norppa & Vaino
1002	Phenylacetaldehyde	Reverse mutation	*S. typhimurium* TA98, TA100, TA104	Not specified	Negative[a]	Kato et al. (1989)
1002	Phenylacetaldehyde	Mutation	*E. coli* WP2uvrA/pkM101	Not specified	Negative[a]	Kato et al. (1989)
1007	Phenylacetic acid	Reverse mutation	*S. typhimurium* TA98, TA100, TA1535, TA1537, TA1538	1000 mg	Negative[a]	Heck et al. (1989)
1007	Phenylacetic acid synthesis	Unscheduled DNA	Rat hepatocytes	500 mg	Negative	Heck et al. (1989)
1007	Phenylacetic acid	Mutation	Mouse lymphoma L5178Y $Tk^{+/-}$ cells	1500 mg	Negative[a]	Heck et al. (1989)
1009	Ethyl phenylacetate	Mutation	*B. subtilis* H17 (rec^{+}) and M45 (rec^{-})	21 mg/disc	Negative	Oda et al. (1979)
1009	Ethyl phenylacetate	Mutation	*B. subtilis* H17 (rec^{+}) and M45 (rec^{-})	20 ml/disc	Positive[b]	Yoo (1986)
1009	Ethyl phenylacetate	Reverse mutation	*S. typhimurium* TA92, TA94, TA98, TA100, TA1535, TA1537	5 mg	Negative[a]	Ishidate et al. (1984)
1009	Ethyl phenylacetate	Chromosomal abberation	Chinese hamster fibroblast cells	1 mg/ml	Negative[a]	Ishidate et al. (1984)
1009	Ethyl phenylacetate	Mutation	*E. coli* WP2uvrA (trp^{-})	200–1600 μg/plate	Negative[b]	Yoo (1986)
1013	Isobutyl phenylacetate	Reverse mutation	*S. typhimurium* TA97, TA102	0–0.1 mg/plate	Negative[a]	Fujita et al. (1994)
1014	Isoamyl phenylacetate	Mutation	*B. subtilis* H17 (rec^{+}) and M45 (rec^{-})	20 mg/disc	Positive[b]	Oda et al. (1979)
1014	Isoamyl phenylacetate	Mutation	*B. subtilis* H17 (rec^{+}) and M45 (rec^{-})	20 ml/disc	Negative[b]	Yoo (1986)
1014	Isoamyl phenylacetate	Reverse mutation	*S. typhimurium* TA98, TA100	10 mg/plate 50 mg/plate	Negative[a] Lethal[a,b]	Oda et al. (1979)
1023	*para*-Tolylacetaldehyde	Reverse mutation	*S. typhimurium* TA100	0.1–1000 μg/plate	Negative	Rapson et al. (1980)

Table 6 (contd)

No.	Agent	End-point	Test object	Maximum concentration	Result	Reference
1023	*para*-Tolylacetaldehyde	Mutation	*E. coli* PQ37	Not specified	Negative	Ohshima et al. (1989)
1027	Ethyl (*para*-tolyloxy) acetate	Reverse mutation	*S. typhimurium* TA1535, TA1537, TA1538, TA100, TA98	≤ 3600 µg/plate	Negative[a]	Wild et al. (1983)
1028	2-Phenoxyethyl isobutyrate	Reverse mutation	*S. typhimurium* TA1535, TA1537, TA1538, TA100, TA98	≤ 3600 µg/plate	Negative[a]	Wild et al. (1983)
1029	Sodium 2-(4-methoxy-phenoxy) propanoate	Reverse mutation	*S. typhimurium* TA1535, TA98, TA100, TA1537	≤ 5000 µg/plate	Negative[a]	Varley & Trenchard-Morgan (1985)
In vivo						
1028	2-Phenoxyethyl propanoate	Micronucleus formation	Mouse bone marrow cells	≤ 1900 mg/kg/bw	Negative[b]	Wild et al. (1983)
1029	Sodium 2-(4-methoxy-	Micronucleus phenoxy)isobutyrate	Mouse bone marrow cells formation	≤ 2000 mg/kg/bw	Negative	Asquith & Pickering (1985)
	Isoeugenol phenylacet-	Micronucleus aldehyde	Mouse bone marrow cells formation	≤ 2800 mg/kg bw	Negative[b]	Wild et al. (1983)
	Isoeugenol phenylacet-aldehyde	Sex-linked recessive lethal mutation	*D. melanogaster*	25 mmol/l	Negative	Wild et al. (1983)

[a] With and without metabolic activation
[b] Administered intraperitoneally

(Wild et al., 1983) or sodium 2-(4-methoxyphenoxy)propanoate (No. 1029) at up to 5000 μg per plate (Varley & Trenchard-Morgan, 1985), with or without metabolic activation.

Tests of the ability of ethyl phenylacetate (No. 1009) and isoamyl phenylacetate (No. 1014) to induce mutation in *Bacillus subtilis* H17 and M45 were inconclusive. In a study in which ethyl phenylacetate was incubated with *B. subtilis* H17 and M45 at 21 μg per disc, the difference in the zone of inhibition (0.8 mm) between the two strains indicated that it was not active (Oda et al., 1979). In a study with a lower concentration, ethyl phenylacetate was incubated at a concentration of 20 μl per disc with *B. subtilis* H17 and M45 in the same assay. The difference in the zone of inhibition (> 8 mm) between the two strains was considered to provide evidence of mutagenicity (Yoo, 1986). Contradictory data have also been reported with isoamyl phenylacetate (No. 1014). When 20 μg per disc were incubated with *B. subtilis* H17 and M45, a weak (2–5 mm difference) positive response was reported by Oda et al. (1979), while Yoo (1986) reported a negative response with 20 μl per disc.

Phenylacetaldehyde (No. 1002) was tested in *E. coli* strain WP2uvrA/pKM101 with preincubation (Kato et al., 1989), and *para*-tolylacetaldehyde (No. 1023) was studied in *E. coli* strain PQ37 (Ohshima et al., 1989), both at unspecified concentrations. There was no evidence of mutagenicity in either assay. In another assay, 200–1600 μg per plate of ethyl phenylacetate showed no evidence of mutagenicity when incubated with *E. coli* WP2uvrA (Yoo, 1986). The contradictory results reported by Yoo (1986) and Oda et al. (1979), the negative result in the WP2 uvrA strain and the fact that phenethyl alcohol is bactericidal in *E. coli* (Treick & Konetzka, 1964; Brunner & Treick, 1982) support the conclusion that the results with *B. subtilis* H17 and M45 should not be used in the overall assessment of the genotoxic potential of these substances.

No increase in sister chromatid exchange frequency was observed when human whole blood lymphocyte cultures were exposed to 2-phenethyl alcohol (No. 987) for 72 h (Norppa & Vainio, 1983). Also, no increase in unscheduled DNA synthesis was noted when rat hepatocytes were incubated with phenylacetic acid (Heck et al., 1989). Incubation of ethyl phenylacetate at 1000 μg/ml with Chinese hamster fibroblasts for 48 h caused chromosomal aberrations in 3% of cells. On the basis of a threshold of positivity of > 10%, ethyl phenylacetate gave negative results in this assay (Ishidate et al., 1984).

(ii) In vivo

The results of tests for genotoxicity *in vivo* with phenylacetate ester and 2-phenoxyethyl isobutyrate (No. 1028) and sodium 2-(4-methoxyphenoxy)propanoate (No. 1029) were negative. No significant increase in the number of micronucleated polychromatic erythrocytes was seen in mice given intraperitoneal injections of 2-phenoxyethyl isobutyrate at 620–1900 mg/kg bw (Wild et al., 1983). In another test, sodium 2-(4-methoxyphenoxy) propanoate was given to mice by gavage at doses of 500–2000 mg/kg bw. An increased frequency of micronucleated polychromatic erythrocytes was found in males at 500 mg/kg bw at 24 h (Asquith & Pickering, 1985).

A phenyl acetate ester that was not included in this group of substances, isoeugenol phenylacetate, was also tested for its ability to induce micronucleus formation. Groups of male and female NMRI mice were given the compound at doses of 1100–2800 mg/kg bw by intraperitoneal injection. After 30 h, they were killed, and the mean number of micronucleated polychromatic erythrocytes per 1000 normochromatic erythrocytes was calculated. No effect was seen at any dose. Furthermore, the frequency of sex-linked lethal mutations was not increased when *Drosophila melanogaster* were fed a solution of isoeugenol phenylacetate at 25 mmol/l for 3 days (Wild et al., 1983).

(d) Reproductive toxicity

(i) Multigeneration study

Male and female CD-1 mice received diets containing a hydrolysis product of the phenoxyethyl alcohol derivatives in the group, phenoxyethyl alcohol, at a concentration of 0.25, 1.2 or 2.5% for 18 weeks, corresponding to doses of 380, 1900 and 3700 mg/kg bw per day. No effect was seen on the number of pairs capable of producing at least one litter or on the sex of pups born alive. Reductions in litter size and the number of live pups per litter were reported at the highest dose when compared with the control group and the two lower doses, and there was a dose-related decrease in live pup weight at the two higher doses. The weights of live male pups (F_1) were decreased, corresponding to a decrease in food consumption through week 18 among F_0 pairs at the lower dose. Necropsy of control and high-dose F_0 mice revealed decreased body weights (6% less) and increased absolute liver weights (14% more) in treated males and increased absolute liver weights among treated females. No effects were reported on the weights of other organs or on sperm indices.

In the second part of this study, phenoxyethyl alcohol was administered after weaning to the last F_1 litter at the same dose provided to their F_0 parents. A dose-related decrease in body weight was seen in the F_1 generation at the two higher doses, from birth to post natal day 74. Increased mortality was reported at the two higher doses from weaning until postnatal day 74, with marked effects at the highest dose, only 6/56 pups surviving until mating. Consequently, this group was terminated. After mating of the F_1 generation, no treatment-related effects were seen on F_2 litter size or sex ratio. The weights of live F_2 pups at the intermediate dose were decreased by 7%. Necropsy of these mice and of the controls revealed decreased body weight (13% less), decreased absolute testis weight (16% less), decreased relative seminal vesicle weight (14% less) in the male mice and decreased body weight (7%) in the female mice in comparison with the controls. No effects were reported on epididymal sperm concentration, motility or morphology. The decrease in body weight reported for live male F_1 pups born to the pairs given the low dose (F_0) was barely statistically significant, and its biological relevance is questionable as it was observed in male pups only. The NOEL for reproductive effects was 0.25%, corresponding to 380 mg/kg bw per day (National Toxicology Program, 1984).

(ii) Developmental toxicity

In screening studies, low doses of phenethyl alcohol (No. 987) and phenylacetic acid (No. 1007) were reported to have teratogenic effects resembling fetal alcohol syndrome (Mankes et al., 1983, 1984, 1985). These results were contradicted by those of a study in which phenethyl alcohol given to pregnant rats at high doses at critical periods of embryogenesis did not cause any visible anomalies in embryonal development (Burdock et al., 1987). More recent, comprehensive studies with high doses of phenethyl alcohol given orally (Bottomley et al., 1987) or dermally (Palmer et al., 1986) indicated that this group of substances has little reproductive or developmental toxicity.

Phenethyl alcohol (No. 987) and phenylacetic acid (No. 1007)

Long-Evans rats were given phenethyl alcohol by gavage at a dose of 4.3, 43 or 430 mg/kg bw per day on days 6–15 of gestation. The average birth weight and pup size in all treated groups were significantly lower than those of the control group, but the changes were not dose-related, and in fact the birth weights were greater in the group at the intermediate dose than in controls. The mean litter size was greater in the high-dose group (13) than at the two lower doses (9) or in controls (12). No embryolethality was seen at the high dose, while the rates were 18% at 43 mg/kg bw per day and 10% at the lowest dose. The authors reported a clear dose-related increase in the percentage of malformations in live offspring (100% at the highest dose, 93% at the intermediate and 50% at the lowest dose). The malformations were mainly ocular changes, neural tube defects, hydronephrosis and limb defects (Mankes et al., 1983). In abstracts of subsequent studies reported by the same authors (Mankes et al., 1984, 1985), doses of phenethyl alcohol equivalent to 0.02% and 24% of the oral LD_{50} were administered to pregnant Long-Evans rats. Intrauterine growth retardation (birth weight reduction) and embryolethality were reported at all doses. These observations are inconsistent with those of the original study.

No teratogenic effects were reported in rats fed diets containing phenethyl alcohol at concentrations exceeding the normal human intake by > 8000 times. Microencapsulated phenethyl alcohol was administered in the diet to nulliparous Sprague-Dawley rats during the period of major organogenesis (days 6–15 post coitum) at a concentration of 1000, 3000 or 10 000 ppm, calculated to provide an average daily intake of 50, 150 or 500 mg/kg bw per day. The effects on the dams were limited to transient suppression of maternal food consumption, resulting in slight weight loss during the first 2 days of treatment with the high dose. The effects on the offspring were minimal, malformations being found in only five pups (three controls and two at the intermediate dose). The number and type of visceral anomalies were comparable in control and treated groups. An increased incidence of incomplete calcification was seen only in fetuses of dams at 500 mg/kg bw per day, and this was considered to be a possible consequence of the earlier impairment of maternal weight gain. There were no obvious differences between control and treated groups in skeletal variants, number of live young, embryolethality, number of implants, litter weight, mean fetal weight or sex ratio (Burdock et al., 1987).

Sprague-Dawley-derived rats received diets containing micro-encapsulated phenethyl alcohol at a nominal concentration of 0, 1000, 3000 or 10 000 ppm, equivalent to 0, 50, 150 and 500 mg/kg bw per day, on days 6–15 of gestation. Spray-dried gum arabic, the microencapsulant, was used as a placebo control and was also added with the lower concentrations so that the total dose remained constant for all groups at 5%. The animals were killed on day 20 post coitum, and development was assessed by determination of litter values and examination of fetuses for structural malformations or anomalies. The achieved intake of phenethyl alcohol by dams during the treatment period was calculated, and the values were adjusted to take account of the assayed content of test material in the microcapsules. The actual intake was thus 83, 270 and 800 mg/kg bw per day for the three groups, respectively. Treatment at the highest dose had a negligible detrimental effect on uterine development. Although there was clear evidence of impaired weight gain in dams after initial treatment, fetal development was virtually unaffected, the only possible exception being a marginal delay in ossification, which is usually considered to be transient and self-correcting during postnatal maturation. At the two lower doses, phenethyl alcohol did not elicit any overt response in the dams, and embryofetal development and morphology were unaffected (Bottomley et al., 1987).

Sprague-Dawley-derived rats received phenethyl alcohol topically at a dose of 0, 0.14, 0.43 or 1.4 ml/kg bw, corresponding to 140, 440 or 1400 mg/kg bw per day, on days 6–15 of gestation. These doses were chosen so that the intermediate dose was roughly equivalent to the highest dose used in a previous study of oral administration (Mankes et al., 1983) and designed to extend the range in case of differential absorption by the dermal route. An occlusive bandage was used to minimize oral ingestion of the test material, and this was put on immediately after dosing and not removed until the following day. The animals were killed on day 20 of gestation, and uterine development was assessed by determining litter values and examining fetuses for soft tissue and skeletal changes. The highest dose clearly induced both maternal toxicity, including lethality, suppression of mean food intake and growth rate, and embryo-fetal toxicity, indicated by resorption, wastage, reduction in mean litter size, depression of fetal weight, a wide range of soft tissue and skeletal changes and incomplete ossification. The pattern of incomplete ossification and the comprehensive nature of the morphological changes were considered by the authors to be more severe than those that would occur merely as a secondary consequence of the maternal response. Thus, the intermediate dose was considered close to the threshold for maternal toxicity. While there was no evidence of an adverse effect on litter values, dose-dependent increases were found in some morphological changes in fetuses. The lowest dose did not elicit any adverse effects in litter values. The slight increases over control values in morphological changes (cervical ribs, thoracic vertebral irregularities) at the lowest dose indicated that 140 mg/kg bw per day was a threshold for developmental toxicity in rats (Palmer et al., 1986).

Phenoxyacetic acid (No. 1026)

Phenoxyacetic acid was administered by gavage to mice on one of days 8–15 of gestation at a dose of 800–900 mg/kg bw or on three consecutive days (7–9, 10–12 or 13–15) at a dose of 250–300 mg/kg bw per day. Treated females were killed on day 18 of gestation, and their uteri were examined to determine the numbes of live, dead and resorbed fetuses. Live fetuses were examined for gross external malformations and weighed. Fetuses from each litter were further examined for visceral abnormalities and skeletal malformations. The authors stated that phenoxyacetic acid was not strongly teratogenic or fetotoxic (Hood et al., 1979).

Sodium 2-(4-methoxyphenoxy)propanoate (No. 1029)

Groups of five time-mated Charles River female rats were given suspensions of sodium 2-(4-methoxyphenoxy) propanoate at a single daily dose of 0, 100, 500, 1000 or 5000 mg/kg bw by gavage on days 6–15 of gestation, in order to determine suitable doses for a second study. The surviving animals were killed on day 20 of gestation and grossly necropsied. No adverse effects were observed at 100 or 500 mg/kg bw per day, and at 1000 and 5000 mg/kg bw per day, there were no treatment-related clinical signs of toxicity, mortality or abnormal findings at necropsy among dams, and embryonic and fetal development was not affected. Maternal body-weight gain and food consumption were adversely affected from the onset of dosing at 1000 mg/kg bw per day, and food consumption was significantly lower ($p < 0.01$, Student's t test) than that of controls throughout dosing. These differences were considered to be treatment-related even though statistical significance was not achieved (probably because of the small group size). However, there were no deaths or adverse findings at necropsy. There was a trend towards slightly lower fetal body weight (not statistically significant), which may have been related to the reduced food consumption of the dams. Two animals at 5000 mg/kg bw per day were found dead on day 8 of gestation, having received two doses of sodium 2-(4-methoxyphenoxy) propanoate; the remaining three animals were reported to be in poor condition on day 7, after having received one dose, and were killed. The authors also reported brown stains on the fur and piloerection in most animals. Additionally, there were lesions of the gastric mucosa, which included inflammation, haemorrhage or ulceration in four animals. The authors concluded that 1000 mg/kg per day was a suitable highest dose for the main study. Although slight toxicity was observed at this dose, there were no maternal deaths (Ridgway, 1985).

The next study was designed to investigate the effects of sodium 2-(4-methoxyphenoxy) propanoate on embryonic and fetal development when administered during the period of organogenesis. Groups of 24 time-mated female CD rats were dosed once daily by gavage with suspensions of sodium 2-(4-methoxyphenoxy) propanoate at 0, 100, 300 or 1000 mg/kg per day on days 6–15 of gestation and were killed and necropsied on day 20 of gestation. The fetuses were subjected to detailed external, visceral and skeletal examinations. No maternal deaths occurred. Treatment-related clinical signs were reported only at the highest dose. A slight, transient retardation in maternal body-weight gain was seen at all doses at the start of treatment, with statistically significant differences of on days 6–7 at 100 and 300 mg/kg bw per day ($p < 0.05$) and on days 6–7 and 6–9 at 1000 mg/kg bw per day ($p < 0.001$). Maternal food consumption was not affected. There were no abnormal findings at necropsy of dams and no evidence of developmental toxicity. Implantation, post-implantation loss, fetal weight, sex ratio and the incidence of fetal abnormalities were not affected by treatment (Ridgway, 1986).

(e) Other studies

(i) Phenethyl alcohol (No. 987) and phenethyl acetate (No. 989)

Male rats were given phenethyl alcohol at a dose of 51 mg/kg bw per day or phenethyl acetate at 73 mg/kg bw per day by gavage for 4 months. On days 40 and 140, cholinesterase activity, the activity of serum enzymes such as aldolases, amylase, sorbitol dehydrogenase and aspartate and alanine aminotransferases, and the content of thiol groups and total serum protein were measured in serum. Treatment with phenethyl alcohol increased cholinesterase and alanine aminotransferase activities and increased the content of thiol groups in the blood at day 40. The serum protein content decreased to 7.2 g/100 ml after 40 days of treatment. The effects on thiol group content and cholinesterase activity persisted at 140 days. The only change reported was an increase in cholinesterase activity 140 days after the start of the experiment (Zaitsev & Rakhmanina, 1974).

(ii) Phenylacetic acid (No. 1007)

A rapid screening protocol was used to evaluate the potential immunotoxicity of flavouring ingredients, including phenylacetic acid. The protocol incorporated key elements of the National Toxicology Program's tier testing strategy, including measurement of body weight, lymphoid organ weight and cellularity, as well as functional tests of the humoral (antibody plaque-forming cells) response to sheep erythrocytes and cell-mediated immunity to *Listeria monocytogenes* bacterial challenge. Decreases in body weight, spleen and/or thymus weight or a decrease in spleen cellularity may be indicative of depressed immune competence. The number of antibody-producing plasma cells, the result of antigen-driven B-cell differentiation, after immunization with a T-cell-dependent antigen such as sheep red blood cells, provides information about the functional integrity of, and communication among, several cell populations important in antibody-mediated immunity, including T cells, B cells and macrophages. The *Listeria* model system was selected because the pathogenesis of this infection and the host's immune response are similar in mice and humans. The model system is useful for assessing immunosuppression since both immunocompetent T cells and macrophages are required to control infection and supply protective immunity. Phenylacetic acid administered orally to groups of 10–20 female CD-1 or $B6C3F_1$ mice, aged 6–8 weeks, at doses as high as 100 mg/kg per day had no effect on spleen weight, thymus weight, spleen cellularity, anti-sheep red blood cell response or *Listeria* mortality (Vollmuth et al., 1989).

Phenylacetic acid was shown to be devoid of immunomodulatory effects in a testing strategy to evaluate the effects of 35 commonly used flavouring ingredients on humoral and cell-mediated immune responses. Female CD-1 or $B6C3F_1$ mice were given phenylacetic acid at a dose of 250, 500 or 1000 mg/kg bw per day intragastrically for 5 days. *L. monocytogenes* bacterial challenge was conducted to assess cell-mediated immunity and the antibody response to sheep erythrocytes was determined as a measure of humoral immunity. Body weights, lymphoid organ weights and spleen cellularity were also measured. Cyclophosphamide served as a positive control. Phenylacetic acid did not modulate the cell-mediated or humoral response at any dose tested (Gaworski et al., 1994).

In a study of the effects of excess L-phenylalanine, L-tyrosine, L-valine and phenylacetic acid on serotonin in brain and liver, 5% or 7% phenylacetic acid in the diet for 1–3 weeks increased brain serotonin and decreased liver serotonin in rats (Boggs et al., 1963).

(iii) Sodium 2-(4-methoxyphenoxy) propanoate (No. 1029)

The effects of sodium 2-(4-methoxyphenoxy) propanoate and ibuprofen on prostaglandin E_2 levels and leukocyte counts were studied in rats in which carrageenan-soaked sponges had been implanted subcutaneously. Oral administration of sodium 2-(4-methoxyphenoxy) propanoate at 3, 10, 30 or 100 mg/kg bw to male and female rats did not alter leukocyte migration into the sponges, and, in general, the test substance had no significant or dose-related effect on the prostaglandin E_2 content of the sponges. However, there was a marked increase in the levels in male rats treated at 30 mg/kg bw and a significant reduction in females at 100 mg/kg bw. The reference standard, ibuprofen, at 100 mg/kg bw did not alter the leukocyte counts but significantly reduced the prostaglandin E_2 content of the sponges in both male and female rats. The authors concluded that this reduction is consistent with the anti-inflammatory action of ibuprofen and that sodium 2-(4-methoxyphenoxy) propanoate could have a slight inhibitory effect on prostaglandin biosynthesis (Algate et al., 1986a).

In a study by the same authors, sodium 2-(4-methoxyphenoxy) propanoate at doses of 10, 30 and 100 mg/kg bw produced statistically significant inhibition of carrageenan-induced oedema, at both the 3-h (30 and 100 mg/kg bw) and 6-h (10, 30 and 100 mg/kg bw) observation times. Ibuprofen caused statistically significant inhibition of oedema at all three intervals, with peak activity recorded at the 6-h observation time. The authors concluded that sodium 2-(4-methoxyphenoxy) propanoate and the reference standard ibuprofen suppressed the inflammatory response induced by the irritant carrageenan; however, the activity of 2-(4-methoxyphenoxy) propanoate was reported to be minor (Algate et al., 1986b).

The effects of sodium 2-(4-methoxyphenoxy) propanoate (lactisole) on glucose tolerance and insulin secretion were investigated in eight healthy male volunteers. The men received either a glucose solution (75 g) or a glucose solution containing 80 mg of the test substance on two separate occasions, at least 7 days apart. Blood samples were collected at 30-min intervals for 3 h and then hourly until 5 h after the glucose load. Glucose, insulin, C-peptide and glucagon levels were assayed. Lactisole had no effect on glucose metabolism or insulin, C-peptide or glucagon secretion under the conditions of the study (Marks, 1988).

3. REFERENCES

Algate, D.R., Lawrence, D.J. & Atterson, P.R. (1986a) An investigation of the anti-inflammatory activity of ORP 178 against carrageenan-induced oedema in the rat. Private communication. Submitted to WHO by Flavor and Extract Manufacturers' Association of the United States.

Algate, D.R., Munt, P.L. & Green, T.D. (1986b) Effects of ORP 178 and ibuprofen on PGE_2 levels and leucocyte counts in subcutaneously implanted sponges in the rat. Private communication. Submitted to WHO by Flavor and Extract Manufacturers' Association of the United States.

Ambrose, A.M., Power, F.W. & Sherwin, C.P. (1933) Further studies on the detoxication of phenylacetic acid. *J. Biol. Chem.*, **101**, 669–675.

Asquith, J.C. & Pickering, K.J. (1985) Mouse micronucleus test on ORP 178. Private communication. Submitted to WHO by Flavor and Extract Manufacturers' Association of the United States.

Augustinsson, K.B. & Ekedahl, G. (1962) On the specificity of arylesterases. *Acta Chem. Scand.*, **16**, 240–241.

Beroza, M., Inscoe, M.N., Schwartz, P.H., Keplinger, M.L. & Mastri, C.W. (1975) Acute toxicity studies with insect attractants. *Toxicol. Appl. Pharmacol.*, **31**, 421–429.

Bierbower, G.W. (1970) Memo to J. Taylor regarding a pathology report on Osborne-Mendel rats fed phenethyl senocioate in rations for four months. Private communication. Submitted to WHO by Flavor and Extract Manufacturers' Association of the United States.

Block, W. (1953) On the physiology of C(14)-radioactive mescaline in animals. Part IV. Comparative studies with C(14)-mescaline and C(14)-beta-phenylethylamine. *Z. Naturforsch.*, **8B**, 440–444.

Boggs, D.E., Rosenberg, R. & Waiman, H.A. (1963) Effects of phenylalanine, phenylacetic acid, tyrosine and valine on brain and liver serotonin in rats. *Proc. Soc. Biol. Med.*, **114**, 356–358.

Bosron, W.F. & Li, T.K. (1980) Alcohol dehydrogenase. In: Jakoby, W.B., ed., *Enzymatic Basis of Detoxication*, Orlando, Florida: Academic Press, Vol. 1, pp. 231–248.

Bottomley, A.M., Ratcliffe, H.E., John, D.M., Anderson, A. & Dawe, I.S. (1987) Effect of dietary administration of micro-encapsulated phenylethyl alcohol on pregnancy of the rat (embryotoxicity study). Unpublished report to RIFM. Submitted to WHO by Flavor and Extract Manufacturers' Association of the United States.

Bray, H.G., Neale, F.C. & Thorpe, W.V. (1946) The fate of certain organic acids and amines in the rabbit. *Biochem. J.*, **40**, 134–139.

Bray, H.G., James, S.P. & Thorpe, W.V. (1958) Metabolism of some omega-halogenoalkylbenzenes and related alcohols in the rabbit. *Biochem. J.*, **70**, 570–579.

Brown, P.M., Rhenius, S.T. & Ryder, J.R. (1986) ORP/178: Absorption, distribution, and excretion in rat. Final report. Private communication. Submitted to WHO by Flavor and Extract Manufacturers' Association of the United States.

Brunner, D.P. & Treick, R.W. (1982) Effects of phenethyl alcohol treatment upon the folded chromosome of *Escherichia coli*. *J. Gen. Appl. Microbiol.*, **28**, 491–498.

Burdock, G.A. & Ford, R.A. (1990) Acute oral toxicity (LD50) study in the rat with phenoxyacetic acid. Acute toxicity data. *J. Am. Coll. Toxicol. B*, **1**, 94–95.

Burdock, G.A., Ford, R.A., Bottomley, A.M. & John, D.M. (1987) An evaluation of the teratogenic potential of orally administered phenylethyl alcohol. *Toxicologist*, **7**, 702.

Caldwell, J. (1987) Human disposition of [^{14}C]-ORP/178. Unpublished report. Private communication. Submitted to WHO by Flavor and Extract Manufacturers' Association of the United States.

Colaianni, L. (1967) Acute toxicity, eye and skin irritation test of fragrance materials. Unpublished report. Private communication. Submitted to WHO by Flavor and Extract Manufacturers' Association of the United States.

Cramer, G.M., Ford, R.A. & Hall, R.L. (1978) Estimation of toxic hazard—A decision tree approach. *Food Cosmet. Toxicol.*,**16**, 255–276.

Cummins, H.A. (1985) Private communication. Submitted to WHO by Flavor and Extract Manufacturers' Association of the United States.

Department of Agriculture (1965) *Food intake and nutritive value of the diets of men, women, and children in the in the United States, Spring 1965*, a preliminary report by the consumer and Food Economics Research Division, Agriculture Research Service, Washington DC.

Erwin, V.G., Kim, J. & Anderson, A.D. (1975) Effects of aldehydes on sodium plus potassium ion-stimulated adenosine triphosphatase of mouse brain. *Biochem. Pharmacol.*, **24**, 2089–2095.

Fassett, D.W. (1963) Toxicity of phenethyl alcohol. *Ind. Hyg. Toxicol.*, Vol. ?, 1476–1477.

Florin, I., Rutberg, L., Curvall, M. & Enzell, C. (1980) Screening of tobacco smoke constituents for mutagenicity using the Ames test. *Toxicology*, **18**, 219–232.

Fujita, H., Aoki, N. & Sasaki, M. (1994) Mutagenicity test of food additives with *Salmonella typhimurium* TA97 and TA102(IX). *Tokyo-toritsu Eisei Kenkysho Kenkyu Nenpo*, **45**, 191–199.

Gaworski, C.L., Vollmuth, T.A., Dosier, M.M., Heck, J.D., Dunn, L.T., Ratajczak, H.V. & Thomas, P.T. (1994) An immunotoxicity assessment of food flavouring ingredients. *Food Chem. Toxicol.*, **32**, 409–415.

Gregus, Z., Fekete, T., Varga, F. & Klaassen, C.D. (1993) Dependence of glycine conjugation on availability of glycine: Role of the glycine cleavage system. *Xenobiotica*, **23**,141–153.

Grundschober, F. (1977) Toxicological assessment of flavouring esters. *Toxicology*, **8**, 387–390.

Hagan, E.C., Hansen, W.H., Fitzhugh, O.G., Jenner, P.M., Jones, W.I., Taylor, J.M., Long, E.L., Nelson, A.A. & Brouwer, J.B. (1967) Food flavourings and compounds of related structure. II. Subacute and chronic toxicity. *Food Cosmet. Toxicol.*, **5**, 141–157.

Hawkins, D.R. & Mayo, B.C. (1986) Plasma kinetics of [^{14}C]-ORP/178 in the rat. Unpublished report. Private communication. Submitted to WHO by Flavor and Extract Manufacturers' Association of the United States.

Heck, J.D., Vollmuth, T.A., Cifone, M.A., Jagannath, D.R., Myhr, B. & Curren, R.D. (1989) An evaluation of food flavouring ingredients in a genetic toxicity screening battery. *Toxicologist*, **9**, 257.

Hijikata, Y. (1922) The influence of putrefaction products on cellular metabolism. II. On the influence of phenylacetic and phenylproionic acids on the distribution of nitrogen in the urine. *J. Biol. Chem.*, **51**, 141–154.

Hill, R.E. (1986) Metabolism phase of 90 day toxicity study in the rat ORP-178. Unpublished report. Private communication. Submitted to WHO by Flavor and Extract Manufacturers' Association of the United States.

Hill, R.E. & Wood, C.M. (1985) 14-day oral toxicity study in the rat ORP-178. Unpublished report. Private communication. Submitted to WHO by Flavor and Extract Manufacturers' Association of the United States.

Hill, R.E. & Wood, C.M. (1986) 90-day oral toxicity study in the rat ORP-178. Unpublished report. Private communication. Submitted to WHO by Flavor and Extract Manufacturers' Association of the United States.

Hood, R.D., Patterson, B.L., Thacker, G.T., Sloan, G.L & Szczech, G.M. (1979) Prenatal effect of 2,4,5-T,2,4,5-trichlorophenol and phenoxyacetic acid in mice. *J. Environ. Sci. Health*, **C13**, 189–204.

Howes, D. (1988) Absorption and metabolism of 2-phenoxyethanol in rat and man. *Cosmet. Toiletries*, **103**, 75.

International Organization of the Flavour Industry (1995) European inquiry on volume use. Private communication. Submitted to WHO by Flavor and Extract Manufacturers' Association of the United States.

Ishidate, M., Sofuni, K., Yoshikawa, K., Hayashi, M., Nohmi, T., Sawada, M. & Matsuoka, A. (1984) Primary mutagenicity screening of food additives currently used in Japan. *Food Chem. Toxicol.*, **22**, 623–636.

James, M.O. & Smith, R.L. (1973) Conjugation of phenylacetic acid in phenylketonurics. *Eur. J. Clin. Pharmacol.*, **5**, 243–246.

James, M.O., Smith, R.L., Williams, R.T. & Reidenberg, M. (1972) The conjugation of phenylacetic acid in man, sub-human primates and some non-primate species. *Proc. R. Soc. London B*, **182**, 25–35.

Jenner, P.M., Hagan, E.C., Taylor, J.M., Cook, E.L. & Fitzhugh, O.G. (1964) Food flavourings and compounds of related structure. I. Acute oral toxicity. *Food Cosmet. Toxicol.*, **2**, 327–343.

Johannsen, E. & Purchase, I.F.H. (1969) Kaffircorn malting and brewing studies. XXI: The effect of the fusel oils of Bantu beer on rat liver. *S.A. Med. J.*, **43**, 326–328.

Kato, F., Araki, A., Nozaki, K. & Matsushima, T. (1989) Mutagenicity of aldehydes and diketones. *Mutat. Res.*, **216**, 366.

Kay, H.D. & Raper, H.S. (1922) Mode of oxidation of fatty acids with branched chains. II. The fate in the body of hydratropic, tropic, atrolactic and atropic acids together with phenylacetaldehyde. *Biochem. J.*, **16**, 465–474.

Keating, J.W. (1972) Acute toxicity study in rats and rabbits. Unpublished report. Private communication. Submitted to WHO by Flavor and Extract Manufacturers' Association of the United States.

Klyosov, A.A. (1996) Kinetics and specificity of human liver aldehyde dehydrogenases toward aliphatic, aromatic and fused polycyclic aldehydes. *Biochemistry*, **35**, 4457–4467.

Ko, G.K.W., Raghupathy, E. & McKean, C.M. (1973) UDP-galactose: Glycoprotein galactosyl transferase and UDP-N-acetylgalactosamine: Protein N-acetylgalactosaminyl transferase activities of human cerebrospinal fluid. *Can. J. Biochem.*, **51**, 1460–1469.

Koss, F.W. & Lamprecht, W. (1968) Experimental investigations in choleresis. *Eur. J. Pharmacol.*, **4**, 215–223.

Levenstein, I. (1973) Acute oral toxicity reports on rats. Unpublished report. Private communication. Submitted to WHO by Flavor and Extract Manufacturers' Association of the United States.

Levenstein, I. (1974) Acute toxicity study in rats and rabbits. Unpublished report to RIFM. Submitted to WHO by Flavor and Extract Manufacturers' Association of the United States.

Levenstein, I. (1975) Acute toxicity studies in rats, mice and rabbits. Unpublished report to RIFM. Submitted to WHO by Flavor and Extract Manufacturers' Association of the United States.

Levey, S. & Lewis, H.B. (1947) The metabolism of phenoxyacetic acid, its homologues, and some monochlorophenoxyacetic acids. New examples of β oxidation. *J. Biol. Chem.*, **168**, 213–221.

Longland, R.C., Shilling, W.H. & Gangolli, S.D. (1977) The hydrolysis of flavouring esters by artificial gastrointestinal juices and rat tissue preparations. *Toxicology*, **8**, 197–204.

Lucas, C.D., Putnam, J.M. & Hallagan, J.B. (1999) *Flavour and Extract Manufacturers Association of the United States 1995 Poundage and Technical Effects Update Survey*, Washington DC: Flavour and Extract Manufacturers Association of the United States.

Maarse, H., Visscher, C.A., Willemsens, L.C. & Boelens, M.H. (1999) *Volatile Components in Food—Qualitative and Quantitative Data*, Zeist: Centraal Instituut Voor Voedingsonderzioek TNO.

Mallory, V.T., Naismith, R.W. & Matthews, R.J. (1982) Acute oral toxicity study in rats (14 days). Unpublished report No. PH 402-IFF-018-82. Private communication. Submitted to WHO by Flavor and Extract Manufacturers' Association of the United States.

Mankes, R.F., LeFevre, R., Bates, H. & Abraham, R. (1983) Effects of various exposure levels of 2-phenylethanol on fetal development and survival in Long-Evans rats. *J. Toxicol. Environ. Health*, **12**, 235–244.

Mankes. R.F., LeFevre, R. & Abraham, R. (1984) Reproductive effects of some solvent alcohols: Evaluation of structure–activity relationships in 2-substituted ethanols. *J. Am. Coll. Toxicol.*, **3**, 166.

Mankes, R.F., LeFevre, R., Renak, R., Fiesher, J. & Abraham, R. (1985) Reproductive effects of some solvent alcohols with differing partition coefficients. *Teratology*, **31**, 67A.

Marks, V. (1988) The effect of lactisole on glucose tolerance and insulin secretion in healthy subjects. Unpublished report. Private communication. Submitted to WHO by Flavor and Extract Manufacturers' Association of the United States.

Martini, R. & Murray, M. (1996) Rat hepatic microsomal aldehyde dehydrogenase. Identification of 3- and 4-substituted aromatic aldehydes as substrates of the enzyme. *Chem. Res. Toxicol.*, **9**, 268–276.

Moldave, K. & Meister, A. (1957) Enzymic acylation of glutamine by phenylacetic acid. *Biochim. Biophys. Acta*, **24**, 654–655.

Moreno, O.M. (1972) Single dose toxicity in rats. Unpublished report by Toxicological Resources, East Millstone, New Jersey, USA. Submitted to WHO by Flavor and Extract Manufacturers' Association of the United States.

Moreno, O.M. (1973) Acute oral toxicity in rats. Unpublished report by MB Research Laboratories, Inc., Perkasie, Pennsylvania, USA. Submitted to WHO by Flavor and Extract Manufacturers' Association of the United States.

Moreno, O.M. (1974) Acute oral toxicity in rats. Unpublished report by MB Research Laboratories, Inc., Perkasie, Pennsylvania, USA. Submitted to WHO by Flavor and Extract Manufacturers' Association of the United States.

Moreno, O.M. (1976) Acute oral toxicity in rats. Unpublished report by MB Research Laboratories, Inc., Perkasie, Pennsylvania, USA. Submitted to WHO by Flavor and Extract Manufacturers' Association of the United States.

Moreno, O.M. (1977) Acute oral toxicity in rats. Unpublished report by MB Research Laboratories, Inc., Perkasie, Pennsylvania, USA. Submitted to WHO by Flavor and Extract Manufacturers' Association of the United States.

Moreno, O.M. (1979) Acute oral toxicity in rats. Unpublished report by MB Research Laboratories, Inc., Perkasie, Pennsylvania, USA. Submitted to WHO by Flavor and Extract Manufacturers' Association of the United States.

Moreno, O.M. (1980) Acute oral toxicity in rats. Unpublished report by MB Research Laboratories, Inc., Perkasie, Pennsylvania, USA. Submitted to WHO by Flavor and Extract Manufacturers' Association of the United States.

Moreno, O.M. (1982) Acute oral toxicity in rats. Unpublished report by MB Research Laboratories, Inc., Perkasie, Pennsylvania, USA. Submitted to WHO by Flavor and Extract Manufacturers' Association of the United States.

Morgareidge, K. (1962) In vitro digestion of four acetals. Unpublished report. Private communication. Submitted to WHO by Flavor and Extract Manufacturers' Association of the United States.

National Toxicology Program (1984) Ethylene glycol monophenyl ether: Reproductive and fertility assessment in CD-1 mice when administered in the feed. Report No. NTP-84-410; NTP-83-FACB022, Washington DC.

Norppa, H. & Vainio, H. (1983) Induction of sister-chromatid exchanges by styrene analogues in cultured human lymphocytes. *Mutat. Res.*, **116**, 379–387.

Nour-Eldin, F. (1968) Phenols and blood coagulation. *J. Biomed. Mater. Res.*, **2**, 23–42.

Oda, Y., Hamono, Y., Inoue, K., Yamamoto, H., Niihara, T. & Kunita, N. (1979) Mutagenicity of food flavours in bacteria. *Shokuhin Eisei Hen*, **9**, 177–181 (in Japanese, with English abstract and tables).

Ohshima, H., Friesen, M., Malaveille, C., Brouet, I., Hautefeuille, A. & Bartsch, H. (1989) Formation of direct-acting genotoxic substances in nitrosated smoked fish and meat products: Identification of simple phenolic precursors and phenyldiazonium ions as reactive products. *Food Chem. Toxicol.*, **27**, 193–203.

Oser, B.L. & Hall, R.L. (1977) Criteria employed by the expert panel of Flavor and Extract Manufacturers' Association for the GRAS evaluation of flavouring substances. *Food Cosmet. Toxicol.*,**15**, 457–466.

Palmer, A.K., Bottomley, A.M., Ratcliffe, H.E., Clark, R. & John, D.M. (1986) Effect of phenylethyl alcohol (PEA) on pregnancy of the rat. Unpublished report from Huntingdon Research Center to RIFM. Submitted to WHO by Flavor and Extract Manufacturers' Association of the United States.

Piccirillo, V.J., Hartman, W.C., Dauvin, E. & Swidersky, ?. (1983) Acute oral toxicity (LD_{50}) study in the rat with phenoxyacetic acid. Unpublished report. Private communication. Submitted to WHO by Flavor and Extract Manufacturers' Association of the United States.

Pietruszko, R., Crawford, K. & Lester, D. (1973) Comparison of substrate specificity of alcohol dehydrogenases from human liver, horse liver, and yeast towards saturated and 2-enoic alcohols ands aldehydes. *Arch. Biochem. Biophys.*, **159**, 50–60.

Posternak, J.M., Linder, A. & Vodoz, C.A. (1969) Summaries of toxicological data. Toxicological tests on flavouring matters. *Food Cosmet. Toxicol.*, **7**, 405–407.

Power, F.W. & Sherwin, C.P. (1927) The detoxication of putrefactive products by the human body. *Arch. Intern. Med.*, **39**, 60–66.

Rabinowitz, T. (1969) Subacute toxicity (90 days) of senecioic acid, phenethyl ester. Unpublished report. Private communication. Submitted to WHO by Flavor and Extract Manufacturers' Association of the United States.

Rapson, W.H., Nazar, M.A. & Butsky, V.V. (1980) Mutagenicity produced by aqueous chlorination of organic compounds. *Bull. Environ. Contam. Toxicol.*, **24**, 590–596.

Richter, D. (1938) Elimination of amines in man. *Biochem. J.*, **32**, 1763–1769.

Ridgway, P. (1985) Rat teratology dose ranging study. Unpublished report. Private communication. Submitted to WHO by Flavor and Extract Manufacturers' Association of the United States.

Ridgway, P. (1986) Rat teratology test. Unpublished report. Private communication. Submitted to WHO by Flavor and Extract Manufacturers' Association of the United States.

Romano, C., Meyers, F.H. & Anderson, H.H. (1954) Pharmacological relationship between aldehydes and arterenol. *Arch. Int. Pharmacodyn.*, **99**, 378–398.

Rumyantsev, G.I., Novikov, S.M., Fursova, T.N.M., Kochetkova, T.A. & Ivanov, Yu.V. (1987) Experimental study of the toxic properties of phenylethyl alcohol and phenylethyl acetate. *Gig. Sanit.*, **10**, 83–84.

Russell, T. (1973) Acute oral and dermal toxicity studies. Unpublished report. Private communication. Submitted to WHO by Flavor and Extract Manufacturers' Association of the United States.

Sangster, S.A. & Lindley, M.G. (1986) Metabolism and excretion of ORP/178 in man. Unpublished report. Private communication. Submitted to WHO by Flavor and Extract Manufacturers' Association of the United States.

Seakins, J.W.T. (1971) The determination of urinary phenylacetylglutamine as phenylacetic acid: Studies on its origin in normal subjects and children with cystic fibrosis. *Clin. Chim. Acta*, **35**, 121–131.

Shelanski, M. & Moldovan, M. (1971) Acute oral toxicity in rats. Unpublished report. Private communication. Submitted to WHO by Flavor and Extract Manufacturers' Association of the United States.

Shiple, G.J. & Sherwin, C.P. (1922) Fate of phenylacetyl derivatives of the amino acids in the animal organism. *J. Biol. Chem.*, **53**, 463.

Simpson, V.J., Baker, R. & Deitrich, R.A. (1985) Inducible aldehyde dehydrogenases from rat liver cytosol. *Toxicol. Appl. Pharmacol.*, **79**, 193–203.

Stein, W.H., Paladini, A.C., Hirs, C.H. & Moore, S. (1954) A phenylacetylglutamine as a constituent of normal human urine (Letter to the Editor). *J. Chem. Soc. Perkin Trans.*, **76**, 2848–2849.

Stewart, G.A. (1962) Pharmacological studies on oral hypoglycaemic agents. *Anglo-Ger. Med. Rev.*, **1**, 334–347.

Stofberg, J. & Grundschober, F. (1987) Consumption ratio and food predominance of flavouring materials. *Perfumer Flavourist*, **12**, 27–68.

Stofberg, J. & Kirschman, J.C. (1985) The consumption ratio of flavouring materials: A mechanism for setting priorities for safety evaluation. *Food Chem. Toxicol.*, **23**, 857–860.

Tabatabaie, T. & Floyd, R.A. (1996) Inactivation of glutathione peroxidase by benzaldehyde. *Toxicol. Appl. Pharmacol.*, **141**, 389–393.

Tachibana, A. & Yonei, S. (1985) Inhibition of excision repair in DNA in UV-irradiated Escherichia coli by phenethyl alcohol. *Int. J. Radiat. Biol.*, **47**, 663–671.

Tachibana, A., Yonei, T., Todo, T. & Kato, M. (1982) Inhibitory effect of phenethyl alcohol on DNA repair in UV-radiated E. coli cells (Abstract No. 23910). *J. Radiat. Res.*, **23**.

Tashian, R.E. (1960) Effect of phenylalanine metabolites on urinary excretion of indole-3-acetic acid and 5-hydroxyindole-3-acetic acid in man. *Proc. Soc. Exp. Biol. Med.*, **103**, 407–410.

Teuchy, H., Quatacker, G. & Van Sumere, C.F. (1971) Quantitative investigation of the hippuric acid formation in the rat after administration of some possible aromatic and hydroaromatic precursors. *Arch. Intern. Physiol. Biochem.*, **79**, 573–587.

Thierfelder, H. & Schempp, E. (1917) Behaviour of benzoylpropionic acid, phenethyl alcohol and phenoxyacetic acid in the body of men and dogs. *Arch. Ges. Physiol.*, **167**, 280–288.

Todo, T. & Yonei, S. (1983) Inhibitory effect of membrane-binding drugs on excision repair of DNA damage in UV-irradiated Escherichia coli. *Mutat. Res.*, **112**, 97–107.

Tomiyama, H., Tachibana, A. & Yonei, S. (1986) Differential effects of procaine and phenethyl alcohol on excision repair in DNA in UV-irradiated *Escherichia coli. Int. J. Radiat. Biol.*, **50**, 973–981.

Treick, R.W. & Konetzka, W.A. (1964) Physiological state of *Escherichia coli* and the inhibition of deoxyribonucleic acid synthesis by phenethyl alcohol. *J. Bacteriol.*, **88**, 1580–1584.

Tulane, V.J. & Lewis, H.B. (1933) Studies in the synthesis of hippuric acid in the animal organism. IX. A comparative study of the rate of synthesis and excretion of hippuric and phenaceturic acids by the rabbit. *J. Biol. Chem.*, **103**, 151–160.

Varley, R. & Trenchard-Morgan, S. (1985) Completion of bacterial reverse mutation tests to on ORP 178 to OECD guideline 471 (Ames test). Tate and Lyle Group Research and Development, Ledbury, Herefordshire, England. Unpublished report. Private communication. Submitted to WHO by Flavor and Extract Manufacturers' Association of the United States.

Vollmuth, T.A., Heck, J.D., Ratajczak, H.V. & Thomas, P.T. (1989) Immunotoxicity assessment of flavouring Ingredients using a rapid and economical screen. *Toxicologist*, **9**, 206.

Wagreich, H., Kamin, H. & Harrow, B. (1940) On the detoxication of phenylacetic acid by glucuronic acid in humans. *Proc. Soc. Exp. Biol. Med.*, **43**, 468–470.

Wild, D., King, M.T., Gocke, E. & Eckhardt, K. (1983) Study of artificial flavouring substances for mutagenicity in the *Salmonella*/microsome, Basc and micronucleus tests. *Food Chem. Toxicol.*, **21**, 707–719.

Williams, R.T. (1959) *Detoxication Mechanisms—The Metabolism and Detoxication of Drugs, Toxic Substances and Other Organic Compounds*, 2nd Ed., London, Chapman & Hall.

Wingard, C., Hitchcock, P. & Teague, R.S. (1955) A survey of aldehydes with respect to their action on the blood pressure. *Arch. Int. Pharmacodyn. Ther.*, **102**, 65–84.

Yonei, S. (1980) Inhibitory effect of membrane-specific drugs on liquid-holding recovery in UV-irradiated E. coli cells. *Int. J. Radiat. Biol.*, **37**, 685–689.

Yoo, Y.S. (1986) Mutagenic and antimutagenic activities of flavouring agents used in foodstuffs. *J. Osaka City Med. Center*, **34**, 267–288.

Zaitsev, A.N. & Rakhmanina, N.L. (1974) Some data on the toxic properties of phenylethyl and cinnamyl alcohols. *Vop. Pitan.*, **6**, 48–53.

SULFUR-CONTAINING HETEROCYCLIC COMPOUNDS

First draft prepared by

Dr P.J. Abbott[1] and Professor A.G. Renwick[2]
[1]Australia New Zealand Food Authority, Canberra, ACT, Australia; and
[2]Clinical Pharmacology Group, University of Southampton, Southampton, England

1. EVALUATION

The Committee evaluated a group of 30 flavouring agents comprising sulfur-containing heterocyclic compounds (see Table 1) using the Procedure for the Safety Evaluation of Flavouring Agents (See Figure 1). The group is composed of both five- and six-member S-containing aromatic and non-aromatic heterocyclic compounds, including thiazole itself and derivatives of thiazole, dithiazine, thiazoline and thiophene. Thiazole is a five-membered aromatic heterocyclic compound containing sulfur and nitrogen atoms in the 1- and 3-ring positions, respectively (Nos 1030–1044 and 1054–1057).

Dithiazine is a six-membered non-aromatic heterocyclic compound containing two sulfur atoms and one nitrogen atom in the 1-, 3- and 5-positions, respectively (Nos 1046–1049). Thiazoline (partially reduced thiazole) is a five-membered non-aromatic heterocyclic compound containing sulfur and nitrogen atoms in the 1- and 3-positions, respectively (Nos 1045, 1058 and 1059). Thiophene is a five-membered aromatic heterocyclic compound containing a sulfur atom in the 1-position (Nos 1050–1053). Except for thiazole (No. 1032), all the flavouring agents in this group

Table 1. *Summary of the results of safety evaluations of sulfur-containing heterocyclic compounds*[a]

Flavouring agent	No.	CAS No. and structure	*Step 2* Predicted to be metabolized to innocuous products?	*Step A3 /B3*[b] Does intake exceed threshold for structural class?	*Step A4* Is the flavouring agent or are its metabolites endogenous?	*Step A5 / B4* Adequate margin of safety for the flavouring agent or related chemical?	Conclusion based on current intake
Structural class II							
Thiamine hydrochloride	1030	67-03-8 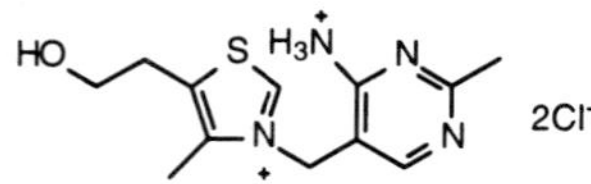	Yes[c] A side	Yes Europe: 2900 USA: 1200	No	A5 Yes. The NOEL of 36 mg/kg bw per day is > 500 times the estimated intake of thiamine hydrochloride when used as a flavouring agent.	No safety concern
4-Methyl-5-thiazoleethanol	1031	137-00-8 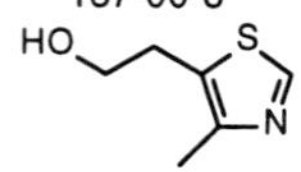	Yes[d] A side	No Europe: 170 USA: 380	N/R	N/R	No safety concern
Thiazole	1032	288-47-1 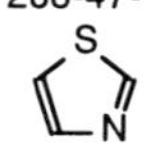	No B side	No Europe: 0.01 USA: 0.07	N/R	B5 Yes The NOEL of 0.92 mg/kg bw per day for the related flavouring agent No. 1039 is > 900 000 times the estimated intake of thiazole when used as a flavouring agent.	No safety concern

Table 1 *(continued)*

Flavouring agent	No.	CAS No. and structure	*Step 2* Predicted to be metabolized to innocuous products?	*Step A3 /B3*[a,b] Does intake exceed threshold for structural class?	*Step A4* Is the flavouring agent or are its metabolites endogenous?	*Step A5 / B4* Adequate margin of safety for the flavouring agent or related chemical?	Conclusion based on current intake
2-(1-Methylpropyl)-thiazole	1033	18277-27-5 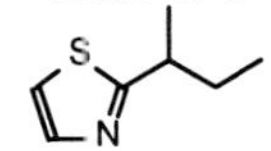	No B side	No Europe: 0.03 USA: 0.01	N/R	B5 Yes The NOEL of 0.92 mg/kg bw per day for the related flavouring agent No. 1039 is > 1 million times the estimated intake of 2-(1-methylpropyl)-thiazole when used as a flavouring agent.	No safety concern
2-Isobutylthiazole	1034	18640-74-9 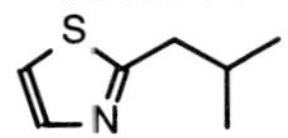	No B side	No Europe: 3 USA: 0.4	N/R	B5 Yes The NOEL of 0.92 mg/kg bw per day for the related flavouring agent No. 1039 is 10 000 times the estimated intake of 2-isobutylthiazole when used as a flavouring agent.	No safety concern
4,5-Dimethylthiazole	1035	3581-91-7 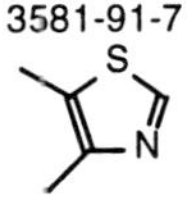	No B side	No Europe: 0.2 USA: 0.4	N/R	B5 Yes The NOEL of 0.92 mg/kg bw per day for the related flavouring agent No. 1039 is > 200 000 times the estimated intake of 4,5-dimethylthiazole when used as a flavouring agent.	No safety concern
2,4,5-Trimethylthiazole	1036	13623-11-5 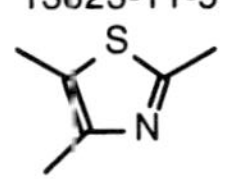	No B side	No Europe: 1 USA: 0.3	N/R	B5 Yes The NOEL of 0.92 mg/kg bw per day for the related flavouring agent No. 1039 is > 90 000 times the estimated intake of 2,4,5-trimethylthiazole when used as a flavouring agent.	No safety concern

Table 1 *(continued)*

Flavouring agent	No.	CAS No. and structure	*Step 2* Predicted to be metabolized to innocuous products?	*Step A3 /B3*[a,b] Does intake exceed threshold for structural class?	*Step A4* Is the flavouring agent or are its metabolites endogenous?	*Step A5 / B4* Adequate margin of safety for the flavouring agent or related chemical?	Conclusion based on current intake
2-Isopropyl-4-methylthiazole	1037	15679-13-7 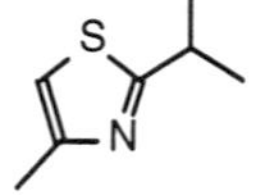	No B side	No Europe: 22 USA: 10	N/R	B5 Yes The NOEL of 0.92 mg/kg bw per day for the related flavouring agent No. 1039 is > 2000 times the estimated intake of 2-isopropyl-4-methylthiazole when used as a flavouring agent.	No safety concern
4-Methyl-5-vinylthiazole	1038	1759-28-0 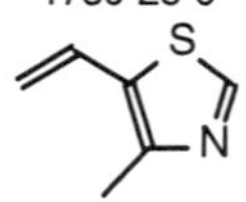	No B side	No Europe: 2 USA: 0.2	N/R	B5 Yes The NOEL of 0.92 mg/kg bw per day for the related flavouring agent No. 1039 is > 20 000 times the estimated intake of 4-methyl-5-vinylthiazole when used as a flavouring agent.	No safety concern
2,4-Dimethyl-5-vinylthiazole	1039	65505-18-2 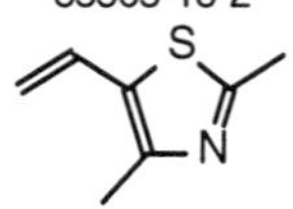	No B side	No Europe: ND USA: 0.007	N/R	B5 Yes The NOEL of 0.92 mg/kg bw per day is > 9 million times the estimated intake of 2,4-dimethyl-5-vinylthiazole when used as a flavouring agent.	No safety concern
2-Acetylthiazole	1041	24295-03-2 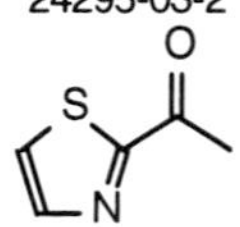	No B side	No Europe: 11 USA: 10	N/R	B5 Yes The NOEL of 50 mg/kg bw per day is > 200 000 times the estimated intake of 2-acetyl-thiazole when used as a flavouring agent.	No safety concern

Table 1 *(continued)*

Flavouring agent	No.	CAS No. and structure	*Step 2* Predicted to be metabolized to innocuous products?	*Step A3 /B3*[a,b] Does intake exceed threshold for structural class?	*Step A4* Is the flavouring agent or are its metabolites endogenous?	*Step A5 / B4* Adequate margin of safety for the flavouring agent or related chemical?	Conclusion based on current intake
2-Propionylthiazole	1042	43039-98-1 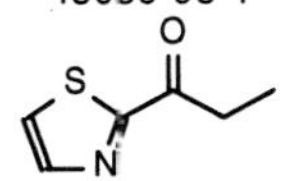	No B side	No Europe: ND USA: 0.2	N/R	B5 Yes The NOEL of 50 mg/kg bw per day for the related flavouring agent No. 1041 is > 10 million times the estimated intake of 2-propionylthiazole when used as a flavouring agent.	No safety concern
4-Methylthiazole	1043	693-95-8 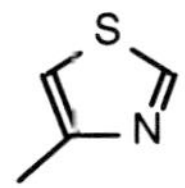	No B side	No Europe: 0.1 USA: 0.05	N/R	B5 Yes The NOEL of 0.92 mg/kg bw per day for the related flavouring agent No. 1039 is > 400 000 times the estimated intake of 4-methylthiazole when used as a flavouring agent.	No safety concern
2-Ethyl-4-methylthiazole	1044	15679-12-6 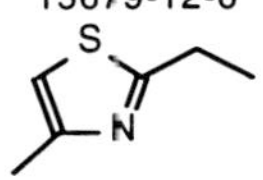	No B side	No Europe: 4 USA: 1	N/R	B5 Yes The NOEL of 0.92 mg/kg bw per day for the related flavouring agent No. 1039 is > 10 000 times the estimated intake of 2-ethyl-4-methylthiazole when used as a flavouring agent.	No safety concern

Table 1 *(continued)*

Flavouring agent	No.	CAS No. and structure	*Step 2* Predicted to be metabolized to innocuous products?	*Step A3 /B3*[a,b] Does intake exceed threshold for structural class?	*Step A4* Is the flavouring agent or are its metabolites endogenous?	*Step A5 / B4* Adequate margin of safety for the flavouring agent or related chemical?	Conclusion based on current intake
2-Isobutyl-4,6-dimethyldihydro-1,3,5-dithiazine and 4-isobutyl-2,6-dimethyldihydro-1,3,5-dithiazine (mixture)	1046	101517-87-7 101517-86-6 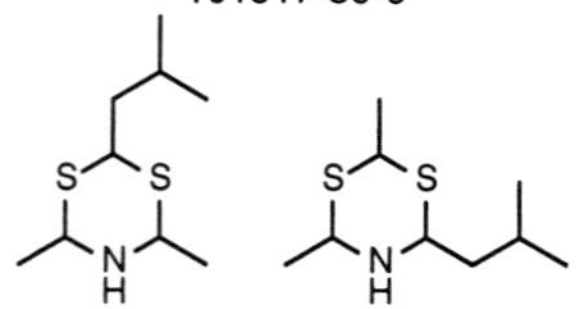 18% 62%	No B side	No Europe: 0.1 USA: 0.05	N/R	B5 Yes The NOEL of 11 mg/kg bw per day is > 5 million times the estimated intake of a mixture of 2-isobutyl-4,6-dimethyl and 4-isobutyl-2,6-dimethyldihydro-1,3,5-dithiazine when used as a flavouring agent.	No safety concern
2-Isopropyl-4,6-dimethyl 2,6-dimethyldihydro-1,3,5-dithiazine and 4-isopropyl-2,6-dimethyldihydro-1,3,5-dithiazine (mixture)	1047	104691-41-0 104691-40-9 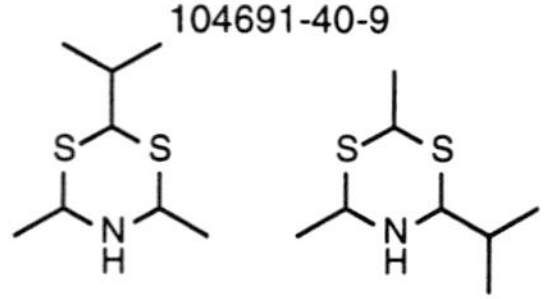 27% 44%	No B side	No Europe: ND USA: 0.07	N/R	B5 Yes The NOEL of 11 mg/kg bw per day is > 10 million times the estimated intake of a mixture of 2-isopropyl-4,6-dimethyl and 4-isopropyl-2,6-dimethyldihydro-1,3,5-dithiazine when used as a flavouring agent.	No safety concern

Table 1 *(continued)*

Flavouring agent	No.	CAS No. and structure	*Step 2* Predicted to be metabolized to innocuous products?	*Step A3 /B3*[a,b] Does intake exceed threshold for structural class?	*Step A4* Is the flavouring agent or are its metabolites endogenous?	*Step A5 / B4* Adequate margin of safety for the flavouring agent or related chemical?	Conclusion based on current intake
2,4,6-Triisobutyl-5,6-dihydro-4*H*-1,3,5-dithiazine	1048	74595-94-1 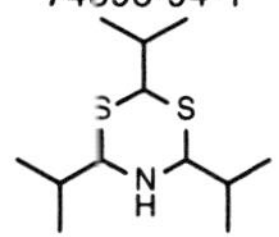	No B side	No Europe: ND USA: 2.6	N/R	B5 Yes The NOEL of 11 mg/kg bw per day for the related flavouring agent No. 1046 is > 200 000 times the estimated intake of 2,4,6-triisobutyl-5,6-dihydro-4*H*-1,3,5-dithiazine when used as a flavouring agent.	No safety concern
2,4,6-Trimethyldihydro-4*H*-1,3,5-dithiazine	1049	638-17-5 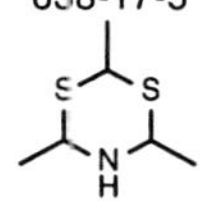	No B side	No Europe: ND USA: 3.3	N/R	B5 Yes The NOEL of 11 mg/kg bw per day for the related flavouring agent No. 1046 is > 5 million times the estimated intake of2,4,6-trimethyldihydro-4*H*-1,3,5-dithiazine when used as a flavouring agent.	No safety concern
5-Methyl-2-thiophene-carboxyaldehyde	1050	13679-70-4 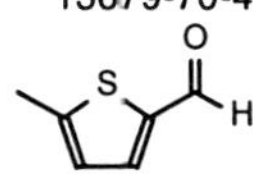	No B side	No Europe: 1 USA: 0.01	N/R	B5 Yes The NOEL of 290 mg/kg bw per day for the related flavouring agent No. 1053 is > 20 million times the estimated intake of 5-methyl-2-thiophene-carboxyaldehyde when used as a flavouring agent.	No safety concern

Table 1 *(continued)*

Flavouring agent	No.	CAS No. and structure	*Step 2* Predicted to be metabolized to innocuous products?	*Step A3 /B3*[a,b] Does intake exceed threshold for structural class?	*Step A4* Is the flavouring agent or are its metabolites endogenous?	*Step A5 / B4* Adequate margin of safety for the flavouring agent or related chemical?	Conclusion based on current intake
3-Acetyl-2,5-dimethylthiophene	1051	2530-10-1 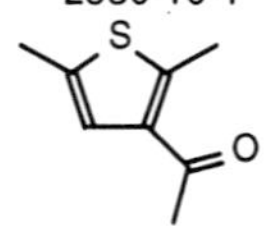	No B side	No Europe: 22 USA: 0.2	N/R	B5 Yes The NOEL of 290 mg/kg bw per day for the related flavouring agent No. 1053 is > 700 000 times the estimated intake of 3-acetyl-2,5-dimethylthiophene when used as a flavouring agent.	No safety concern
4-Methyl-5-thiazoleethanol acetate	1054	656-53-1 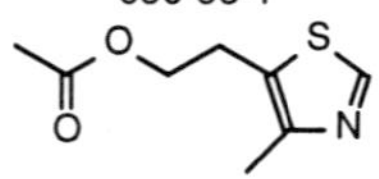	Yes[c] A side	No Europe: 10 USA: 3	N/R	N/R	No safety concern
Structural class III							
Benzothiazole	1040	95-16-9 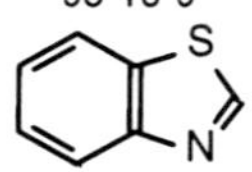	No B side	No Europe: 1 USA: 0.2	N/R	B5 Yes The NOEL of 5.1 mg/kg bw per day is > 20 000 times the estimated intake of benzothiazole when used as a flavouring agent.	No safety concern
4,5-Dimethyl-2-isobutyl-3-thiazoline	1045	65894-83-9 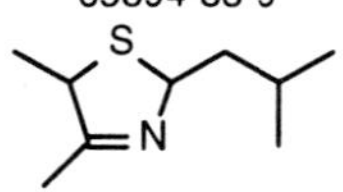	No B side	No Europe: 0.01 USA: 4	N/R	B5 Yes The NOEL of 1.2 mg/kg bw per day for the related flavouring agent No. 1059 is > 10 000 times the estimated intake of 4,5-dimethyl-2-isobutyl-3-thiazoline when used as a flavouring agent.	No safety concern

Table 1 *(continued)*

Flavouring agent	No.	CAS No. and structure	*Step 2* Predicted to be metabolized to innocuous products?	*Step A3 /B3*[a,b] Does intake exceed threshold for structural class?	*Step A4* Is the flavouring agent or are its metabolites endogenous?	*Step A5 / B4* Adequate margin of safety for the flavouring agent or related chemical?	Conclusion based on current intake
2-Thienyl mercaptan	1052	7774-74-5 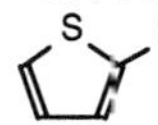	No B side	No Europe: 0.01 USA: 0.03	N/R	B5 Yes The NOEL of 290 mg/kg bw per day for the related flavouring agent No. 1053 is > 700 million times the estimated intake of 2-thienyl mercaptan when used as a flavouring agent.	No safety concern
2-Thienyl disulfide	1053	6911-51-9 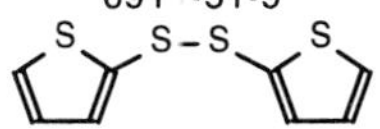	No B side	No Europe: ND USA: 0.07	N/R	B5 Yes The NOEL of 290 mg/kg bw per day is > 200 million times the estimated intake of 2-thienyl disulfide when used as a flavouring agent.	No safety concern
2,4-Dimethyl-5-acetylthiazole	1055	38205-60-6 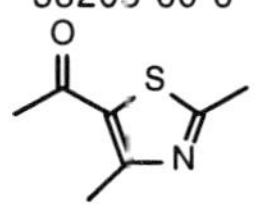	No B side	No Europe: 0.01 USA: 2	N/R	B5 Yes The NOEL of 24 mg/kg bw per day is > 10 million times the estimated intake of 2,4-dimethyl-5-acetylthiazole when used as a flavouring agent.	No safety concern
2-Ethoxythiazole	1056	15679-19-3 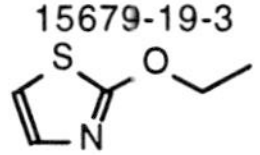	No B side	No Europe: 0.01 USA: 0.12	N/R	B5 Yes The NOEL of 50 mg/kg bw per day for the related flavouring agent No. 1041 is > 20 million times the estimated intake of 2-ethoxythiazole when used as a flavouring agent.	No safety concern

Table 1 *(continued)*

Flavouring agent	No.	CAS No. and structure	*Step 2* Predicted to be metabolized to innocuous products?	*Step A3 /B3*[a,b] Does intake exceed threshold for structural class?	*Step A4* Is the flavouring agent or are its metabolites endogenous?	*Step A5 / B4* Adequate margin of safety for the flavouring agent or related chemical?	Conclusion based on current intake
2-Methyl-5-methoxythiazole	1057	38205-64-0 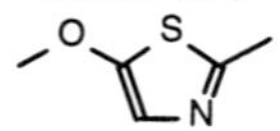	No B side	No Europe: ND USA: 0.01	N/R	B5 Yes The NOEL of 8.6 mg/kg bw per day is > 40 million times the estimated intake of 2-methyl-5-methoxythiazole when used as a flavouring agent.	No safety concern
4,5-Dimethyl-2-ethyl-3-thiazoline	1058	76788-46-0 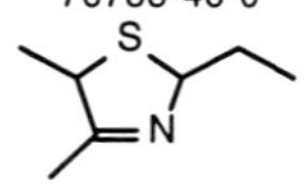	No B side	No Europe: ND USA: 0.01	N/R	B5 Yes The NOEL of 1.2 mg/kg bw per day for the related flavouring agent No. 1059 is > 6 million times the estimated intake of 4,5-dimethyl-2-ethyl-3-thiazoline when used as a flavouring agent.	No safety concern
2-(2-Butyl)-4,5-dimethyl-3-thiazoline	1059	65894-82-8 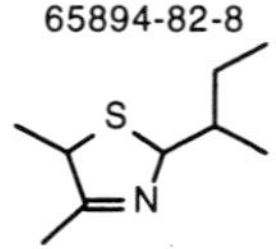	No B side	No Europe: ND USA: 5	N/R	B5 Yes The NOEL of 1.2 mg/kg bw per day is > 10 000 times the estimated intake of 2-(2-butyl)-4,5-dimethyl-3-thiazoline when used as a flavouring agent.	No safety concern

CAS: Chemical Abstracts Service; ND: no intake data reported; N/R: not required for evaluation

[a] Step 1. Twenty-one flavouring agents are in structural class II and nine flavouring agents are in structural class III.

[b] The thresholds for human intake for structural classes II and III are 540 μg/day and 90 μg/day, respectively. All intake values are expressed in μg/day. The combined per capita intake of flavouring agents in structural class II is 3100 μg/day in Europe and 1600 μg/day in the USA. The combined per capita intake of flavouring agents in structural class III is 1 μg/day in Europe and 11 μg/day in the USA.

[c] Enzymatically cleaved to yield 4-methyl-5-thiazoleethanol (No. 1031) and 2-methyl-4-amino-5-hydroxymethylpyrimidine. The thiazole and pyrimidine fragments are further oxidized to yield 4-methylthiazole-4-acetic acid and the 5-pyrimidine carboxylic acid derivative, respectively, which, together with thiamine, are excreted in the urine. Can also be converted to 2-methyl-4-amino-5-formylaminopyrimidine and thiamine acetic acid.

[d] Oxidized to yield 4-methylthiazole-4-acetic acid and excreted in the urine (See note c)

[e] Hydrolysed to yield 4-methylthiazole-4-acetic acid and excreted in urine (See note c)

are ring-substituted with one or more of the following functional groups: alkyl, alkenyl, aryl, alcohol, keto, thiol and disulfide.

The Committee has not previously evaluated any of the members of this group.

Eighteen flavouring agents in the group have been detected as natural components of food, and quantitative data have been reported on the natural occurrence of eight agents (Nos 1032, 1034, 1036, 1040, 1041, 1043 and 1050). The foods in which one or more of these flavouring agents can be found include lean meat, beans, nuts, whole-grain cereals, fish, coffee, milk, beer, peanuts, popcorn, pork liver, shrimp, tomato, potato, grapes and apples (Maarse et al., 1999).

1.2 Estimated daily intake

The total annual volume of production of the 30 sulfur-containing heterocyclic compounds in this group was reported to be 22 000 kg in Europe (International Organization of the Flavor Industry, 1995) and 12 000 kg in the USA (Lucas et al., 1999). These values are equivalent to total daily per capita intakes of 3100 µg in Europe and 1600 µg in the USA.

In both Europe and the USA, thiamine hydrochloride (vitamin B1; No. 1030) and its principal metabolite, 4-methyl-5-thiazoleethanol (No. 1031), accounted for approximately 98% of the total per capita dietary intake of the flavouring agents in this group. In Europe, thiamine hydrochloride and 4-methyl-5-thiazoleethanol accounted for approximately 92% (2900 µg/day) and 6% (170 µg/day), respectively, of the total per capita intake. In the USA, thiamine hydrochloride and 4-methyl-5-thiazoleethanol accounted for approximately 75% (1200 µg/day) and 23% (380 µg/day), respectively, of the total per capita intake of the flavouring agents in this group.

The estimated intakes of the remaining flavouring agents are between 0.01 and 22 µg/day in both Europe and the USA, with most intakes below 4 µg/day.

1.3 Absorption, distribution, metabolism and elimination

Thiazole and its derivatives are metabolized primarily by side-chain oxidation or oxidation of the ring sulfur or nitrogen atoms (Rance, 1989); however, other routes of metabolism, involving ring cleavage, are possible. Derivatives of dithiazine, which are cyclic sulfides, are expected to be metabolized primarily by S-oxidation to yield the corresponding sulfoxides and sulfones. Thiazoline is predicted to be similarly metabolized. Thiamine hydrochloride (No. 1030) at an intake of less than 5 mg/day is readily absorbed in the small intestine by an active transport system, followed by conversion to the coenzyme, thiamine pyrophosphate (Hegsted, et al., 1976; Wilson, 1982; Tietz, 1986). Thiamine is metabolized primarily to a pyrimidine and a thiazole (4-methyl-5-thiazoleethanol, No. 1031) fragment. These fragments are further oxidized and excreted in the urine.

Thiophene derivatives are metabolized primarily by *S*-oxidation, followed by conjugation with glutathione; however, other routes of metabolism, involving ring cleavage, are also possible. The resulting mercapturic acid derivative is eliminated in the urine (Dansette et al., 1992; Valadon et al., 1996). The aromatic-fused thiazole, 2-acetylthiazole (benzothiazole; No. 1041), which lacks ring substituents, is metabolized by thiazole ring cleavage, yielding a series of free and conjugated *ortho*-aminophenyl sulfone and sulfoxide metabolites (Wilson et al., 1991).

1.3 Application of the Procedure for the Safety Evaluation of Flavouring Agents

The stepwise evaluation of the 30 sulfur-containing heterocyclic compounds is described in detail below and summarized in Table 1.

Step 1. All the flavouring agents in this group are heterocyclic compounds, and five of these (Nos 1040, 1050, 1051, 1052 and 1053) are also aromatic 1[1]. One of the aromatic heterocyclic flavouring agents, benzothiazole (No. 1040) is unsubstituted and was therefore placed in structural class III (Cramer et al., 1978). Of the other four substituted compounds, two (Nos 1050 and 1051) are common constituents of food and were therefore classified in structural class II, while the other two (Nos 1052 and 1053) are not common constituents of food and were therefore placed in structural class III.

Of the 25 non-aromatic heterocyclic flavouring agents, 19 (Nos 1030–1039, 1041–1044, 1046–1049 and 1054) are or are structurally closely related to common constituents of food and were therefore placed in structural class II. The remaining six flavouring agents (Nos 1045 and 1055–1059) were classified in structural class III.

Step 2. The data on most individual members of the group were, in most cases, insufficient to allow conclusions about their probable metabolic fate. The metabolism of thiamine hydrochloride (No. 1030) and its metabolite, 4-methyl-5-thiazoleethanol (No. 1031), is well characterized, and both are metabolized to innocuous products. 4-Methyl-5-thiazoleethanol acetate (No. 1054) is expected to be readily hydrolysed to 4-methyl-5-thiazoleethanol. The evaluation of all three flavouring agents therefore proceeded via the A (left-hand) side of the scheme. For all other flavouring agents in the group, the evaluation proceeded via the B (right-hand) side of the scheme.

Step A3. With respect to those flavouring agents evaluated via the A side of the scheme, the estimated daily per capita intakes of 4-methyl-5-thiazoleethanol (No. 1031) and 4-methyl-5-thiazoleethanol acetate (No. 1054) are below the threshold for human intake for structural class II (540 μg), and therefore these flavouring agents are not expected to raise any safety concern. The daily per capita intake of thiamine hydrochloride (No. 1030) is 2900 μg in Europe and 1200 μg in the USA. The intake of this flavouring agent therefore exceeds the threshold for human intake of compounds in structural class II, and its evaluation proceeded to step A4.

Step A4. Thiamine hydrochloride (No. 1030) is not endogenous; therefore, its evaluation proceeded to step A5.

[1] According to the decision-tree process, an 'aromatic' compound has at least one benzene, furan, thiophene, pyridine or pyrole ring, however substituted and regardless of whether it is fused to another ring.

Step A5. The NOEL of 36 mg/kg bw per day for thiamine hydrochloride (No. 1030) in a 90-day dietary study in rats is more than 500 times the estimated intake of this substance from its use as a flavouring agent in Europe (48 μg/kg bw per day) and more than 1000 times the estimated intake in the USA (22 μg/kg bw per day). Thiamine hydrochloride would therefore not be expected to be a safety concern.

Step B3 With regard to those flavouring agents evaluated via the B side of the scheme, the estimated daily per capita intakes of all 18 flavouring agents in structural class II (Nos 1032–1039, 1041–1044 and 1046–1051) are below the threshold for human intake for structural class II (540 μg per person per day). Similarly, the estimated daily per capita intakes of all nine flavouring agents in structural class III (Nos 1040, 1045, 1052, 1053 and 1055–1059) are below the threshold for human intake for compounds in this class (90 μg per person per day). The evaluation of all of these flavouring agents therefore proceeded to step B4.

Step B4 The NOEL for 2,4-dimethyl-5-vinylthiazole (No. 1039) in a 90-day dietary study in rats was 0.92 mg/kg bw per day, and this NOEL is appropriate for the structurally related flavouring agents, thiazole (No. 1032), 2-(1-methylpropyl)thiazole (No. 1033), 2-isobutylthiazole (No. 1034), 4,5-dimethylthiazole (No. 1035), 2,4,5-trimethylthiazole (No. 1036), 2-isopropyl-4-methylthiazole (No. 1037), 4-methyl-5-vinylthiazole (No. 1038), 4-methylthiazole (No. 1043) and 2-ethyl-4-methylthiazole (No. 1044).

The NOEL for 2-acetylthiazole (No. 1041) in a 90-day dietary study in rats was 50 mg/kg bw per day, and this NOEL is appropriate for the structurally related flavouring agent, 2-propionylthiazole (No. 1042).

The NOEL for 2-(2-butyl)-4,5-dimethyl-3-thiazoline (No. 1059) in a 90-day dietary study in rats was 1.2 mg/kg bw per day, and this NOEL is appropriate for the structurally related flavouring agents, 4,5-dimethyl-2-isobutyl-3-thiazoline (No. 1045) and 4,5-dimethyl-2-ethyl-3-thiazoline (No. 1058).

The NOEL for a mixture of 2-isobutyl-4,6-dimethyl and 4-isobutyl-2,6-dimethyldihydro-1,3,5-dithiazine (No. 1046) and of a mixture of 2-isopropyl-4,6-dimethyl and 4-isopropyl-2,6-dimethyldihydro-1,3,5-dithiazine (No. 1047) in a 14-day dietary study in rats was 11 mg/kg bw per day, and this NOEL is appropriate for the structurally related agents, 2,4,6-triisobutyl-5,6-dihydro-4H-1,3,5-dithiazine (No. 1048) and 2,4,6-trimethyldihydro-4H-1,3,5-dithiazine (No. 1049).

The NOEL for 2-thienyl disulfide (No. 1053) in a 90-day dietary study in rats was 290 mg/kg bw per day, and this NOEL is appropriate for the structurally related agents 5-methyl-2-thiophene carboxyaldehyde (No. 1050), 3-acetyl-2,5-dimethylthiophene (No. 1051) and 2-thienyl mercaptan (No. 1052).

The NOEL for benzothiazole (No. 1040) in a 90-day dietary study in rats was 5.1 mg/kg bw per day.

The NOEL for 2,4-dimethyl-5-acetylthiazole (No. 1055) in a dietary study in rats was 24 mg/kg bw per day.

The NOEL for 2-methyl-5-methoxythiazole (No. 1057) in a 90-day dietary study in rats was 8.6 mg/kg bw per day.

For all of the above-mentioned flavouring agents (Nos 1032–1053, 1055–1059), therefore, a NOEL exists either for the substance itself, which provides an adequate margin of safety under the intended conditions of use, or for a structurally related substance, which is high enough to accommodate any perceived difference in toxicity between the substance itself and the related substance.

The NOEL used to evaluate each flavouring agent and the margins of safety they provide on the basis of current levels of intake are summarized in Table 2. On the basis of these data, these flavouring agents would not be expected to be a safety concern.

1.4 Consideration of combined intake from use as flavouring agents

Although the 30 flavouring agents in this group are unlikely to produce common metabolites, consideration of their combined intake is appropriate since all are likely to be conjugated with glutathione before excretion. In the unlikely event that all 21 flavouring agents in structural class II were to be consumed simultaneously on a daily basis, the estimated combined per capita intakes (3100 μg/day in Europe and 1600 μg/day in the USA) would exceed the threshold of human intake for their structural class (540 μg per person). Since, at these levels of intake, the conjugation capacity of the glutathione pool is unlikely to be depleted, this level of intake is not anticipated to be a safety concern. In the unlikely event that the nine flavouring agents in structural class III were to be consumed simultaneously on a daily basis, the estimated combined per capita intakes (1 μg/day in Europe and 11 μg/day in the USA) would not exceed the threshold for human intake for their structural class (90 μg per person).

1.5 Conclusions

On the basis of the predicted metabolism and the NOELs for some members of the group, the Committee concluded that the 30 sulfur-containing heterocyclic compounds in this group would not raise safety concerns at current levels of intake.

The Committee noted that, for the three flavouring agents that were predicted to be metabolized to innocuous products, the available data on toxicity were consistent with the results of the safety evaluation.

2. RELEVANT BACKGROUND INFORMATION

2.1 Explanation

This monograph summarizes the key data relevant to the evaluation of the 30 flavouring substances in this group. The group included both five- and six-member sulfur-containing aromatic and non-aromatic heterocycles, namely, thiazole and derivatives of thiazole, dithiazine, thiazoline and thiophene. Thiazole derivatives are five-membered aromatic heterocycles containing sulfur and nitrogen atoms in the 1- and 3-ring positions, respectively (Nos 1030–1044 and 1054–1057). Thiazoline derivatives (containing partially reduced thiazole) are five-membered non-aromatic

Table 2. Margins of safety for the flavouring agents proceeding via the B side of the Scheme

Flavouring agent	No.	Related flavouring agent(s)	No.	NOEL (mg/kg bw per day)	Reference	Highest estimated intake (mg/kg bw per day)	Margin of safety
Structural class II							
Thiazole	1032	2,4-Dimethyl-5-vinylthiazole	1039	0.92[a]	Posternak et al. (1969)	0.001	> 900 000
2-(1-Methylpropyl)-thiazole	1033	2,4-Dimethyl-5-vinylthiazole	1039	0.92[a]	Posternak et al. (1969)	0.0005	> 1 million
2-Isobutylthiazole	1034	2,4-Dimethyl-5-vinylthiazole	1039	0.92[a]	Posternak et al. (1969)	0.05	> 10 000
4,5-Dimethylthiazole	1035	2,4-Dimethyl-5-vinylthiazole	1039	0.92[a]	Posternak et al. (1969)	0.004	> 200 000
2,4,5-Trimethylthiazole	1036	2,4-Dimethyl-5-vinylthiazole	1039	0.92[a]	Posternak et al. (1969)	0.01	> 90 000
2-Isopropyl-4-methylthiazole	1037	2,4-Dimethyl-5-vinylthiazole	1039	0.92[a]	Posternak et al. (1969)	0.4	> 2000
4-Methyl-5-vinylthiazole	1038	2,4-dimethyl-5-vinylthiazole	1039	0.92[a]	Posternak et al. (1969)	0.04	> 20 000
2,4-Dimethyl-5-vinylthiazole	1039	NA	NA	0.92[a]	Posternak et al. (1969)	0.0001	> 9 million
2-Acetylthiazole	1041	NA	NA	50[a]	Wheldon et al. (1970)	0.2	> 200 000
2-Propionylthiazole	1042	2-Acetylthiazole	1041	50[a]	Wheldon et al. (1970)	0.003	> 10 million
4-Methylthiazole	1043	2,4-Dimethyl-5-vinylthiazole	1039	0.92[a]	Posternak et al. (1969)	0.002	> 400 000
2-Ethyl-4-methylthiazole	1044	2,4-Dimethyl-5-vinylthiazole	1039	0.92[a]	Posternak et al. (1969)	0.06	> 10 000
2-Isobutyl-4,6-dimethyldihdro-1,3,5-dithiazine and 4-isobutyl-2,6-dimethyldihydro-1,3,5-dithiazine (mixture)	1046	NA	NA	11[b]	Rush (1989a,b)	0.002	> 5 million
2-Isopropyl-4,6-dimethyldihydro-1,3,5-dithiazine and 4-isopropyl-2,6-dimethyldihydro-1,3,5-dithiazine (mixture)	1047	NA	NA	11[b]	Rush (1989a,b)	0.001	> 10 million
2,4,6-Triisobutyl-5,6-dihydro-4H-1,3,5-dithiazine	1048	2-Isobutyl-4,6-dimethyl and 4-isobutyl-2,6-dimethyldihydro-1,3,5-dithiazine (mixture)	1046	11[b]	Rush (1989a,b)	0.04	> 200 000
2,4,6-Trimethyldihydro-4H-1,3,5-dithiazine	1049	2-Isobutyl-4,6-dimethyl and 4-isobutyl-2,6-dimethyldihydro-1,3,5-dithiazine	1046	11[b]	Rush (1989a,b)	0.002	> 5 million

Table 2 (contd)

Flavouring agent	No.	Related flavouring agent(s)	No.	NOEL (mg/kg bw per day)	Reference	Highest estimated intake (mg/kg bw per day)	Margin of safety
5-Methyl-2-thiophene-carboxyaldehyde	1050	2-Thienyl disulfide	1053	290[a]	Morgareidge & Oser (1970)	0.01	> 20 million
3-Acetyl-2,5-dimethylthio-phene	1051	2-Thienyl disulfide	1053	290[a]	Morgareidge & Oser (1970)	0.4	> 700 000
Structural class III							
Benzothiazole	1040	NA	NA	5.1[a]	Margareidge (1971)	0.02	> 20 000
4,5-Dimethyl-2-isobutyl-3-thiazoline	1045	2-(2-Butyl)-4,5-dimethyl-3-thiazoline	1059	1.2[a]	Babish (1978)	0.07	> 10 000
2-Thienyl mercaptan	1052	2-Thienyl disulfide	1053	290[a]	Morgareidge & Oser (1970)	0.0004	> 700 million
2-Thienyl disulfide	1053	NA	NA	290[a]	Morgareidge (1971)	0.001	> 200 million
2,4-Dimethyl-5-acetylthiazole	1055	NA	NA	24[a]	Shellenberger (1971)	0.03	> 10 million
2-Ethoxythiazole	1056	2-Acetylthiazole	1041	50[a]	Wheldon et al. (1970)	0.002	> 20 million
2-Methyl-5-methoxythiazole	1057	NA	NA	8.6[a]	Posternak et al. (1975)	0.0002	> 40 million
4,5-Dimethyl-2-ethyl-3-thiazoline	1058	2-(2-Butyl)-4,5-dimethyl-3-thiazoline	1059	1.2[a]	Babish (1978)	0.0002	> 6 million
2-(2-Butyl)-4,5-dimethyl-3-thiazoline	1059	NA	NA	1.2[a]	Babish (1978)	0.09	> 10 000

NA, not applicable
[a] 90-day study in rats
[b] 14-day study in rats

heterocycles containing sulfur and nitrogen atoms in the 1- and 3-ring positions, respectively (Nos 1045, 1058 and 1059). Dithiazine derivatives are six-membered non-aromatic heterocycles containing two sulfur atoms and one nitrogen (Nos 1046–1049). Thiophene derivatives are five-membered aromatic heterocycles containing a sulfur atom in the 1-ring position (Nos 1050–1053). Except for thiazole (No. 1032), all the substances in this group are ring-substituted with one or more of the following functional groups: alkyl, alkenyl, aryl, alcohol, keto, thiol and disulfide.

2.2 Additional considerations on intake

The total annual volume of production of the 30 sulfur-containing heterocyclic compounds in this group was reported to be 22 000 kg in Europe (International Organization of the Flavor Industry, 1995) and 12 000 kg in the USA (Lucas et al., 1999). In both Europe and the USA, 98% of the total volume of use was accounted for by thiamine hydrochloride (vitamin B1) and its metabolite, 4-methyl-5-thiazole ethanol (see Table 3).

Eighteen of the 30 flavouring agents in this group have been reported to occur naturally in foods (Maarse et al., 1999). Quantitative data on natural occurrence and consumption ratios have been reported for eight substances in the group (Nos 1032, 1034–1036, 1040, 1041,1043 and 1050), which indicate that the consumptionof six substances (Nos 1032, 1034, 1036, 1040, 1043 and 1050) derives predominantly from traditional foods (i.e., consumption ratio > 1) (Stofberg & Kirschman, 1985; Stofberg & Grundschober, 1987) (see Table 1). Thiamine hydrochloride (No. 1030) is vitamin B1 and is found in most plants and in animal tissues, the main sources being lean meat, beans, nuts, whole-grain cereals and fish (Lehninger, 1987). Other thiazole and thiophene derivatives have been detected in foods including coffee, milk, beer, peanuts, popcorn, beef, pork liver, fish, shrimp, tomatoes, potatoes, grapes and apples (Maarse et al., 1999).

The daily per capita intake of this group of flavouring substances is 3100 μg/day in Europe (52 μg/kg bw per day) and 1600 μg/day in the USA (27 μg/kg bw per day). Thiamine hydrochloride accounted for approximately 92% of the total per capita intake in Europe, while in the USA, thiamine hydrochloride accounted for approximately 75% of the total per capita intake. The high per capita intake of thiamine hydrochloride is likely to be the result of its use to fortify food products rather than of its use as a flavour.

2.3 Biological data

2.3.1 Biochemical data

(a) Absorption and excretion

Little specific information was available on the absorption and transformation of individual members of this group of flavouring substances, apart from thiamine hydrochloride (No. 1030). The absorption and excretion of thiamine hydrochloride is well understood (Informatics, Inc., 1974; Tietz, 1986; Lehninger, 1987). It is readily absorbed in the small intestine *via* an active transport process at an intake of < 5 mg/day (Tietz, 1986). Passive diffusion across the small intestine becomes significant at intakes > 5 mg/day (Wilson, 1982). In the jejunal mucosa, thiamine is phosphorylated to yield thiamine pyrophosphate. Both free and phosphorylated

Table 3. Annual volumes of production of sulfur-containing heterocyclic compounds used as flavouring agents in Europe and the USA

Substance (No.)	Most recent annual production volume (kg)[a]	Intake[b]		Consumption ratio[c]
		μg/day	μg/kg bw per day	
Thiamine hydrochloride (1030)				
Europe	20 000	2900	48	NA
USA	9 300	1200	20	NA
4-Methyl-5-thiazoleethanol (1031)				
Europe	1 200	170	3	NA
USA	2 800	380	6	NA
Thiazole (1032)				
Europe	0.1	0.01	0.0002	180
USA	0.5	0.07	0.001	35
2-(1-Methylpropyl)thiazole (1033)				
Europe	0.2	0.03	0.0005	NA
USA	0.1	0.01	0.0002	NA
2-Isobutylthiazole (1034)				
Europe	19	3	0.05	37
USA	2.7	0.4	0.006	261
4,5-Dimethylthiazole (1035)				
Europe	1.5	0.2	0.004	1
USA	2.7	0.4	0.006	0.4
2,4,5-Trimethylthiazole (1036)				
Europe	5	1	0.01	214
USA	2.3	0.3	0.005	465
2-Isopropyl-4-methylthiazole (1037)				
Europe	160	22	0.4	NA
USA	73	10	0.2	NA
4-Methyl-5-vinylthiazole (1038)				
Europe	17	2	0.04	NA
USA	1.4	0.2	0.003	NA
2,4-Dimethyl-5-vinylthiazole (1039)				
Europe	N/D	N/D	N/D	NA
USA	0.05	0.007	0.0001	NA
Benzothiazole (1040)				
Europe	10	1	0.02	0.4
USA	1.4	0.2	0.003	3
2-Acetylthiazole (1041)				
Europe	80	11	0.2	0.2
USA	73	10	0.2	0.2

Table 3 (contd)

Substance (No.)	Most recent annual production volume (kg)[a]	Intake[b]		Consumption ratio[c]
		μg/day	μg/kg bw per day	
2-Propionylthiazole (1042)				
Europe	N/D	N/D	N/D	NA
USA	1.5	0.2	0.003	NA
4-Methylthiazole (1043)				
Europe	0.8	0.1	0.002	1 200
USA	0.4	0.05	0.001	2 400
2-Ethyl-4-methylthiazole (1044)				
Europe	26	4	0.06	NA
USA	5.4	1	0.01	NA
4,5-Dimethyl-2-isobutyl-3-thiazoline (1045)				
Europe	0.1	0.01	0.0002	NA
USA	30	4	0.07	NA
Mixture of 2-isobutyl-4,6-dimethyl- and 4-isobutyl-2,6-dimethyldihydro-1,3,5-dithiazine (1046)				
Europe	1	0.1	0.002	NA
USA	0.4	0.05	0.001	NA
Mixture of 2-isobutyl-4,6-dimethyl and 4-isobutyl-2,6-dimethyldihydro-1,3,5-dithiazine (1047)				
Europe	N/D	N/D	N/D	NA
USA	0.5	0.07	0.001	NA
2,4,6-Triisobutyl-5,6-dihydro-4*H*-1,3,5-dithiazine (1048)				
Europe	N/D	N/D	N/D	NA
USA	20	2.6	0.04	NA
2,4,6-Trimethyldihydro-4*H*-1,3,5-dithiazine (1049)				
Europe	N/D	N/D	N/D	NA
USA	25	3.3	0.05	NA
5-Methyl-2-thiophenecarboxyaldehyde (1050)				
Europpc	6	1	0.01	76
USA	0.05	0.01	0.0002	9 100
3-Acetyl-2,5-dimethylthiophene (1051)				
Europe	150	22	0.4	NA
USA	1.8	0.2	0.004	NA
2-Thienyl mercaptan (1052)				
Europe	0.1	0.01	0.0002	NA
USA	0.2	0.03	0.0004	NA

Table 3 (contd)

Substance (No.)	Most recent annual production volume (kg)[a]	Intake[b]		Consumption ratio[c]
		µg/day	µg/kg bw per day	
2-Thienylsulfide (1053)				
Europe	N/D	N/D	N/D	NA
USA	0.5	000	0.001	NA
4-Methyl-5-thiazoleethanol acetate (1054)				
Europe	71	10	0.2	NA
USA	25	3	0.03	NA
2,4-Dimethyl-5-acetylthiazole (1055)				
Europe	0.1	0.01	0.0001	NA
USA	13	2	0.03	NA
2-Ethoxythiazole (1056)				
Europe	0.1	0.01	0.0002	NA
USA	0.9	0.12	0.002	NA
2-Methyl-5-methoxythiazole (1057)				
Europe	N/D	N/D	N/D	NA
USA	0.1	0.01	0.0002	NA
4,5-Dimethyl-2-ethyl-3-thiazoline (1058)				
Europe	N/D	N/D	N/D	NA
USA	0.1	0.01	0.0002	NA
2-(2-Butyl)-4,5-dimethyl-3-thiazoline (1059)				
Europe	N/D	N/D	N/D	NA
USA	40	5	0.09	NA
Total annual volume				
Europe	22 000			
USA	12 000			

NA, not available; N/D, no intake data reported

[a] From International Organization of the Flavor Industry (1995) and Lucas et al. (1999)

[b] Intake (µg/person per day) calculated as follows: [(annual volume, kg) × (1 × 10^9 µg/kg)/(population × survey correction factor × 365 days)], where population (10%, 'eaters only') = 32 × 10^6 for Europe and 26 × 10^6 for the USA. The correction factor = 0.6 for Europe and 0.8 for the USA, representing the assumption that only 60% and 80% of the annual production volume of the flavour, resepctively, was reported in the poundage surveys. Intake (µg/kg bw per day) calculated as follows: [(µg/person per day)/body weight], where body weight = 60 kg. Slight variations may occur from rounding.

[c] Calculated as follows: (annual consumption in food, kg)/(most recently reported volume as a flavouring agent, kg)

thiamine occur in the plasma. Thiamine pyrophosphate functions as a Mg(II)-coordinated coenzyme involved in several biochemical processes in which transfer of an aldehyde group occurs. It also has functions in energy production and the biosynthesis of lipids and acetylcholine (Lehninger, 1987).

A thiazole derivative of possible clinical importance, 1-(2,6-difluorophenyl)-1*H*,3*H*-thiazolo[3,4-*a*]benzimidazole, was rapidly absorbed when administered orally at a dose of 50 mg/kg bw to male $CD2F_1$ mice, and was detected in the plasma up to 30 min after dosing; a single elimination phase with a half-life of 6 min was noted. Of the administered dose, only 0.4% was recovered in the urine as the intact compound. The major and minor urinary metabolites were the thiazole ring sulfoxide and sulfone, respectively (El Dareer et al., 1993).

Four compounds (Nos 1050–1053) in the group are derivatives of thiophene. In a study in male Sprague-Dawley rats, thiophene tritiated at the C-1 and C-4 positions was injected intraperitoneally at a dose of 200 mg/kg bw. About 31% of the radioactivity was excreted between 0 and 15 h, while 4% was excreted between 15 and 50 h. More than 94% of the urinary radiolabel was accounted for by a single metabolite, the 2-mercapturic acid derivative of 1,4-dihydrothiophene-*S*-oxide (Dansette et al., 1992).

Excretion of a related thiophene derivative, 3-benzothiophene, was examined in male Sprague-Dawley rats. After an intraperitoneal injection of 30 mg/kg bw of [^{14}C-keto]3-benzoylthiophene, 20% of the radiolabel was recovered in the urine within 24 h. About 15% of the urinary radiolabel could be accounted for by a mixture of two diastereoisomers of the mercapturic acid conjugate of 4,5-dihydro-3-benzolythiophene (Valadon et al., 1996).

(b) Biotransformation

(i) Metabolism of thiazole and its derivatives (Nos 1030–1044 and 1054–1059)

Thiazole is a five-membered heterocycle containing sulfur and nitrogen atoms in the 1- and 3-ring positions. Derivatives of thiazole in this group of flavouring substances are substituted in various combinations at the 2-, 4- or 5-position.

Two primary metabolic routes are potentially available to unsubstituted thiazole (No. 1032). The first is oxidation of the sulfur to form the corresponding sulfoxide and sulfone, and this route is the most favoured. The second, electrophilic attack at ring carbons (usually C-5), is unlikely because thiazoles do not readily undergo electrophilic substitution reactions (Rance, 1989). Addition of alkyl and acyl substituents to the thiazole ring increases the potential metabolic routes available. Alkyl- and acyl-substituted thiazole derivatives (Nos 1033–1039, 1041–1044 and 1054–1059) are primarily metabolized via side-chain oxidation and ring S- and N-oxidation. The major metabolites are readily excreted in the urine either free or as glutathione conjugates (see Figure 1). A minor metabolic pathway (< 0.5%) is ring cleavage followed by C-oxidation to yield α-diketone and thioamide fragments.

(ii) Metabolism of thiamine hydrochloride (No. 1030) and a metabolite (4-methyl-5-thiazole ethanol) (No. 1031)

Thiamine is metabolized via two major pathways. In one, thiamine is cleaved to yield 4-methyl-5-thiazoleethanol (No. 1031) and 2-methyl-4-amino-5-hydroxymethyl-

Figure 1. Metabolism of alkylthiazoles

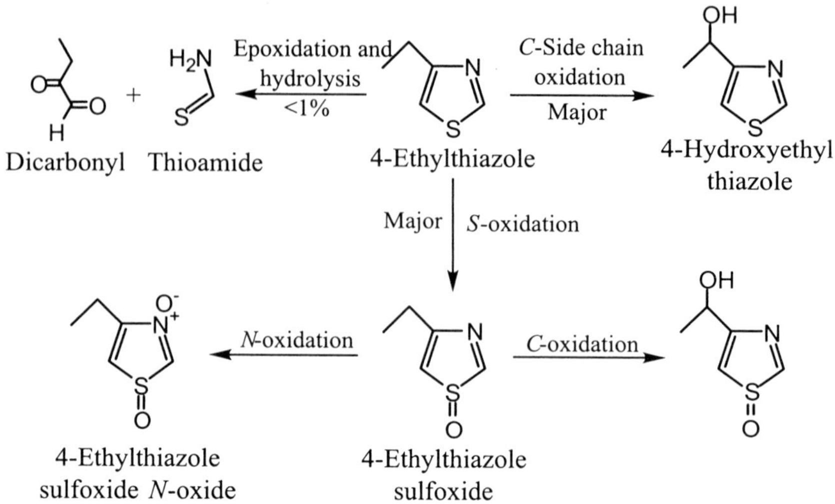

pyrimidine. The thiazole and pyrimidine fragments are further oxidized to yield 4-methylthiazole-5-acetic acid and the 5-pyrimidine carboxylic acid derivative, respectively, which, together with thiamine, are excreted in the urine (Tietz, 1986). In other pathways, thiamine is converted to 2-methyl-4-amino-5-formylamino-pyrimidine and thiamine acetic acid (Hanley & Fenwick, 1989) (see Figure 2).

(iii) Metabolism of benzothiazole (No. 1040)

The fused-ring thiazole derivative, benzothiazole, is metabolized in guinea-pigs primarily by thiazole ring cleavage. A dose of 30 mg/kg bw given by intraperitoneal injection was metabolized to the free and conjugated forms of *ortho*-aminophenyl methyl sulfide, *ortho*-aminophenyl methyl sulfoxide and *ortho*-aminophenyl methyl sulfone. Small amounts of the *N*-hydroxy derivative of the sulfoxide and sulfone were also detected (Wilson et al., 1991).

(iv) Metabolism of related (non-flavouring) thiazole derivatives

The therapeutic agent chlormethiazole is an alkyl- and chloroalkyl-substituted thiazole derivative [5-(2-chloroethyl)-4-methylythiazole] that is metabolized by oxidation of the alkyl and chloroalkyl substituents and by S- and N-oxidation. In volunteers treated orally, chlormethiazole was metabolized via side-chain oxidation of the ethyl C-1 position. The major urinary metabolite is 5-(acetic acid)-4-methylthiazole (see Figure 3). Methyl oxidation at C-4 yields 5-(hydroxyethyl)-4-thiazolecarboxylic acid lactone. The ring sulfoxide, the ring N-oxide and the combined ring S- and N-oxides of the side-chain were also found in urine. This was the first reported instance of sulfoxidation of a thiazole sulfur and oxidation of two different heteroatoms in the same heterocyclic aromatic ring in vivo (Offen et al., 1985).

Figure 2. Metabolism of thiamine and 4-methyl-5-thiazoleethanol

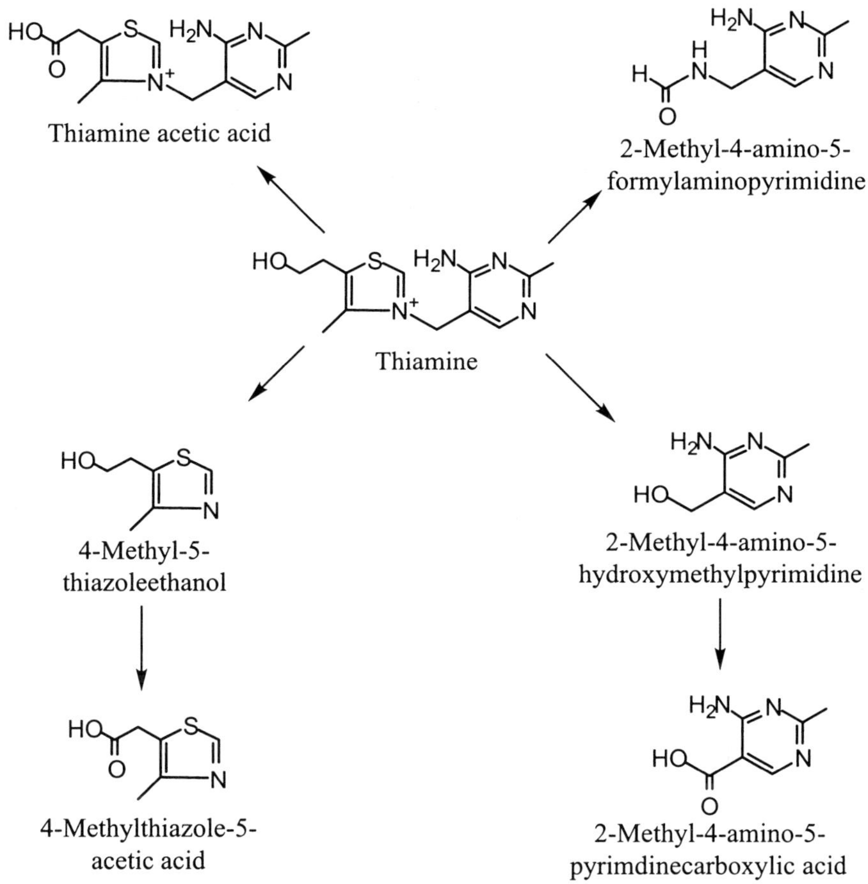

Chlormethiazole has also been reported to undergo nucleophilic attack at C_2 (Grupe & Spiteller, 1982).

Oral administration of the thiazole derivative, 4-*tert*-butyl-2-methylthiazole, to mice resulted in limited ring C-oxidation. Of the administered dose, 0.25% was excreted as ring fragmentation products in the urine within 24 h. These products included an α-dicarbonyl metabolite (3,3-dimethyl-2-oxobutanal) and a thioamide metabolite (thioacetamide). On the basis of the structure of the dicarbonyl fragment, it was postulated that the 4,5-double bond of the thiazole ring undergoes epoxidation, then hydrolysis to yield 4,5-diol. The diols then undergo hydrolytic cleavage to yield the corresponding carbonyl derivatives and thioacetamide (Mizutani et al., 1994).

Figure 3. Metabolism of chlormethiazole

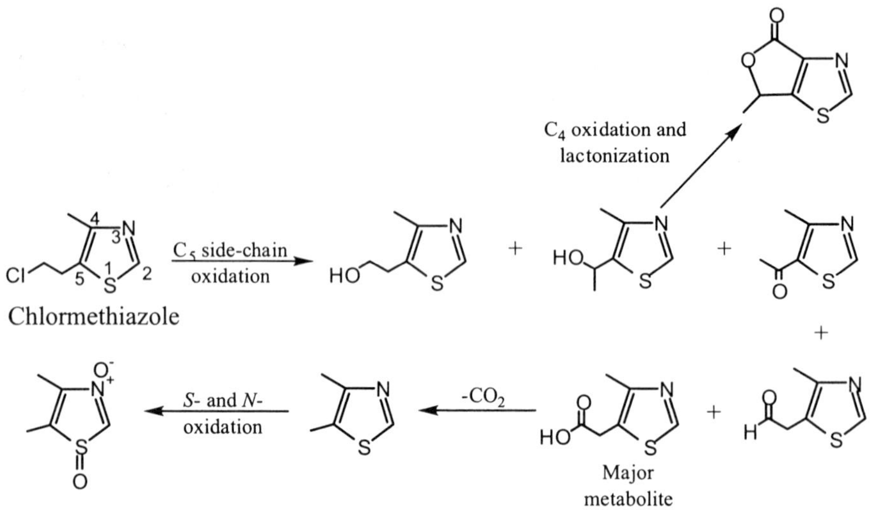

The thioamide metabolite is postulated to be responsible for the nephrotoxicity of 4-*tert*-butyl-2-methylthiazole (Mizutani et al., 1993).

(v) Metabolism of thiazoline derivatives (Nos 1045, 1058 and 1059)

Thiazoline (containing a partially reduced thiazole ring) is a five-membered, non-aromatic hererocycle containing sulfur and nitrogen atoms in the 1- and 3-ring positions. Derivatives of thiazoline in this group of flavouring substances are substituted in the 2-, 4- and 5-positions. Substances in this group are expected to be metabolized primarily by oxidation of the ring sulfur.

(vi) Metabolism of dithiazine derivatives (Nos 1046–1049)

Dithiazine is a six-membered, non-aromatic heterocycle containing two sulfur atoms (at the 1- and 3-positions) and one nitrogen (at the 5-position). The derivatives of diathiazine in this group of flavouring substances are substituted at the 2-, 4- and 6-positions.

No specific information was available on the metabolism of the dithiazine derivatives, but it is expected to be similar to that of thiazole derivatives. Because of the presence of alkyl substituents, metabolism is expected to be primarily via side-chain oxidation and ring S- and N-oxidation.

(vii) Metabolism of thiophene derivatives (Nos 1050–1053)

Thiophene is a five-membered aromatic heterocycle containing a sulfur atom in the 1-position. Derivatives of thiophene in this group of flavouring substances are substituted in the 1-, 2- and 4- positions.

Thiophene and its derivatives follow a similar metabolic pathway to that of thiazole and its derivatives, namely, S-oxidation and side-chain C-oxidation. In a study in Sprague-Dawley rats given an intraperitoneal injection of radiolabelled thiophene at a dose of 200 mg/kg bw, 94% of the radiolabel in the urine was accounted for by the 2-mercapturic acid derivative of 1,4-dihydrothiophene-*S*-oxide. The authors concluded that it was formed by S-oxidation, yielding the thiophene sulfoxide, which underwent further conjugation with glutathione at position 2 (Dansette et al., 1992).

Ring-substituted thiophenes also undergo S-oxidation and glutathione conjugation (see Figure 4). In a study in which radiolabelled 3-benzoylthiophene was injected intraperitoneally into Sprague-Dawley rats, 15% of the urinary radiolabel was accounted for by a mixture of two diastereomers of the mercapturic acid conjugate of 4,5-dihydro-3-benzoylthiophene. The N-acetylcysteine residue (mercapturate) is bonded to the 4-position of the thiophene ring. Experiments with rat liver microsomes confirmed that thiophene sulfoxide is a reactive intermediate in the conversion of 3-benzothiophene to the dihydromercapturic acid metabolite (Valadon et al., 1996).

2.3.2 Toxicological studies

(a) Acute toxicity

Oral LD_{50} values have been reported for 18 of the 30 substances in this group. With the exception of thiamine hydrochloride, the values in rats ranged from 460 to 1400 mg/kg bw (Sprince et al., 1974; Posternak et al., 1975; Babish, 1977; Mondino, 1981; Moreno, 1981; Piccirillo, 1982a,b; Reddy & Mayhew, 1986). The LD_{50} of thiamine hydrochloride in rats was 3700 mg/kg bw. The LD_{50} values in mice were in the range 400–4800 mg/kg bw (Calvery & Nelson, 1944; Oser, 1970; Shellenberger, 1970; Babish, 1978; Moran & Easterday, 1980).

Figure 4. Metabolism of alkylthiophenes

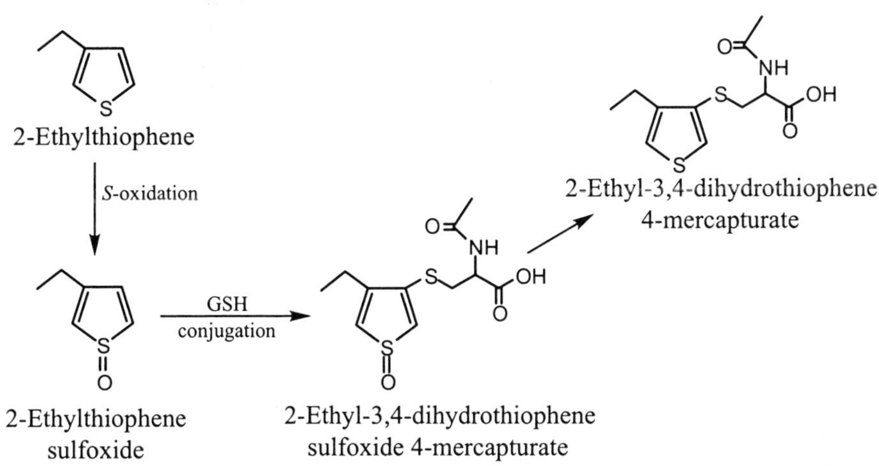

(b) Short-term studies of toxicity

The results of short-term studies of toxicity with substances in this group are summarized in Table 4.

(i) Thiamine hydrochloride (No. 1030) and 4-methyl-5-thiazoleethanol (No. 1031)

In a 90-day feeding study, groups of 15 FDRL strain rats of each sex were maintained on a diet which provided a daily intake of 5 g/kg bw of an imitation meat flavour containing thiamine hydrochloride (No. 1030) at a concentration of 13 mg/kg. Analysis of the meat flavour revealed that 40–50% of the thiamine hydrochloride had degraded to 4-methyl-5-thiazoleethanol (No. 1031). Thus, the daily intake of thiamine hydrochloride was 33–44 mg/kg bw, and the daily intake of 4-methyl-5-thiazoleethanol was 11–14 mg/kg bw. The animals were observed daily for behaviour, appearance and survival. Body weights and food intake were reported weekly. At 6 and 12 weeks, the following clinical parameters were measured: blood haemoglobin and haematocrit, total and differential leukocyte counts, blood glucose and urea nitrogen and urinary pH, albumin and glucose. All animals were necropsied at termination of the study. A range of tissues was examined microscopically.

Daily observations revealed no treatment-related effects, and weekly measurement of body weights and food consumption revealed no significant differences between test and control animals. Haematological examinations, blood chemical determinations and urine analysis performed during weeks 6 and 12 showed normal values. The gross appearance was unremarkable, and liver and kidney weights were normal. Histopathological examination did not reveal any evidence of treatment-related effects. The increase in tubular dilatation observed in the kidneys of 2/15 males and 4/15 females was considered to be the result of a high intake of protein and salt and not treatment-related. The NOEL was 36 mg/kg bw per day (median value) for thiamine hydrochloride and 12 mg/kg bw per day (median value) for 4-methyl-5-thiazoleethanol (Oser, 1964).

(ii) 2,4-Dimethyl-5-vinylthiazole (No. 1039)

In a 90-day study, groups of 16 Charles River CD rats of each sex were maintained on diets containing 2,4-dimethyl-5-vinylthiazole at concentrations adjusted to provide an intake of approximately 0.92 mg/kg bw per day for males and 1.0 mg/kg bw per day for females. Clinical observations were recorded daily, and food consumption and body weights were determined weekly. During weeks 7 and 13 of the study, haematological and clinical chemical (blood urea) determinations were conducted on 50% of the animals. At the end of the study, all animals were necropsied, and the liver and kidneys were weighed. A wide range of tissues and organs from each animal was preserved, and major organs and tissues were examined histologically.

There were no clinical signs of toxicity, and body weights and food consumption were normal. There were no treatment-related changes in clinical chemical parameters. Organ weights were normal, and histopathological examination did not reveal any treatment-related changes. The NOEL was 0.92 mg/kg bw per day (Posternak et al., 1969).

Table 4. Results of short-term studies of the toxicity of sulfur-containing heterocyclic compounds administered in the diet

No.	Substance	Species, sex	No. test groups[a]/ no. per group[b]	Duration (days)	NOEL (mg/kg bw per day)	Reference
1030	Thiamine hydrochloride	Rat; M/F	1/30	90	33–40	Oser (1964)
1031	4-Methyl-5-thiazole ethanol	Rat; M/F	1/30	90	11-14	Oser (1964)
1039	2,4-Dimethyl-5-vinylthiazole	Rat; M/F	1/32	90	0.92 (M) 1.0 (F)	Posternak et al. (1969)
1040	Benzothiazole	Rat; M/F	1/30	90	5.1	Morgareidge (1971)
1041	2-Acetylthiazole	Rat; M/F	1/30	90	50	Wheldon et al. (1970)
	Mixture of 2-isobutyl-4,6-dimethyl and 4-isobutyl-2,6-dimethyldihydro-1,3,5-dithaizole	Rat; M/F	1/5	14	11	Rush (1989a,b)
	Mxture of 2-isopropyl-4,6-dimethyl and 4-isopropyl-2,6-dimethyldihydro-1,3,5-dithaizole	Rat, M/F	1/5	14	11	Rush (1989a,b)
1053	2-Thienyldisulfide	Rat; M/F	1/30	90	9300	Morgareidge & Oser (1970)
1055	2,4-Dimethyl-5-acetylthiazole	Rat; M/F	1/46	90	25 (M) 24 (F)	Shellenberger (1971)
1057	2-Methyl-5-methoxythiazole	Rat; M/F	1/32	90 90	8.8 (M) 8.6 (F)	Posternak et al. (1975)
1059	2-(2-Butyl)-4,5-dimethyl-3-thiazoline	Rat; M/F	1/30	90	1.2 (M) 1.3 (F)	Babish (1978)

[a] Does not include control animals
[b] Includes both male and female animals

(iii) Benzothiazole (No. 1040)

In a 90-day study, groups of 15 FDRL rats of each sex were maintained on a diet which provided a daily intake of benzothiazole (No. 1040) at 5.1 mg/kg bw. All animals were observed daily for behaviour, appearance and survival; body weights and food intake were recorded weekly. At weeks 6 and 12, the following clinical parameters were measured: blood haemoglobin and erythrocyte volume fraction, total and differential leukocyte counts, blood glucose and urea nitrogen concentrations and urinary pH, albumin and glucose. All animals were necropsied at the termination of the study, and a range of tissues was examined microscopically.

Daily observations reveaed no treatment-related effects, and weekly measurements of body weights and food consumption revealed no significant differences between test and control animals. Haematological, blood chemical and urine analyses showed normal values. Gross appearance was unremarkable, and the liver and kidney weights were normal. Histopathological examination showed no evidence of treatment-related effects. The NOEL was 5.1 mg/kg bw per day (Morgareidge, 1971).

(iv) 2-Acetylthiazole (No. 1041)

In a 28-day study, groups of five weanling Cfy Wistar male rats were maintained on diets containing 0, 5000 or 10 000 ppm of 2-acetylthiazole, calculated to provide an average daily intake of 0, 250 or 500 mg/kg bw per day, for 4 weeks. The high dose was reduced to 2500 ppm (125 mg/kg bw per day) on days 4–10. Decreased body weights were found at both the low dose (87%) and the high dose (78%), and slightly increased food consumption and reduced efficiency of food use were reported. At autopsy, no treatment-related changes were observed (Wheldon et al., 1970).

In a 90-day study, groups of 15 male Cfy Wistar rats were maintained on diets containing 0, 100, 1000 or 10 000 ppm of 2-acetylthiazole, equivalent to 0, 5, 50 and 500 mg/kg bw per day, for 13 weeks. The highest dose was maintained for 6 weeks and then raised to 1000 mg/kg bw per day for 7 weeks. Clinical signs were monitored daily, and body weights and food consumption were monitored weekly. Haematological parameters were measured weekly. At the end of the study, all animals were necropsied; the weights of the adrenals, heart, kidneys, liver, lungs, spleen, testes, and thyroid were recorded; and a range of tissues was taken for histopathology.

Reduced body weight and food intake were seen at doses of 50 mg/kg bw per day and above, and decreased food conversion efficiency at the highest dose. Haematological parameters were similar in treated and untreated animals. At necropsy, the relative weights of the liver, adrenals and thyroid were increased in animals at the highest dose. Gross examination showed occasional pallor and mottling of kidneys in some rats, but these findings were not considered significant. Histopathological examination of 13 tissues from five animals at the highest dose revealed minimal fatty changes in the liver. Histopathological examination of the livers of five animals at the low and intermediate doses showed no evidence of treatment-related changes. The NOEL was 50 mg/kg bw per day (Wheldon et al., 1970).

(v) 2- and 4-Isobutyl (No. 1046) and 2- and 4-isopropyl (No. 1047) derivatives of 2,6-dimethyldihydro-1,3,5-dithiazine

In a 14-day study, groups of five male and five female Sprague-Dawley rats were maintained on a diet containing either 2- and 4-isobutyl or 2- and 4-isopropyl derivatives of 2,6-dimethyldihydro-1,3,5-dithiazine. The diets were calculated to provide an average daily intake of 12 mg/kg bw of the isobutyl or isopropyl derivative for males and 11 mg/kg bw of isobutyl or isopropyl derivative for females. During the study, the rats were observed daily for clinical signs of toxicity, and body weight and food consumption were measured weekly. All animals were subjected to a complete gross necroscopy at termination (day 15). No deaths or treatment-related toxic effects were observed. Body-weight gain and food consumption were normal. Histopathological evaluation of the liver and kidneys did not reveal any treatment-related effects. The NOEL for both substances was 11 mg/kg per day (Rush, 1989a,b).

(vi) 5-Methyl-2-thiophenecarboxyaldehyde (No. 1050) and 3-acetyl-2,5-dimethylthiophene (No. 1051)

In a 14-day study, groups of five Fischer 344 rats of each sex were maintained on a diet containing either 5-methyl-2-thiophenecarboxyaldehyde (No. 1050) or 3-acetyl-2,5-dimethylthiophene (No. 1051) at concentrations calculated to provide an estimated average daily intake of 10 mg/kg bw for 14 days. Daily clinical examinations and weekly measurements of body weight and food consumption revealed decreased food consumption (7%, $p \leq 0.01$) by males on both substances during the first week. No significant decrease in food consumption was observed among females or males during the second week of the study. On the basis that similar changes were seen with other substances in this study and that the changes in body weight were not accompanied by other signs of toxicity, this was considered not to be a treatment-related effect. At day 14, decreased body-weight gain was observed in males and females given 5-methyl-2-thiophenecarboxyaldehyde (11% and 20%, respectively) and 3-acetyl-2,5-dimethylthiophene (17% and 11%, respectively). At necropsy, decreased absolute (11–14%) and relative liver weights (7–10%, $p \leq 0.05$) and decreased absolute kidney weights (8–11%, $p \leq 0.05$) were reported for treated males. No difference in absolute or relative liver or kidney weights was observed between control and test females. Macroscopic and histopathological examination did not reveal any gross or microscopic lesions that could be attributed to treatment. A NOEL could not be identified in this study (Gill & Van Miller, 1987).

(vii) 2-Thienyl disulfide (No. 1053)

In a 90-day study, groups of 15 FDRL rats of each sex were maintained on a diet containing 2-thienyl disulfide (No. 1053) which provided a daily intake of 0.29 mg/kg bw. All animals were observed daily for behaviour, appearance and survival; body weights and food intake were recorded weekly. At weeks 6 and 12, the following clinical parameters were measured: blood haemoglobin and erythrocyte volume fraction, total and differential leukocyte counts, blood glucose and urea nitrogen concentrations and urine pH, albumin and glucose. All animals were necropsied at termination of the study, and a range of tissues was examined microscopically.

Daily observations did not reveal any treatment-related effects, and weekly measurement of body weights and food consumption showed no significant differences between test and control animals. Haematological, blood chemical and urine analyses gave normal values. Gross appearance was unremarkable, and the liver and kidney weights were normal. Histopathological examination did not reveal any evidence of treatment-related effects. The NOEL was 0.29 mg/kg bw per day (Morgareidge & Oser, 1970).

(viii) 2,4-Dimethyl-5-acetylthiazole (No. 1055)

In a 90-day study, groups of 23 Sprague-Dawley rats of each sex were maintained on a diet containing 2,4-dimethyl-5-acetylthiazole at a concentration calculated to result in a daily intake of 24 mg/kg bw. Animals were examined daily for clinical appearance and survival and weekly forbody weight, food consumption and food use efficiency. Haematological, blood chemical and urine analyses were performed on eight animals of each sex at weeks 6 and 13. At necropsy, all animals were examined grossly, and the heart, kidneys, liver, spleen, testes and ovaries were weighed. Histopathology was carried out on 25 tissues from eight males and eight females and on the liver and kidneys of seven additional males and females.

There were no deaths or clinical signs of toxicity. Body weights and food consumption were normal, and there were no treatment-related changes in clinical parameters. At necropsy, there was an apparent increase in absolute liver and kidney weights in both males and females, which was considered to be due to unusually low average organ weights in the control group. Histopathology did not reveal any treatment-related changes in the liver or kidney or in any other organs. The NOEL was 24 mg/kg bw per day (Shellenberger, 1971).

(ix) 2-Methyl-5-methoxythiazole (No. 1057)

In a 90-day study, groups of 16 Charles River CD rats of each sex were maintained on a diet containing 2-methyl-5-methoxythiazole (No. 1057) at a concentration adjusted to provide an intake of approximately 8.8 mg/kg bw per day for males and 8.6 mg/kg bw per day for females. Clinical observations were recorded daily, and food consumption and body weights were determined weekly. During weeks 7 and 13 of the study, haematological and clinical chemical (blood urea) determinations were conducted on 50% of the animals. At the end of the study, all animals were necropsied, and the liver and kidneys were weighed. A wide range of tissues and organs from each animal was preserved, and major organs and tissues were examined histologically.

There were no clinical signs of toxicity, and body weights and food consumption were normal. No treatment-related changes were found in clinical chemical parameters. The organ weights were normal, and histopathological examination did not reveal any treatment-related changes. The NOEL was 8.6 mg/kg bw per day (Posternak et al., 1975).

(x) 2-(2-Butyl)-4,5-dimethyl-3-thiazoline (No. 1059)

In a 90-day study, groups of 15 FDRL rats of each sex were maintained on a diet containing 2-(2-butyl)-4,5-dimethyl-3-thiazoline (No. 1059), which provided a daily intake of 1.2 mg/kg bw for males and 1.3 mg/kg bw for females. All animals were observed daily for behaviour, appearance and survival; body weights and food intake were reported weekly. At 6 and 12 weeks, the following clinical parameters were measured: blood haemoglobin and erythrocyte volume fraction, total and differential leukocyte counts, blood glucose and urea nitrogen concentrations and urinary pH, albumin and glucose. All animals were necropsied at termination of the study.

Daily observations did not reveal any treatment-related effects, and weekly measurement of body weights and food consumption revealed no significant differences between test and control animals. Haematological, blood chemical and urine analyses revealed normal values. The weights of the liver and kidney were normal. The NOEL was 1.2 mg/kg bw per day (Babish, 1978).

(c) Genotoxicity

Four substances, thiazole (No. 1032), 4,5-dimethylthiazole (No. 1035), 4-methylthiazole (No. 1043) and 5-methyl-2-thiophenecarboxyaldehyde (No. 1050), in this group of flavouring substances were testedfor their ability to induce reverse mutation in *Salmonella typhimurium* strains TA98 and TA100 at doses of 4–100 μmol/plate. The purity of the chemicals was not stated. Positive results were obtained with thiazole in strain TA100 only at a minimum concentration of 4 μmol/plate; however, the mutagenicity that was observed in the absence of an exogenous metabolic activation system was less marked in the presence of such a system, indicating that thaizole does not undergo metabolic activation. The other substances gave uniformly negative results in both strains (Lee et al., 1994). No other tests have been reported.

(d) Other relevant studies

4-Methylthiazole (No. 1043)

In a study to determine the effect of glutathione on the metabolism of thiazoles and, in particular, 4-methylthiazole (No. 1043), groups of mice were given a single oral dose of 495 mg/kg bw alone or 24 h after pretreatment with DL-buthionine sulfoximine, a glutathione depletion agent. In those animals in which glutathione was depleted, treatment with 4-methylthiazole resulted in an increase in relative kidney weight and a significant increase in serum urea nitrogen. In animals treated with 4-methylthiazole alone at doses up to 1.0 mmol/kg bw, there was no effect on kidney weight or serum urea nitrogen concentration (Mizutani et al., 1992).

3. REFERENCES

Babish, J.G. (1977) Acute oral toxicity of 2-isopropyl-4-methylthiazole in rats. Unpublished report from Food and Drug Research Laboratories, Inc., Maspeth, New York, USA. Submitted to WHO by Flavor and Extract Manufacturers' Association of the United States.

Babish, J.G. (1978) 90-day feeding study of 2-(2-butyl)-4,5-dimethyl-3-thiazoline in rats. Unpublished report from Food and Drug Research Laboratories, Inc., Maspeth, New York, USA. Submitted to WHO by Flavor and Extract Manufacturers' Association of the United States.

Calvery, H.O. & Nelson, A.A. (1944) Acute toxicity study for 4-methyl-5-thiazoleethanol in mice. Unpublished report. Submitted to WHO by Flavor and Extract Manufacturers' Association of the United States.

Cramer, G.M. & Ford, R.A. (1978) Estimation of toxic hazard—A decision tree approach. *Food Cosmet. Toxicol.*, **16**, 255–276.

Dansette, P.M., Thang, D.C., El Amri, H. & Mansuy, D. (1992) Evidence for thiophene S-oxide as a primary reactive metabolite of thiophene in vivo: Formation of a dihydrothiophene sulphoxide mercapturic acid. *Biochem. Biophys. Res. Communications*, **186**, 1624–1630.

Decouvelaere, B., Gaillard, C., Kettler, J.P., Martin, S., Gires, P. & Gaillot, J. (1986) Biotransformation of 48740 RP. Unpublished report presented at the 10th European Drug Metabolism Workshop, Guildford, Surrey, England.

El Dareer, S.M., Tillery, K.F., Rose, L.M., Posey, C.F., Struck, R.F., Stiller, S.W. & Hill, D.L (1993) Metabolism and disposition of a thiazolobenzimidazole active against human immunodeficiency virus-1. *Drug Metab. Disposition*, **21**, 231–235.

Gill, M.W. & Van Miller, J.P. (1987) Fourteen-day dietary minimum toxicity (MTS) in albino rats. Unpublished report from Bushy Run Research Centre, Export, Pennsylvania, USA. Submitted to WHO by Flavor and Extract Manufacturers' Association of the United States.

Grupe, A. & Spiteller, G. (1982) Unexpected metabolites produced from clomethiazole. *J. Chromatogr.*, **230**, 335–344.

Hanley, A.B. & Fenwick, G.R. (1989) Naturally occuring sulphur compounds. In: Damani, L.A., ed., *Sulphur Containing Drugs and Related Organic Compounds*, Chichester: Ellis Horwood, Vol. 1A, pp. 41–44.

Hegsted, D.M., Chichester, C.O. & Darby, W.J., eds (1976) *Present Knowledge in Nutrition*, 4th Ed., Washington DC, The Nutrition Foundation Inc.

Informatics, Inc. (1974) *Scientific Literature Reviews on Generally Recognized as Safe (GRAS) Ingredients—Thiamine*, Washington DC: US Food & Drug Administration.

International Organization of the Flavor Industry (1995) European usage data. Unpublished report. Submitted to WHO by Flavor and Extract Manufacturers' Association of the United States.

Lee, H., Bian, S.S. & Chen, Y.L. (1994) Genotoxicity of 1,3-dithiane and 1,4-dithiane in the CHO/SCE assay and in the Salmonella/microsomal test. *Mutat. Res.*, **321**, 213–218.

Lehninger, A.L. (1987) *Principles of Biochemistry*, Worth Publishing Inc., pp. 252–254, 768.

Lucas, C.D., Putnam, J.M. & Hallagan, J.B. (1999) *Flavor and Extract Manufacturers' Association of the United States 1995 Poundage and Technical Effects Update Survey*. Washington DC: Flavor and Extract Manufacturers' Association of the United States.

Maarse, H., Visscher, C.A., Willemsens, L.C. & Boelens, M.H. (1999) *Volatile Components in Food—Qualitative and Quantitative Data*, Zeist, Centraal Instituut Voor Voedingsonderzioek TNO.

Mizutani, T., Yoshida, K. & Ito, K. (1992) Nephrotoxicity of thiazoles structurally related to thiabendazole in mice depleted of glutathione by treatment with bithionine sulfoximine. *Res. Communications Chem. Pathol. Pharmacol.*, **75**, 29–38.

Mizutani, T., Yoshida, K. & Kawazoe, S. (1993) Possible role of thioformamide as a proximate toxicant in the nephrotoxicity of thiabendazole and related thiazoles in the glutathione-depleted mice: Structure–toxicity and metabolism studies. *Chem. Res. Toxicol.*, **6**, 174–179.

Mizutani, T., Yoshida, K. & Kawazoe, S. (1994) Formation of toxic metabolites from thiabendazole and other thiazoles in mice. *Drug Metab. Disposition*, **22**, 750–755.

Mondino, A. (1981) Acute toxicity study on methyl-4-thiazol. Unpublished report from Instituto di Ricerche Biomedicine, Rome. Submitted to WHO by Flavor and Extract Manufacturers' Association of the United States.

Moran, E.J. & Easterday, O.D. (1980). Acute oral toxicity on selected flavor chemicals. *Food Chem. Toxicol.*, **3**, 249–258.

Moreno, O.M. (1981). Single dose oral toxicity/LD50 in rats. Unpublished report by MB Research Laboratories Inc., Spinerstown, Pennsylvania, USA. Submitted to WHO by Flavor and Extract Manufacturers' Association of the United States.

Morgareidge, K. (1971) 90-day feeding study with benzothiazole in rats. Unpublished report from Food and Drug Research Laboratories, Inc., Maspeth, New York, USA. Submitted to WHO by Flavor and Extract Manufacturers' Association of the United States.

Morgareidge, K. & Oser, B.L. (1970) 90-day feeding studies in rats with 2-thienyldisulfide. Unpublished report from Food and Drug Research Laboratories, Inc., Maspeth, New York, USA. Submitted to WHO by Flavor and Extract Manufacturers' Association of the United States.

Offen, C.P., Frearson, M.J., Wilson, K. & Burnett, D. (1985) 4,5-Dimethylthiazole-N-oxide-S-oxide: A metabolite of chloromethiazole in man. *Xenobiotica*, **15**, 503–511.

Oser, B.L. (1964) Subacute (90-day) feeding study with imitation meat flavor IT-2341 in rats. Unpublished report from Food and Drug Research Laboratories, Inc., Maspeth, New York, USA. Submitted to WHO by Flavor and Extract Manufacturers' Association of the United States.

Oser, B.L. (1970) The acute oral toxicity of ten compounds. Unpublished report from the Food and Drug Research Laboratories, Inc., Maspeth, New York, USA. Submitted to WHO by Flavor and Extract Manufacturers' Association of the United States.

Piccirillo, V.J. (1982a) Final report on acute oral toxicity (LD50) study in the rat with 4,5-dimethylthiazole. Unpublished report from Borriston Laboratories, Inc., Temple Hills, Maryland, USA. Submitted to WHO by Flavor and Extract Manufacturers' Association of the United States.

Piccirillo, V.J. (1982b) Acute oral toxicity (LD50) study in the rat with 2-ethoxythiazole. Unpublished report from Borriston Laboratories, Inc., Temple Hills, Maryland, USA. Submitted to WHO by Flavor and Extract Manufacturers' Association of the United States.

Posternak, J.M., Linder, A. & Vpodos, C.A. (1969) Summaries of toxicological data: Toxicological tests on flavouring matters. *Food Cosmet. Toxicol.*, **7**, 405–407.

Posternak, J.M., Dufour, J.J., Rogg, C. & Vodoz, C.A. (1975) Toxicology tests on flavouring matters. II. Pyrazines and other compounds. *Food Cosmet. Toxicol.*, **13**, 487–490.

Rance, D.J. (1989) Sulphur heterocycles. In: Damani, L.A., ed., *Sulfur-containing Drugs and Related Organic Compounds. Chemistry, Biochemistry and Toxicology*, Chichester, Ellis Horwood, Vol. 1, pp. 217–259.

Reddy, G. & Mayhew, D.A. (1986) Acute oral toxicity (LD50) study in rats with benzothiazole. *J. Am. Coll. Toxicol.*, **4**, 666.

Rush, R. (1989a) 14-day dietary toxicity study in rats with 2(4)-isobutyl-4(2),6-dimethyldihydro-1,3,5-dithiazine. Unpublished report from Springborn Laboratories, Inc., Spencerville, Ohio, USA. Submitted to WHO by Flavor and Extract Manufacturers' Association of the United States.

Rush, R. (1989b) 14-day dietary toxicity study in rats with 2(4)-isopropyl-4(2),6-dimethyldihydro-1,3,5-dithiazine. Unpublished report from Springborn Laboratories, Inc., Spencerville, Ohio, USA. Submitted to WHO by Flavor and Extract Manufacturers' Association of the United States.

Sprince, H., Parker, C.M., Smith, G.G. & Gonzales, L.J. (1974) Protection against acetaldehyde toxicity in the rat by L-cysteine, thiamin and 1,2-methylthiazolidine-4-carboxylic acid. *Agents Actions*, **4**, 125–130.

Shellenberger, T.E. (1970) Acute toxicological evaluations of chemicals with mice, Unpublished report from the Gulf South Research Institute, Iszeria, Louisiana, USA. Submitted to WHO by Flavor and Extract Manufacturers' Association of the United States.

Shellenberger, T.E. (1971) Subacute toxicity evaluation of 2,4-dimethyl-5-acetylthiazole with rats. Unpublished report from the Gulf South Research Institute, Iszeria, Louisiana, USA. Submitted to WHO by Flavor and Extract Manufacturers' Association of the United States.

Stofberg, J. & Grundschober, F. (1987) Consumption ratio and food predominance of flavoring materials. *Perfumer Flavorist*, **12**, 27.

Stofberg, J. & Kirschman, J.C. (1985) The consumption ratio of flavouring materials: A mechanism for setting priorities for safety evaluations. *Food Chem. Toxicol.*, **23**, 857–860.

Tietz, N.W., ed. (1986) B-vitamins. In: *Textbook of Clinical Chemistry*, New York, W.B. Saunders Co., pp. 940–945.

Valadon, P., Dansette, P.M., Girault, J-P., Amar, C. & Mansuy, D. (1996) Thiophene sulfoxides as reactive metabolites: Formation upon microsomal oxidation of a 3-aroylthiophenes and fate in the presence of nucleophiles *in vitro* and *in vivo*. *Chem. Res. Toxicol.*, **9**, 1403–1413.

Wilson, J.D. (1982) Disorder of vitamins—Deficiency, excess and errors of metabolism. In: *Harrison's Principles of Internal Medicine*, 10th ed., New York, McGraw-Hill, pp. 461–470.

Wilson, K., Chissick, A.M., Fowler, A.M., Frearson, F.J., Gittins, M. & Swinbourne, F.J. (1991) Metabolism of benzothiazole. 1. Identification of ring-cleavage products. *Xenobiotica*, **21**, 1179–1183.

Wheldon, G.H., Amyes, S.J., Street, A.E., Hague, P.H. & Mawdesley-Thomas, L.E. (1970) Toxicity of Wa4295, Sa 927, SH3048, and Wa 3328 in dietary administration to rats over a period of 13 weeks. Unpublished report from Huntingdon Research Centre, United Kingdom. Submitted to WHO by Flavor and Extract Manufacturers' Association of the United States.

SULFUR-SUBSTITUTED FURAN DERIVATIVES

First draft prepared by

Dr A. Mattia[1] and Professor A.G. Renwick[2]

[1]Division of Petition Review, Office of Food Additive Safety, Center for Food Safety and Applied Nutrition, Food and Drug Administration, College Park, Maryland, USA; and [2]Clinical Pharmacology Group, University of Southampton, Southampton, England

1. EVALUATION

1.1 Introduction

The Committee evaluated a group of 33 flavouring agents that includes thiofuran and thiofurfuryl derivatives (see Table 1), using the Procedure for the Safety Evaluation of Flavouring Agents (see Figure 1). The Committee has not previously evaluated any member of the group. Eighteen of the 33 flavouring agents in this group have been reported to occur naturally in foods (Stofberg & Grundschober, 1987; Maarse et al., 1999). Thiofurfuryl and thiofuran derivatives have been detected mainly in coffee and cooked meats.

1.2 Estimated daily intake

The total annual volume of production of the 33 sulfur-substituted furan derivatives is approximately 370 kg in Europe (international Organization of the Falvor Industry, 1995) and 210 kg in the USA (Lucas et al., 1999). Approximately 65% of the total annual volume in Europe and 40% of that in the USA arises from use of one

Table 1. Summary of results of safety evaluations of sulfur-substituted furan derivatives used as flavouring agents[a]

Flavouring agent	No.	CAS No. and structure	*Step B3*[b] Does intake exceed threshold for human intake?	*Step B4* Adequate margin of safety for the flavouring agent or related chemical?	Conclusion based on current intake
Structural class II					
2-Methyl-3-furanthiol	1060	28588-74-1 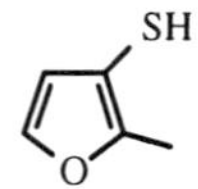	No Europe: 0.6 USA: 0.9	Yes. The NOEL of 5 mg/kg bw per day is > 100 000 times the estimated intake of 2-methyl-3-furanthiol when used as a flavouring agent.	No safety concern
Ethanoic acid, S-(2-methyl-3-furanyl) ester	1069	55764-25-5 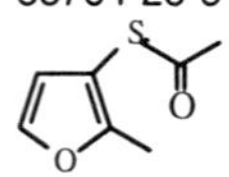	No Europe: 0.01 USA: 0.07	Yes. The NOEL of 5 mg/kg bw per day for the related substance No. 1060 is > 1 million times the estimated intake of *S*-ethanoic acid when used as a flavouring agent.	No safety concern
2-Methyl-3-(methylthio)furan	1061	63012-97-5 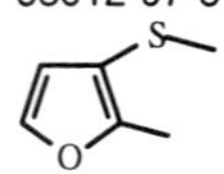	No Europe: 1 USA: 0.1	Yes. The NOEL of 1.3 mg/kg bw per day for the related substance No. 1077 is > 10 000 times the estimated intake of 2-methyl-3-(methylthio)furan when used as a flavouring agent.	No safety concern
2-Methyl-5-(methylthio)furan	1062	13678-59-6 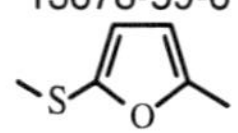	No Europe: 1 USA: 0.02	Yes. NOEL of 1.3 mg/kg bw per day for the related substance No. 1077 is > 10 000 times the estimated intake of 2-methyl-5-(methylthio)furan when used as a flavouring agent.	No safety concern
Methyl 2-methyl-3-furyl disulfide	1064	65505-17-1 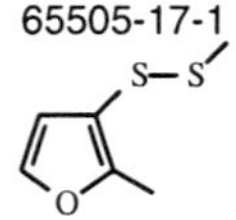	No Europe: 0.9 USA: 0.05	Yes. The NOEL of 1.2 mg/kg bw per day is > 10 000 times the estimated intake of methyl 2-methyl-3-furyl disulfide when used as a flavouring agent.	No safety concern

Table 1 *(continued)*

Flavouring agent	No.	CAS No. and structure	*Step B3*[a,b] Does intake exceed threshold for human intake?	*Step B4* Adequate margin of safety for the flavouring agent or related chemical?	Conclusion based on current intake
Propyl 2-methyl-3-furyl disulfide	1065	61197-09-9 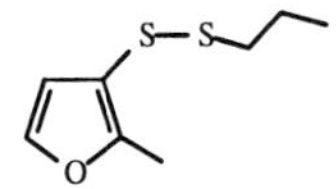	No Europe: ND USA: 0.7	Yes. The NOEL of 1.2 mg/kg bw per day for the related substance (No. 1064 is > 100 000 times the estimated intake of propyl 2-methyl-3-furyl disulfide when used as a flavouring agent.	No safety concern
Bis(2-methyl-3-furyl) disulfide	1066	28588-75-2 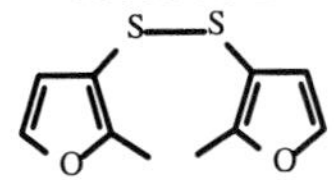	No Europe: 0.3 USA: 0.7	Yes. The NOEL of 0.45 mg/kg bw per day is > 10 000 times the estimated intake of bis(2-methyl-3-furyl) disulfide when used as a flavouring agent.	No safety concern
2,5-Dimethyl-3-furanthiol	1063	55764-23-3 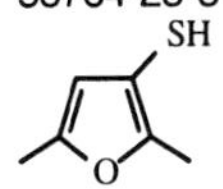	No Europe: 0.03 USA: 0.7	Yes. The NOEL of 5 mg/kg bw per day for the related substance No. 1060 is > 100 000 times the estimated intake of 2,5-dimethyl-3-furanthiol when used as a flavouring agent.	No safety concern
Furfuryl mercaptan	1072	98-02-2 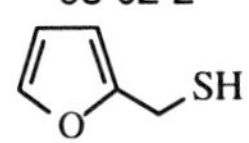	No Europe: 40 USA:11	Yes. The NOEL of 3 mg/kg bw per day is > 1000 times the estimated intake of furfuryl mercaptan when used as a flavouring agent.	No safety concern
Furfuryl thioacetate	1074	13678-68-7 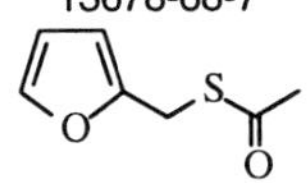	No Europe: 0.5 USA: 0.05	Yes. The NOEL of 0.83 mg/kg bw per day is > 100 000 times the estimated intake of furfuryl thioacetate when used as a flavouring agent.	No safety concern

Table 1 (continued)

Flavouring agent	No.	CAS No. and structure	*Step B3*[a,b] Does intake exceed threshold for human intake?	*Step B4* Adequate margin of safety for the flavouring agent or related chemical?	Conclusion based on current intake
Furfuryl methyl sulfide	1076	1438-91-1 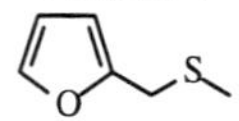	No Europe: 1 USA: 0.1	Yes. The NOEL of 1.3 mg/kg bw per day for the related substance No. 1077 is > 10 000 times the estimated intake of furfuryl methyl sulfide when used as a flavouring agent.	No safety concern
Methyl furfuryl disulfide	1078	57500-00-2 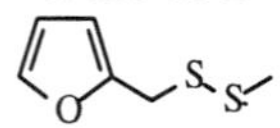	No Europe: 1 USA: 0.04	Yes. The NOEL of 1.2 mg/kg bw per day for the related substance No. 1064 is > 10 000 times the estimated intake of methyl furfuryl disulfide when used as a flavouring agent.	No safety concern
Furfuryl propyl disulfide	1079	2527366-36-0 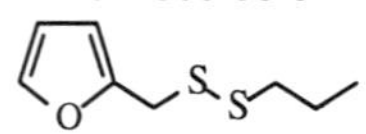	No Europe: ND USA: 3	Yes. The NOEL of 1.2 mg/kg bw per day for the related substance No. 1064 is > 10 000 times the estimated intake of furfuryl propyl disulfide when used as a flavouring agent.	No safety concern
2,2´-(Dithiodimethylene) difuran	1081	4437-20-1 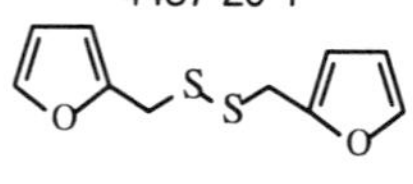	No Europe: 4 USA: 0.7	Yes. The NOEL of 3 mg/kg bw per day for the related substance No. 1072 is > 10 000 times the estimated intake of 2,2´-(dithiodimethylene) difuran when used as a flavouring agent.	No safety concern
Methyl thiofuroate	1083	13679-61-3 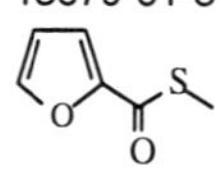	No Europe: 0.4 USA: 0.1	Yes. The flavouring agent is expected to hydrolyse to methanethiol and furoic acid. The NOEL of 0.74 mg/kg bw per day for the related substance No. 1071, which also hydrolyses to furoic acid, is > 100 000 times the estimated intake of methyl thiofuroate when used as a flavouring agent.	No safety concern

Table 1 *(continued)*

Flavouring agent	No.	CAS No. and structure	*Step B3*[a,b] Does intake exceed threshold for human intake?	*Step B4* Adequate margin of safety for the flavouring agent or related chemical?	Conclusion based on current intake
4-[(2-Methyl-3-furyl)thio]-2-pentanone	1084	180031-78-1 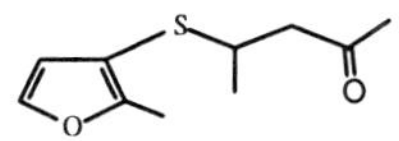	No Europe: ND USA: 0.6	Yes. The NOEL of 3.8 mg/kg bw per day for the related substance No. 1085 is > 100 000 times, respectively, the estimated intake of 4-[(2-methyl-3-furyl)thio]-2-pentanone when used as a flavouring agent.	No safety concern
2-Methyl-3-tetrahydrofuranthiol	1090	57124-87-5 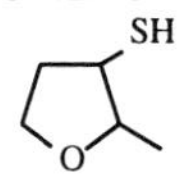	No Europe: 4 USA: 0.7	Yes. The NOEL of 8.3 mg/kg bw per day for the related substance No. 1089 is > 100 000 times the estimated intake of 2-methyl-3-tetrahydrofuranthiol when used as a flavouring agent.	No safety concern
2,2´-(Thiodimethylene) difuran	1080	13678-67-6 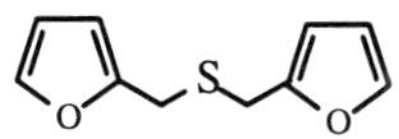	No Europe: 0.9 USA: 0.005	Yes. The NOEL of 10 mg/kg bw per day is > 100 000 times the estimated intake of 2,2´-(thiodimethylene) when used as a flavouring agent.	No safety concern
Structural class III					
Bis(2,5-dimethyl-3-furyl)disulfide	1067	28588-73-0 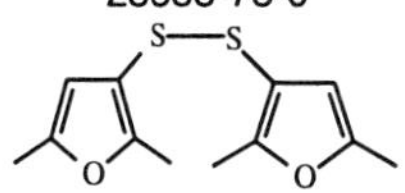	No Europe: 0.01 USA: 0.7	Yes. The NOEL of 0.45 mg/kg bw per day for the related substance No. 1066 is > 10 000 times the estimqted intake of bis(2,5-dimethyl-3-furyl)disulfide when used as a flavouring agent.	No safety concern
Bis(2-methyl-3-furyl) tetrasulfide	1068	28588-76-3 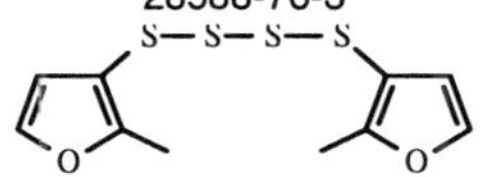	No Europe: ND USA: 0.7	Yes. The NOEL of 0.56 mg/kg bw per day is > 10 000 times the estimated intake of bis(2-methyl-3-furyl) tetrasulfide when used as a flavouring agent.	No safety concern

Table 1 *(continued)*

Flavouring agent	No.	CAS No. and structure	*Step B3*[a,b] Does intake exceed threshold for human intake?	*Step B4* Adequate margin of safety for the flavouring agent or related chemical?	Conclusion based on current intake
2,5-Dimethyl-3-furan thioisovalerate	1070	55764-28-8 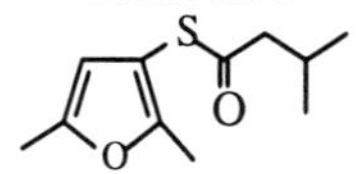	No Europe: ND USA: 0.7	Yes. The NOEL of 0.73 mg/kg bw per day is > 10 000 times the estimated intake of 2,5-dimethyl-3-furan thioisovalerate when used as a flavouring agent.	No safety concern
S-2,5-Dimethyl-3-thiofuroyl furan	1071	65505-16-0 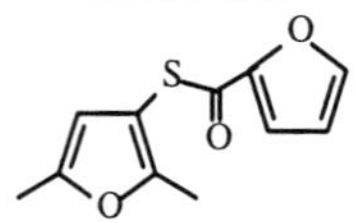	No Europe: ND USA: 0.01	Yes. The NOEL of 0.74 mg/kg bw per day is > 1 million times the estimated intake of *S*-2,5-dimethyl-3-thiofuroyl furan when used as a flavouring agent.	No safety concern
Furfuryl thioformate	1073	59020-90-5 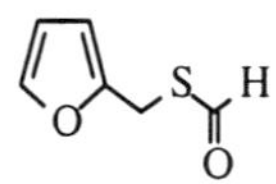	No Europe: 2 USA: 0.02	Yes. The NOELs of 3 and 0.83 mg/kg bw per day for the related substances Nos 1072 and 1074 are > 1 million and > 10 000 times, respectively, the estimated intake of furfuryl thioformate when used as a flavouring agent.	No safety concern
Furfuryl thiopropionate	1075	59020-85-8 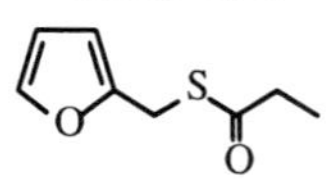	No Europe: 0.01 USA: 0.005	Yes. The NOELs of 3 and 0.83 mg/kg bw per day for the related substances Nos 1072 and 1074 are > 1 million times the estimated intake of furfuryl thiopropionate when used as a flavouring agent.	No safety concern
Furfuryl isopropyl sulfide	1077	1883-78-9 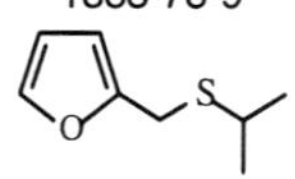	No Europe: 0.001 USA: 0.1	Yes. The NOEL of 1.3 mg/kg bw per day is > 100 000 times the estimated intake of furfuryl isopropyl sulfide when used as a flavouring agent.	No safety concern

Table 1
(continued)

Flavouring agent	No.	CAS No. and structure	*Step B3*[a,b] Does intake exceed threshold for human intake?	*Step B4* Adequate margin of safety for the flavouring agent or related chemical?	Conclusion based on current intake
2-Methyl-3-, 5- or 6-(furfurylthio)pyrazine	1082	65530-53-2 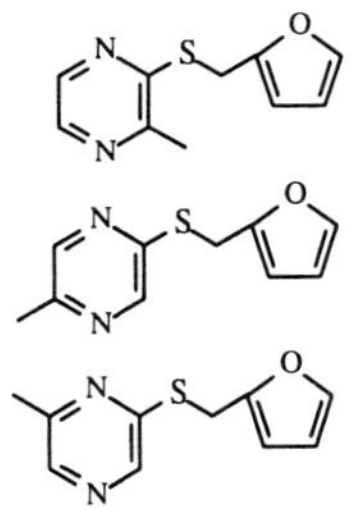	No Europe: 0.4 USA: 0.7	Yes. The NOEL of 1.7 mg/kg bw per day is > 100 000 times the estimated intake of 2-methyl-3-, 5- or 6-(furfurylthio)pyrazine when used as a flavouring agent.	No safety concern
3-[(2-Methyl-3-furyl)thio]-4-heptanone	1085	61295-41-8 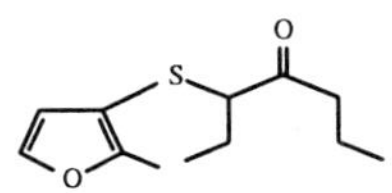	No Europe: ND USA: 0.7	Yes. The NOEL of 3.8 mg/kg bw per day is > 100 000 times the estimated intake of 3-[(2-methyl-3-furyl)thio]-4-heptanone when used as a flavouring agent.	No safety concern
2,6-Dimethyl-3-[(2-methyl-3-furyl)thio]-4-heptanone	1086	61295-51-0 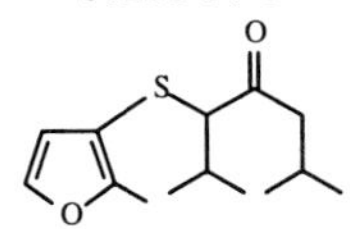	No Europe: ND USA: 0.7	Yes. The NOEL of 3.8 mg/kg bw per day for the related substance No. 1085 is > 100 000 times the estimated intake of 2,6-dimethyl-3-[(2-methyl-3-furyl)thio]-4-heptanone when used as a flavouring agent.	No safety concern

Table 1
(continued)

Flavouring agent	No.	CAS No. and structure	*Step B3*[a,b] Does intake exceed threshold for human intake?	*Step B4* Adequate margin of safety for the flavouring agent or related chemical?	Conclusion based on current intake
4-[(2-Methyl-3-furyl)thio]-5-nonanone	1087	61295-50-9 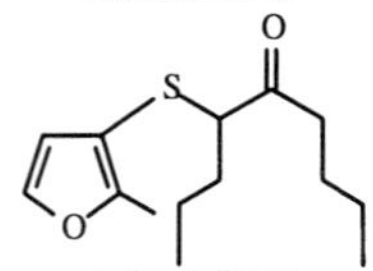	No Europe: ND USA: 0.7	Yes. The NOEL of 3.8 mg/kg bw per day for the related substance No. 1085 was > 100 000 times the estimqted intake of 4-[(2-methyl-3-furyl)thio]-5-nonanone when used as a flavouring agent.	No safety concern
Ethyl 3-(furfurylthio)propionate	1088	94278-27-0 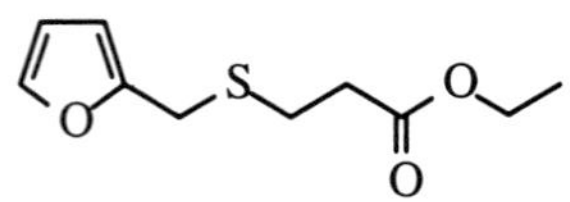	No Europe: 0.01 USA: 0.2	Yes. The NOEL of 17 mg/kg bw per day is > 1 million times the estimated intake of ethyl 3-(furfurylthio)propionate when used as a flavouring agent.	No safety concern
2-Methyl-3-thioacetoxy-4,5-dihydrofuran	1089	26486-14-6 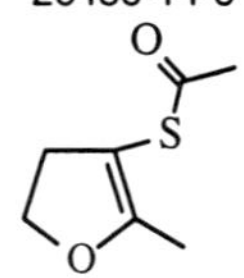	No Europe: ND USA: 0.7	Yes. The NOEL of 8.3 mg/kg bw per day is > 100 000 times the estimated intake of 2-methyl-3-thioacetoxy-4,5-dihydrofuran at when used as a flavouring agent.	No safety concern
cis- and *trans*-2,5-Dimethyl-3-tetrahydrofuranthiol	1091	26486-21-5 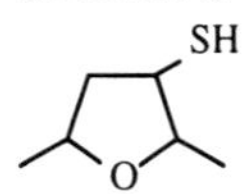	No Europe: ND USA: 0.9	Yes. The NOEL of 8.3 mg/kg bw per day for the related substance No. 1089 is > 100 000 times the estimated intake of *cis*- and *trans*-2,5-dimethyl-3-tetrahydrofuranthiol when used as a flavouring agent.	No safety concern

Table 1 *(continued)*

Flavouring agent	No.	CAS No. and structure	*Step B3*[a,b] Does intake exceed threshold for human intake?	*Step B4* Adequate margin of safety for the flavouring agent or related chemical?	Conclusion based on current intake
cis- and *trans*-2,5-Dimethyltetrahydro-3-furyl thioacetate	1092	252736-39-3 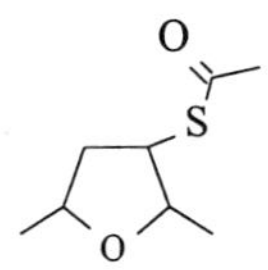	No Europe: ND USA: 2	Yes. The NOEL of 8.3 mg/kg bw per day for the related substance No. 1089 is > 100 000 times the estimated intake of *cis*- and *trans*-2,5-dimethyltetrahydro-3-furyl thioacetate when used as a flavouring agent.	No safety concern

CAS: Chemical Abstracts Service; ND: no intake data reported

[a] Step 2: None of the flavouring agents in this group was predicted to be metabolized to innocuous products.

[b] The human intake threshold is 540 μg/person per day for class II and 90 μg/person per day for class III. All intake levels are expressed in μg/person per day. The combined intake of flavouring agents in class II is 56 and 20 μg/person per day in Europe and the USA, respectively. The combined intake of flavouring agents in class III is 2.4 and 8.8 μg/person per day in Europe and the USA, respectively.

substance, furfuryl mercaptan (No. 1072), with annual volumes of 240 kg in Europe and 82 kg in the USA. The estimated per capita intake of furfuryl mercaptan is 40 μg/day in Europe and 11 μg/day in the USA. The estimated daily intakes of each of the other 32 flavouring agents in this group are less than 5 μg per day in Europe and in the USA.

1.3 Absorption, distribution, metabolism and elimination

No data on the metabolism of these flavouring agents were available; however, all the members of the group contain a reactive divalent sulfur atom attached to the heteroaromatic or heterocyclic ring. Therefore, these flavouring agents are probably metabolized via reactions of the divalent sulfur, like other heteroaromatic ring systems containing sulfur as a side-chain substituent. However, in the absence of data on the disposition of these substances, the metabolic fate of this group of flavouring agents is unknown.

1.4 Application of the Procedure for the Safety Evaluation of Flavouring Agents

Step 1. Twenty-nine of the 33 flavouring agents in this group are aromatic heterocyclic compounds bearing ring substituents. Twenty-two of the 29 heteroaromatic S-substituted furan derivatives containing only one aromatic ring were assigned to structural class II (Cramer et al., 1978) if they were naturally occurring or to structural class III if they did not occur naturally in foods. The seven remaining heteroaromatic substances contain two aromatic rings. Of these, five agents (Nos 1066, 1067, 1068, 1071 and 1081) would be hydrolysed (thioester) or readily reduced (di- or tetrasulfides) to mononuclear residues. They were assigned to structural class II or III depending on their natural occurrence. The two remaining substances (Nos 1080 and 1082), which are not readily converted to mononuclear residues, were placed in structural class III.

The remaining four flavouring agents are heterocyclic but not heteroaromatic (Nos 1089–1092). One agent, 2-methyl-3-tetrahydrofuranthiol (No. 1090), occurs naturally in foods and was placed in structural class II. The other three agents were placed in structural class III.

In summary, 18 of the 33 flavouring agents were assigned to structural class II because they are common components of food, while 15 that are not common components of food were assigned to structural class III.

Step 2. None of the flavouring agents in this group was predicted to be metabolized to innocuous products. The evaluation of these agents therefore proceeded down the B (right-hand) side of the decision tree.

Step B3. The current estimated daily per capita intakes of each of the 33 flavouring agents in this group is below the threshold for human intake for the respective structural classes (540 mg/day for structural class II and 90 mg/day for structural class III). Accordingly, the evaluation of all 33 substances in the group proceeded to step B4.

Step B4. The NOEL for 2-methyl-3-furanthiol (No. 1060) in a 90-day dietary study in rats was 5 mg/kg bw per day (Oser, 1970a). This NOEL is appropriate for ethanoic acid, S-(2-methyl-3-furanyl) ester (No. 1069) because the acetate ester would be hydrolysed to 2-methyl-3-furanthiol (No. 1060). This NOEL is also appropriate for the closely structurally related agent, 2,5-dimethyl-3-furanthiol (No. 1063).

The NOEL for furfuryl isopropyl sulfide (No. 1077) in a 90-day dietary study in rats was 1.3 mg/kg bw per day (Posternak et al., 1969). This NOEL is also appropriate for three structurally related sulfides, 2-methyl-3-(methylthio)furan (No. 1061), 2- methyl-5-(methylthio)furan (No. 1062) and furfuryl methyl sulfide (No. 1076), which would be expected to participate in the same metabolic pathways as furfuryl isopropyl sulfide.

The NOEL for methyl 2-methyl-3-furyl disulfide (No. 1064) in a 90-day dietary study in rats was 1.2 mg/kg bw per day (Gallo et al., 1976a). This NOEL is also appropriate for three structurally related disulfides, propyl 2-methyl-3-furyl disulfide (No. 1065), methyl furfuryl disulfide (No. 1078) and furfuryl propyl disulfide (No. 1079).

The NOEL for bis(2-methyl-3-furyl) disulfide (No. 1066) in a 90-day dietary study in rats was 0.45 mg/kg bw per day (Morgareidge & oser, 1970a). This NOEL is also appropriate for a structurally related bis-disulfide, bis(2,5-dimethyl-3-furyl) disulfide (No. 1067).

The NOEL for furfuryl mercaptan (No. 1072) in a multiple-dose, 13-week study in rats treated by gavage was 3 mg/kg bw per day (Phillips et al., 1977); and the NOEL for furfuryl thioacetate (No. 1074) in a 90-day dietary study in rats was 0.83 mg/kg bw per day (Posternak et al., 1969). These NOELs are also appropriate for the esters furfuryl thioformate (No. 1073) and furfuryl thiopropionate (No. 1075), because they are either close structural relatives of furfuryl thioacetate (No. 1074) or are expected to be hydrolysed to furfuryl mercaptan (No. 1072). The NOEL of 3 mg/kg bw per day for furfuryl mercaptan (No. 1072) is also appropriate for 2,2´-(dithiodimethylene) difuran (No. 1081), because this chemical is anticipated to be readily reduced to furfuryl mercaptan.

The NOEL for 2,5-dimethyl-3-thiofuroyl furan (No. 1071) in a 90-day dietary study in rats was 0.74 mg/kg bw per day (Morgareidge et al., 1974b). This NOEL is appropriate for methyl thiofuroate (No. 1083), because both 2,5-dimethyl-3-thiofuroyl furan and methyl thiofuroate would be hydrolysed to furoic acid.

The NOEL for 3-[(2-methyl-3-furyl)thio]-4-heptanone (No. 1085) in a 90-day dietary study in rats was 3.8 mg/kg bw per day (Gallo et al., 1976b). This NOEL is also appropriate for three structurally related thioketones, 4-[(2-methyl-3-furyl)thio]-2-pentanone (No. 1084), 2,6-dimethyl-3-[(2-methyl-3-furyl)thio]-4-heptanone (No. 1086) and 4-[(2-methyl-3-furyl)thio]-5- nonanone (No. 1087).

The NOEL for 2,2´-(thiodimethylene) difuran (No. 1080) in a 14-day study in rats treated in the diet was 10 mg/kg bw per day (Gill & van Miller, 1987); that for bis(2-methyl-3-furyl) tetrasulfide (No. 1068) in a 90-day study was 0.56 mg/kg bw per day (Morgareidge & Oser, 1970b); that for

2,5-dimethyl-3-furan thioisovalerate (No. 1070) in a 90-day study was 0.73 mg/kg bw per day (Morgareidge et al., 1974a); that for 2-methyl-3-, 5- or 6-(furfurylthio)pyrazine (No. 1082) in a 90-day study was 1.7 mg/kg bw per day (Posternak et al., 1975); and that for ethyl 3-(furfurylthio)propionate (No. 1088) in a 90-day study was 17 mg/kg bw per day (Bio-Research Laboratory, 1980).

The NOEL for 2-methylthioacetoxy-4,5-dihydrofuran (No. 1089) in a 1-year dietary study in rats was 8.3 mg/kg bw per day (Munday & Gellatly, 1974a). This NOEL is also appropriate for three structurally related analogues, 2-methyl-3-tetrahydrofuranthiol (No. 1090), cis- and trans-2,5-dimethyl-3-tetrahydrofuranthiol (No. 1091) and cis-and trans-2,5-dimethyltetrahydro-3-furyl thioacetate (No. 1092), because these chemicals are expected to be hydrolysed to the corresponding dihydro- or tetrahydrofuranthiol.

The NOEL(s) used to evaluate each flavouring agent and the margins of safety they provide, on the basis of current intake levels, are summarized in Table 2.

1.5 Consideration of combined intake from use as flavouring agents

The hydrolysis of furfuryl thioesters (Nos 1073, 1074 and 1075) or reduction of the disulfides (Nos 1078, 1079 and 1081) in this group would yield the common metabolite furfuryl mercaptan (No. 1072). In the unlikely event that these seven flavouring agents were to be consumed simultaneously on a daily basis, the estimated combined intakes (48 and 15 μg per person per day in Europe and the USA, respectively) would not exceed the threshold for daily human intake of compounds in structural class II (540 μg per person) or III (90 μg per person). Likewise, hydrolysis of 2-methyl-3-furan thioester (No. 1069) or reduction of the corresponding disulfides (Nos 1064, 1065 and 1066) or tetrasulfide (No. 1068) would yield the common metabolite 2-methyl-3-furanthiol (No. 1060). In the unlikely event that these six flavouring agents were to be consumed simultaneously on a daily basis, the estimated combined daily intake would be less than 5 μg per person in Europe and the USA, which would not exceed the threshold for human daily intake of compounds in structural class II (540 μg per person) or III (90 μg per person). On the basis of the evaluation of the collective data, the Committee identified no safety concern from combined intake.

1.6 Conclusions

The Committee concluded that use of flavouring agents in this group of 33 thiofurfuryl and thiofuran derivatives at current levels of intake would not present a safety concern. In the Procedure, data on toxicity were required for all the evaluations. The Committee noted the absence of data on the metabolic fate of these substances but considered that their potential metabolism via the reactive divalent sulfur atom and their very low levels of use as flavouring agents were consistent with the outcome of the evaluations.

Table 2. No-observed-effect levels and margins of safety for sulfur-substituted furan derivatives used as flavouring agents

Flavouring agent	No.	Related flavouring agent(s)	No.	NOEL (mg/kg bw per day)	Reference	Highest estimated intake (mg/kg bw per day)	Margin of safety
Methyl-3-furanthiol	1060			5[a]	Oser (1970a)	0.015	> 100 000
		Ethanoic acid, S-(2-methyl-3-furanyl)ester	1069			0.001	> 1 000 000
		2,5-Dimethyl-3-furanthiol	1063			0.01	> 100 000
Methyl 2-methyl-3-furyl disulfide	1064			1.2[a]	Gallo et al. (1976a)	0.014	> 10 000
		Propyl 2-methyl-3-furyl disulfide	1065			0.01	> 100 000
		Methyl furfuryl disulfide	1078			0.017	> 10 000
		Furfuryl propyl disulfide	1079			0.04	> 10 000
Bis(2-methyl-3-furyl) disulfide	1066			0.45[a]	Morgareidge & Oser (1970a)	0.01	> 10 000
		Bis(2,5-dimethyl-3-furyl) disulfide	1067			0.01	> 10 000
Furfuryl mercaptan	1072			3[a]	Phillips et al. (1977)	0.58	> 1 000
		2,2'-(Dithiodimethylene) difuran	1081			0.064	> 10 000
		Furfuryl thioformate	1073			0.026	> 100 000
		Furfuryl thiopropionate	1075			0.0002	> 1 000 000
Furfuryl thioacetate	1074			0.83[a]	Posternak et al. (1969)	0.008	> 100 000
		Furfuryl thioformate	1073			0.026	> 10 000
		Furfuryl thiopropionate	1075			0.0002	> 1 000 000
4-[(2-Furanmethyl)thio]-2-pentanone	1084			54[b, c]	Wnorowski (1997)	0.01	> 1 000 000
		2-Methyl-3-tetrahydro-furanthiol	1090	12[b, c]	Rush (1991)	0.07	> 100 000

Table 2 (contd)

Flavouring agent	No.	Related flavouring agent(s)	No.	NOEL (mg/kg bw per day)	Reference	Highest estimated intake (mg/kg bw per day)	Margin of safety
2,2'-(Thiodimethylene) difuran	1080			10 [b]	Gill & Van Miller, (1987)	0.014	> 100 000
Bis(2-methyl-3-furyl) tetrasulfide	1068			0.56 [a]	Morgareidge & Oser (1970b)	0.01	> 10 000
2,5-Dimethyl-3-furan thio- isovalerate	1070			0.73 [a]	Morgareidge et al. (1974a)	0.01	> 10 000
2,5-Dimethyl-3-thiofuroyl- furan	1071			0.74 [a]	Morgareidge et al. (1974b)	0.0002	> 1 000 000
		Methyl thiofuroate	1083			0.002	> 100 000
Furfuryl isopropyl sulfide	1077			1.3 [a]	Posternak et al. (1969)	0.002	> 100 000
		2-Methyl-3-(methylthio)- furan	1061			0.024	> 10 000
		2-Methyl-5-(methylthio)furan	1062			0.021	> 10 000
		Furfuryl methyl sulfide	1076			0.02	> 10 000
2-Methyl-3-, 5-, or 6-(furfuryl- thio)pyrazine		1082		1.7 [a]	Posternak & Vodoz (1975)	0.01	> 100 000
3-[(2-Methyl-3-furyl)thio]-4- heptanone		1085		3.8 [a]	Gallo et al. (1976b)	0.01	> 100 000
		4-[(2-Methyl-3-furyl)thiol]-2- pentanone	1084			0.01	> 100 000
		2,6-Dimethyl-3-[(2-methyl-3- furyl)thio]-4-heptanone	1086			0.01	> 100 000
		4-[(2-Methyl-3-furyl)thio]-5- nonanone	1087			0.01	> 100 000
Ethyl 3-(furfurylthio) propionate	1088			17 [a]	Bio-Research Laboratory (1980)	0.003	> 1 000 000

Table 2 (contd)

Flavouring agent	No.	Related flavouring agent(s)	No.	NOEL (mg/kg bw per day)	Reference	Highest estimated intake (mg/kg bw per day)	Margin of safety
2-Methyl-3-thioacetoxy-4,5-dihydrofuran	1089			8.3[d]	Munday & Gellatly (1974a)	0.01	> 100 000
		2-Methyl-3-tetrahydro-furanthiol	1090			0.07	> 100 000
		cis- and *trans*-2,5-Dimethyl-3-tetrahydrofuranthiol	1091			0.02	> 100 000
		cis- and *trans*-2,5-Dimethyl-tetrahydrofuryl thioacetate	1092			0.03	> 100 000

[a] 90-day study
[b] 14-day study
[c] This study was not used in the evaluation as a 90-day study on a structurally related flavouring agent was available.
[d] 1-year study

2. RELEVANT BACKGROUND INFORMATION

2.1 Explanation

The background information provides key data relevant to the safety evaluation of 33 thiofuran derivatives used as flavouring agents. The S-containing functional groups in this review include thiols, mono-, di- and tetra-sulfides and thioesters. Fifteen are either 2- or 3-thiofuran derivatives, 14 are furfurylthio derivatives, and four are dihydro or tetrahydro-3-thiofuran derivatives.

All the substances in the group contain a divalent S atom. They are expected to be metabolized by biotransformations at the reactive S atom. They participate in common routes of absorption, distribution and metabolic detoxication and have similar toxicological end-points. Saturation of metabolic pathways is unlikely, given the extremely low levels of intake of thiofurans from use as flavouring agents or from natural occurrence in food (see section 2.2).

2.2 Additional considerations on intake

Annual use volumes and intake values for each of the sulfur-substituted furan derivatives used as flavouring agents in this group are reported in Table 3. As noted, more than 65% of the total annual volume for flavouring agents in this group in Europe and 40% in the USA comes solely from the use of furfuryl mercaptan (No. 1072) as a flavouring agent (i.e., 244/374 kg in Europe and 82/207 kg in the USA). It is the only substance in this group for which the estimated daily *per capita* intake[1] ("eaters only") from use as a flavouring agent is > 0.1 μg/kg bw. In four national surveys conducted in the USA in 1970, 1975, 1982 and 1987 (National Academy of Sciences, 1989), the highest annual volume of use of furfuryl mercaptan was 160 kg, indicating that the results of surveys for furfuryl mercaptan have been consistent over time.

Quantitative data on natural occurrence and consumption ratios have been reported for five substances in the group. They are consumed predominantly from coffee (i.e., consumption ratio > 1) (Stofberg & Kirschman, 1985; Stofberg & Grundschober, 1987) (see Table 3). On the basis of the mean daily per capita consumption of coffee in the USA (Department of Agriculture, 1997), the estimated daily per capita intake of furfuryl mercaptan or furfuryl methyl sulfide from consumption of coffee by men aged ≥ 20 is 0.81 mg. By comparison, the estimated daily per capita intake ("eaters only') of furfuryl mercaptan from its use as a flavouring agent in the USA is 0.011 mg. Therefore, intake of furfuryl mercaptan as a result of consumption of coffee is approximately 80 times its consumption from use as a flavouring agent (Stofberg & Kirschman, 1985; Stofberg & Grundschober, 1987).

Thiofuran derivatives are known for their powerful aromas in cooked, browned and roasted foods. Their strong organoleptic properties make a significant contribution to the taste and smell of various foods, even when they are present at extremely low concentrations. For instance, at 100 ppm, furfuryl mercaptan (No. 1072) has an extremely powerful, diffusive odor. After dilution to < 2 ppm, it contributes an agreeable, coffee-like, burnt caramel, sweet aroma. The thresholds of aroma detection for thiofurans range from 0.01 ppm for 2-methyl-3-(methylthio)furan (No. 1061) to 0.1 ppm for ethyl 3-(furfurylthio)propionate (No. 1088). In sensory tolerance

Table 3. Annual volumes of use of S-substituted furan derivatives used as flavouring agents in Europe and the USA

Substance (No.)	Most recent annual volume (kg)[a]	Intake[b] μg/day	Intake[b] μg/kg bw per day	Annual volume in naturally occurring foods (kg)[c]	Consumption ratio[d]
Structural class II					
2-Methyl-3-furanthiol (1060)					
Europe	4.3	0.6	0.010	+	NA
USA	6.8	0.9	0.015		
Ethanoic acid, *S*-(2-methyl-3-furanyl) ester (1069)					
Europe	0.1	0.01	0.0002	+	NA
USA	0.5	0.07	0.001		
2-Methyl-3-(methylthio)furan (1061)					
Europe	10	1.4	0.024	6700	670
USA	0.8	0.1	0.002		>10 000
2-Methyl-5-(methylthio)furan (1062)					
Europe	9	1.3	0.021	+	
USA	0.14	0.02	0.0001		NA
Methyl 2-methyl-3-furyl disulfide (1064)					
Europe	6	0.9	0.014	11	1.4
USA	0.4	0.05	0.0009		28
Propyl 2-methyl-3-furyl disulfide (1065)					
Europe	NR	NA	NA	+	NA
USA	4	0.7	0.01		
Bis(2-methyl-3-furyl) disulfide (1066)					
Europe	2.2	0.31	0.005	+	NA
USA	4	0.7	0.01		
2,5-Dimethyl-3-furanthiol (1063)					
Europe	0.2	0.029	0.0005	+	NA
USA	4	0.7	0.01		
Furfuryl mercaptan (1072)					
Europe	240	35	0.6	1700	7
USA	82	11	0.18		21
Furfuryl thioacetate (1074)					
Europe	3.5	0.5	0.008	+	NA
USA	0.4	0.05	0.0009		
Furfuryl methyl sulfide (1076)					
Europe	8	1.1	0.019	1800	230
USA	0.9	0.11	0.002		2000
Methyl furfuryl disulfide (1078)					
Europe	7	1.0	0.017	430	61
USA	0.28	0.037	0.0006		1500

***Table 3* (contd)**

Substance (No.)	Most recent annual volume (kg)[a]	Intake[b] µg/day	Intake[b] µg/kg bw per day	Annual volume in naturally occurring foods (kg)[c]	Consumption ratio[d]
Furfuryl propyl disulfide (1079)					
Europe	NR	NA	NA	+	NA
USA	20	2.6	0.04		
2,2'-(Dithiodimethylene)difuran (1081)					
Europe	27	3.9	0.064	+	NA
USA	5	0.66	0.011		
Methyl thiofuroate (1083)					
Europe	3	0.43	0.007	+	NA
USA	0.9	0.11	0.002		
4-[(2-Furanmethyl)thio]-2-pentanone (1084)					
Europe	NR	NA	NA	+	NA
USA	4.5	0.6	0.01		
2-Methyl-3-tetrahydrofuranthiol (1090)					
Europe	29	4.1	0.07	+	NA
USA	4	0.7	0.01		
2,2'-(Thiodimethylene)difuran (1080)					
Europe	6	0.86	0.014	+	NA
USA	0.04	0.0053	0.0001		
Structural class III					
Bis(2,5-dimethyl-3-furyl) disulfide (1067)					
Europe	0.1	0.01	0.0002	–	NA
USA	4	0.7	0.01		
Bis(2-methyl-3-furyl) tetrasulfide (1068)					
Europe	NR	NA	NA	–	NA
USA	4	0.7	0.01		
2,5-Dimethyl-3-furyl thioisovalerate (1070)					
Europe	NR	NA	NA	–	NA
USA	4	0.7	0.01		
2,5-Dimethyl-3-thiofuroylfuran (1071)					
Europe	NR	NA	NA	–	NA
USA	0.1	0.013	0.0002		
Furfuryl thioformate (1073)					
Europe	11	1.6	0.026	–	NA
USA	0.14	0.02	0.0003		
Furfuryl thiopropionate (1075)					
Europe	0.1	0.014	0.0002	–	NA
USA	0.04	0.005	0.0001		

Table 3 **(contd)**

Substance (No.)	Most recent annual volume (kg)[a]	Intake[b] µg/day	Intake[b] µg/kg bw per day	Annual volume in naturally occurring foods (kg)[c]	Consumption ratio[d]
Furfuryl isopropyl sulfide (1077)					
Europe	0.01	0.0014	0.00002	–	NA
USA	1	0.1	0.002		
2-Methyl-3-, 5- or 6-(furfurylthio)pyrazine (1082)					
Europe	3	0.43	0.007	–	NA
USA	4	0.7	0.01		
3-[(2-Methyl-3-furyl)thio]-4-heptanone (1085)					
Europe	NR	NA	NA	–	NA
USA	4	0.7	0.01		
2,6-Dimethyl-3-[(2-methyl-3-furyl)thio]-4-heptanone (1086)					
Europe	NR	NA	NA	–	NA
USA	4	0.7	0.01		
4-[(2-Methyl-3-furyl)thio]-5-nonanone (1087)					
Europe	NR	NA	NA	–	NA
USA	4	0.7	0.01		
Ethyl 3-(furfurylthio)propionate (1088)					
Europe	0.1	0.01	0.0002	–	NA
USA	1.4	0.2	0.0031		
2-Methyl-3-thioacetoxy-4,5-dihydrofuran (1089)					
Europe	NR	NA	NA	–	NA
USA	4	0.7	0.01		
cis- and *trans*-2,5-Dimethyl-3-tetrahydrofuranthiol (1091)					
Europe	NR	NA	NA	–	NA
USA	7	0.9	0.02		
cis- and *trans*-2,5-Dimethyltetrahydro-3-furyl thioacetate (1092)					
Europe	NR	NA	NA	–	NA
USA	15	2	0.03		
Total					
Europe	370	53	0.9		NA
USA	210	27	0.49		

NR, no data reported; NA, not applicable; +, reported to occur naturally in foods (Maarse et al., 1999) but no quantitative data available; –, not reported to occur naturally in foods

[a] From International Organization of the Flavor Industry (1995) and Lucas et al. (1999)

[b] Intake (µg/person per day) was calculated as follows: [(annual volume, kg) × (1 × 10^9 µg/kg)/(population × survey correction factor × 365 days)], where population (10%, 'eaters only') = 32 × 10^6 for Europe and 26 × 10^6 for the USA. The correction factor = 0.6 for Europe and 0.8 for the USA, representing the assumption that only 60% and 80% of the annual production volume of the flavour, resepctively, was reported in the poundage surveys. Intake (µg/kg bw per day) calculated as follows: [(µg/person per day)/body weight], where body weight = 60 kg. Slight variations may occur from rounding.

[c] Quantitative data from Stofberg & Grundschober (1987)

[d] Calculated as follows: (annual consumption in food, kg)/(most recently reported volume as a flavouring agent, kg)

[e] Anticipated annual volume in the USA as reported by the Flavor and Extract Manufacturers Association

tests, most panelists reported that atmospheres containing thiofurans were intolerable in the range 0.5–200 ppm (Flavor and Extract Manufacturers' Association, 1996).

A direct result of their low odour thresholds is that thiolfurans and sulfides have a compelling, self-limiting effect in food. In the vast majority of food categories, < 1 ppm is used. Even in foods in which substantial evaporation of the sulfur derivative occurs during processing (e.g., hard sweets and baked goods), use is predominantly < 20 ppm. The low levels of use in food are reflected in the low reported annual volume of use of thiofurans as flavouring agents.

2.3 Biological data

2.3.1 Biochemical data

Studies have been reported on the hydrolysis in vitro of two substances in this group. Approximately 96% ester hydrolysis was observed when ethyl 3-(furfurylthio) propionate (No. 1088) was incubated in simulated intestinal fluid for 1 h (Bio-Research Laboratory, 1980). When 2-methyl-3-tetrahydrofuranthiol (No. 1090) was incubated in intestinal fluid for 4 h, 18% bis(2-methyl-3-tetrahydrofuryl)disulfide was formed (Salzer, 1991)

2.3.2 Biotransformation

Although no studies in vivo are available, thiofurans and thiofurfuryl derivatives are probably metabolized via reactions of the divalent sulfur atom, like other heteroaromatic ring systems containing a sulfur substituent (Damani, 1987). Thiols are highly reactive in vivo mainly because most exist in the ionized form at physiological pH. Thiols are oxidized to unstable sulfenic acids (RSOH), which are further oxidized to the corresponding sulfinic (RSO_2H) and sulfonic acids (RSO_3H). Methylation of thiols, primarily by *S*-adenosyl methionine, yields methyl sulfides, which are then readily oxidized to sulfoxides and sulfones. Thiols may react with physiological thiols to form mixed disulfides or conjugate with glucuronic acid. Oxidation of the α-carbon results in desulfuration and formation of an aldehyde, which oxidizes to the corresponding acid (McBain & Menn, 1969; Dutton & Illing, 1972; Maiorino et al., 1989; Richardson et al., 1991).

The labile nature of the S–S bond also presents di- and tetrasulfides with a variety of metabolic options for detoxication, as the disulfide bond is rapidly reduced to the corresponding thiol (i.e. mercaptan) in a reversible reaction in vivo. Thiol–disulfide exchange reactions occur in vivo, resulting from nucleophilic substitution by sulfur. Such reactions with cellular concentrations of endogenous thiols (GSH) and disulfides (GSSG) produce mixed disulfides, which may also undergo reduction (Brigelius, 1985; Sies et al., 1987; Cotgreave et al., 1989).

On the basis of the metabolism of simple aliphatic and aromatic monosulfides, furfuryl and furan monosulfides are expected to undergo oxidation, mainly to the sulfoxide and sulfone. Sulfoxides and sulfones are physiologically stable and are excreted unchanged in the urine (McBain & Menn, 1969; Nickson & Mitchell, 1994; Nickson et al., 1995; Nnane & Damani 1995).

2.4 Toxicological studies

2.4.1 Acute toxicity

LD_{50} values after oral administration have been reported for 13 of the 33 sulfur compounds in this group, and ranged from 100 mg/kg bw for 2-methyl-3-furanthiol (No. 1060) to > 5000 mg/kg bw for ethyl 3-(furfurylthio)propionate (No. 1088) (see Table 4). Most values were in the range 100–500 mg/kg bw.

2.4.2 Short-term studies of toxicity

Short-term studies (lasting 1 year or less) on toxicity have been performed with 16 thiofuran derivatives, and are summarized in Table 5. Studies on representative thiols (Nos 1060, 1072 and 1090), monosulfides (Nos 1077, 1080, 1082, 1084,

Table 4. Studies of the acute toxicity of S-substituted furan derivatives used as flavouring agents used as flavouring agents

No.	Agent	Species	Sex[a]	Route	LD_{50} (mg/kg bw)	Reference
1060	2-Methyl-3-furanthiol	Mouse	M,F	Gavage	100	Oser (1969a)
1060	2-Methyl-3-furanthiol	Mouse	M,F	Gavage	100	Moran et al. (1980)
1064	Methyl 2-methyl-3-furyl disulfide	Mouse	M,F	Gavage	140	Moran et al. (1980)
1065	Propyl 2-methyl-3-furyl disulfide	Mouse	M,F	Gavage	280	Moran et al. (1980)
1066	Bis(2-methyl-3-furyl) disulfide	Mouse	M,F	Gavage	110	Oser (1969b)
1063	2,5-Dimethyl-3-furanthiol	Mouse	M,F	Gavage	< 540	Fogleman & Suppers (1973c)
1063	2,5-Dimethyl-3-furanthiol	Mouse	M,F	Gavage	360	Moran et al. (1980)
1072	Furfuryl mercaptan	Mouse	M	Intraperitoneal	100–200	Doull et al.(1962)
1090	2-Methyl-3-tetrahydro-furanthiol	Mouse	NR	Gavage	1900	Oser (1970b)
1068	Bis(2-methyl-3-furyl) tetrasulfide	Mouse	M,F	Gavage	220	Oser (1970b)
1068	Bis(2-methyl-3-furyl) tetrasulfide	Mouse	NR	Gavage	220	Moran et al. (1980)
1070	2,5-Dimethyl-3-furan thioisovalerate	Mouse	M,F	Gavage	580	Fogleman & Suppers (1974a)
1070	2,5-Dimethyl-3-furan thioisovalerate	Mouse	M,F	Gavage	720	Fogleman & Suppers (1973b)
1071	2,5-Dimethyl-3-thio-furoylfuran	Mouse	M,F	Gavage	620	Fogleman & Suppers (1973a,b)
1082	2-Methyl-3-, 5- or 6-(furfurylthio)pyrazine	Rat	M,F	Oral	1000	Posternak et al. (1975)
1085	3-[(2-Methyl-3-furyl)thio]-4-heptanone	Mouse	M,F	Gavage	420	Moran et al. (1980)
1088	Ethyl 3-(furfurylthio) propionate	Mouse	M,F	Oral	> 5000	Griffiths & Babish (1977)

[a] M, male; F, female; NR, not reported

1086 and 1088), disulfides (Nos 1064 and 1066), a tetrasulfide (No. 1068) and thioesters (Nos 1070, 1071, 1074 and 1089) are described below. Most of the studies were carried out in two laboratories, each using a standardized test protocol. The protocol and results are discussed below.

(a) 2-Methyl-3-furanthiol (No. 1060), methyl 2-methyl-3-furyl disulfide (No. 1064), bis(2-methyl-3-furyl) disulfide (No. 1066), bis(2-methyl-3-furyl)tetrasulfide (No. 1068), 2,5-dimethyl-3-furan thioisovalerate (No. 1070), 2,5-dimethyl-3-thiofuroyl furan (No. 1071) and 3-[(2-methyl-3-furyl)thio]-4-heptanone (No. 1085)

A series of 90-day studies was performed with seven of the substances in this group, given at a single target level of intake that was 100 times the estimated possible average daily intake from use of the substance as a flavouring agent. The possible average daily intake was determined by multiplying usual use levels in each of 33 food categories (e.g., baked goods and meat products) by the average amount of that food category consumed daily and summing the intake over all 33 food categories (Department of Agriculture, 1965). The possible average daily intake of the vast majority of flavouring agents with low reported annual volumes of use (National Academy of Sciences, 1987; Lucas et al., 1999) is a gross exaggeration of the average daily intake. The calculation is based on the assumption that all foods in a category always contain the substance and that the food category is consumed each day (Oser & Hall, 1977). Therefore, the dietary concentrations in these studies were many orders of magnitude greater than actual levels of intake of thiofurans derivatives used as flavouring agents.

Each study was performed according to the protocol described below. The results are summarized in Table 5. The only study that showed effects (decreased growth rates in both sexes and decreased food use efficiency and increased liver and kidney weights in males) was conducted with bis(2-methyl-3-furyl) disulfide (No. 1066) at 4 mg/kg bw per day (Morgareidge & Oser, 1970b) and was repeated at a lower level of intake (Morgareidge & Oser, 1970a). This study is discussed separately.

In each study, groups of 15 male and 15 female weanling Wistar rats were individually housed and given their respective diet and tap water *ad libitum*. The substances were dissolved in acetone and blended into a basal laboratory diet to yield the required daily intake for each substance. Controls received the basal laboratory diet admixed with acetone. The acetone was evaporated from the diets before presentation to the animals. Samples of each diet were taken weekly to assess the stability and concentration of the test material.

Daily observations of appearance, behaviour, appetite, elimination, gross signs of adverse effects and mortality showed no differences among test and control animals. Weekly measurement of body weight and food consumption also showed no significant differences in any of the seven studies. Haematological, blood chemical and urine analyses performed during weeks 6 and 12 on eight males and eight females from each group gave normal values. No differences in absolute or relative weightsof the liver or kidneys were found at necropsy. No evidence of gross or histological alteration was seen in tissues from major organs of eight male and eight female rats or in the livers and kidneys of the remaining seven animals. The levels of intake of these substances that had no effect are listed in Table 5.

Table 5. Results of short-term studies of the toxicity of S-substituted furan derivatives used as flavouring agents

No.	Substance	Species, sex	No. test groups[a]/ no. per group[b]	Route	Duration	NOEL (mg/kg bw per day)	Reference
1060	2-Methyl-3-furanthiol	Rat, M, F	1/30	Diet	90 days	5[c]	Oser (1970a)
1064	Methyl 2-methyl-3-furyl disulfide	Rat, M, F	1/30	Diet	90 days	1.2[c]	Gallo *et al.* (1976a)
1066	Bis(2-methyl-3-furyl) disulfide	Rat, M, F	1/30	Diet	90 days	0.45[c]	Morgareidge & Oser (1970a)
1066	Bis(2-methyl-3-furyl) disulfide	Rat, M, F	1/30	Diet	90 days	< 4.0	Oser (1970b)
1072	Furfuryl mercaptan	Rat, M, F	3/30	Gavage	91 days	3	Phillips et al. (1977)
1074	Furfuryl thioacetate	Rat, M, F	1/32	Diet	90 days	0.81 (F)[c] 0.83 (M)[c]	Posternak et al. (1969)
1084	4-[(2-Furanmethyl)thio]-2-pentanone	Rats, M,F	1/10	Diet	14 days	51 (F)[c] 54 (M)	Wnorowski (1997)
1090	2-Methyl-3-tetrahydrofuranthiol	Rat, M, F	1/10	Diet	14 days	11 (F) 12 (M)	Rush (1991)
1080	2,2'-(Thiodimethylene)difuran	Rat, M, F	1/10	Diet	14 days	10	Gill & Van Miller (1987)
1068	Bis(2-methyl-3-furyl) tetrasulfide	Rat, M, F	1/30	Diet	90 days	0.56[c]	Morgareidge & Oser (1970b)
1070	2,5-Dimethyl-3-furan thioisovalerate	Rat, M, F	1/30	Diet	90 days	0.73[c]	Morgareidge et al. (1974a)

Table 5 (contd)

No.	Substance	Species, sex	No. test groups[a]/ no. per group[b]	Route	Duration	NOEL (mg/kg bw per day)	Reference
1071	2,5-Dimethyl-3-thiofuroylfuran	Rat, M, F	1/30	Diet	90 days	0.74[c]	Morgareidge et al. (1974b)
1077	Furfuryl isopropyl sulfide	Rat, M, F	1/32	Diet	90 days	1.3 (F)[c] 1.3 (M)[c]	Posternak et al. (1969)
1082	2-Methyl-3-, 5-, or 6-(furfurylthio)-pyrazine	Rat, M, F	1/32	Diet	90 days	1.6 (F)[c] 1.7 (M)[c]	Posternak et al. (1975)
1085	3-[(2-Methyl-3-furyl)thio]-4-heptanone	Rat, M, F	1/30	Diet	90 days	3.8[c]	Gallo et al. (1976b)
1088	Ethyl 3-(furfurylthio)propionate	Rats, M,F	2/20	Diet	90 days	17[c]	Bio-Research Laboratory (1980)
1089	2-Methyl-3-thioacetoxy-4,5-dihydrofuran	Rat, M	1/6	Diet	7 days	50	Munday & Gellatly (1972)
1089	2-Methyl-3-thioacetoxy-4,5-dihydrofuran	Rats, M,F	6/6	Diet	3-weeks	2.5	Munday & Gellatly (1972)
1089	2-Methyl-3-thioacetoxy-4,5-dihydrofuran	Rats, M,F	6/16	Diet	13 weeks	1.4	Munday & Gellatly (1973a)
1089	2-Methyl-3-thioacetoxy-4,5-dihydrofuran	Rats, M,F	2/16	Diet	16 weeks	< 11	Munday & Gellatly (1973b)
1089	2-Methyl-3-thioacetoxy-4,5-dihydrofuran	Rats, M,F	4/16	Diet	1 year	8.3[c,d]	Munday & Gellatly (1974a)

[a] Does not include control animals
[b] Includes both male and female animals
[c] Performed at either a single or multiple doses that produced no adverse effects. Therefore, this dose level is not a true NOEL but the highest dose tested that produced no adverse effects. The actual NOEL would be higher.
[d] For the first 16 weeks, animals were maintained on diets containing lower doses of 1.4, 2.8, 5.5 or 8.3 mg/kg bw of test material.

In the only study in which effects were found, bis(2-methyl-3-furyl) disulfide (No. 1066) was added to the diet of groups of 15 male and 15 female Wistar rats at a concentration providing a dose of 4 mg/kg bw per day, for 90 days. The body-weight gain of treated males was 37% less than that of controls, and treated males had significantly less efficient food use than control males. Females gained 15% less weight but showed no difference in food use efficiency . The absolute weights of the liver and kidney were lower in treated males, but the relative weights of these organs were significantly greater than in control males. Haematological, blood chemical and urine analyses and gross and histopathological examination showed no observable difference between test and control animals. The authors concluded that treatment was associated with retarded growth rates in animals of each sex, accompanied by decreased food use efficiency and increased relative liver and kidney weights in males. No effects were observed in a 90-day study with the same protocol but at a lower dose (0.45 mg/kg bw per day) (Morgareidge & Oser, 1970a).

(b) Furfuryl mercaptan (No. 1072)

Furfuryl mercaptan is the flavouring agent with the highest reported annual volume of use in Europe and the USA. Groups of 15 weanling Wistar rats of each sex were given a dose of 0, 1, 3 or 30 mg/kg bw per day in corn oil by stomach tube for 13 weeks. Additional groups of five rats of each sex were given 0, 3 or 30 mg/kg bw per day for 2 or 6 weeks. Clinical observations were made daily. Body weights were measured initially on days 1, 6 and 9 and then weekly up to week 12. Food and water intake was measured 1 day before the weight measurements.

At the highest dose, a decrease in body-weight gain associated with reduced food intake was observed, and many differences in organ weights were found at the end of the study. Significant differences in the absolute and relative weights of the brain, kidneys and stomach of males and the brain, heart, liver, kidneys and stomach of females were reported at 30 mg/kg bw per day at 13 weeks; these were considered to be associated with the reduced body weights. Increased relative heart weight in males and reduced absolute and relative kidney weight in females were reported in the group at 3 mg/kg bw per day that was terminated at 6 weeks. These changes were not present in the group at 3 mg/kg bw per day terminated at 13 weeks and were considered not to be related to treatment. Urine analysis, including concentration tests, performed at 6 and 13 weeks revealed no significant differences between test and control groups. Haematological examinations at study termination showed increases in haemoglobin concentration and packed cell volume at the highest dose, but these were not considered to be adverse effects. At termination, macroscopic and microscopic examination showed no evidence of abnormalities related to treatment. As the changes in organ weights were seen only at the highest dose, the authors concluded that 3 mg/kg bw per day had no untoward effects. No NOEL could be identified (Phillips et al., 1977).

(c) Furfuryl thioacetate (No. 1074), furfuryl isopropyl sulfide (No. 1077) and 2-methyl-3-,5- or 6-(furfurylthio)pyrazine (No. 1082)

Three flavouring agents were studied with a similar 90-day protocol. A control and a test group, each consisting of 16 male and 16 female Charles River CD rats, were housed in pairs of the same sex and given access to water and food *ad libitum.* The concentration of the test material in the diet was adjusted during the study to maintain a constant intake of furfuryl thioacetate of about 0.83 mg/kg bw per day for males and 0.81 mg/kg bw per day for females (Posternak et al., 1969), furfuryl isopropyl sulfide at 1.3 mg/kg bw per day for both males and females (Posternak et al., 1969) and 2-methyl-3-,5- or 6-(furfurylthio)pyrazine at 1.7 mg/kg bw per day for males and 1.6 mg/kg bw per day for females (Posternak et al., 1975). The doses were calculated to be > 100 times the possible average daily intakes. Clinical observations were recorded daily, and food consumption and body weights were determined weekly. During weeks 7 and 13 of the study, haematological and clinical chemical(blood urea) determinations were conducted on 50% of the animals. After 90 days, all animals were killed and subjected to a detailed necropsy, and the liver and kidneys were weighed. A wide range of tissues and organs from each animal were preserved, and major organs and tissues were examined histologically. No differences were observed between groups of test and control animals

(d) 2,2'-(Thiodimethylene)difuran (No. 1080)

Groups of five Fischer 344 rats of each sex were given 2,2'-(thiodimethylene)-difuran at a dose of 10 mg/kg bw per day for 14 days and observed for deaths, body weight, toxic signs and food intake; each underwent gross necropsy, and the liver and kidney were examined microscopically. No biologically significant physical signs of toxicity were seen, and no differences were found in absolute body weight or food consumption, gross observations at necropsy or histological appearance of the liver or kidneys. The treated group showed a statistically significant decrease in body-weight gain, and all male rats showed decreased absolute and relative weights of the liver and decreased absolute kidney weight. The authors considered that the biological significance of the decreases seen was questionable in the absence of other treatment-related effects (Gill & Van Miller, 1987).

(e) 4-[(2-Methyl-3-furyl)thio]-2-pentanone (No. 1084)

In a 14-day study, groups of five Sprague-Dawley rats of each sex were maintained on diets containing 630 ppm of 4-[(2-furanylmethyl)thio]-2-pentanone for 7 days; the concentration was reduced to 530 ppm during the second week to achieve a targeted intake of 41 mg/kg bw per day. Weekly measurement of body weight and food consumption indicated that the average daily intake was 54 mg/kg bw for males and 51 mg/kg bw for females. Animals were observed daily for gross signs of toxicity. On day 15, gross necropsies were performed on all animals, and the kidneys and liver of each animal were removed, weighed and prepared for histological evaluation.

All animals survived and appeared healthy. Although a statistically significant ($p < 0.05$) increase in relative liver weight was reported in males, no pathological changes were seen microscopically. A significant ($p < 0.05$) reduction in food consumption was reported for treated males, which was presumed to be related to the poor palatability of the sulfide-containing diet. The change in food consumption was not accompanied by a significant difference in body-weight gain between treated and control groups. Gross necropsy and microscopic examination of the kidneys and liver revealed no effects related to treatment (Wnorowski, 1997).

(f) Ethyl 3-(furfurylthio) propionate (No. 1088)

Ethyl 3-(furfurylthio) propionate was added to the diet of groups of 10 male and 10 female rats (strain not given) at concentrations calculated to result in an average daily intake of 5.8 or 17 mg/kg bw for 90 days. Body weight, food intake and food use efficiency were recorded weekly. No significant difference was found between treated and control animals at either dose. Haematological, blood chemical and urine analyses conducted at weeks 7 and 13 gave normal values. The weights of the liver and kidneys at necropsy were normal, and gross and microscopic examination revealed no dose-related effects (Bio-Research Laboratory, 1980).

(g) 2-Methyl-3-thioacetoxy-4,5-dihydrofuran (No. 1089)

Groups of three male and three female Wistar rats received diets containing 2-methyl-3-thioacetoxy-4,5-dihydrofuran spray-dried in maltodextrin at a concentration of 50, 130, 250, 500, 750 or 1000 ppm for 3 weeks, corresponding to calculated intakes of 2.5, 6.5, 12, 25, 38 and 50 mg/kg bw per day (Food and Drug Administration, 1993). Five male and six female rats were fed a control diet.. The test diets also contained a heat-reaction flavour prepared by heating 4-hydroxy-5-methyl-2,3-dihydrofuran-3-one and 1-cysteine in aqueous solution. The corresponding dietary concentrations of the reaction flavour were 4000, 10 000, 20 000, 40 000, 60 000 and 80 000 ppm. Body-weight gain, food intake, food use efficiency, water intake, activities of serum alanine and aminotransferases, lactic dehydrogenase and alkaline phosphatase, and serum sodium, potassium, chloride, urea nitrogen, creatinine and total protein were measured; haematological parameters (haemoglobin concentration, erythrocyte volume fraction, total leukocyte count and blood film examination), organ weights (heart, liver, spleen, kidneys, brain, testes, adrenals, pituitary and thyroid), and macroscopic and microscopic appearance *post mortem* were recorded.

Haemolysis of red blood cells occurred at doses ≥ 6.5 mg/kg bw per day, as evidenced by a reduced haemoglobin concentration and erythrocyte volume fraction, increased spleen weights and microscopic changes in the spleen and liver. No other biological effects could be attributed to feeding of the test diets, but it was not possible to determine whether 2-methyl-3-thioacetoxy-4,5-dihydrofuran or the heat-reaction flavour was the haemolytic agent (Munday & Gellatly, 1972).

Groups of eight male and eight female Wistar rats were fed diets containing 28, 56, 110, 170, 280 or 550 ppm of 2-methyl-3-thioacetoxy-4,5-dihydrofuran spray-dried with maltodextrin for 13 weeks, corresponding to calculated intakes of 1.4, 2.8, 5.6, 8.3, 14 and 28 mg/kg bw per day (Food and Drug Administration, 1993). Sixteen male and 16 female rats were fed a control diet, and groups of six rats of each sex per dose and a control group of 12 rats of each sex were fed the same diets for 6 weeks. The parameters studied were the same as those in the 3-week study (Munday & Gallatly, 1972). Haemolysis of red blood cells, as evidenced by reduced erythrocyte volume fraction, generally accompanied by decreased haemoglobin concentrations, increased spleen weights and microscopic changes in the spleen and liver, were observed. The NOEL was 1.4 mg/kg bw per day (Munday & Gellatly, 1973a).

Groups of eight male and eight female Wistar rats were fed diets containing 230 or 550 ppm of 2-methyl-3 thioacetoxy-4,5-dihydrofuran for 16 weeks and then transferred to a control diet for a recovery period of 12 weeks. These concentrations corresponded to calculated intakes of 11 and 28 mg/kg bw per day (Food and Drug Administration, 1993). Sixteen male and 16 female rats were fed the control diet. Half the animals were killed after a further 6 weeks and the remainder after the 12-week recovery period to investigate the reversibility of the haemolytic effects. At necropsy, the spleen, liver and kidneys were weighed and examined histologically. All the changes in the spleen and liver associated with the breakdown and production of red blood cells were reversible after cessation of treatment, except for the levels of splenic haemosiderin, which remained elevated. Haemosiderin, an iron-containing breakdown product of haemoglobin, is a normal constituent of the mammalian organism. Rapid elimination of this material cannot therefore be expected, particularly from the spleen, which is a major depot of haemosiderin in healthy animals (Munday & Gellatly, 1973b).

In the 13-week study of Munday & Gallatly (1973a), an additional eight male and eight female rats were maintained on diets containing 2-methyl-3-thioacetoxy-4,5-dihydrofuran at a concentration of 28, 56, 110 or 170 ppm for the 13 weeks of the study, plus an additional 3 weeks while the results of the 13-week study were being evaluated. All rats were then given the diet containing 170 ppm for a further 36 weeks, for a total duration of 1 year. Observation of general condition and survival, body-weight gain, haematological end-points and macroscopic and microscopic examination at necropsy showed no adverse effects. In particular, there was no evidence of the accelerated splenic erythropoiesis that had been reported in the 13-week study (Munday & Gellatly, 1974).

The haemolysis observed in the study of Munday & Gellatly (1973a) was a result of oxidative stress due to active oxygen species formed via intra-erythrocytic redox cycling of disulfides. Disulfides are reduced to their corresponding thiols by thiol–disulfide exchange with glutathione. Although generally recognized as anti-oxidants, thiols act as pro-oxidants after one electron oxidation to a thyil radical. In erythrocytes, such oxidation is mediated by oxyhaemoglobin and results in the formation of methaemoglobin, hydrogen peroxide and a thiolate anion. The hydrogen peroxide and the more active oxygen species formed from the thyil radical are responsible for initiating the cellular damage that leads to haemolysis (Munday & Mann, 1994). Haemolysis was observed only when very high doses of disulfides were administered. The authors noted that human erythrocytes are usually resistant to oxidative damage, although certain individuals are susceptible because of hereditary deficiencies (Munday et al., 1976, 1990; Munday & Mann, 1994).

(h) 2-Methyl-3-tetrahydrofuranthiol (No. 1090)

2-Methyl-3-tetrahydrofuranthiol was added to the diet of groups of five male and five female Sprague-Dawley rats for 14 days at a concentration calculated to provide a target dose of 10 mg/kg bw per day. Weekly measurements of body weight and food consumption indicated that the average daily intake was 12 mg/kg for males and 11 mg/kg for females. The kidney and liver weights at termination on day 15 didi not differ significant between treated and control groups. Gross necropsy on day 15 and histopathological examination of the kidneys, liver and gross lesions revealed no dose-related adverse effects (Rush, 1991).

3. REFERENCES

Bio-Research Laboratory (1980) Ethyl-beta-furfural-alpha-thiopropionate. Unpublished report. Submitted to WHO by Flavor and Extract Manufacturers' Association of the United States.

Brigelius, R. (1985) Mixed disulfides: Biological function and increase in oxidative stress. In: Sies, H., ed., *Oxidative Stress,* New York, Academic Press, pp. 243–271.

Cotgreave, I.A., Atzori, L. & Moldéus, N. (1989) Thiol–disulphide exchange: Physiological and toxicological aspects. In: Damani, L.A., ed., *Sulphur-containing Drugs and Related Organic Compounds. Chemistry, Biochemistry and Toxicology,* Vol. 2, Part B, *Analytical, Biochemical and Toxicological Aspects of Sulphur Xenobiochemistry,* New York: John Wiley & Sons, pp. 101–119.

Damani, L.A. (1987) Metabolism of sulphur-containing drugs. In: Benford, D.J., Bridges, J.W. & Gibson, G.G., eds, *Drug Metabolism—From Molecules to Man,* New York, Taylor & Francis, pp. 581–603.

Department of Agriculture (1965) *Food Intake and Nutritive Value of the Diets of Men, Women, and Children in the United States, Spring 1965,* Preliminary report, Washington DC: Consumer and Food Economics Research Division, Agricultural Research Service.

Department of Agriculture (1997) *Data Tables: Results from USDA's 1994–96 Continuing Survey of Food Intakes by Individuals and 1994–96 Diet and Health Knowledge Survey,* Beltsville, Maryland: Food Surveys Research Group.

Doull, J., Plzak, V. & Brois, S.J. (1962) *A Survey of Compounds for Radiation Protection,* Arlington, Virginia: United States Air Force, Armed Services Technical Information Agency.

Dutton, G.J. & Illing, H.P.A. (1972) Mechanism of biosynthesis of thio-beta-D-glucosides. *Biochem. J.,* **129**, 539–550.

Flavor and Extract Manufacturers' Association of the United States (1996) Sensory tolerance test protocol. Private communication. Submitted to WHO by Flavor and Extract Manufacturers' Association of the United States.

Fogleman, R.W. & Suppers, L. (1973a) Acute oral toxicity in mice with 2,5-dimethyl-3-thiofuroylfuran. Private communication. Submitted to WHO by Flavor and Extract Manufacturers' Association of the United States.

Fogleman, R.W. & Suppers, L. (1973b) Acute oral toxicity in mice with 2,5-dimethyl-3-thioisovalerate. Private communication. Submitted to WHO by Flavor and Extract Manufacturers' Association of the United States.

Fogleman, R.W. & Suppers, L. (1973c) Acute oral toxicity in mice with 2,5-dimethyl-3-furanthiol. Private communication. Submitted to WHO by Flavor and Extract Manufacturers' Association of the United States.

Fogleman, R.W. & Suppers, L. (1974a) Acute oral toxicity in mice with 2,5-dimethyl-3-thioisovalerate. Private communication. Submitted to WHO by Flavor and Extract Manufacturers' Association of the United States.

Fogleman, R.W. & Suppers, L. (1974b) Acute oral toxicity in mice with 2,5-dimethyl-3-thiofuroylfuran. Private communication. Submitted to WHO by Flavor and Extract Manufacturers' Association of the United States.

Food and Drug Administration (1993) *Priority-based Assessment of Food Additives (PAFA) Database,* Washington DC: Center for Food Safety and Applied Nutrition, p. 58.

Gallo, M.A., Cox, G.E. & Babish, J.G. (1976a) 90-day feeding studies in rats with compound 75-31065 (methyl benzyl disulfide). Unpublished report from Food and Drug Research Laboratories, Inc. Submitted to WHO by Flavor and Extract Manufacturers' Association of the United States.

Gallo, M.A., Cox, G.E. & Babish, J.G. (1976b) 90-day feeding study in rats with compound 75-15963 (3-[(2-methyl-3-furyl)-thio]-4-heptanone). Unpublished report from Food and Drug Research Laboratories, Inc. Submitted to WHO by Flavor and Extract Manufacturers' Association of the United States.

Gill, M.W. & Van Miller, J.P. (1987) Fourteen-day dietary minimum toxicity screen (MTS) in albino rats. Unpublished report No. 50-527 from Bushy Run Research Center, Export, Pennsylvania, USA. Submitted to WHO by Flavor and Extract Manufacturers' Association of the United States.

Griffiths, J. & Babish, J.G. (1977) Acute oral toxicity (LD50) of ethyl furfuryl alpha-thiopropionate in mice. Private communication. Submitted to WHO by Flavor and Extract Manufacturers' Association of the United States.

International Organization of the Flavor Industry (1995) European inquiry on volume of use. Private communication. Submitted to WHO by Flavor and Extract Manufacturers' Association of the United States.

Lucas, C.D., Putnam, J.M. & Hallagan, J.B. (1999) *Flavor and Extract Manufacturers Association of the United States 1995 Poundage and Technical Effects Update Survey*, Washington DC: Flavor and Extract Manufacturers' Association of the United States.

Maarse, H., Visscher, C.A., Willemsens, L.C. & Boelens, M.H. (1999) *Volatile Components in Food—Qualitative and Quantitative Data*, Zeist: Centraal Instituut Voor Voedingsonderzioek TNO.

Maiorino, R.M., Bruce, D.C. & Aposhian, H.V. (1989) Determination and metabolism of dithiol chelating agents. VI. Isolation and identification of the mixed disulfides of *meso*-2,3-dimercaptosuccinic acid with L-cysteine in human urine. *Toxicol. Appl. Pharmacol.*, **97**, 338.

McBain, D.A. & Menn, J.J. (1969) S-Methylation, oxidation, hydroxylation, and conjugation of thiophenol in the rat. *Biochem. Pharmacol.*, **18**, 2282–2285.

Moran, E.J., Easterday, O.D. & Oser, B.L. (1980) Acute oral toxicity of selected flavor chemicals. *Drug Chem. Toxicol.*, **3**, 249–258.

Morgareidge, K. & Oser, B.L. (1970a) 90-day feeding studies in rats with bis-(2-methyl-3-furyl)-disulfide. Unpublished report. Submitted to WHO by Flavor and Extract Manufacturers' Association of the United States.

Morgareidge, K. & Oser, B.L. (1970b) 90-day feeding studies in rats with bis-(2-methyl-3-furyl)-tetrasulfide. Unpublished report. Submitted to WHO by Flavor and Extract Manufacturers' Association of the United States.

Morgareidge, K., Cox, G.E. & Bailey, D.E. (1974a) 90-day feeding study in rats with 2,5-dimethyl-3-thioisovalerylfuran. Unpbulished report. Submitted to WHO by Flavor and Extract Manufacturers' Association of the United States.

Morgareidge, K., Cox, G.E. & Bailey, D.E. (1974b) 90-day feeding study in rats with 2,5-dimethyl-3-thiofuroylfuran. Unpbulished report. Submitted to WHO by Flavor and Extract Manufacturers' Association of the United States.

Munday, R. & Gellatly, J.B. (1972) Biological evaluation of dusmelline and DUS-5. I. Palatability study in rats. Unpublished report. Submitted to WHO by Flavor and Extract Manufacturers' Association of the United States.

Munday, R. & Gellatly, J.B. (1973a) Biological evaluation of feeding trial with DUS-5. 4. 13-week rat. Unpublished report. Submitted to WHO by Flavor and Extract Manufacturers' Association of the United States.

Munday, R. & Gellatly, J.B. (1973b) Biological evaluation of DUS-5. Part 5. Reversibility of toxic effects induced by feeding of DUS-5. Unpublished report. Submitted to WHO by Flavor and Extract Manufacturers' Association of the United States.

Munday, R. & Gellatly, J.B. (1974) Biological evaluation of DUS-5. Part 7. 1-year test with DUS-5 in rats. Unpublished report. Submitted to WHO by Flavor and Extract Manufacturers' Association of the United States.

Munday, R. & Manns, E. (1994) Comparative toxicity of prop(en)yl disulfides derived from Alliaceae. Possible involvement of 1-propenyl disulfides in onion-induced hemolytic anemia. *J. Agric. Food Chem.*, **42**, 959–962.

Munday, R., Robinson, J.A., Salmond, G.W. & Gellatly, J.B. (1976) Biochemical effects of haemolytic thiols. Unpublished report. Submitted to WHO by Flavor and Extract Manufacturers' Association of the United States.

Munday, R., Manns, E. & Fowke, E.A. (1990) Steric effects on the haemolytic activity of aromatic disulphides in rats. *Food Chem. Toxicol.*, **28**, 561–566.

National Academy of Sciences (1989) *1987 Poundage and Technical Effects Update of Substances Added to Food*, Washington DC.

Nickson, R.M. & Mitchell, S.C. (1994) Fate of dipropyl sulfide and dipropyl sulfoxide in rat. *Xenobiotica*, **24**, 157–168.

Nickson, R.M., Mitchell, S.C. & Zhang, A.Q. (1995) Fate of dipropyl sulfone in rat. *Xenobiotica*, **25**, 1391–1398.

Nnane, P. & Damani, L.A. (1995) The involvement of rat liver CYP2B1 and CYP2D1 in the microsomal sulphoxidation of 4-chlorophenyl methyl sulphide. *ISSX Proc.* , **8**, 110.

Oser, B.L. (1969a) The acute oral toxicity (LD50) to mice of 2-methyl-3-furanthiol. Unpublished report. Submitted to WHO by Flavor and Extract Manufacturers' Association of the United States.

Oser, B.L. (1969b) The acute oral toxicity (LD50) to mice of bis-(2-methyl-3-furyl) disulfide. Unpublished report. Submitted to WHO by Flavor and Extract Manufacturers' Association of the United States.

Oser, B.L. (1970a) 90-day feeding studies with 2-methyl-3-furanthiol in rats. Unpublished report. Submitted to WHO by Flavor and Extract Manufacturers' Association of the United States.

Oser, B.L. (1970b) The acute oral toxicity (LD50) to mice of ten compounds. Unpublished report. Submitted to WHO by Flavor and Extract Manufacturers' Association of the United States.

Oser, B.L. & Hall, R.L. (1977) Criteria employed by the expert panel of Flavor and Extract Manufacturers' Association of the United States for the GRAS evaluation of flavoring substances. *Food Cosmet. Toxicol.*, **15**, 457–466.

Phillips, J.C., Gaunt, I.F., Hardy, J., Kiss, I.S., Gangolli, S.D. & Butterworth, K.R. (1977) Short-term toxicity of furfuryl mercaptan in rats. *Food Cosmet. Toxicol.*, **15**, 383–387.

Posternak, J.M., Linder, A. & Vodoz, C.A. (1969) Summaries of toxicological data. Toxicological tests on flavoring matters. *Food Cosmet. Toxicol.*, **7**, 405–407.

Posternak, J.M., Dufour, J.J., Rogg, C. & Vodoz, C.A. (1975) Toxicology tests on flavouring materials. II. Pyrazines and other compounds. *Food Cosmet. Toxicol.*, **13**, 487–490.

Richardson, K.A., Edward, V.T., Jones, B.C. & Hutson, D.H. (1991) Metabolism in the rat of a model xenobiotic plant metabolite S-benzyl-N-malonyl-L-cysteine. *Xenobiotica*, **21**, 371.

Rush, R.E. (1991) 2-Methyl-3-tetrahydrofuranthiol. Unpublished report. Submitted to WHO by Flavor and Extract Manufacturers' Association of the United States.

Salzer, U. (1991) In vitro hydrolysis test for 2-methyl-3-tetrahydrofuranthiol. Unpublished report. Submitted to WHO by Flavor and Extract Manufacturers' Association of the United States.

Sies, H., Brigelius, R. & Graf, P. (1987) Hormones, glutathione status and protein S-thiolation. *Adv. Enzyme Regul.*, **26**, 175–184.

Stofberg, J. & Grundschober, F. (1987) Consumption ratio and food predominance of flavoring materials. *Perfumer Flavorist*, **12**, 27.

Stofberg, J. & Kirschman, J.C. (1985) The consumption ratio of flavoring materials: A mechanism for setting priorities for safety evaluation. *Food Chem. Toxicol.*, **23**, 857–860.

Wnorowski, G. (1997) 14-day dietary study in rats. Study No. 5212. Unpublished report No. 088816-3206 from Product Safety Labs, East Brunswick, New Jersey, USA. Submitted to WHO by Flavor and Extract Manufacturers' Association of the United States.

ALICYCLIC KETONES, SECONDARY ALCOHOLS AND RELATED ESTERS

First draft prepared by

Professor I.G. Sipes[1] and Professor A.G. Renwick[2]
[1]Department of Pharmacology, College of Medicine, University of Arizona, Tucson, Arizona, USA; and [2]Clinical Pharmacology Group, University of Southampton, Southampton, England

1. EVALUATION

1.1 Introduction

The Committee evaluated a group of 25 flavouring agents consisting of alicyclic ketones, secondary alcohols and related esters (see Table 1). The evaluations were conducted according to the Procedure for the Safety Evaluation of Flavouring Agents (see Figure 1). None of these agents has been evaluated previously by the Committee.

Twelve of the 25 substances (Nos 1093, 1098, 1100–1104, 1107, 1108, 1110, 1112 and 1114) have been reported to occur naturally in foods. They have been detected in fruits, vegetables, cheese, meats, seafood, grains, alcoholic beverages, coffee and tea.

Table 1. Summary of results of the safety evaluation of alicyclic ketones, secondary alcohols and related esters used as flavouring agents[a]

Flavouring agent	No.	CAS No. and structure	*Step A3*[b] Does intake exceed the threshold for human intake?	Comments	Conclusion based on current intake
Structural class I					
Cyclohexyl acetate	1093	622-45-7 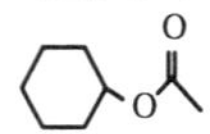	No Europe: 14 USA: 10	See note 1.	No safety concern
Cyclohexyl butyrate	1094	1551-44-6 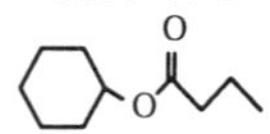	No Europe: ND USA: 0.1	See note 1.	No safety concern
Cyclohexyl formate	1095	4351-54-6 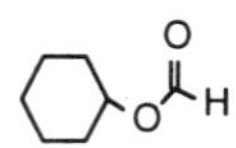	No Europe: 0.01 USA: 0.2	See note 1.	No safety concern
Cyclohexyl isovalerate	1096	7774-44-9 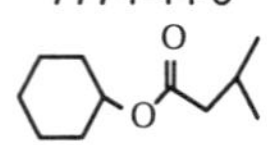	No Europe: 0.3 USA: 0.05	See note 1.	No safety concern
Cyclohexyl propionate	1097	6222-35-1 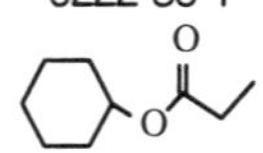	No Europe: 0.01 USA: 0.05	See note 1.	No safety concern
3,3,5-Trimethyl cyclohexanol	1099	116-02-9 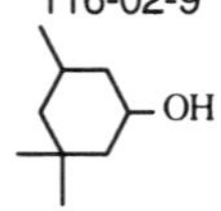	No Europe: 0.1 USA: 0.1	See note 2.	No safety concern

Table 1 *(continued)*

Flavouring agent	No.	CAS No. and structure	*Step A3*[a,b] Does intake exceed the threshold for human intake?	Comments	Conclusion based on current intake
Structural class II					
cis- and *trans-para*-1(7),8-Menthadien-2-yl acetate	1098	71660-03-2 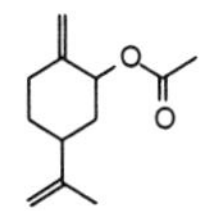	No Europe: ND USA: 0.6	See note 1.	No safety concern
Cyclohexanone	1100	108-94-1 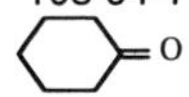	No Europe: 0.1 USA: 0.1	See note 2.	No safety concern
Cyclopentanone	1101	120-92-3 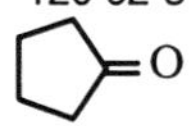	No Europe: 0.02 USA: 0.02	See note 2.	No safety concern
2-Methylcyclohexanone	1102	583-60-8 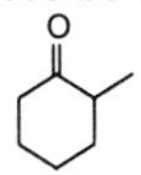	No Europe: 0.1 USA: 0.1	See note 2.	No safety concern
3-Methylcyclohexanone	1103	591-24-2 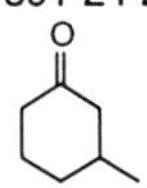	No Europe: 0.1 USA: 0.1	See note 2.	No safety concern
4-Methyl cyclohexanone	1104	589-92-4 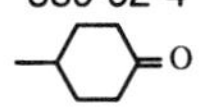	No Europe: 0.1 USA: 0.1	See note 2.	No safety concern

Table 1 *(continued)*

Flavouring agent	No.	CAS No. and structure	*Step A3*[a,b] Does intake exceed the threshold for human intake?	Comments	Conclusion based on current intake
1-Methyl-1-cyclopenten-3-one	1105	2758-18-1 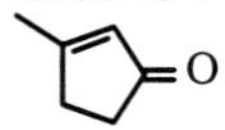	No Europe: 0.07 USA: ND	See note 3.	No safety concern
Structural class II					
2-Hexylidene cyclopentanone	1106	17373-89-6 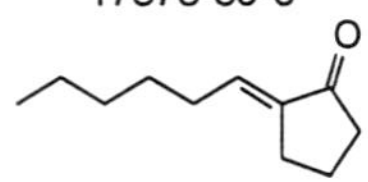	No Europe: 0.3 USA: 0.01	See notes 3 and 4.	No safety concern
3-Methyl-2-cyclohexen-1-one	1107	1193-18-6 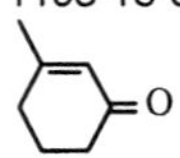	No Europe: 0.01 USA: 0.1	See note 3.	No safety concern
2,2,6-Trimethylcyclohexanone	1108	2408-37-9 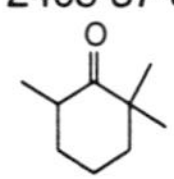	No Europe: 2.4 USA: 0.04	See note 2.	No safety concern
2-*sec*-Butylcyclohexanone	1109	14765-30-1 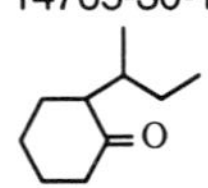	No Europe: 6 USA: ND	See note 4.	No safety concern
4-Isopropyl-2-cyclohexenone	1110	500-02-7 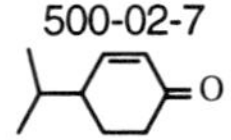	No Europe: 0.001 USA: 0.001	See note 3.	No safety concern

Table 1 *(continued)*

Flavouring agent	No.	CAS No. and structure	*Step A3*[a,b] Does intake exceed the threshold for human intake?	Comments	Conclusion based on current intake
Tetramethyl ethyl cyclohexenone (mixture of isomers)	1111	17369-60-7 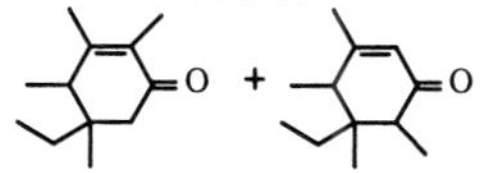	No Europe: 9 USA: 0.2	See note 3.	No safety concern
Isophorone	1112	78-59-1 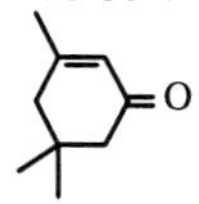	No Europe: 5 USA: 0.1	See note 3.	No safety concern
3-Methyl-5-propyl-2-cyclohexen-1-one	1113	3720-16-9 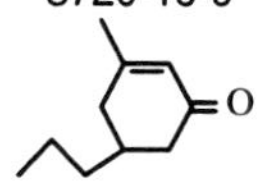	No Europe: 0.1 USA: 4	See note 3.	No safety concern
3-Methyl-2-(2-pentenyl)-2-cyclopenten-1-one	1114	488-10-8 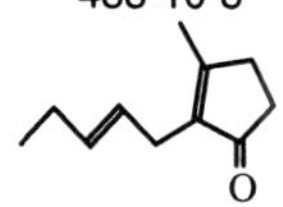	No Europe: 16 USA: 7	See notes 3 and 4.	No safety concern
Isojasmone	1115	11050-62-7 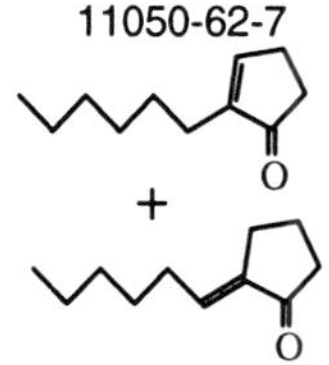	No Europe: 0.4 USA: 0.01	See notes 3 and 4.	No safety concern

Table 1 *(continued)*

Flavouring agent	No.	CAS No. and structure	*Step A3*[a,b] Does intake exceed the threshold for human intake?	Comments	Conclusion based on current intake
(E)-2-(2-Octenyl) cyclopentanone	1116	65737-52-2 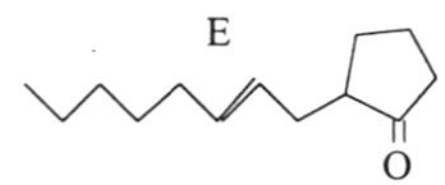	No Europe: 7 USA: 7	See note 4.	No safety concern
2-(3,7-Dimethyl-2,6-octadienyl) cyclopentanone	1117	68133-79-9 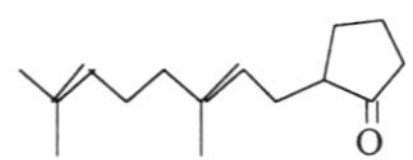	No Europe: 7 USA: 7	See note 4.	No safety concern

CAS: Chemical Abstracts Service; ND: No intake data reported

[a] Step 2: All of the flavouring agents in this group were predicted to be metabolized to innocuous products.

[b] The threshold for human intake is 1800 mg/person per day for class I and 540 mg/person per day for class II. All intake levels expressed in mg/person per day. The combined intake of flavouring agents in class I is 14 and 10 mg per person per day in Europe and the USA, respectively. The combined intake of flavouring agents in class II is 54 and 26 mg per person per day in Europe and the USA, respectively.

Notes:

1. Detoxicated by hydrolysis of ester and glucuronic acid conjugation of the resulting alicyclic alcohol and complete oxidation of the carboxylic acid
2. Detoxicated by reduction of the ketone followed by glucuronic acid conjugation of the corresponding alcohol
3. Detoxicated by reduction of the ketone functional group followed by glucuronic acid conjugation of the resulting alcohol and glutathione conjugation of the parent ketone
4. Detoxicated by reduction of the ketone and alkyl side-chain oxidation and excretion

1.2 Estimated daily per capita intake

The total annual volume of production of the 25 alicyclic ketones, secondary alcohols and related esters in this group is approximately 520 kg in Europe (International Organization of the Flavor Industry, 1995) and 310 kg in the USA (Lucas et al., 1999). The reported annual production of only one agent, 3-methyl-2-(2-pentenyl)-2-cyclopenten-1-one (No. 1114), is greater than 100 kg (110 kg in Europe). The daily per capita intake of each of the agents in this group is < 15 μg in Europe and in the USA.

1.3 Absorption, distribution, metabolism and elimination

Esters (Nos 1093–1098) in this group are hydrolysed to their corresponding alcohols and carboxylic acids by carboxylesterases, which are found abundantly in hepatocytes. The resulting alicyclic secondary alcohols are conjugated with glucuronic acid and excreted mainly in the urine. Side-chain oxidation of methyl groups may also occur.

Alicyclic ketones (Nos 1100–1117) are reduced to the corresponding secondary alcohol and excreted primarily as glucuronic acid conjugates. If a double bond is present, the chemical may be reduced to the corresponding dihydro derivative. Reduction of the double-bond in metabolites excreted in the bile is probably associated with the action of gut microflora. Endocyclic double bonds (Nos 1105, 1107 and 1110–1115) are more prone to reduction than exocyclic double bonds (Nos 1098 and 1106). Alicyclic ketones containing an alkyl side-chain (Nos 1098, 1099 and 1102–1117) can not only follow reductive pathways but can undergo oxidation of the side-chain to form poly-oxygenated metabolites, which are excreted as the glucuronic acid or sulfate conjugates in the urine and, to a lesser extent, in the faeces. The nine chemicals that are α,β-unsaturated ketones (Nos 1105–1107 and 1110–1116) are subject to glutathione conjugation, with subsequent elimination in the urine as mercapturic acids.

As the α,β-unsaturated carbonyl group is a structural alert for toxicity, the Committee, at previous meetings, has devoted considerable attention to the safety of flavouring agents containing this reactive moiety. The Committee concluded at its fifty-seventh meeting (Annex 1, reference *154*) that the presence of protective processes in cells provides adequate detoxication capacity at the low doses associated with use of such compounds as flavouring agents. With respect to α,β-unsaturated ketones, these protective processes include reduction of the ketone to the corresponding alcohol (followed by conjugation of the alcohol with glucuronic acid) and conjugation with glutathione. These processes operate for the aliphatic ketones used as flavouring agents.

1.4 Application of the Procedure for the Safety Evaluation of Flavouring Agents

Step 1. In applying the Procedure for the Safety Evaluation of Flavouring Agents to this group of flavouring agents, the Committee assigned six (Nos 1093–1097 and 1099) of the 25 substances to structural class I (Cramer et al., 1978). These agents are either alicyclic secondary alcohols or are readily hydrolysed to secondary alcohols and simple short-chain carboxylic acids. They are simple aliphatic substances containing secondary hydroxyl groups, which have low toxicity. The remaining 19 compounds, which are monocyclic alkanones (Nos 1100–1117) or have a secondary alcohol attached to a vinyl group (No. 1098), were assigned to structural class II.

Step 2. All the flavouring agents in this group were predicted to be metabolized to innocuous products. The evaluation of these agents therefore proceeded via the A (left-hand) side of the decision tree.

Step A3. The estimated daily per capita intake of each of the six flavouring agents in structural class I and each of the 19 agents in structural class II is below the threshold for daily human intake of compounds in the respective structural class (i.e., 1800 μg per person for class I and 540 μg per person for class II). According to the Procedure, the safety of these 25 flavouring agents raises no concern when they are consumed at the currently estimated levels.

Table 1 summarizes the evaluations of this group of acyclic ketones, secondary alcohols and related esters used as flavouring agents.

1.5 Consideration of combined intake from use as flavouring agents

In the unlikely event that all the flavouring agents in structural class I or II were to be consumed concurrently on a daily basis, the estimated combined intake would not exceed the threshold for human intake of either class.

1.6 Conclusions

The Committee concluded that none of the flavouring agents in this group of alicyclic ketones, secondary alcohols and related esters would raise a safety concern at current estimated levels of intake. Other data on the toxicity of alicyclic ketones, secondary alcohols and related esters were consistent with the results of the safety evaluation.

2. RELEVANT BACKGROUND INFORMATION

2.1 Explanation

This section summarizes the key data relevant to the safety evaluation of 25 alicyclic ketones, secondary alcohols and esters derived from secondary alcohols (see Table 1). The group includes both terpenoid and non-terpenoid substances and contains 18 saturated and unsaturated alicyclic ketones (Nos 1100–1117), one secondary alcohol (No. 1099) and six esters (Nos 1093–1098). Groups of structurally related menthol and carvone derivatives that are also alicyclic secondary alcohols and ketones were previously evaluated for use as flavouring agents (Annex 1, reference *138*).

The flavouring agents in this group contain a ketone functional group or are esters of a related secondary alcohol. Before absorption, the esters are readily hydrolysed to an alicyclic secondary alcohol. Once they have been absorbed, secondary alcohols and ketones are readily interconverted in vivo (McMahon, 1982). Therefore, all the substances in this group are expected to participate in common routes of absorption, metabolism, distribution and excretion and show similar toxicological profiles.

2.2 Additional considerations on intake

Quantitative data on natural occurrence and consumption ratios have been reported for two substances, isophorone (No. 1112) and 3-methyl-2-(2-pentenyl)-2-cyclopenteny-1-one (No. 1114). These show that intake is predominantly from the consumption of traditional foods (i.e., consumption ratio > 1) (Stofberg & Kirschman, 1985; Stofberg & Grundschober, 1987; see Table 2).

2.3 Biological data

2.3.1 Biochemical data

(a) Hydrolysis

The unsubstituted monocyclic esters (e.g., cyclohexyl acetate, No. 1093) hydrolyse rapidly to cyclohexanol and the component aliphatic carboxylic acids by classes of enzymes recognized as carboxylesterases (Ford & Moran, 1978; Heymann, 1980; White et al., 1990), the most important of which are the β-esterases. These enzymes occur in most mammalian tissues (Heymann, 1980; Anders, 1989) but predominate in hepatocytes (Heymann, 1980).

cis and *trans-para*-1(7),8-Menthadien-2-yl acetate (No. 1098) was rapidly hydrolysed in vitro in the presence of a rat liver homogenate (Salzer, 1998). Incubation of the ester resulted in 92% hydrolysis after 15 min and 100% after 60 min (Salzer, 1998). The structurally related (–)-menthol carbonate esters, (–)-menthol ethylene glycol carbonate and (–)-menthol propylene glycol carbonate, were almost completely hydrolysed after incubation for 1 h with rat liver homogenate (Emberger, 1994).

Sterically hindered esters were also hydrolysed in rat liver homogenate. The sterically hindered ester 3,5,5-trimethyl-[2,3-^{3}H]-cyclohexanyl-[^{14}C]mandelate (cyclandelate) was completely hydrolysed to 3,5,5-trimethyl-[2,3-^{3}H]-cyclohexanol and [^{14}C]mandelic acid within 5 min of incubation with rat hepatocytes. By 20 min, 80% of the alcohol had disappeared, with a concomitant linear increase in a β-glucuronidase-reactive substance. The resulting alcohol was presumably conjugated with glucuronic acid (White et al., 1990).

In urine collected 18 h after administration of cyclohex-1-en-1-yl acetate at a dose of 350 mg/kg bw to rabbits, 39% of the dose was hydrolysed and then conjugated with glucuronic acid (Elliott et al., 1959). This information indicates that all the esters in this group will be hydrolysed in vivo to yield the corresponding cyclohexanol derivative.

(b) Absorption, distribution and excretion

Once formed, the alicyclic terpenoid secondary alcohols and ketones are rapidly absorbed, metabolized and excreted, mainly in the urine as glucuronic acid and, in the case of α,β-unsaturated ketones, as glutathione conjugates (Williams, 1959; Portoghese et al., 1989).

Male Sprague-Dawley rats were exposed to atmospheres containing cyclohexanone (No. 1100) at either 400 ppm [approximately 240 mg/kg bw (Fassett, 2001)] or 1600 ppm [approximately 980 mg/kg bw (Fassett, 2001)] for 6 h (unpublished data). Blood samples obtained at termination showed average plasma concentrations of cyclohexanone and cyclohexanol of 26 and 20 μg/ml, respectively,

Table 2. Annual usage volume of alicyclic ketones, secondary alcohols and related esters used as flavouring agents in Europe and the USA

Substance (No.)	Most recent annual volume (kg)[a]	Intake ('eaters only')[b]		Annual volume in naturally occurring foods (kg)[c]	Consumption ratio[d]
		μg/day	μg/kg bw per day		
Cyclohexyl acetate (1093)					
Europe	99	14	0.24		
USA	77	10	0.17	+	NA
Cyclohexyl butyrate (1094)					
Europe	NR	NA	NA		
USA	0.90	0.12	0.002	–	NA
Cyclohexyl formate (1095)					
Europe	0.10	0.01	0.0002		
USA	1.4	0.18	0.003	–	NA
Cyclohexyl isovalerate (1096)					
Europe	2.3	0.33	0.01		
USA	0.40	0.05	0.001	–	NA
Cyclohexyl propionate (1097)					
Europe	0.10	0.01	0.0002		
USA	0.40	0.05	0.001	–	NA
cis- and *trans-para-*1(7),8-Menthadien-2-yl acetate (1098)					
Europe	NR	NA	NA		
USA	4.5	0.59	0.01	+	NA
3,5,5-Trimethylcyclohexanol (1099)					
Europe	1.0	0.14	0.002		
USA	1.0	0.13	0.002	–	NA
Cyclohexanone (1100)					
Europe	1.0	0.14	0.002		
USA	1.0	0.13	0.002	+	NA
Cyclopentanone (1101)					
Europe	0.15	0.02	0.0003		
USA	0.15	0.02	0.0003	+	NA
2-Methylcyclohexanone (1102)[e]					
Europe	1.0	0.14	0.002		
USA	1.0	0.13	0.002	+	NA
3-Methylcyclohexanone (1103)[e]					
Europe	1.0	0.14	0.002		
USA	1.0	0.13	0.002	+	NA
4-Methylcylcohexanone (1104)[e]					
Europe	1.0	0.14	0.002		
USA	1.0	0.13	0.002	+	NA

Table 2 (contd)

Substance (No.)	Most recent annual volume (kg)[a]	Intake ('eaters only')[b]		Annual volume in naturally occurring foods (kg)[c]	Consumption ratio[d]
		µg/day	µg/kg bw per day		
1-Methyl-1-cyclopenten-3-one (1105)					
Europe	0.50	0.07	0.001		
USA	NR	NA	NA	–	NA
2-Hexylidenecyclopentanone (1106)					
Europe	2.0	0.29	0.005		
USA	0.05	0.01	0.0002	–	NA
3-Methyl-2-cyclohexen-1-one (1107)					
Europe	0.10	0.01	0.0002		
USA	0.90	0.12	0.002	+	NA
2,2,6-Trimethylcyclohexanone (1108)					
Europe	17	2.4	0.04		
USA	0.30	0.04	0.001	+	NA
2-*sec*-Butylcyclohexanone (1109)					
Europe	42	6.0	0.10		
USA	NR	NA	NA	–	NA
4-Isopropyl-2-cyclohexenone (1110)[e]					
Europe	0.01	0.001	0.00002		
USA	0.01	0.001	0.00002	+	NA
Tetramethylethylcyclohexenone (mixture of isomers) (1111)					
Europe	64	9.1	0.15		
USA	1.4	0.18	0.003	–	NA
Isophorone (1112)					
Europe	38	5.4	0.09		
USA	0.90	0.12	0.002	30	34
3-Methyl-5-propyl-2-cyclohexen-1-one (1113)					
Europe	0.80	0.11	0.002		
USA	31	4.1	0.07	–	NA
3-Methyl-2-(2-pentenyl)-2-cyclopenten-1-one (1114)					
Europe	110	16	0.26		
USA	54	7.2	0.12	970	18
Isojasmone (1115)					
Europe	3.0	0.43	0.01		
USA	0.05	0.01	0.0002	–	NA
(E)-2-(2-Octenyl)cyclopentanone (1116)[e]					
Europe	50	7.1	0.12		
USA	50	6.6	0.11	–	NA

Table 2 (contd)

Substance (No.)	Most recent annual volume (kg)[a]	Intake ('eaters only')[b]		Annual volume in naturally occurring foods (kg)[c]	Consumption ratio[d]
		µg/day	µg/kg bw per day		
2-(3,7-Dimethyl-2,6-octadienyl) cyclopentanone (1117)[e]					
Europe	50	7.1	0.12		
USA	50	6.6	0.11	–	NA
Total					
Europe		480			
USA		280			

NR, no data reported; NA, not applicable; +, reported to occur naturally in foods (Maarse et al., 1999) but no quantitative data available; –, not reported to occur naturally in foods

[a] From International Organization of the Flavor Industry (1995) and Lucas et al. (1999)
[b] Intake (µg/person per day) was calculated as follows: [(annual volume, kg) × (1 × 10^9 µg/kg)/(population × survey correction factor × 365 days)], where population (10%, 'eaters only') = 32 × 10^6 for Europe and 26 × 10^6 for the USA. The correction factor = 0.6 for Europe and 0.8 for the USA, representing the assumption that only 60% and 80% of the annual production volume of the flavour, resepctively, was reported in the poundage surveys. Intake (µg/kg bw per day) calculated as follows: [(µg/person per day)/body weight], where body weight = 60 kg. Slight variations may occur from rounding.
[c] Quantitative data from Stofberg & Grundschober (1987)
[d] Calculated as follows: (annual consumption in food, kg)/(most recently reported volume as a flavouring agent, kg)
[e] Anticipated annual volume in the USA as reported by the Flavor and Extract Manufacturers Association of the United States

at 400 ppm and 120 and 140 µg/ml, respectively, at 1600 ppm. Total urinary excretion of cyclohexanol in samples collected 24, 48 and 72 h after exposure was at least 10 times greater than that of cyclohexanone, with 16 and 15 µg/ml cyclohexanone and 140 and 260 µg/ml cyclohexanol at 400 and 1600 ppm, respectively; 13 and 72 mg/ml of conjugated cyclohexanol were excreted within 72 h after exposure to 400 and 1600 ppm, respectively (Topping et al., 1994).

Menthol (5-methyl-2-isopropylcyclohexanol), which is structurally similar to 4-isopropyl-2-cyclohexanone (No. 1110), is rapidly absorbed and excreted in rabbits and rats. Urine and faeces were collected 24 and 48 h after dosing from rats given 3-[^{3}H]menthol at a dose of 500 mg/kg bw (128 µCi/mg) by gavage. Both intact and bile duct-cannulated rats excreted > 70% of the dose within 48 h. Several metabolites of menthol were detected in urine and faecal extracts, but no unchanged menthol was found. The glucuronic acid conjugate of menthol was the main metabolite in bile, while the glucuronic acid conjugate and metabolites formed by side-chain oxidation (e.g., 3,8-menthanediol) were detected in urine (Yamaguchi et al., 1994). Rabbits given a maximum toxic dose of 3.5 g of menthol over 10 h by gavage eliminated 86% of the dose as a glucuronic acid conjugate in the urine (Annex 1, reference *138*).

Isophorone (3,5,5-trimethyl-2-cyclohexenone, No. 1112) was partially excreted as the parent compound in exhaled air and urine of rabbits. Rabbits and rats given isophorone at a dose of 1000 mg/kg bw by gavage in water and olive oil, respectively, had untransformed isophorone in expired air and glucuronic acid conjugates in the urine 24 h later (Dutertre-Catella, 1978).

Polar alicyclic ketones of low relative molecular mass and secondary alcohols [e.g., cyclopentanone (No. 1101] were rapidly absorbed, metabolized, conjugated and excreted, mainly in the urine. Treatment with β-glucuronidase of urine collected from rabbits given cyclopentanone at a dose of 190 mg/kg bw orally showed that the major urlnary component was a glucuronic acid conjugate of cyclopentanol (James & Waring, 1971). Similar results were obtained with cyclohexanone (No. 1100). Urine collected from rabbits 18 h after dosing showed that 66% of the oral dose of 250 mg/kg bw had been excreted as the glucuronic acid conjugate of cyclohexanol. The authors concluded that cyclohexanone is first reduced to cyclohexanol and then conjugated with glucuronic acid prior to excretion in the urine (Elliott et al., 1959).

Four groups of male beagles were given cyclohexanone at a dose of 280 mg/kg bw per day by intravenous injection at various concentrations and injection rates for 18–21 days. When given at the highest concentration, cyclohexanol was detected in plasma within 30 min of injection, with mean distribution and elimination half-lives of 6.6 and 81 min, respectively. The mean steady-state volume of distribution of cyclohexanone was 2.6 l/kg, and the mean total body clearance was 27 ml/kg per min, indicating that cyclohexanone is rapidly cleared from the body. To clarify the pharmacokinetics of cyclohexanol, one dog was given 330 mg/kg bw by intravenous injection in a separate study. The plasma half-life was 99 min, the apparent distribution volume was 1.2 l/kg, and the total body clearance was 8.8 ml/kg per min; therefore, cyclohexanol was rapidly cleared. About 60% of an administered dose of cyclohexanone was recovered in 24-h urine as a glucuronide conjugate of cyclohexanol. Direct renal clearance of unmodified cyclohexanone and cyclohexanol is a minor route of elimination, accounting for < 1% of an administered dose. It has been proposed that 74–100% of cyclohexanone is converted to cyclohexanol and further metabolized before elimination. The authors proposed that some cyclohexanone is exhaled through the lungs (Martis et al., 1980; Koeferl et al., 1981).

Four men and four women volunteers were exposed to cyclohexanone at an atmospheric concentration of 100, 210 or 410 mg/m^3 for 8 h. Urine collected at 2-h intervals during exposure and for 72 h afterwards showed the presence of glucuronic acid conjugates of cyclohexanediol, with a peak excretion rate about 16 h after exposure. About 60% of the dose was excreted within 72 h (Mraz et al., 1994).

Premature neonates fed a solution of dextrose parenterally showed the presence of *trans*-1,2-cyclohexanediol, with lesser amounts of *trans*-1,4-cyclohexanediol and *cis*-1,3-cyclohexanediol in their urine. It was determined that cyclohexanone had leached from the containers used to deliver the dextrose solution at an average of 0.89 g/24 h (Mills & Walker, 1990). The absence of glucuronide conjugates in the urine of these neonates was explained by decreased activity of microsomal glucuronosyltransferase in premature neonatal liver tissue (Boreus, 1982).

After consuming 720 ml of saké (ethanol content, 10% w/v), a man ingested 100 ml of liquid adhesive containing 39% cyclohexanone. The cyclohexanone was rapidly absorbed as cyclohexanol, presumably by conversion of cyclohexanone catalysed by alcohol dehydrogenase. The plasma and urine concentrations of cyclohexanone and its metabolites were unaffected by gastric lavage (5.5 l of saline), two plasma exchanges (2.4 l each) and haemoperfusion, as compared with pre-treatment values. Cyclohexanol and cyclohexanone were detected in plasma for up to 25 h after ingestion. The concentration of cyclohexanone was at the lower limit of detection but was highest (220 μg/ml) 5 h after ingestion and had decreased to < 10 μg/ml after 20 h. High levels of cyclohexanol glucuronide were detected in the urine for up to 47 h. Urinary excretion of the parent ketone was described as minimal. The elimination half-life of cyclohexanone in human plasma was determined to be 4.8 h, and the rate of elimination (K_e) was 0.14 μg/ml per h. The mechanism of elimination in humans thus involves conversion of cyclohexanone to cyclohexanol, followed by conjugation with glucuronic acid (Sakata et al., 1989).

In summary, the esters of alicyclic secondary alcohols are readily hydrolysed. The resulting secondary alcohols are interconvertible with their corresponding ketones. In the principal excretion pathway, the secondary alcohols are conjugated with glucuronic acid and excreted primarily in the urine.

(c) Metabolism

As indicated above, the major metabolic pathway involves reduction of the alicyclic ketones to yield the corresponding secondary alcohols, which are subsequently excreted primarily as the glucuronic acid conjugates (Williams, 1959; Lington & Bevan, 1994; Topping, 1994). If a double-bond is present, it may be reduced to the corresponding dihydro derivative (Krasavage et al., 1982), presumably by the gut microflora for metabolites excreted into the bile. Endocyclic double-bonds are more prone to reduction than exocyclic double-bonds. In addition to reductive pathways, alicyclic ketones and secondary alcohols containing an alkyl side-chain also undergo oxidation of the side-chain to form polar metabolites, which are excreted as the glucuronide or sulfate conjugates mainly in the urine. In the case of substances of higher relative molecular mass, more lipophilic ketones or sterically hindered functional groups, oxidation of a ring position by cytochrome P450 may compete with reduction of the ketone function or oxidation of the alcohol function (Hildebrandt, 1902; Nelson et al., 1992).

(i) Alicyclicketones and secondary alcohols (Nos 1100 and 1101)

The metabolic fate of cyclohexanol is similar to that of simple acyclic aliphatic secondary alcohols (see Figure 1) (Lington & Bevan, 1994; Topping, 1994).

In rabbits, 66% of a dose of cyclohexanone of 190 mg/kg bw and 47% of a dose of cyclopentanone of 190 mg/kg bw administered by gavage were reduced to the corresponding secondary alcohol and excreted in urine as the glucuronic acid conjugate (Elliot et al., 1959; James & Waring, 1971). Trace amounts of mercapturic acid conjugate of the 2-hydroxycyclohexyl derivative were also detected, indicating that the epoxide may be the intermediary metabolite formed from dehydration (James & Waring, 1971). Cyclohexene is formed by dehydration of cyclohexanol, a known contaminant of commercially available cyclohexanone. Hepatic cytochrome P450 monooxygenases catalyse the epoxidation of cycloalkenes. The epoxide may hydrolyse to form diols or undergo nucleophilic attack by mercapturic acid to form the 2-hydroxycyclohexyl mercapturic acid conjugate.

Figure 1. Metabolic fate of cyclohexyl esters in humans

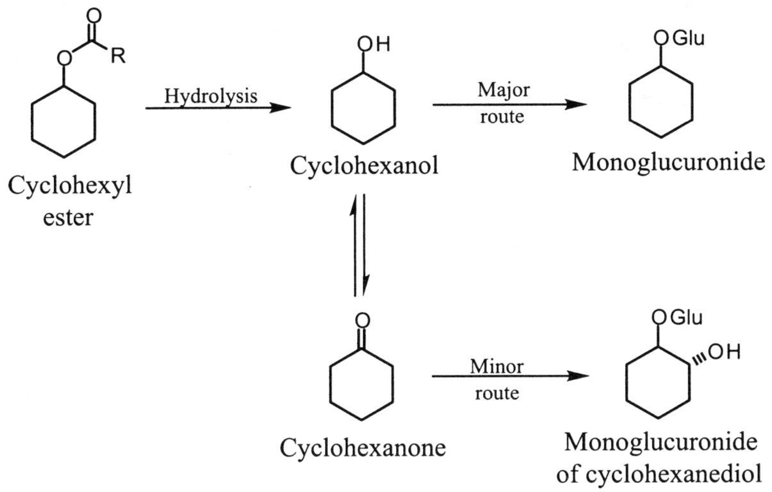

Glu, glucuronic acid

Eighteen-hour urine samples from rabbits given 1500 mg of cyclohexanone by gavage contained 65% cyclohexanol and 6% *trans*-cyclohexane-1,2-diol as monoglucuronide conjugates (Elliott et al., 1959). Presumably, the diol forms by hydroxylation at the α position and subsequent reduction of cyclohexanone. When cyclohexane at 460 mg/kg bw, cyclohexanol at 260 mg/kg bw or cyclohex-1-en-1-yl acetate at 350 mg/kg bw was fed to rabbits, the alcohol and to a lesser extent the diol were common urinary metabolites (Elliott et al., 1959).

The urine of rabbits given a dose of 1200 mg/kg bw of cyclohexanol orally showed the presence of glucuronic acid conjugates. Similar results were obtained in rabbits given cyclohexanone at a dose of 890 mg/kg bw orally (Treon et al., 1943). The glucuronic acid conjugates of cyclohexanol (1.6 mg/l) and cyclohexanone (0.23 mg/l) were found in the urine of workers exposed to a mixture of atmospheric hexanes including 460 mg/m^3 of cyclohexane (Perbellini et al., 1980; Governa et al., 1987). The authors concluded that cyclohexane is transformed to cyclohexanol, which subsequently forms glucuronic acid and sulfate conjugates.

(ii) Alkyl substituted alicyclic ketones and secondary alcohols (Nos 1099–1117)

When rabbits were given 2-, 3- or 4-methylcyclohexanone orally at doses of 200–3200 mg/kg bw, the glucuronic acid and sulfate conjugates of the corresponding secondary alcohols were excreted in their urine (Treon et al., 1943; Elliott et al., 1959; Tao & Elliott, 1962).

When a secondary alcohol or ketone function is located on alicyclic-containing alkyl substituents, as in 2-methylcyclohexanone (No. 1102), oxidation of the alkyl substituents competes with oxidation of the alcohol or reduction of the ketone function. When the substance contains allylic or tertiary hydrogen (e.g., 2-hexylidene cyclopentanone, No. 1106), the rate of side-chain oxidation increases, often leading to polyoxygenated metabolites (Dutertre-Catella, 1978, Nelson et al., 1992). The number of possible polyoxygenated metabolites increases with the number of types of alkyl ring substituents (e.g., methyl and isopropyl substituents) and the presence of an alkene (endocyclic and/or exocyclic), in which hydrogenation may precede or succeed oxidation of the substituents (Madyastha & Raj, 1990; Nelson et al., 1992; Yamaguchi et al., 1994). More lipophilic substances may undergo oxidation of the secondary alcohol function to the corresponding ketone, in addition to oxidation of alkyl substituents (Asakawa et al., 1986).

Urine samples collected over up to 4 days from rabbits given isophorone (No. 1112, 3,5,5-trimethyl-2-cyclohexen-1-one) by gavage at a dose of 1000 mg/kg bw contained several metabolites: the major one was 5,5-dimethyl-1-cyclohexene-3-one-1-carboxylic acid, formed by methyl-group oxidation at an exocyclic allylic position; 3,5,5-trimethyl-2-cyclohexen-1-ol (isophorol) was formed by reduction of the ketone group and then conjugation with glucuronic acid; 3,5,5-trimethylcyclohexanone (dihydroisophorone) was formed by hydrogenation of the endocyclic double-bond; and *cis*- and *trans*-3,5,5-trimethylcyclohexanol were formed by hydrogenation of the endocyclic double-bond and reduction of the ketone group (see Figure 2) (Truhaut et al., 1970; Dutertre-Catella, 1978).

Carvone (2-methyl-5-(1-methylethenyl)-2-cyclohexen-1-one), a structurally related α,β-unsaturated ketone, was partially excreted as the parent compound in both humans and rats (Tamura et al., 1962; Zlatkis et al., 1973). Allylic oxidation products, namely 9-hydroxycarvone, have also been detected in rabbits (see Figure 3) (Williams, 1959; Ishida et al., 1989). Carvone has been shown to induce cytosolic glutathione transferase activity in mice (Zheng et al., 1992), suggesting that it undergoes conjugation with glutathione at the β position (Portoghese et al., 1989). In rabbits, carvone was mainly reduced to yield carveol, which was then converted to the glucuronic acid conjugate and excreted in urine (Fisher & Bielig, 1940). Unchanged dihydrocarveol I (Fisher & Bielig, 1940) and its glucuronic acid conjugate (Hamalainen, 1912) were detected in the urine of rabbits given carvone.

Figure 2. Recognized metabolic fates of isophorone in rabbits

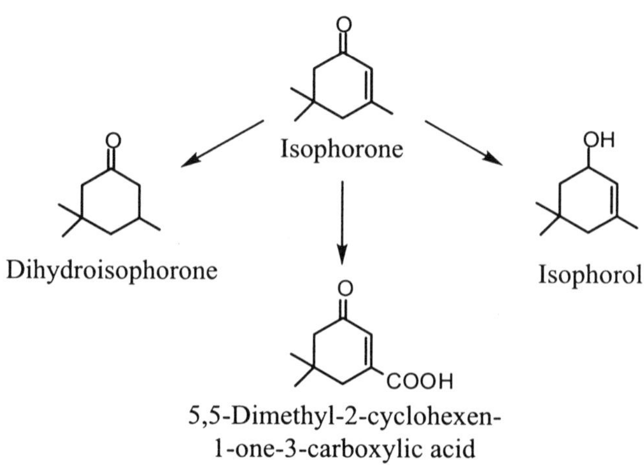

Figure 3. Metabolic fate of carvone in animals

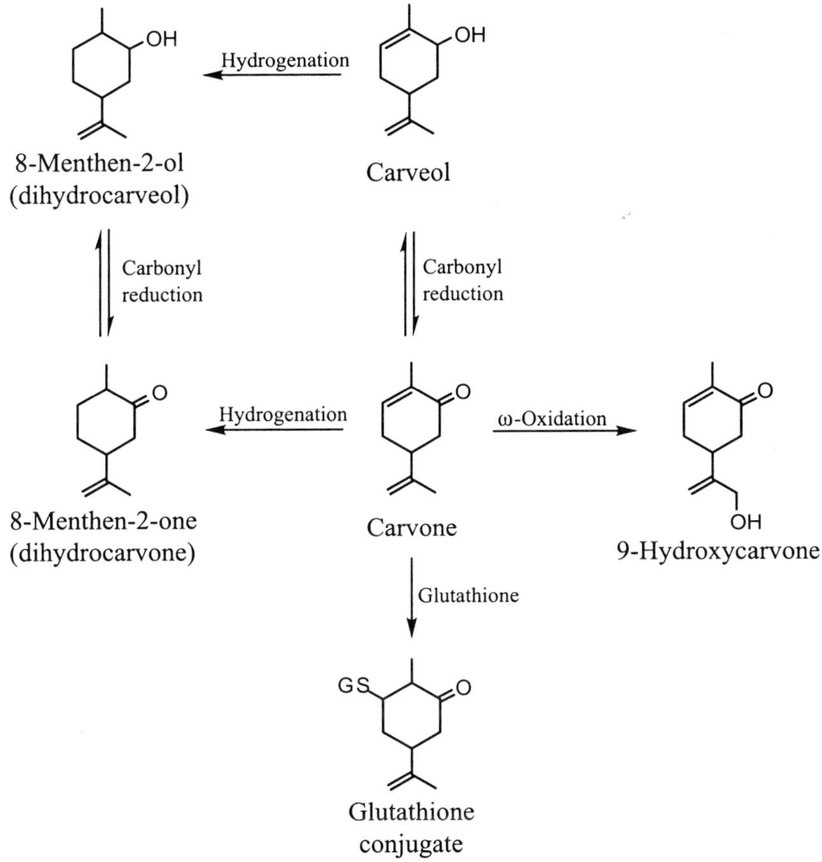

In summary, these 25 alicyclic ketones, secondary alcohols and related esters are metabolized primarily by reduction of the ketone to yield the corresponding secondary alcohol, followed by conjugation with glucuronic acid and excretion in urine. There is also evidence that small amounts are excreted as sulfate conjugates. If side-chains are present, side-chain oxidation may occur. If the ketone is α,β-unsaturated, it may be conjugated with glutathione or undergo allylic oxidation to yield polar oxygenated metabolites that can be excreted in urine.

2.3.2 *Toxicological studies*

(a) Acute toxicity

Oral LD_{50} values have been reported for 10 of the 25 substances in this group (see Table 3). In rats, the values ranged from 1200 mg/kg bw for cyclopentanone (No. 1101) to > 5000 mg/kg bw for cyclohexyl acetate (No. 1093), 2-hexylidenecyclopentanone (No. 1106), isojasmone (No. 1115) and 2-(3,7-dimethyl-2,6-octadienyl)-

Table 3. Studies of acute toxicity of alicyclic ketones, secondary alcohols and related esters used as flavouring agents

No.	Agent	Species	Sex[1]	Route	LD_{50} (mg/kg bw)	Reference
1093	Cyclohexyl acetate	Rat	M,F	Oral	> 5000	Moreno (1977)
1094	Cyclohexyl butyrate	Mouse	M,F	Oral	> 5000	Moreno (1982)
1100	Cyclohexanone	Rat	M,F	Oral	1700	Kohli et al. (1967)
1100	Cyclohexanone	Rat	M,F	Oral	1800	Deichmann & LeBlanc (1943)
1100	Cyclohexanone	Rat	M,F	Oral	1500	Smyth et al. (1969)
1100	Cyclohexanone	Rat	M,F	Oral	1800	Gupta et al. (1979)
1100	Cyclohexanone	Mouse	M	Oral	2100	Gupta et al. (1979)
1100	Cyclohexanone	Mouse	F	Oral	2100	Gupta et al. (1979)
1101	Cyclopentanone	Rat	M,F	Oral	1200	Levenstein (1976)
1101	Cyclopentanone	Rat	M	Oral	2700	Exxon Chemical Americas (1982)
1106	2-Hexylidene cyclopentanone	Rat	M,F	Oral	> 5000	Moreno (1980)
1109	2-*sec*-Butylcyclohexanone	Rat	NR	Oral	2300	Leberco Laboratories (1965)
1109	2-*sec*-Butylcyclohexanone	Rat	M,F	Oral	2400	Moreno (1978)
1112	Isophorone	Rat	M,F	Oral	3400	Exxon Chemical Americas (1982)
1113	3-Methyl-5-propyl-2-cyclohexen-1-one	Rat	M,F	Oral	2700	Moreno (1982)
1115	Isojasmone	Rat	M,F	Oral	> 5000	Moreno (1974)
1117	2-(3,7-Dimethyl-2,6-octadienyl)cyclopentanone	Rat	M,F	Oral	> 5000	Gabriel (1991)

M, male; F, female; NR, not reported

cyclopentanone (No. 1117), demonstrating that the acute oral toxicity of alicyclic ketones, secondary alcohols and related esters is extremely low.

In mice, the oral LD_{50} values were reported to range from 2100 mg/kg bw for cyclohexanone (No. 1100) to > 5000 mg/kg bw for cyclohexyl butyrate (No. 1094), confirming the extremely low acute toxicity of these alicyclic ketones, secondary alcohols and related esters.

(b) Short-term studies of toxicity

Ninety-day or 13-week studies were available for five of the 25 alicyclic ketones and secondary alcohols in this group (Table 4). Two of the five studies were performed at a single target level of intake that was at least 100 times the estimated possible average daily intake from use of the substance as a flavouring agent (Oser & Hall, 1977). The possible average daily intake is determined by multiplying usual use levels in each of 33 food categories (e.g., baked goods and meat products) by the average amount of that food category consumed daily and summing the intake over all 33 food categories (Department of Agriculture, 1965). The possible average daily

Table 4. Results of short-term, long-term and carcinogenicity studies of aicyclic ketones, secondary alcohols and related esters used as flavouring substances

No.	Substance	Species, sex	No. test groups[a]/ no. per group[b]	Route	Duration (days)	NOEL (mg/kg bw per day)	Reference
1100	Cyclohexanone	Mouse, M,F	7/20	Drinking-water	91	1600 (M) 3200 (F)	Lijinsky & Kovatch (1986)
1100	Cyclohexanone	Mouse, M,F	3[c]/96–102	Drinking-water	730	ND (M) 1600 (F)	Lijinsky & Kovatch (1986)
1100	Cyclohexanone	Rat, M,F	7/10	Drinking-water	175	500 (M) 720 (F)	Lijinsky & Kovatch (1986)
1100	Cyclohexanone	Rat, M,F	2/104	Drinking-water	730	650	Lijinsky & Kovatch (1986)
1106	2-Hexylidene cyclopentanone	Rat, M,F	1/10–32	Diet	90	2.9 (M) 3.4 (F)	Posternak et al. (1969)
1109	2-*sec*-Butyl cyclohexanone	Rat	3/NR	Diet	91	370	Hummler (1969)
1111	Tetramethyl ethyl cyclohexenone (mixed isomers)	Rat, M,F	1/10–32	Diet	90	40 (M) 48 (F)	Posternak et al. (1969)
1112	Isophorone	Mouse, M, F	5/20	Gavage	91	1000 (M) 500 (F)	Bucher et al. (1986); National Toxicology Program (1986)
1112	Isophorone	Mouse, M,F	2/100	Gavage	730	250[d]	Bucher et al. (1986); National Toxicology Program (1986)

[a] Does not include control animals
[b] Includes both male and female animals
[c] At the highest dose (6.2 mg/kg bw per day), only females (41) were tested
[d] Performed at either a single or multiple doses that produced no adverse effects. Therefore, this dose level is not a true NOEL but the highest dose tested that produced no adverse effects. The actual NOEL would be higher.

intake of the vast majority of flavouring agents with low reported annual volumes of use (National Academy of Sciences, 1987; Lucas et al., 1999) is a gross exaggeration of the average daily intake. The calculation is based on the assumption that all foods in a category always contain the substance and that the food category is consumed each day (Oser & Hall, 1977). Therefore, the dietary concentrations in these studies were many orders of magnitude greater than actual levels of intake of alicyclic ketone, secondary alcohol and related esters as flavouring agents.

An alicyclic ketone, cyclohexanone (No. 1100), was used in two studies (Lijinsky & Kovatch, 1986); alkyl-substituted ketones and secondary alcohols, 2-*sec*-butylcyclohexanone (No. 1109), tetramethylethylcyclohexenone (isomeric mixture) (No. 1111) and 2-hexylidenecyclopentanone (No. 1106) in two (Hummler, 1969; Posternak et al., 1969); and an α,β-unsaturated alicyclic ketone, isophorone (No. 1112) in one (Bucher et al., 1986; National Toxicology Program, 1986). Additionally, 13-week studies were performed with structurally related alicyclics, DL-menthol [(1*S*-1α,2β,5α)-5-methyl-2-(1-methylethyl)cyclohexanol] (National Cancer Institute, 1978) and D-carvone (2-methyl-5-(1-methylethenyl)-2-cyclohexenone) (Hagan et al., 1967; National Toxicology Program, 1990). No histopathological changes were found. The only consistent effect found at high doses in these studies was a decrease in body-weight gain in treated groups.

(i) Alicyclic ketones and secondary alcohols

Cyclohexanone (No. 1100) and related secondary alcohols: A 13-week study was performed in which groups of 10 B6C3F$_1$ mice of each sex received drinking-water containing cyclohexanone (No. 1100) at a concentration of 0, 400, 2300, 6500, 13 000, 25 000 or 47 000 ppm, corresponding to 0, 100, 580, 1600, 3200, 6200, 8500 and 12 000 mg/kg bw per day, respectively (Food and Drug Administration, 1993). The drinking-water was acidified to pH 2.5 to suppress bacterial growth. Food and water were provided *ad libitum*. The mice were monitored daily for survival and weekly for changes in body weight. At necropsy, at least 28 tissues were examined histologically.

No effects were reported at 100, 580 or 1600 mg/kg bw per day. Male mice given 3200 mg/kg bw per day gained slightly less weight, and those at 6200 mg/kg bw per day dose showed a 19% decrease in body-weight gain when compared with controls. Among animals at 8500 mg/kg bw per day, one male died before 13 weeks, and weight gain was depressed by 15% in females and 24% in males. At the highest dose, 6/10 males and 3/10 females died, some mice showed focal coagulative liver necrosis (number and sex not reported), and two females showed hyperplasia of the thymus. The maximal tolerated dose was estimated by the authors to be 6200 mg/kg bw per day for females and 3200 mg/kg bw per day for males (Lijinsky & Kovach, 1986). The concentration of 3200 mg/kg bw per day is more than 10 000 000 times the daily per capita intake ("eaters only") of 0.002 μg/kg bw of cyclohexanone from its use as a flavouring agent in Europe and the USA (Table 2).

A 25-week study was performed in which groups of five Fischer 344 rats of each sex were given drinking-water containing cyclohexanone (No. 1100) at a concentration of 0, 190, 400, 800, 1600, 3300, 4700 or 6500 ppm, corresponding to 0, 30, 60, 120, 240, 500, 720 and 1000 mg/kg bw per day, respectively (Food and Drug Administration, 1993). The drinking-water was acidified to pH 2.5 to suppress bacterial growth. Food and water were provided *ad libitum*. The rats were monitored daily for survival and weekly for changes in body weight. At necropsy, major organs and tissues were examined histologically, and the observations were used to determine the maximal tolerated dose for this species. All the rats survived to termination of the study at 25 weeks. No observable effects were reported in male or female rats at doses up to 500 mg/kg bw per day. Two males at 720 mg/kg bw per day developed degenerative changes of the thyroid gland that were not seen in other animals in this study. Animals at the highest dose had a 10% depression in weight gain. No other adverse effects were observed. The maximal tolerated dose was calculated to be 1000 mg/kg bw per day (Lijinsky & Kovatch, 1986). This concentration is more than 5 000 000 times the daily per capita intake[2] ("eaters only") of 0.002 μg/kg bw of cyclohexanone from its use as a flavouring agent in Europe and the USA (Table 2).

(ii) Alkyl substituted ketones and secondary alcohols

2-sec-Butylcyclohexanone (No. 1109): In a 13-week study, an unspecified number of rats of an unspecified strain were given diets containing 2-*sec*-butylcyclohexanone (No. 1109) at a concentration calculated to provide an average daily intake of 0, 160, 370 or 900 mg/kg bw per day. Some dose-dependent depression in body-weight gain was reported, which was attributed in part by the authors to the presumed unpalatability of the food. No adverse effects were observed at the two lower doses. An increased mortality rate was reported among rats at 900 mg/kg bw per day, but no details were given (Hummler, 1969). The dose of 370 mg/kg bw per day that resulted in no adverse effects is more than 1 000 000 times the daily per capita intake ("eaters only") of 0.1 μg/kg bw from its use as a flavouring agent in Europe and the USA (Table 2).

Tetramethylethylcyclohexenone (isomeric mixture) (No. 1111) and 2-hexylidene-cyclopentanone (No. 1106): These two alicyclic ketones were studied by the same protocol in two 90-day studies. A control and a test group, each consisting of 10–16 male and 10–16 female Charles River CD rats, were housed in pairs of the same sex and given access to water and food *ad libitum*. The concentration of the test material in the diet was adjusted during the study to maintain constant levels of intake of tetramethylethylcyclohexenone (isomeric mixture) (No. 1111) of 40 mg/kg bw per day for males and 48 mg/kg bw per day for females, and levels of 2-hexylidenecyclopentanone (No. 1106) of 2.9 mg/kg bw per day for males and 3.4 mg/kg bw per day for females. These dietary levels were calculated to be > 100 times an exaggerated estimate of possible average daily intake from use as flavouring agents. Clinical observations were recorded daily, and food consumption and body weights were determined weekly. Haematological and clinical chemical determinations (blood urea) performed at weeks 7 and 13 revealed elevated blood urea at week 7 in rats given 2-hexylidenecyclopentanone but depressed blood urea at week 14. When the values were compared with those for a composite rather than individual control groups, the differences were not significant. At necropsy, liver and kidney weights were measured; a wide range of tissues and organs from each animal were preserved, and major organs and tissues were examined microscopically. Observations of growth, food intake, haematological and clinical chemical parameters, organ weights and histopathological examination revealed no significant differences between control animals and treated groups given tetramethylethylcyclohexenone (isomeric mixture) or 2-hexylidenecyclopentanone (Posternak et al., 1969). The dose of 48 mg/kg bw per day of tetramethylethylcyclohexenone that resulted in no adverse effects is more than 100 000 times the estimated total daily per capita intakes ("eaters only") of up to 0.15 μg/kg bw from use of this substance as a flavouring agent in Europe and the USA (Table 2).

Isophorone (No. 1112): In a 13-week study, groups of 10 B6C3F$_1$ mice and Fischer 344/N rats of each sex received isophorone (No. 1112) in corn oil by gavage at a dose of 0, 62, 120, 250, 500 or 1000 mg/kg bw per day for 5 days per week. Food and water were provided *ad libitum*. Moribund animals were killed. Body weights were recorded weekly. At termination of the experiment, survivors were killed and necropsied. Three female mice receiving the highest dose died before the end of the study. The body weights of male mice receiving 250, 500 or 1000 mg/kg bw per day were reported to be approximately 10% less than those of male controls; however, no such difference was observed in females, and there was no correlation between final mean body weights and dose. No compound-related gross or microscopic changes were observed. The kidneys of animals at the highest dose were reviewed on two separate occasions, and no evidence of nephropathy was found. One female rat given the highest dose died, but no other adverse effects or dose-related responses were noted. A comprehensive examination of the kidneys of animals at the highest dose revealed no evidence of nephrotoxicity (Bucher et al., 1986; National Toxicology Program, 1986).

(c) *Long-term studies of toxicity and carcinogenicity*

(i) *Alicyclic ketones and secondary alcohols*

Cyclohexanone (No. 1100): Groups of 41–52 B6C3F$_1$ mice of each sex were given drinking-water containing cyclohexanone at a concentration of 0, 6500, 13 000 or 25 000 ppm (females only), corresponding to dietary intakes of 0, 1600, 3200 and 6200 mg/kg bw per day, respectively. The pH of the drinking-water was maintained at 2.5 to suppress bacterial growth. The mice were monitored daily for survival and weekly for body-weight changes. At death or termination, they were subjected to complete necropsy and histopathology. Mice at the lowest dose and females at 3200 mg/kg bw per day showed no significant changes in weight when compared with controls. The body weights of male mice at 3200 mg/kg bw per day and of females at 6200 mg/kg bw per day were 15–20% lower than those of controls during most of the study. No differences in survival rates were reported for animals receiving 1600 mg/kg bw per day; 80% of those at 3200 mg/kg bw per day were alive at week 90. The survival of female mice at the two higher doses was poor: only half the females at the highest dose survived to 50 weeks and < 20% to 75 weeks, and of females at 3200 mg/kg bw per day, only 50% survived to 75 weeks and only 40% to 90 weeks. Histopathological examination of male mice at 1600 mg/kg bw per day revealed an increased incidence of proliferative lesions of the liver and lung, combined with a statistically significant ($p = 0.041$) increase in the incidence of benign and malignant hepatocellular neoplasms (25/51, 49%) when compared with controls (16/52, 31%). No benign or malignant neoplasms were seen at the two higher concentrations. Female mice at 1600 mg/kg bw per day had an increased incidence of malignant lymphomas (17/50, 34%), which was significant in the life-table (p = 0.036) and incidental tumour tests (p = 0.040), suggesting a weak carcinogenic effect (Lijinsky & Kovatch, 1986). The increased incidence of malignant lymphomas among females at the lowest dose may be attributable to the high, variable background incidence of this lesion in control B6C3F$_1$ mice used for National Toxicology Program studies, which has been increasing with time (Haseman et al., 1986, 1997, 1998). The incidence in female B6C3F$_1$ mice used as controls in studies sponsored by the National Toxicology Program (2001) was 6–30%. The background incidence in female control mice in this study (15%, 8/52) (Lijinsky & Kovatch, 1986) was thus within the range for this tumour type in control B6C3F$_1$ female mice.

In a separate 2-year study with the same protocol, groups of 52 Fischer 344/N rats of each sex were given drinking-water containing cyclohexanone at a concentration of 0, 3300 or 6500 ppm, corresponding to 0, 330 or 650 mg/kg bw per day (Food and Drug Administration, 1993). Rats in both treated groups failed to reach the mean body weight of the control group. More than 85% of the rats at 650 mg/kg bw per day survived to week 90, and 70% survived to the end of the study. The authors noted that the incidences and types of many neoplasms were similar in treated and control groups. A statistically significant increase in the incidence of adrenal cortex adenomas (7/52, 13%; p = 0.030) was reported in males at 330 mg/kg bw per day when compared with controls (1/52, 2%), but no increase was found at 650 mg/kg bw per day (1/51, 2%). The authors noted that the incidence of adrenal cortex adenomas in Fischer 344/N rats in National Toxicology Program laboratories was approximately 1%. They concluded that, in the absence of a dose–response relationship, the increased incidence of benign neoplasms was not indicative of a carcinogenic response (Lijinsky & Kovatch, 1986).

Thus, no dose-dependent effects were observed in either mice or rats, and histopathological examination revealed no lesions that could be associated with administration of the test material. The authors concluded that the doses of 3200 mg/kg bw per day for mice of each sex and 330 mg/kg bw per day for male rats were weakly carcinogenic. These conclusions should be evaluated in relation to the finding of no evidence of carcinogenicity with higher doses in female mice (6200 mg/kg bw per day) and in rats (650 mg/kg bw per day). These data were evaluated independently by a working group convened by the International Agency for Research on Cancer (1989), which determined that cyclohexanone was "not classifiable as to its carcinogenicity to humans".

Isophorone (No. 1112): A 2-year bioassay was conducted in which isophorone (No. 1112) was tested in mice and rats with the standardized National Toxicology Program protocol. The doses were determined from the results of a 13-week study. In the 2-year study, groups of 50 B6C3F$_1$ mice and Fischer 344/N rats of each sex received isophorone in corn oil by gavage at a dose of 0, 250 or 500 mg/kg bw per day on 5 days per week for 103 weeks. Food and water were provided *ad libitum*. Moribund animals were killed. Body weights were recorded weekly. At the end of the experiment, survivors were killed and necropsied. No clinical signs of toxicity were reported, but significant numbers of animals (36/300) died due to gavage errors. The survival of male rats at the higher dose was reported to be lower than that of controls, and the mean body weights of rats at this dose were reported to be significantly lower than those of controls. No other treatment-related clinical signs were reported. In mice, no treatment-related clinical signs were reported, and isophorone did not adversely affect survival rates. The mean body weights of females at the higher dose were reported to be 5% lower than those of controls during the second year of the study (Bucher et al., 1986; National Toxicology Program, 1986).

Neoplastic and non-neoplastic lesions associated with administration of isophorone to male mice developed principally in the liver (see Table 5). Male mice at 500 mg/kg bw per day showed a significantly increased incidence of coagulative necrosis and hepatocytomegaly, but females showed a decrease. Male mice at the higher dose showed a statistically significant increase in the incidence of combined hepatocellular adenomas and carcinomas (29/50, 58%; $p = 0.033$) over that in controls (18/48, 38%). No significant difference in the incidence of hepatocellular neoplasms was found in the group at the lower dose. Female $B6C3F_1$ mice showed no evidence of hepatocellular neoplasms at either dose. The authors concluded that there was equivocal evidence of carcinogenic activity in male B6C3F1 mice at 500 mg/kg bw per day, as shown by an increased incidence of hepatocellular adenoma and carcinoma (combined), but no evidence of carcinogenicity in female $B6C3F_1$ mice at 250 or 500 mg/kg per day (National Toxicology Program, 1986).

As stated previously, the primary neoplastic effects observed in male mice in the National Toxicology Program study were in the liver. The incidences of hepatocellular adenoma in control and high-dose groups of male mice (6/48 and 13/50, respectively) and of carcinomas (14/48 and 22/50, respectively) demonstrate the susceptibility and sensitivity of the $B6C3F_1$ male mouse liver to carcinogenic agents. This conclusion is supported by the fact that the combined incidence of adenomas and carcinomas in control males was greater than in any group of treated females (6/50 and 8/50, respectively). These responses are consistent with the high incidences of hepatocellular neoplasms found in control $B6C3F_1$ male mice in other studies (Maronpot & Boorman, 1982; Maronpot et al., 1987).

The observation of hepatic neoplasms in mice in the National Toxicology Program bioassay are not relevant to the safety evaluation of isophorone in humans at low levels of intake from use as a flavour ingredient. This conclusion is based on the

Table 5. Incidences of hepatocellular neoplasms associated with administration of isophorone to mice by gavage for 103 weeks

Lesion	Dose (mg/kg bw per day)		
	Control	250	500
Males			
Hepatocellular adenoma	6/48	7/50	13/50
Hepatocellular carcinoma	14/48	13/50	22/50
Combined*	18/48 (38%)	18/50 (36%)	29/50 (58%) $p = 0.033$
Females			
Hepatocellular adenoma	2/50	4/50	6/50
Hepatocellular carcinoma	2/50	2/50	2/50
Combined*	4/50 (8%)	6/50 (12%)	8/50 (16%)

*Historical incidence rate: 335/1034 (32% ± 9.4%)

high incidence of spontaneous hepatocellular neoplasms (adenomas and carcinomas) in the strain of mice studied, the absence of a dose–response relationship, the lack of hepatocellular neoplastic effects in rats and the relatively high doses administered, as compared with intake levels from use as a flavour ingredient. The dose of 250 mg/kg bw per day used in the 2-year rat bioassay is > 1 000 000 times the estimated daily *per capita* intake ("eaters only") of 2 ng/kg of isophorone from its use as a flavour ingredient.

A statistically significant increase in the incidence of mesenchymal neoplasms (14/50, 28%; $p = 0.011$), primarily subcutaneous fibrosarcomas, was reported in male mice at the higher dose (500 mg/kg bw per day) when compared with the control group (6/48, 13%) (Bucher et al., 1986; National Toxicology Program, 1986). More recent bioassays in mice in which corn oil was used as the vehicle showed a high background incidence of subcutaneous fibrosarcomas. The incidence of combined subcutaneous fibrosarcomas, fibromas and fibrosarcomas in control groups of male mice has been reported to be as high as 20% (National Toxicology Program, 1992b). The observation of mesenchymal neoplasms in mice in this bioassay is also not relevant to the safety evaluation of isophorone in humans at low levels of intake. This conclusion is based on the high background incidence of spontaneous fibrosarcomas in the strain and sex of mice studied, the lack of similar neoplastic effects in rats and the relatively high dose (500 mg/kg bw per day) administered as compared with intake levels from use as a flavour ingredient.

Nephropathy was seen in both treated and control rats of each sex after natural death or at termination. In male rats at 250 and 500 mg/kg bw per day, increased incidences of mineral deposits in renal collecting ducts (31/50, 62%, and 20/50, 40%), tubular-cell hyperplasia (1/50, 2%, and 4/50, 8%), adenomas (0/50 and 2/50, 8%) and adenocarcinomas (3/50, 6%, and 1/50, 2%) were observed, but these were not found not in female rats (see Table 6). Tubule mineralization was characterized by basophilic aggregates in the medullary collecting ducts, often occurring coincidentally with lesions of chronic nephropathy. The authors of the report concluded that there was some evidence for the carcinogenicity of isophorone in male Fischer 344/N rats, as shown by the occurrence of renal tubule cell adenomas and adenocarcinomas in animals given 250 or 500 mg/kg per day; however, there was no evidence of carcinogenicity in female rats at these doses (Bucher et al., 1986; National Toxicology Program, 1986).

It has been shown that renal lesions result from the accumulation of aggregates of α-2u-globulin (a protein of low relative molecular mass synthesized in the liver) and isophorone or its metabolites in the P2 segment of the renal proximal tubule. This phenomenon has been observed only in male Fischer 344/N rats and has been referred to as α-2u-globulin nephropathy (Strasser et al., 1988; Borghoff et al., 1990). The gene that encodes α-2u-globulin has been isolated and the sequence deduced (Untermann et al., 1981). These proteins are expressed in the liver under hormonal control and are the major urinary protein in adult male rats (Roy & Neuhaus, 1967; Wang & Hodgetts, 1998). α-2u-Globulin belongs to a superfamily of proteins characterized by a unique hydrophobic binding pocket (Swenberg et al., 1989). The characteristic renal lesions have not been reported in female Fischer 344/N rats or in humans (Bucher et al., 1986). Subsequent investigations have shown that the α-2u-globulin nephropathy found in male Fischer 344/N rats does not develop in mammals that do not express the hepatic form of α-2u-globulin (Swenberg et al.,

Table 6. Incidences of renal neoplasms associated with administration of isophorone to rats by gavage for 103 weeks

Lesion	Dose (mg/kg bw per day)		
	Control	250	500
Males			
Nephropathy	49/50	47/50	46/50
Tubule mineralization	1/50	31/50	20/50
Renal tubule hyperplasia	0/50	1/50	4/50
Renal tubule adenoma*	0/50	0/50	2/50
Renal tubule adenocarcinoma*	0/50	3/50	1/50
Females			
Nephropathy	21/50	39/50	32/50
Tubule mineralization	10/50	4/50	2/50
Renal tubule hyperplasia	0/50	0/50	1/50
Renal tubule adenoma*	0/50	0/50	0/50
Renal tubule adenocarcinoma*	0/50	0/50	0/50

* Incidence of tubule cell adenoma or adenocarcinoma in other controls: 4/1091 (0.4%); $p < 0.05$

1989), such as other strains of rats (Dietrich & Swenberg, 1991), mice (Bucher et al., 1986; National Toxicology Program, 1986; Lehman-McKeeman & Caudill, 1994) and dogs (Webb et al., 1990).

While humans produce serum proteins of low relative molecular mass which are resorbed by the kidney, there is no evidence that α-2u-globulin is produced (Swenberg et al., 1989; Olson et al., 1990). Comparison of urine collected from adult male Fischer 344 rats and humans revealed no evidence of α-2u-globulin production in humans (Olson et al., 1990). The authors concluded that the very low protein content of human urine, the relatively small proportion of cationic to total proteins and the high relative molecular mass of the most abundant human urinary proteins form a biological basis for suggesting that humans are not at risk for developing the α-2u-globulin nephropathy observed in male Fischer 344/N rats.

It is not known whether any human serum proteins have a binding site similar to that of α-2u-globulin. Although this is a possibility, it appears remote, since female rats, mice and dogs do not show the renal changes seen in male rats exposed to isophorone. There is a class of human proteins referred to as the α-2u-globulin-related proteins, which appear to have no functional relationship to adult male rat urine proteins. The human protein has a higher relative molecular mass, 25 kDa, and is a component of a neutrophil gelatinase complex (Triebel et al., 1992; Kjeldsen et al., 2000). An extensive review of the current scientific literature and genome databases revealed no native protein or biological entity that has the nephropathic effects of α-2u-globulin in mature male rats. The accumulated evidence indicates that the unique anatomical, physiological and biochemical properties of the male rat kidney, especially the proximal convoluted tubule, allow isophorone to interfere with renal processing of the strain-specific α-2u-globulin. Therefore, this process is not predictive of human carcinogenicity. In a comprehensive review of α-2u-globulin nephropathy and associated renal tubule tumours produced in male Fischer 344/N rats exposed to isophorone and other simple chemical substances (e.g., limonene, decalin and methyl isobutyl ketone), it was concluded that the Fischer 344/N rat is not an appropriate model for assessing human renal carcinogenic risk (Environmental Protection Agency, 1991). It has also been concluded that the mechanisms leading to renal carcinogenicity in male Fischer 344/N rats are largely known and strongly indicate that the α-2u-globulin nephropathy associated with compounds such as isophorone have no significance for human risk (Burdock et al., 1990).

Preputial gland carcinomas were observed in 5/50 (10%) male rats at the higher dose, and clitoral gland adenomas were reported in 2/50 (4%) female rats at the lower dose, as compared with the incidence of 0/50 in vehicle control groups of each sex. The preputial lesions were believed to be significant, in the apparent absence of lesions in vehicle controls (0/50), males at the lower dose (0/50) and females at the higher dose (0/50) and the low incidence of this lesion in controls given corn oil in other experiments (12/1094). However, the authors emphasized that the actual incidence of lesions of the prepuce and clitoris was unknown since only animals with visible lesions were examined histologically (Bucher et al., 1986).

Macroscopic preputial tumours have been reported sporadically in vehicle controls in previous studies of the National Toxicology Program (1986). In more recent studies, in which histopathological examination of the prepuce was performed on all male rats, the incidence of preputial neoplasms in control rats exceeded the incidence reported in the isophorone-treated male rats (5/50; $p < 0.05$). For example, the combined incidence of preputial adenomas and carcinomas in control male rats treated only with corn oil by gavage was as high as 23% (11/47; $p = 0.218$) (National Toxicology Program, 1993a), while the incidence of preputial carcinomas in other control males was reported to be as high as 12% (6/50; $p = 0.092$) (National Toxicology Program, 1993b). In addition, the high background incidence of clitoral gland adenomas, 6% (3/47; $p = 0.346$) to 14% (7/49; $p = 0.093$), in female rats serving as vehicle controls in studies with corn oil by gavage (National Toxicology Program, 1993a,b, 1994) supports the conclusion that the clitoral gland adenomas reported in two females at the lower dose in the study of isophorone (National Toxicology Program, 1986) are of no relevance to humans.

This conclusion was confirmed in a similar study to determine whether there is a link between preputial gland carcinomas and α-2u-globulin and/or isophorone (No. 1112). A series of experiments was performed that included measurements of binding of α-2u-globulin to DNA, with [^{14}C]isophorone as tracer. [^{32}P]DNA post-labelling was used to detect adduct formation and cell proliferation. No DNA adducts were detected. An increase in cell proliferation was observed after 1 week of exposure in both male and female rats, but only a slight increase was seen in the males after 4 weeks (data not shown). The authors concluded that preputial carcinomas are specific to rats and are not relevant to human risk assessment (Morishita et al., 1997).

(c) Genotoxicity

The studies of genotoxicity conducted with this group of substances are summarized in Table 7.

Table 7. Studies of genotoxicity with alicyclic ketones, secondary alcohols and related esters used as flavouring agents

No.	Agent	End-point	Test object	Maximum concentration	Results	Reference
In vitro						
1093	Cyclohexyl acetate	DNA damage	*B. subtilis* H17(*rec*+), M45 (*rec*-)	19 mg/disc	Negative[a]	Yoo (1986)
1094	Cyclohexyl butyrate	DNA damage	*B. subtilis* H17(*rec*+), M45 (*rec*-)	19 mg/plate	Negative[a]	Oda *et al.* (1979
1100	Cyclohexanone	Reverse mutation	*S. typhimurium* TA98, TA100, TA1535, TA1537	33–10 000 mg/plate	Negative[a]	Haworth et al. (1983)
1100	Cyclohexanone	Reverse mutation	*S. typhimurium* TA98, TA100, TA1535, TA1537	2.9–2900 mg/plate	Negative[a]	Florin et al. (1980)
1100	Cyclohexanone	Chromosomal aberration	Chinese hamster ovary cells	7.5 μl/ml	Negative[a]	Aaron et al. (1985)
1100	Cyclohexanone	Chromosomal aberration	Human lymphocytes	9.8–980 mg/ml	Positive[a]	Lederer et al. (1971)
1100	Cyclohexanone	Chromosomal aberration	Human lymphocytes	0.005–0.1 mg/ml	Positive[a]	Dyshlovoi et al. (1981)
1100	Cyclohexanone	Sister chromatid exchange	Chinese hamster ovary cells	7.5 μl/ml	Negative[b] Positive[c]	Aaron et al. (1985)
1101	Cycopentanone	Reverse mutation	*S. typhimurium* TA98, TA100, TA1535, TA1537	2.5–2500 mg/plate	Negative[a]	Florin et al. (1980)
1106	2-Hexylidene cyclo-pentanone	Reverse mutation	*S. typhimurium* TA98, TA100, TA1535, TA1537, TA1538	≤ 3600 mg/plate	Negative[a]	Wild et al. (1983)
1108	2,2,6-Trimethyl cyclo-hexanone	Reverse mutation	*S. typhimurium* TA98, TA100, TA1535,TA1537	4.2–3600 mg/plate	Negative[a]	Florin et al. (1980)
1111	Tetramethyethyl cyclo-hexanone (mixed isomers)	Reverse mutation	*S. typhimurium* TA98, TA100, T A1535, TA1537, TA1538	≤ 3600 mg/plate	Negative[a]	Wild et al. (1983)
1112	Isophorone	Reverse mutation	*S. typhimurium* TA97, TA98, TA100, TA1535, TA1537	33–10 000 mg/plate	Negative[a]	Mortelmans *et al.* (1986)

Table 7 (contd)

No.	Agent	End-point	Test object	Maximum concentration	Results	Reference
1112	Isophorone	Mutation	*S. typhimurium* TA98, TA100, TA1535, TA1537	33–10 000 mg/plate	Negative[a]	National Toxicology Program (1986)
1112	Isophorone	Mutation	L5178YTk$^{+/-}$ mouse lymphoma cells	67–810 mg/ml 130–1300 mg/ml	Negative[b] Negative[c]	McKee et al. (1987)
1112	Isophorone	Mutation	L5178YTk$^{+/-}$ mouse lymphoma cells	0.089–0.89 ml/ml 0.13–1.3 ml/ml	Negative[b] Negative[c]	O'Donoghue et al. (1988)
1112	Isophorone	Mutation cells	L5178YTk$^{+/-}$ mouse lymphoma	≤ 800 mg/ml[d]	Positive[c]	McGregor et al. (1988)
1112	Isophorone	Mutation	L5178YTk$^{+/-}$ mouse lymphoma cells	≤ 1200 mg/ml	Negative[b] Positive[c]	National Toxicology Program (1986)
1112	Isophorone	Chromosomal aberration	Chinese hamster ovary cells	5–1600 mg/ml	Negative[a]	Gulati et al. (1989)
1112	Isophorone	Chromosomal aberration	Chinese hamster ovary cells	250–1600 mg/ml	Negative[a]	National Toxicology Program (1986)
1112	Isophorone	Chromosomal aberration	Chinese hamster lung fibroblasts	1200[c], 1500[b] mg/ml 250–1000 mg/ml	Positive[a] Negative[a]	Matsuoka et al. (1996)
1112	Isophorone	Sister chromatid exchange	Chinese hamster ovary cells	5–1600 mg/ml	Positive[b,e]	Gulati et al. (1989)
1112	Isophorone	Sister chromatid exchange	Chinese hamster ovary cells	160–1000 mg/ml	Negative[a]	National Toxicology Program (1986)
1112	Isophorone	Unscheduled DNA synthesis	Rat hepatocytes	0.005–0.4 ml/ml	Negative[a]	O'Donoghue et al. (1988)
1112	Isophorone	Unscheduled DNA synthesis	Rat hepatocytes	200 ml/ml	Negative[a]	McKee et al. (1987)

Table 7 (contd)

No.	Agent	End-point	Test object	Maximum concentration	Results	Reference
In vivo						
1100	Cyclohexanone	Sex-linked recessive lethal mutation	*D. melanogaster*	0.1 ml/100 ml	Negative	Goncharova (1970)
1106	2-Hexylidene cyclo-pentanone	Sex-linked recessive lethal mutation	*D. Melanogaster*	5 mmol/l	Negative	Wild et al. (1983)
1106	2-Hexylidene cyclo-pentanone	Micronucleus formation	Mouse bone marrow	170, 330, 500 mg/kg bw	Negative	Wild et al. (1983)
1111	Tetramethyethyl cyclo-hexanone (mixed isomers)	Sex-linked recessive lethal mutation	*D. melanogaster*	10 mM	Negative	Wild et al. (1983)
1111	Tetramethyethyl cyclo-hexanone (mixed isomers)	Micronucleus formation	Mouse bone marrow	180, 307, 450 mg/kg bw	Negative	Wild et al. (1983)
1112	Isophorone	Sex-linked recessive lethal mutation	*D. melanogaster*	2000[f] and 12 500[g] ppm	Negative	Foureman et al. (1994)
1112	Isophorone	Micronucleus formation	CD-1 mice	≤ 540 mg/kg bw	Negative	McKee et al. (1987)
1112	Isophorone	Micronucleus formation	CD-1 mice	0.54 ml/kg bw	Negative	O'Donoghue et al. (1988)

[a] With and without metabolic activation
[b] Without metabolic activation
[c] With metabolic activation
[d] Cytotoxic at next highest dose tested (1600 mg/ml)
[e] A positive response was obtained only in the absence of metabolic activation and only after additional culture time (6–13 h)
[f] Oral administration
[g] Injection

(i) In vitro

Eight alicyclic ketones, secondary alcohols and related esters have been tested for genotoxicity. Overall, negative results were reported in the standard assay for reverse mutation when various strains of *Salmonella typhimurium* (TA98, TA100, TA1535, TA1537 and TA1538) were incubated with up to 10 000 μg/plate of cyclohexanone (No. 1100) or isophorone (No. 1112), 2.5–2500 μg/plate of cyclopentanone (No. 1101), up to 4200 μg/plate of 2,2,6-trimethyl cyclohexanone (No. 1108) or up to 3600 μg/plate of 2-hexylidene cyclopentanone (No. 1106) or tetramethylethylcyclohexanone (No. 1111), with or without metabolic activation (Florin et al., 1980; Haworth et al., 1983; Wild et al., 1983; Mortelmans et al., 1986). In another test for reverse mutation with *S. typhimurium* TA98, TA100, TA1535 and TA1537 (only an abstract), cyclohexanone was reported to produce 'a large number of revertants' in TA98, with no further elaboration and no results for the other strains. The concentrations and test conditions used were not specified (Massoud et al., 1980).

Both cyclohexyl acetate (No. 1093) and cyclohexyl butyrate (No. 1094) gave negative results for mutation in *Bacillus subtilis* M45 (*rec*$^-$) and H17 (*rec*$^+$) (Oda et al., 1979; Yoo, 1986). Positive results were reported with cyclohexanone in an assay for forward mutation assay in *B. subtilis* (Massoud et al., 1980); however, as previously stated, no concentrations or test conditions were reported in the abstract.

The results for forward mutation in mouse lymphoma cells were generally negative with isophorone, with or without metabolic activation (National Toxicology Program, 1986; McKee et al., 1987; O'Donoghue et al., 1988). An increased mutation frequency was reported in L5178Y Tk$^{+/-}$ mouse lymphoma cells without metabolic activation at concentrations of 400 and 800 μg/ml. Isophorone was lethal at 1600 μg/ml (McGregor et al., 1988).

Cyclohexanone (No. 1100) at concentrations up to 980 μg/ml induced chromosomal aberrations in human lymphocytes with or without metabolic activation (Collin et al., 1971; Lederer et al., 1971; Dyshlovoi et al., 1981). It did not induce chromosomal aberrations in Chinese hamster ovary cells at a concentration of 7.5 μl/ml, with or without metabolic activation (Aaron et al., 1985). Isophorone (No. 1112) gave equivocal results in Chinese hamster ovary cells. In one study, no chromosomal aberrations were induced with or without metabolic activation at concentrations up to 1600 μg/ml (Gulati et al., 1989), whereas in another study isophorone at a concentration of 1200 μg/ml without metabolic activation or at a concentration of 1500 μg/ml with metabolic activation induced chromosomal aberrations (Matsuoka et al., 1996); however, lower concentrations of 250–1000 μg/ml tested without metabolic activation did not. In an assay for sister chromatid exchange, cyclohexanone at a concentration of 7.5 μl/ml gave weakly positive results in Chinese hamster ovary cells in the absence of metabolic activation and negative results in the presence of metabolic activation (Aaron et al., 1985). Similarly, isophorone induced sister chromatid exchange in Chinese hamster ovary cells only when tested without metabolic activation at concentrations of 500–1000 μg/ml and then only after delayed harvesting due to the cytostatic effect of isophorone (Gulati et al., 1989). At lower concentrations tested without metabolic activation or at concentrations up to 1600 μg/mL tested with metabolic activation, isophorone did not induce sister chromatid exchange (National Toxicology Program, 1986; Gulati et al., 1989). In an assay for unscheduled DNA synthesis in rat hepatocytes, isophorone showed no sign of genotoxicity at concentrations up to 200 μl/ml (McKee et al., 1987; O'Donoghue et al., 1988).

(ii) In vivo

When cyclohexanone (No. 1100), 2-hexylidene cyclopentanone (No. 1106), tetramethylethyl cyclohexanone (No. 1111) or isophorone (No. 1112) was fed to adult *Drosophila melanogaster* for 3 days, no mutations were observed (Goncharova, 1970; Wild et al., 1983; Foureman et al., 1994). In addition, negative results were obtained when *D. melanogaster* were injected with a single dose of 12 500 μg of isophorone (Foureman et al., 1994).

There was no increase in the frequency of micronucleated polychromatic erythrocytes in the bone marrow of male or female CD-1 mice given isophorone (No. 1112) at a dose of 540 μg/kg bw by intraperitoneal injection (McKee et al., 1987; O'Donoghue et al., 1988) or in NMRI mice injected intraperitoneally with 2-hexylidenecyclopentanone at a dose of 170, 330 or 500 mg/kg bw or tetramethylethylcyclopentenone at a dose of 180, 310 or 450 mg/kg bw (Wild et al., 1983).

(iii) Conclusion

Cyclohexyl acetate (1093), cyclohexyl butyrate (No. 1094), cyclopentanone (No. 1101), 2-hexylidene cyclopentanone (No. 1106), 2,2,6-trimethyl cyclohexanone (No. 1108) and tetramethylethylcyclohexanone (mixed isomers) (No. 1111) gave negative results in assays for genotoxicity in vitro. The results reported for the genotoxicity of cyclohexanone (No. 1100) and isophorone (No. 1112) are conflicting. Most of the assays were conducted before 1986, when the pH and ionic strength of test media were often not adequately maintained. Mammalian cells *in situ* rely on complex regulatory mechanisms to maintain homeostatic conditions, and those in culture are not equipped to respond to environmental changes; therefore, it is important that the culture media used in mammalian cell assays be maintained at a pH of approximately 6.8–7.5. A lower pH or changes in osmolality due to the test agents can give rise to false-positive results, especially when metabolic activation systems are added. Acidity facilitates the breakdown of the components of such systems into mutagenic agents (Brusick, 1986).

The equivocal results of the assays for genotoxicity with cyclohexanone in vitro can be interpreted in terms of physiochemical properties. Compounds that are structurally similar to cyclohexanone have excellent membrane permeability and hydrogen bonding potential (Slater, 1963, 1970; Moreland, 1994). When cyclohexanone and related substances are tested in vitro, they may induce membrane expansion, leading to multiple effects on membrane-related processes. Membrane expansion may increase cell volume and lipid storage vacuoles, block ionic conductance channels, limit the availability of ATP and alter ion fluxes and metabolite distribution between the cytoplasm and organelles. Given these physiochemical properties, it is highly unlikely that any consistent pattern of genotoxicity would result from a battery of assays in bacterial and mammalian cells.

Overall, the tests for genotoxicity yielded mainly negative results. Positive results were reported in mammalian cells at cytotoxic concentrations, usually in the absence of biotransformation enzymes. The results of assays in vivo were negative.

(d) Reproductive toxicity

(i) Cyclohexanone (No. 1100)

Postnatal behaviour of rats was examined on the basis of activity in a figure-eight maze. CD-1 mice (number not specified) were given cyclohexanone at a dose of 800 mg/kg bw per day orally on days 8–12 of gestation, and the offspring were tested for motor activity in the maze on days 22, 58 and 200 after parturition. No effect on motor activity was seen (Gray et al., 1986).

A group of 28 ICR/SIM mice were given cyclohexanone orally at a dose of 2200 mg/kg bw per day on days 8–12 of gestation, and all treated animals and their offspring were examined for toxic effects. Six of the 28 mice died, and the remaining mice showed a significant decrease in body-weight gain. The litter size and number and the 2-day survival of the neonates were unaffected by treatment, but there was a significant decrease in live birth weight. The authors reported that the treatment was maternally toxic, as evidenced by > 7% mortality, overt clinical signs of toxicity or significantly reduced body-weight gain (Seidenberg et al., 1986; Seidenberg & Becker, 1987). The dose of 2200 mg/kg bw per day that was maternally toxic is > 1 100 000 000 times the daily per capita ("eaters only") intake of 2 ng/ kg from use of cyclohexanone as a flavouring agent.

In another study, female CF1 mice were given cyclohexanone at a dose of 50 mg/day intraperitoneally for 28 days and were mated on day 10 of this treatment. The numbers of pregnancies and viable fetuses were similar to those of controls. The authors concluded that cyclohexanone had no effect on fertility in mice (Hall et al., 1974).

(ii) Cyclopentanone (No. 1101)

Female CF1 mice were given cyclopentanone at a dose of 50 mg/day intraperitoneally for 28 days and were mated on day 10 of this treatment. The numbers of pregnancies, viable fetuses and resorptions were similar to those of controls. The authors concluded that cyclopentanone had no effect on fertility in mice (Hall et al., 1974).

Groups of 25 COBS CD female rats received cyclopentanone at a dose of 0, 50 or 300 mg/kg bw per day by gavage in corn oil on days 6–15 of gestation. On day 20 of gestation, the fetuses were removed surgically, and intrauterine survival, fetal development and skeletal and visceral morphology were recorded. No maternal, embryonal or fetal effects were seen at either dose. A slight decrease in mean fetal body weight was reported in the group at the higher dose but was not statistically significant. Although an increase in the number of litters with the fetal variant malaligned sternebrae was reported at 50 mg/kg bw per day, no such effect was seen at 300 mg/kg bw per day. The authors concluded that the NOEL was 50 mg/kg bw per day (Rusch et al., 1988).

(iii) Isophorone (No. 1112)

Groups of 12 Fischer 344 rats and 12 CD-1 mice were exposed to atmospheres containing isophorone at a dose of up to 150 ppm (260 mg/kg bw per day) on days 6–15 of gestation. Slight maternal toxicity, in the form of increased fetal resorptions and decreased food consumption resulting in decreased mean body weight, was observed in both species only at the highest concentration of 150 ppm. No developmental effects were seen (Phillips, 1985). In a similar study, rats were exposed to 500 ppm (860 mg/kg bw per day) of isophorone for 6 h/day for 5 days per week. No reproductive or developmental effects were reported (Dutertre-Catella, 1989).

(e) Special studies: Neurotoxicity

Cyclohexanol (No. 1100)

Seven Sprague-Dawley rats received cyclohexanol (No. 1100) as a 20% solution in peanut oil in two intraperitoneal injections of 200 mg/kg bw each, 5 days per week for up to 6 weeks. A second group of seven rats received the same total dose (400 mg/kg bw per day) on 5 days/ per week for 13 weeks. A control group of five rats received the peanut oil vehicle only. The animals were monitored for changes in body weight and examined by electrophysiology during the study and by neuropathology at the end. The group that received cyclohexanol showed a slight decrease in weight, but no other effects were observed. No evidence of neurotoxicity or neuropathology was reported (Perbellini et al., 1981).

3. REFERENCES

Aaron, C.S., Brewen, J.G., Stetka, D.G., Bleicher, W.T. & Spahn, M.C. (1985) Comparative mutagenesis in mammalian cells (CHO) in culture: Multiple genetic endpoint analysis of cyclohexanone *in vitro* (Abstract). *Environ. Mutag.*, **7** (Suppl. 3), 60–61.

Anders, M.W. (1989) Biotransformation and bioactivation of xenobiotics by the kidney. In: Hutson, D.H., Caldwell, J. & Paulson, G.D., eds, *Intermediary Xenobiotic Metabolism in Animals*, New York: Taylor & Francis, pp. 81–97.

Asakawa, Y., Ishida, T., Toyota, M. & Takemoto, T. (1986) Terpenoid transformation in mammals. IV. Biotransformation of (+)-longifolene, (–)-caryophyllene, (–)-caryophyllene oxide, (–)-cyclocolorenone, (+)-nootkatone, (–)-elemol, (–)-abietic acid, and (+)-dehydroabietic acid in rabbits. *Xenobiotica*, **16**, 753–767.

Boreus, L.O. (1982) Biotransformation. In: Boreus, L.O., ed., *Principles of Pediatric Pharmacology*, New York: Churchill Livingstone, pp. 1–114.

Borghoff, S.J., Short, B.G. & Swenberg, J.A. (1990) Biochemical mechanisms and pathobiology of nephropathy. *Annu. Rev. Pharmacol. Toxicol.*, **30**, 349–367.

Brusick, D (1986) Genotoxic effects in cultured mammalian cells produced by low pH treatment conditions and increased ion concentrations. *Environ. Mutag.*, **8**, 879–886.

Bucher, J.R., Huff, J. & Kluwe, W.M. (1986) Toxicology and carcinogenesis studies of isophorone in F344 rats and B6C3F1 mice. *Toxicology*, **39**, 207–219.

Burdock, G.A., Wagner, B.M., Smith, R.L., Munro, I.C. & Newberne, P.M. (1990) Recent progress in the consideration of flavoring ingredients under the food additives amendment. 15. GRAS substances. *Food Technol.*, **44**, 78–86.

Cramer, G.M., Ford, R.A. & Hall, R.L. (1978) Estimation of toxic hazard—A decision tree approach. *Food Cosmet. Toxicol.*, **16**, 255–276.

Deichmann, W.B. & LeBlanc, T.J. (1943) Determination of the approximate lethal dose with about six animals. *J. Ind. Hyg. Toxicol.*, **25**, 415–417.

Dietrich, D.R. & Swenberg, J.A. (1991) NCI-Reiter (NBR) male rats fail to develop renal disease following exposure to agents that induce α-2u-globulin (α_{2u}) nephropathy. *Fundam. Appl. Toxicol.*, **16**, 749–762.

Dutertre-Catella, H. (1989) *Contributions to the Analytical Toxicological and Biochemical Study of Isophorone*, Thesis for Doctorate in Pharmacology, Université Réné Descartes, Paris.

Dutertre-Catella, H., Nguyen, P., Dang, Q. & Truhaut, R. (1978) Metabolic transformations of the trimethyl-3,5,5, cyclohexene-2, one-1 (isophorone). *Toxicol. Eur. Res.*, **1**, 209–216.

Dyshlovoi, V.D., Boiko, N.L., Shemetun, A.M. & Kharchenko, T.I. (1981) Cytogenetic action of cyclohexanone. *Gig. Sanit.*, **5**, 76–77 (in Russian with English abstract).

Elliott, T.H., Parke, D.V. & Williams, R.T. (1959) Studies in detoxication. The metabolism of cyclo[^{14}C]hexane and its derivatives. *Biochem. J.*, **72**, 193–200.

Emberger, D. (1994) In vitro hydrolysis test on menthyl propyleneglycol carbonate. Private communication. Submitted to WHO by Flavor and Extract Manufacturers' Association of the United States.

Environmental Protection Agency (1991) *Risk Assessment Forum. Alpha $_{2u}$-globulin: Association with Chemically Induced Renal Toxicity and Neoplasia in the Male Rat*, Publication No. EPA/625/3-91019F, Washington DC.

Exxon Chemical Americas (1982) Unpublished report submitted to Environmental Protection Agency.

Felsted, R.L. & Bachur, N.R. (1980) Ketone reductases. In: Jakoby, W.B., ed., *Enzymatic Basis of Detoxication*, 2nd Ed., New York: Academic Press, pp. 281–293.

Fischer, F.G. & Bielig, H.J. (1940) Biochemical hydrogenations. VII. Hydrogenation of unsaturated compounds in the animal body. *Hoppe-Seyler's Z. Physiol. Chem.*, **266**, 73–98.

Florin, I., Rutberg, L., Curvall, M. & Enzell, C.R. (1980) Screening of tobacco smoke constituents for mutagenicity testing using Ames test. *Toxicology*, **18**, 219–232.

Food and Drug Administration (1993) Conversion table for test chemical treatment doses used in PAFA. In: *Priority-based Assessment of Food Additives (PAFA) Database*, Washington DC: Center for Food Safety and Applied Nutrition, p. 58.

Ford, D.M. & Moran, E.J. (1978) Preliminary indications of *in vitro* hydrolysis of two flavor chemical esters. Private communication. Submitted to WHO by Flavor and Extract Manufacturers' Association of the United States.

Foureman, P., Mason, J.M., Valencia, R. & Zimmering, S. (1994) Chemical mutagenesis testing in Drosophila. X. Results of 70 coded chemicals tested for the National Toxicology Program. *Environ. Mol. Mutag.*, **23**, 208–227.

Gabriel, D. (1991) Acute oral toxicity study of 2-(3,7-dimethyl-2,6-octadienyl)cyclopentanone in rats. Unpublished report from Bedoukian Research Inc.

Goncharova, R.I. (1970) Genetic activity of some cyclohexane derivatives. *Genet. Tsitol.*, ?, 137–142.

Governa, M., Calisti, R., Coppa, G., Tagliavento, G., Colombi, A. & Troni, W. (1987) Urinary excretion of 2,5-hexanedione and peripheral polyneuropathies in workers exposed to hexane. *J. Toxicol. Environ. Health*, **20**, 219.

Gray, L.E., Kavlock, R.J., Ostby, J., Ferrell, J., Rogers, J. & Gray, K. (1986) An evaluation of figure-eight maze activity and general behavioral development following prenatal exposure to forty chemicals: Effects of cytosine arabinoside, dinocap, nitrofen, and vitamin A. *Neurotoxicology*, **7**, 449–462.

Gulati, D.K., Witt, K., Anderson, B., Zeiger, E. & Shelby, M.D. (1989) Chromosomal aberration and sister chromatid exchange tests in Chinese hamster ovary cells in vitro. III: Results with 27 chemicals. *Environ. Mol. Mutag.*, **13**, 133–193.

Gupta, P.K., Lawrence, W.H., Turner, J.E. & Autian, J. (1979) Toxicological aspects of cyclohexanone. *Toxicol. Appl. Pharmacol.*, **49**, 525–533.

Hall, I.H., Carlson, G.L., Abernethy, G.S. & Piantadosi, C. (1974) Cycloalkanones. 4. Antifertility activity. *J. Med. Chem.*, **17**, 1253–1257.

Hagan, E.C., Hansen, W.H., Fitzhugh, O.G., Jenner, P.M., Jones, W.I., Taylor, J.M., Long, E.L., Nelson, A.M. & Brouwer, J.B. (1967) Food flavorings and compounds of related structure. II. Subacute and chronic toxicity. *Food Cosmet. Toxicol.*, **5**, 141–157.

Hamalainen, J. (1912) Concerning the behavior of acyclic compounds with the glucuronic acid pairing in the organism. *Skand. Arch. Physiol.*, **27**, 141–226.

Haseman, J.K., Winbush, J.S. & O'Donnell, M.W., Jr (1986) Use of dual control groups to estimate false positive rates in laboratory animal carcinogenicity studies. *Fundam. Appl. Toxicol.*, **7**, 573–584.

Haseman, J.K., Boorman, G.A. & Huff, J. (1997) Value of historical control data and other issues related to the evaluation of long-term rodent carcinogenicity studies. *Toxicol. Pathol.*, **25**, 524–527.

Haseman, J.K., Hailey, J.R. & Morris, R.W. (1998) Spontaneous neoplasm incidences in Fischer 344 rats and B6C3F1 mice in two-year carcinogenicity studies: A National Toxicology Program update. *Toxicol. Pathol.*, **26**, 428–441.

Haworth, S., Lawlor, T., Mortelmans, K., Speck, W. & Zeigler, E. (1983) Salmonella mutagenicity test results for 250 chemicals. *Environ. Mutag.*, **5** (Suppl. 1), 3–142.

Heymann, E. (1980) Carboxylesterases and amidases. In: Jakoby, W.B., ed., *Enzymatic Basis of Detoxication*, 2nd Ed., New York: Academic Press, pp. 291–323.

Hildebrandt, H. (1902) On the behavior of carvone and santalol in the animal body. *Hoppe Selye's Z. Physiol. Chem.*, **36**, 452–461.

Hummler, H. (1969) Private communication. Submitted to WHO by Flavor and Extract Manufacturers' Association of the United States.

International Agency for Research on Cancer (1989) *IARC Monographs on the Evaluatiion of the Carcinogenic Risk of Chemicals to Humans*, Vol. 47, pp. 157–169.

International Organization of the Flavor Industry (1995) European inquiry on volume use. Private communication. Submitted to WHO by Flavor and Extract Manufacturers' Association of the United States.

Ishida, T., Toyota, M. & Asakawa, Y. (1989) Terpenoid biotransformation in mammals. V. Metabolism of (+)-citronellal, (±)-7-hydroxycitronellal, citral, (–)-perillaldehyde, (–)-myrtenal, cuminalehyde, thujone, and (±)-carvone in rabbits. *Xenobiotica*, **19**, 843–855.

James, S.P. & Waring, R.H. (1971) The metabolism of alicyclic ketones in the rabbit and rat. *Xenobiotica*, **1**, 573–580.

Kjeldsen, L., Cowland, J.B. & Borregaard, N. (2000) Human neutrophil gelatinase-associated lipocalin and homologous proteins in rat and mouse. *Biochim. Biophys. Acta*, **482**, 272–283.

Koeferl, M.T., Miller, T.R., Fisher, J.D., Martis, L., Garvin, P.J. & Dorner, J.L. (1981) Influence of concentration and rate of intravenous administration on the toxicity of cyclohexanone in beagle dogs. *Toxicol. Appl. Pharmacol.*, **59**, 215–229.

Kohli, R.P., Kishor, K., Dua, P.R. & Saxena, R.C. (1967) Anticonvulsant activity of some carbonyl containing compounds. *Indian J. Med. Res.*, **55**, 1221–1225.

Krasavage, W.J., O'Donoghue, J.L. & Divincenzo, G.D. (1982) Ketones. In: Clayton, G.D. & Clayton, F.E., eds, *Patty's Industrial Hygiene and Toxicology*, Vol. IIC, *Toxicology*, New York: John Wiley and Sons, pp. 4709–4800.

Leberco Laboratories (1965) Unpublished report. Submitted to WHO by Flavor and Extract Manufacturers' Association of the United States.

Lederer, J., Collin, J.P., Pottier-Arnould, A.M. & Gondry, E. (1971) Cytogenetic and teratogenetic effect of cyclamate and its metabolites *Therapeutique*, **47**, 357–363 (in French).

Lehman-McKeeman, L.D. & Caudill, D. (1994) d-Limonene induced hyaline droplet nephropathy in α-2u-globulin transgenic mice. *Fundam. Appl. Toxicol.*, **23**, 562–568.

Levenstein, I. (1976) Acute oral toxicity study in rats and acute dermal toxicity studies in rabbits. Unpublished report to RIFM. Submitted to WHO by Flavor and Extract Manufacturers' Association of the United States.

Lijinsky, W. & Kovatch, R.M. (1986) Chronic toxicity study of cyclohehanone in rats and mice. *J. Natl Cancer Inst.*, **77**, 941–949.

Lington, A.W. & Bevan, C. (1994) Alcohols. In: Clayton, G.D. & Clayton, F.E., eds, *Patty's Industrial Hygiene and Toxicology*, 4th Ed., New York: John Wiley & Sons, Vol. IID, pp. 2585–2760.

Lucas, C.D., Putnam, J.M. & Hallagan, J.B. (1999) *Flavor and Extract Manufacturers Association of the United States 1995 Poundage and Technical Effects Update Survey*, Washington DC: Flavor and Extract Manufacturers' Association of the United States..

Maarse, H., Visscher, C.A., Willemsens, L.C. & Boelens, M.H. (1999) *Volatile Components in Food—Qualitative and Quantitative Data*, Zeist: Centraal Instituut Voor Voedingsonderzioek TNO.

Madyastha, K.M. & Raj, C.P. (1990) Biotransformations of R-(+)-pulegone and menthofuran *in vitro*: Chemical basis for toxicity. *Biochem. Biophys. Res. Communications*, **3**, 1086–1092.

Maronpot, R.R. & Boorman, G.A. (1982) Interpretation of rodent hepatocellular proliferative alterations and hepatocellular tumors in chemical safety assessment. *Toxicol. Pathol.*, **10**, 71–80.

Maronpot, R.R., Haseman, J.K., Boorman, G.A., Eustis, S.E., Rao, G.N. & Huff, J.E. (1987) Liver lesions in B6C3F1 mice: The National Toxicology Program, experience, and position. *Arch. Toxicol.*, **10** (Suppl.), 10–26.

Martis, L., Tolhurst, T, Koeferl, M.T., Miller, T.R. & Darby, T.D. (1980) Disposition kinetics of cyclohexanone in beagle dogs. *Toxicol. Appl. Pharmacol.*, **59**, 215–229.

Matsuoka, A., Yamakage, K., Kusakabe, H., Wakuri, S., Asakura, M., Noguchi, T., Sugiyama, T., Shimada, H., Nakayama, S., Kasahara, Y., Takahashi, Y., Miura, K.F., Hatanaka, M., Ishidate, M., Jr, Morita, T., Watanabe, K., Hara, M., Odawara, K., Tanaka, N., Hayashi, M. & Sofuni, T. (1996) Re-evaluation of chromosomal aberration induction on nine mouse lymphoma assay 'unique positive' NTP carcinogens. *Mutat. Res.*, **369**, 243–252.

McGregor, D.B., Brown, A., Cattanach, P., Edwards, I., McBride, D., Riach, C. & Caspary, W.J. (1988) Responses of the L5178Y tk+/tk– mouse lymphoma cell forward mutation assay: III. 72 coded chemicals. *Environ. Mol. Mutag.*, **12**, 85–153.

McKee, R.H., Phillips, R.D., Lerman, S.A., Slesinski, R.S., Rogers-Back, A.M., Curren, R.D. & Putman, D.L. (1987) The genotoxic potential of isophorone. *Environ. Mutag.*, **9**, 71.

McMahon, R.E. (1982) Alcohols, aldehydes and ketones. In: Jakoby, W.B., Bend, J.R. & Caldwell, J., eds, *Metabolic Basis for Detoxification. Metabolism of Functional Groups*, New York: Academic Press, p. 91.

Mills, G. & Walker, V. (1990) Urinary excretion of cyclohexanediol, a metabolite of the solvent cyclohexanone, by infants in a special care unit. *Clin. Chem.*, **36**, 870–874.

Moreland, D.E. (1994) Effects of toxicants on oxidative phosphorylation and photophosphorylation. In: Hodgson, E. & Levi, P.E., eds, *Introduction to Biochemical Toxicology*, 2nd Ed., Norwalk, Connecticut: Appleton & Lange, pp. 345–366.

Moreno, O.M. (1974) Acute toxicity studies. Unpublished report to RIFM. Submitted to WHO by Flavor and Manufacturers' Association of the United States.

Moreno, O.M. (1977) Acute toxicity studies in rats for cyclohexyl acetate. Unpublished report No. 1692 (07/22) to RIFM. Submitted to WHO by Flavor and Manufacturers' Association of the United States.

Moreno, O.M. (1978) Acute toxicity studies in rats, mice, rabbits and guinea pigs. Unpublished report to RIFM. Submitted to WHO by Flavor and Manufacturers' Association of the United States.

Moreno, O.M. (1980) Acute toxicity studies. Unpublished report to RIFM. Submitted to WHO by Flavor and Manufacturers' Association of the United States.

Moreno, O.M. (1982) Acute toxicity studies. Unpublished report to RIFM. Submitted to WHO by Flavor and Manufacturers' Association of the United States.

Morishita, K., Schoonhoven, R. & Swenberg, J.A. (1997) Mechanistic studies on isophorone and preputial gland carcinomas. *Toxicologist*, **36**, 94.

Mortelmans, K., Haworth, S., Lawlor, T., Speck, W., Tainer, W. & Zeiger, E. (1986) Salmonella mutagenicity tests: II. Results from the testing of 270 chemicals. *Environ. Mutag.*, **8**, 1–119.

Mraz, J., Galova, E., Nohova, H. & Vitkova, D. (1994) Uptake, metabolism and elimination of cyclohexanone in humans. *Int. Arch. Occup. Environ. Health*, **66**, 203–208.

National Cancer Institute (1978) *Bioassay of dl-Menthol for Possible Carcinogenicity*, Report No. NCI-CG-TR-98; NIH publication 78-2665, Research Triangle Park, North Carolina.

National Toxicology Program (1986) *Toxicology and Carcinogenesis Studies of Isophorone in F344/N Rats and B6C3F1 Mice (Gavage Studies)*, Report No. NTP-TR-291; NIH Publication No. 86-2547, Research Triangle Park, North Carolina.

National Toxicology Program (1990a) *Toxicology and Carcinogenesis Studies of d-Limonene in F344/N Rats and B6C3F1 Mice (Gavage Studies)*, Report No. NTP-TR-347; NIH Publication No. 88-2902, Research Triangle Park, North Carolina.

National Toxicology Program (1990b) *National Toxicology Program Technical Report on the Toxicology and Carcinogenesis Studies of d-Carvone in B6C3F1 Mice (Gavage Studies)*, Report No. NTP TR 381, Research Triangle Park, NC.

National Toxicology Program (1992b) *Toxicology and Carcinogenesis Studies of γ-Butyrolactone in F344/N Rats and B6C3F1 Mice (Gavage Studies)*, Report No. NTP-TR-406; NIH Publication No. 92-3137, Research Triangle Park, North Carolina.

National Toxicology Program (1993a) *Toxicology and Carcinogenesis Studies of 3,4-Dihydrocoumarin in F344/N Rats and B6C3F1 Mice (Gavage Studies)*, Report No. NTP-TR-423; NIH Publication No. 93-3154, Research Triangle Park, North Carolina.

National Toxicology Program (1993b) *Toxicology and Carcinogenesis Studies of Pentachloroanisole in F344/N Rats and B6C3F1 Mice (Gavage Studies)*, Report No. NTP-TR-414; NIH Publication No. 93-3145, Research Triangle Park, North Carolina.

National Toxicology Program (1994) *Toxicology and Carcinogenesis Studies of* o*-Benzyl-*p*-chlorophenol in F344/N Rats and B6C3F1 Mice (Gavage Studies)*, Report No. NTP-TR-424; NIH Publication No. 94-3155, Research Triangle Park, North Carolina.

National Toxicology Program (2001) *Toxicology and Carcinogenesis Studies of Citral (Microencapsulated) in F344/N Rats and B6C3F1 Mice (Feed Studies)*, Report No. NTP-TR-505; NIH Publication No. 01-4439, Research Triangle Park, North Carolina.

Nelson, S.D., McClanahan, R.H., Thomassen, D., Gordon, W.P. & Knebel, N. (1992) Investigations of mechanisms of reactive metabolite formation from (R)-(+)-pulegone. *Xenobiotica*, **22**, 1157–1164.

Oda, Y., Hamano, Y., Inoue, K., Yamamoto, H. Nihara, T. & Kunita, N. (1979) Mutagenicity of food flavors in bacteria. *Shokuhin Eisei Hen*, **9**, 177–181 (in Japanese).

O'Donoghue, J.L., Haworth, S.R., Curren, R.D., Kirby, P.E., Lawlor, T., Moran, E.J., Phillips, R.D., Putnam, D.L., Rogers-Back, A.M., Slesinski, R.S. & Thilagar, A. (1988) Mutagenicity studies on ketone solvents: Methyl ethyl ketone, methyl isobutyl ketone, and isophorone. *Mutat. Res.*, **206**, 149–161.

Olson, M.J., Johnson, J.T. & Reidy, C.A. (1990) A comparison of male rat and human urinary proteins: Implications for human resistance to hyaline droplet nephropathy. *Toxicol. Appl. Pharmacol.*, **102**, 524–536.

Oser, B.L. & Hall, R.L. (1977) Criteria employed be the expert panel of Flavor and Extract Manufacturers' Association of the United States for the GRAS evaluation of flavouring substances. *Food Cosmet. Toxicol.*, **15**, 457–466.

Perbellini, L., Brugnone, F. & Pavan, I. (1980) Identification of the metabolites of n-hexane, cyclohexane and their isomers in men's urine. *Toxicol. Appl. Pharmacol.*, **53**, 220–229.

Perbellini, L., De Grandis, D., Semenzato, F. & Bongiovanni, L.G. (1981) [Experimental study on the neurotoxicity of cyclohexanol and cyclohexanone.] *Med. Lav.*, **2**, 102–107 (in Italian).

Phillips, R.D. (1985) Ketone program panel. Inhalation teratology study in rats and mice—Isophorone. Report No. USEPA/OTS Doc. 408555049. TSCA negiotated testing agreement submissions. Submitted to WHO by Flavor and Manufacturers' Association of the United States.

Portoghese, P.S., Kedziora, G.S., Larson, D.L., Bernard, B.K. & Hall, R.L. (1989) Reactivity of glutathione with α,β-unsaturated ketone flavouring substances. *Food Chem. Toxicol.*, **27**, 773–776.

Posternak, J.M., Linder, A. & Vodoz, C.A. (1969) Summaries of toxicological data. Toxicoligical tests on flavoring matters. *Food Cosmet. Toxicol.*, **7**, 405–407.

Roy, A.K. & Neuhaus, O.W. (1967) Androgenic control of a sex dependent protein in rat. *Nature*, **214**, 618–620.

Rusch, G.M., Rodwell, D.E., Nemec, M.D. & Tasker, E.J. (1988) A teratology screening study in rats with cyclopentanone. *Toxicologist*, **8**, 213.

Sakata, M., Kikuchi, J. & Haga, M. (1989) Disposition of acetone, methyl ethyl ketone and cyclohexanone in acute poisoning. *Clin. Toxicol.*, **27**, 67–77.

Salzer, D. (1998) In vitro hydrolysis test cis and trans-p-1(7),8-menthadien-2-yl acetate. Private communication. Submitted to WHO by Flavor and Manufacturers' Association of the United States.

Seidenberg, J.M. & Becker, R.A. (1987) A summary of the results of 55 chemicals screened for developmental toxicity in mice. *Teratog. Carcinog. Mutag.*, **7**, 17–28.

Seidenberg, J.M., Anderson, D.G. & Becker, R.A. (1986) Validation of an in-vivo developmental toxicity screen in the mouse. *Teratog. Carcinog. Mutag.*, **6**, 361–374.

Slater, E.C. (1963) Uncouplers and inhibitors of oxidative phosphorylation. In: Hochester, R.M. & Quaste, J.J., eds, *Metabolic Inhibitors*, Vol. 2, New York: Academic Press, pp. 503–516.

Slater, E.C. (1970) Application of inhibitors and uncouplers for a study of oxidative phosphorylation. *Meth. Enzymol.*, **10**, 48.

Smyth, H.F., Jr, Carpenter, C.P., Weil, C.S., Pozzani, U.C., Striegel, J.A. & Nycum, J.S. (1969) Range-finding toxicity data: List VII. *Am. Ind. Hyg.Assoc. J.*, **30**, 470–476.

Stofberg, J. & Grundschober, F. (1987) Consumption ratio and food predominance of flavoring materials. *Perfumer Flavorist*, **12**, 27.

Stofberg, J. & Kirschman, J.C. (1985) The consumption ratio of flavoring materials: A mechanism for setting priorities for safety evaluation. *Food Chem. Toxicol.*, **23**, 857–860.

Strasser, J., Jr, Charbonneau, M., Borghoff, S.J., Turner, M.J. & Swenberg, J.A. (1988) Renal protein droplet formation in male Fischer 344 rats after isophorone (IPH) treatment. *Toxicologist*, **8**, 136.

Swenberg, J.A., Short, B., Borghoff, S., Strasser, J. & Charbonneau, M. (1989) The comparative pathobiology of nephropathy. *Toxicol. Appl. Pharmacol.*, **97**, 35–46.

Tamura, S., Tsustumi, S. & Kizu, K. (1962) Studies on glucuronic acid metabolism. I. The influence of borneol, ionone and carvone on the urinary excretion of glucuronic acid and ascorbic acid. *Fol. Pharmacol. Jpn.*, **58**, 323–336.

Tao, C.C. & Elliot, T.H. (1962) The metabolism of ^{14}C-methylcyclohexane. *J. Biochem.*, **84**, 38–39.

Topping, D.C, Morgott, D.A., David, R.M. & O'Donoghue, J.L. (1994) Ketones. In: Clayton, G.D. & Clayton, F.E., eds, *Patty's Industrial Hygiene and Toxicology*, 4th Ed., New York, John Wiley & Sons, Vol. IIC, pp. 1739–1878.

Triebel, S., Blaser, J., Reinke, H. & Tschesche, H. (1992) A 25 kDa alpha-2-microglobulin-related protein is a component of the 125 kDa form of gelatinase. *FEBS Lett.*, **314**, 386–388.

Treon, J.F., Crutchfield, W.E., Jr & Kitzmiller, K.V.. (1943) *J. Ind. Hyg. Toxicol.*, **25**, 199–214.

Truhaut, R., Dutertre-Catella, H. & Nguyen, P. (1970) Metabolic study of an industrial solvent, isophorone, in the rabbit. *C.R. Hebd. Seances Acad. Sci. Ser. D: Sci. Nat.*, **271**, 1333–1336.

Unterman, R.D., Lynch, K.R., Nakhasi, H.L., Dolan, K.P., Hamilton, J.W., Cohn, D.V. & Feigelson, P. (1981) Cloning and sequencing of several alpha-2-microglobulin cDNAs. *Proc. Natl Acad. Sci. USA*, **78**, 3478–3482.

Wang, K.S. & Hodgetts, R.B. (1998) The analysis of regulatory sequences required for the high level induction of rat alpha-2u-globulin gene expression by dexamethasone. *J. Mol. Endocrinol.*, **20**, 129–139.

Webb, D.R., Kanerva, R.L., Hysell, D.K., Alden, C.L. & Lehman-McKeeman, L.D. (1990) Assessment of the subchronic oral toxicity of *d*-limonene in dogs. *Food Chem. Toxicol.*, **28**, 669–675.

White, D.A., Heffron, A., Miciak, B., Middleton, B., Knights, S. & Knights, D. (1990) Chemical synthesis of dual radiolabelled cyclandelate and its metabolism in rat hepatocytes and mouse J774 cells. *Xenobiotica*, **20**, 71.

Wild, D., King, M.T., Gocke, E. & Eckhardt, K. (1983) Study of artificial flavouring substances for mutagenicity in the Salmonella/microsome, Basc, and micronucleus tests. *Food Chem. Toxicol.*, **21**, 707–719.

Williams, R.T. (1959) *Detoxification Mechanisms. The Metabolism and Detoxification of Drugs, Toxic Substances and Other Organic Compounds*, 2nd Ed., London: Chapman & Hall, pp. 4, 114–126.

Yamaguchi, T., Caldwell, J. & Farmer, P.B. (1994) Metabolic fate of [^{3}H]–*l*-menthol in the rat. *Drug Metab. Disposition*, **22**, 616–624.

Yoo, Y.S. (1986) Mutagenic and antimutagenic activities of flavoring agents used in foodstuffs. *J. Osaka City Med. Center*, **34**, 267–288 (in Japanese).

Zheng, G.-Q., Kenney, P.M. & Lam, K.L.T. (1992) Effects of carvone compounds on glutathione S-transferase activity in A/J mice. *J. Agric. Food Chem.*, **40**, 751–755.

Zlatkis, A., Wolfgang, B., Lichtenstein, H.A., Tishbee, A. & Shunbo, F. (1973) Profile of volatile metabolites in urine by gas chromatography–mass spectrometry. *Anal. Chem.*, **45**, 783–787.

ALIPHATIC SECONDARY ALCOHOLS, KETONES AND RELATED ESTERS

First draft prepared by

Professor I.G. Sipes[1] and Professor A.G. Renwick[2]
[1]Department of Pharmacology, College of Medicine, University of Arizona, Tucson, Arizona, USA; and [2]Clinical Pharmacology Group, University of Southampton, Southampton, England

1. EVALUATION

1.1 Introduction

The Committee evaluated a group of 39 flavouring agents that includes aliphatic acyclic secondary alcohols and ketones and esters derived from aliphatic secondary alcohols (see Table 1), using the Procedure for the Safety Evaluation of Flavouring Agents (see Figure 1). The Committee had not previously evaluated any of these agents.

Twenty-six of the 39 substances in this group of flavouring agents (Nos 1118, 1120, 1122–1132, 1134–1136, 1139, 1142, 1144, 1147, 1148, 1150–1152, 1154 and 1156) have been reported to occur naturally in foods. They have been found in fruits, juices, spices, vegetables, cocoa, coffee and tea.

1.2 Estimated daily intake

The total annual volume of production of the 39 aliphatic secondary alcohols and ketones in this group is approximately 3900 kg in Europe (International

Table 1. Summary of the results of safety evaluations of aliphatic secondary alcohols, ketones and related esters used as flavouring agents[a]

Flavouring agent	No.	CAS No. and structure	Step A3[b] Does intake exceed the threshold for human intake?	Comments	Conclusion based on current intake
Structural class I					
3-Penten-2-one	1124	625-33-2 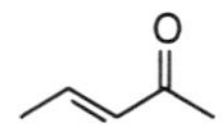	No Europe: 0.3 USA: ND	See notes 1, 3 and 4.	No safety concern
4-Hexen-3-one	1125	2497-21-4 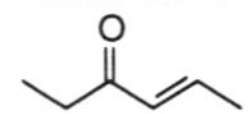	No Europe: 15 USA: 1	See notes 1 and 3.	No safety concern
2-Hepten-4-one	1126	4643-25-8 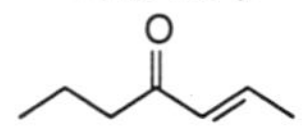	No Europe: 0.01 USA: ND	See notes 1 and 3.	No safety concern
5-Methyl-2-hepten-4-one	1133	81925-81-7 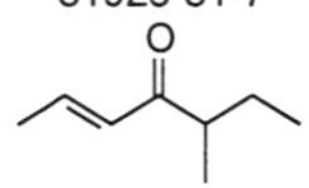	No Europe: 7 USA: 1	See notes 1 and 3.	No safety concern
3-Octen-2-ol	1140	76649-14-4 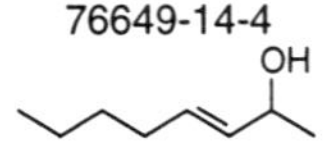	No Europe: 1 USA: ND	See note 1.	No safety concern
(E)-2-Octen-4-ol	1141	20125-81-9 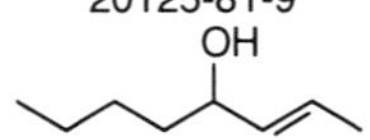	No Europe: ND USA: 1	See note 1.	No safety concern

Table 1 *(continued)*

Flavouring agent	No.	CAS No. and structure	Step A3[a,b] Does intake exceed the threshold for human intake?	Comments	Conclusion based on current intake
2-Pentyl butyrate	1142	60415-61-4 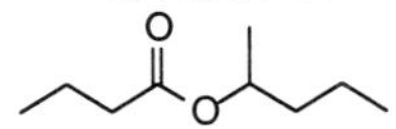	No Europe: ND USA: 1	See note 2.	No safety concern
(±) Heptan-3-yl acetate	1143	5921-83-5 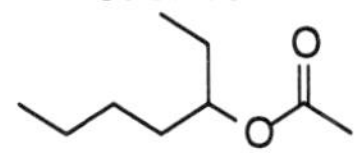	No Europe: 4 USA: 3	See note 2.	No safety concern
(±) Heptan-2-yl butyrate	1144	39026-94-3 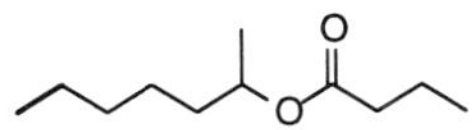	No Europe: 4 USA: 3	See note 2.	No safety concern
(±) Nonan-3-yl acetate	1145	60826-15-5 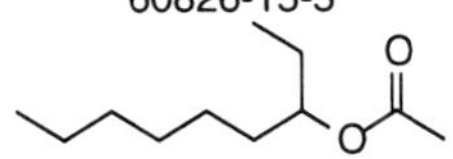	No Europe: 4 USA: 3	See note 2.	No safety concern
2-Pentyl acetate	1146	626-38-0 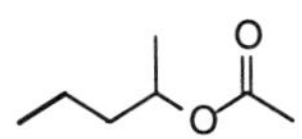	No Europe: 3 USA: 3	See note 2.	No safety concern
Structural class II					
3-Decanone	1118	928-80-3 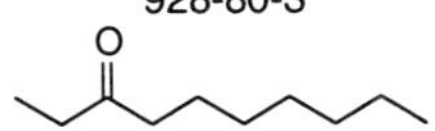	No Europe: 4 USA: 3	See notes 1 and 3.	No safety concern

Table 1 *(continued)*

Flavouring agent	No.	CAS No. and structure	Step A3[a,b] Does intake exceed the threshold for human intake?	Comments	Conclusion based on current intake
5-Methyl-5-hexen-2-one	1119	3240-09-3 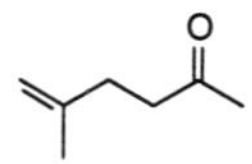	No Europe: ND USA: 0.3	See notes 1, 3 and 4.	No safety concern
6-Methyl-5-hepten-2-one	1120	110-93-0 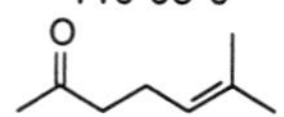	No Europe: 120 USA: 40	See notes 1 and 4.	No safety concern
3,4,5,6-Tetrahydro-pseudo-ionone	1121	4433-36-7 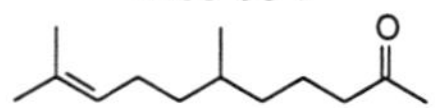	No Europe: ND USA: 0.01	See notes 1 and 4.	No safety concern
6,10-Dimethyl-5,9-undecadien-2-one	1122	3796-70-1 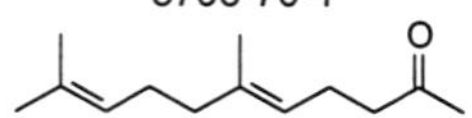	No Europe: 50 USA: 2	See notes 1 and 4.	No safety concern
2,6,10-Trimethyl-2,6,10-pentadecatrien-14-one	1123	762-29-8 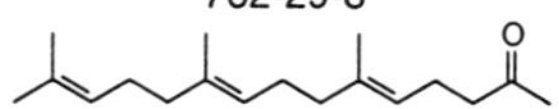	No Europe: 0.1 USA: ND	See notes 1 and 4.	No safety concern
3-Hepten-2-one	1127	1119-44-4 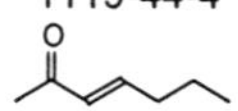	No Europe: 0.2 USA: 0.07	See notes 1, 3 and 4.	No safety concern
3-Octen-2-one	1128	1669-44-9 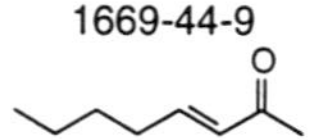	No Europe: 1 USA: 1	See notes 1, 3 and 4.	No safety concern

Table 1 *(continued)*

Flavouring agent	No.	CAS No. and structure	Step A3[a,b] Does intake exceed the threshold for human intake?	Comments	Conclusion based on current intake
2-Octen-4-one	1129	4643-27-0 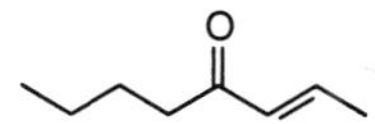	No Europe: 1 USA: 3	See notes 1 and 3.	No safety concern
3-Decen-2-one	1130	10519-33-2 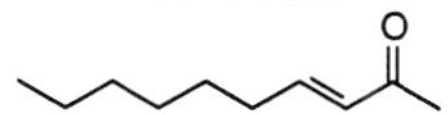	No Europe: 0.01 USA: ND	See notes 1, 3 and 4.	No safety concern
4-Methyl-3-penten-2-one	1131	141-79-7 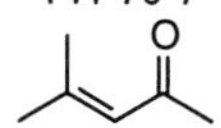	No Europe: 0.4 USA: ND	See notes 1 and 4.	No safety concern
5-Methyl-3-hexen-2-one	1132	5166-53-0 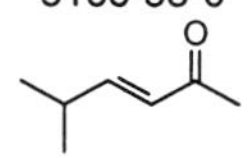	No Europe: ND USA: 0.1	See notes 1, 3 and 4.	No safety concern
6-Methyl-3,5-heptadien-2-one	1134	1604-28-0 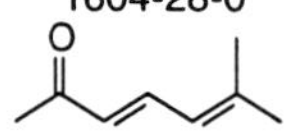	No Europe: 15 USA: 5	See notes 1, 3 and 4.	No safety concern
(E)-7-Methyl-3-octen-2-one	1135	33046-81-0 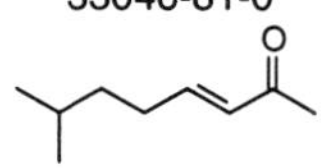	No Europe: ND USA: 2	See notes 1, 3 and 4.	No safety concern
3-Nonen-2-one	1136	14309-57-0 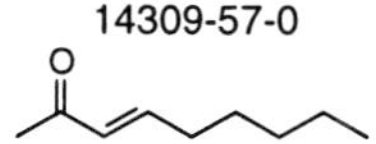	No Europe: 14 USA: 13	See notes 1, 3 and 4.	No safety concern

Table 1 (continued)

Flavouring agent	No.	CAS No. and structure	Step A3[a,b] Does intake exceed the threshold for human intake?	Comments	Conclusion based on current intake
(E) & (Z)-4,8-Dimethyl-3,7-nonadien-2-one	1137	817-88-9 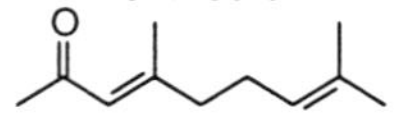	No Europe: 7 USA: 7	See notes 1, 3 and 4.	No safety concern
(E)-6-Methyl-3-hepten-2-one	1138	20859-10-3 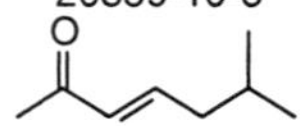	No Europe: 4 USA: 3	See notes 1, 3 and 4.	No safety concern
(E,E)-3,5-Octadien-2-one	1139	30086-02-3 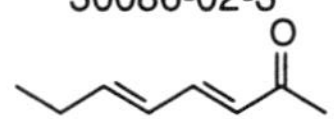	No Europe: 4 USA: 3	See notes 1, 3 and 4.	No safety concern
1-Penten-3-one	1147	1629-58-9 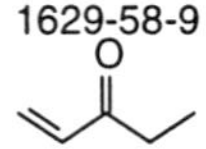	No Europe: 0.3 USA: 0.1	See notes 1 and 3.	No safety concern
1-Octen-3-one	1148	4312-99-6 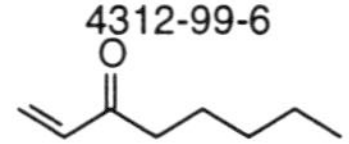	No Europe: 1 USA: 0.1	See notes 1 and 3.	No safety concern
2-Pentyl-1-buten-3-one	1149	63759-55-7 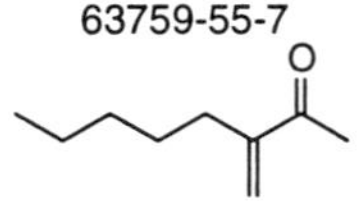	No Europe: 0.2 USA: ND	See notes 1, 3 and 4.	No safety concern
1-Penten-3-ol	1150	616-25-1 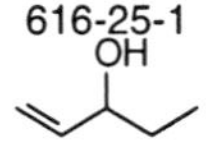	No Europe: 2 USA: 1	See note 1.	No safety concern

Table 1 *(continued)*

Flavouring agent	No.	CAS No. and structure	Step A3[a,b] Does intake exceed the threshold for human intake?	Comments	Conclusion based on current intake
1-Hexen-3-ol	1151	4798-44-1 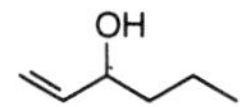	No Europe: 1 USA: 2	See note 1.	No safety concern
1-Octen-3-ol	1152	3391-86-4 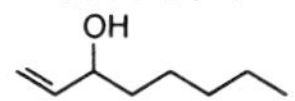	No Europe: 290 USA: 23	See note 1.	No safety concern
1-Decen-3-ol	1153	51100-54-0 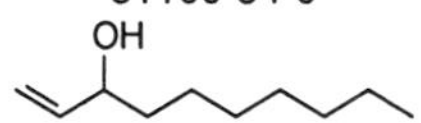	No Europe: ND USA: 0.1	See note 1.	No safety concern
(E,R)-3,7-Dimethyl-1,5,7-octatrien-3-ol	1154	20053-88-7 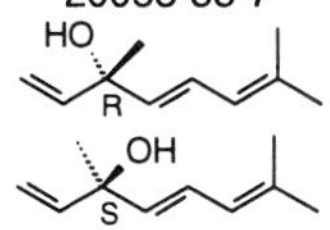	No Europe: ND USA: 6	See note 1.	No safety concern
6-Undecanone	1155	927-49-1 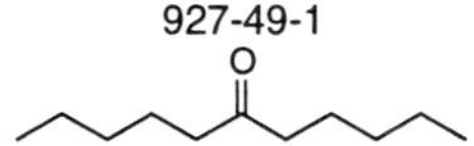	No Europe: 4 USA: 3	See note 1.	No safety concern
2-Methylheptan-3-one	1156	13019-20-0 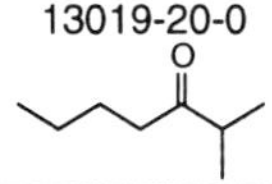	No Europe: 4 USA: 3	See note 1.	No safety concern

CAS: Chemical Abstracts Service; ND: No intake data reported
[a] Step 2: All the flavouring agents in this group were predicted to be metabolized to innocuous products.

Table 1 *(continued)*

[b] The human intake threshold is 1800 μg per person per day for class I and 540 μg/person per day for class II. All intake levels are expressed in μg/person per day. The combined intake of flavouring agents in class I is 58 and 16 μg per person per day in Europe and the USA, respectively. The combined intake of flavouring agents in class II is 520 and 120 μg/person per day in Europe and the USA, respectively.

Notes:

1. Detoxicated by reduction of the ketone followed by glucuronic acid conjugation of the corresponding alcohol or direct glucuronic acid conjugation of the secondary alcohol
2. Detoxicated by hydrolysis of the ester and glucuronic acid conjugation of the resulting alicyclic alcohol and complete oxidation of the carboxylic acid
3. Detoxicated by reduction of the ketone functional group followed by glucuronic acid conjugation of the resulting alcohol and glutathione conjugation of the parent ketone
4. Detoxicated by reduction of the ketone and alkyl side-chain oxidation and excretion

Organization of the Flavor Industry, 1995) and 1100 kg in the USA (Lucas et al., 1999). Approximately 73% and 47% of the total annual production volume in Europe and the USA, respectively, is accounted for by 6-methyl-5-hepten-2-one (No. 1120) and 1-octen-3-ol (No. 1152). The daily per capita intake of 6-methyl-5-hepten-2-one (No. 1120) and 1-octen-3-ol (No. 1152) was 120 μg and 290 μg in Europe and 44 μg and 23 μg, respectively, in the USA, respectively.

1.3 Absorption, distribution, metabolism and elimination

In general, the aliphatic esters in this group hydrolyse to the corresponding secondary alcohols. The secondary alcohols and their corresponding ketones are interconvertible under physiological conditions. In the principal excretion pathway, the ketones are reduced to the corresponding secondary alcohols, which are subsequently conjugated with glucuronic acid and excreted.

When the ketone carbonyl function is located at the 2-position (i.e. a methyl ketone), the methyl group may undergo α,β-hydroxylation and subsequent oxidation, to yield a corresponding ketocarboxylic acid. The ketoacids are intermediary metabolites (e.g. α,β-ketoacids), which can undergo oxidative decarboxylation to yield carbon dioxide and simple aliphatic carboxylic acids. The acid may be completely metabolized in the fatty acid pathway and citric acid cycle. When the substance is an α,β-unsaturated ketone (Nos 1124–1139 and 1147–1149) or secondary alcohol (Nos 1140–1141 and 1150–1154) which is oxidized to an α,β-unsaturated ketone, it can conjugate with glutathione. The glutathione conjugates are converted to the corresponding mercapturic acid derivatives (*N*-acetylcysteine derivative) and excreted.

As the α,β-unsaturated carbonyl group is a structural alert for toxicity, the Committee, at previous meetings, has devoted considerable attention to the safety of flavouring agents containing this reactive moiety. The Committee concluded at its fifty-seventh meeting (Annex 1, reference 154) that the presence of protective processes in cells provides adequate detoxication capacity at the low doses associated with use of such compounds as flavouring agents. With respect to α,β-unsaturated ketones, these protective processes include reduction of the ketone to the corresponding alcohol (followed by conjugation of the alcohol with glucuronic acid) and conjugation with glutathione. These processes operate for the aliphatic ketones used as flavouring agents.

1.4 Application of Procedure for the Safety Evaluation of Flavouring Agents

Step 1. In applying the Procedure for the Safety Evaluation of Flavouring Agents to these chemicals, the Committee assigned 11 of the 39 agents to structural class I (Cramer et al., 1978) on the basis that they are acyclic saturated and unsaturated aliphatic ketones with no other functional groups and three or fewer carbons on either side of the keto group (Nos 1124–1126 and 1133) and secondary alcohols (Nos 1140–1141) or esters of secondary alcohols (Nos 1142–1146). The remaining 28 flavouring agents, which are acyclic aliphatic ketones with four or more carbons on either side of the keto group (Nos 1118–1123, 1127– 1132, 1134–1139, 1148, 1155 and 1156) and either a ketone (Nos 1147 and 1149) or a secondary alcohol (Nos 1150–1154) attached to a terminal vinyl group, were assigned to structural class II.

Step 2. All the flavouring agents in this group were predicted to be metabolized to innocuous products. The evaluation of these substances therefore proceeded via the A (left-hand) side of the decision tree.

Step A3. The estimated daily per capita intakes of the 11 flavouring agents in structural class I and the 28 flavouring agents in structural class II are below the threshold for human intake for each class (i.e., 1800 μg per person for structural class I and 540 μg per person for structural class II). According to the Procedure, the safety of these 39 flavouring agents raises no concern when they are used at their current estimated levels of intake.

Table 1 summarizes the evaluation of aliphatic secondary alcohols, ketones and related esters used as flavouring agents.

1.5 Consideration of combined intake from use as flavouring agents

In the unlikely event that all the flavouring agents in this group were to be consumed concurrently on a daily basis, the estimated combined intake would not exceed the human intake threshold for structural class I or II.

1.6 Conclusions

The Committee concluded that none of the flavouring agents in this group of 39 aliphatic secondary alcohols, ketones and related esters would raise a safety concern at the current estimated levels of intake. Other data on the toxicity of aliphatic secondary alcohols, ketones and related esters are consistent with the results of the safety evaluation.

2. RELEVANT BACKGROUND INFORMATION

2.1 Explanation

This section summarizes the data relevant to the safety evaluation of 39 aliphatic ketones, secondary alcohols and related esters (see Table 1). The group of flavouring agents includes three saturated ketones (Nos 1118, 1155 and 1156), 24 unsaturated ketones (Nos 1119–1139 and 1147–1149), seven secondary alcohols (Nos 1140, 1141 and 1150–1154) and five esters formed from saturated secondary alcohols (Nos 1142–1146). The esters in this group are readily hydrolysed to the corresponding secondary alcohols. After hydrolysis, the resulting secondary alcohols, ketones and other secondary alcohols in the group are readily absorbed. Secondary alcohols are interconvertible with the corresponding ketones in vivo (McMahon, 1982). Therefore, all the substances in this group are expected to participate in common routes of absorption, metabolism, distribution and excretion and have similar toxicological profiles.

2.2 Additional considerations on intake

The total annual volume of use of the 39 secondary alcohols, ketones and related esters is approximately 3900 kg in Europe (International Organization of the Flavor

Table 2. Annual volumes of production and intake of aliphatic secondary alcohols, ketones and related esters used as flavouring agents in Europe and the USA

Substance (No.)	Most recent annual volume (kg)[a]	Intake ('eaters only')[b] µg/day	Intake ('eaters only')[b] µg/kg bw per day	Annual volume in naturally occurring foods (kg)[c]	Consumption ratio[d]
3-Decanone (1118)[e]					
Europe	25	3.6	0.059	+	NA
USA	25	3.3	0.55		
5-Methyl-5-hexen-2-one (1119)[f]					
Europe	NR	NA	NA	–	NA
USA	2.0	0.26	0.0044		
6-Methyl-5-hepten-2-one (1120)					
Europe	820	120	1.9	4300	5
USA	340	44	0.74		13
3,4,5,6-Tetrahydropseudoionone (1121)					
Europe	NR	NA	NA	–	NA
USA	0.1	0.01	0.0002		
6,10-Dimethyl-5,9-undecadien-2-one (1122)					
Europe	340	49	0.82	3200	9
USA	16	2.2	0.036		200
2,6,10-Trimethyl-2,6,10-pentadecatrien-14-one (1123)					
Europe	0.7	0.1	0.002	460	660
USA	NR	NA	NA		NA
3-Penten-2-one (1124)					
Europe	2.1	0.30	0.0050	660	320
USA	NR	NA	NA		NA
4-Hexen-3-one (1125)					
Europe	110	15	0.25	+	NA
USA	6.4	0.84	0.014		
2-Hepten-4-one (1126)					
Europe	0.1	0.01	0.0002	+	NA
USA	NR	NA	NA		NA
3-Hepten-2-one (1127)					
Europe	1.3	0.19	0.0031	+	NA
USA	0.5	0.07	0.001		
3-Octen-2-one (1128)					
Europe	5.2	0.74	0.012	1	0.2
USA	6.4	0.84	0.014		0.2

Table 2 (contd)

Substance (No.)	Most recent annual volume (kg)[a]	Intake ('eaters only')[b]		Annual volume in naturally occurring foods (kg)[c]	Consumption ratio[d]
		µg/day	µg/kg bw per day		
2-Octen-4-one (1129)					
Europe	7.0	1.0	0.016	+	NA
USA	20	2.6	0.044		
3-Decen-2-one (1130)					
Europe	0.1	0.01	0.0002	+	NA
USA	NR	NA	NA		
4-Methyl-3-penten-2-one (1131)					
Europe	2.8	0.40	0.0067	3	1
USA	NR	NA	NA		
5-Methyl-3-hexen-2-one (1132)					
Europe	NR	NA	NA	+	NA
USA	0.9	0.1	0.002		
5-Methyl-2-hepten-4-one (1133)					
Europe	48	6.8	0.11	–	NA
USA	6.4	0.84	0.014		
6-Methyl-3,5-heptadien-2-one (1134)					
Europe	110	15	0.26	+	NA
USA	34	4.5	0.076		
(E)-7-Methyl-3-octen-2-one (1135)					
Europe	NR	NA	NA	+	NA
USA	14	1.8	0.030		
3-Nonen-2-one (1136)[e]					
Europe	100	14	0.24	+	NA
USA	100	13	0.22		
(E) and (Z)-4,8-Dimethyl-3,7-nonadien-2-one (1137)[e]					
Europe	50	7.1	0.12	–	NA
USA	50	6.6	0.11		
(E)-6-Methyl-3-hepten-2-one (1138)[e]					
Europe	25	3.6	0.059	–	NA
USA	25	3.3	0.055		
(E,E)-3,5-Octadien-2-one (1139)[e]					
Europe	25	3.6	0.059	+	NA
USA	25	3.3	0.055		
3-Octen-2-ol (1140)					
Europe	10	1.4	0.024	–	NA
USA	NR	NA	NA		

Table 2 (contd)

Substance (No.)	Most recent annual volume (kg)[a]	Intake ('eaters only')[b]		Annual volume in naturally occurring foods (kg)[c]	Consumption ratio[d]
		μg/day	μg/kg bw per day		
(E)-2-Octen-4-ol (1141)					
Europe	NR	NA	NA	–	NA
USA	11	1.4	0.024		
2-Pentyl butyrate (1142)					
Europe	NR	NA	NA	+	NA
USA	4.5	0.59	0.0099		
(±) Heptan-3-yl acetate (1143)[e]					
Europe	25	3.6	0.059	–	NA
USA	25	3.3	0.055		
(±) Heptan-2-yl butyrate (1144)[e]					
Europe	25	3.6	0.059	+	NA
USA	25	3.3	0.055		
(±) Nonan-3-yl acetate (1145)[e]					
Europe	25	3.6	0.059	–	NA
USA	25	3.3	0.055		
2-Pentyl acetate (1146)[e]					
Europe	24	3.4	0.057	–	NA
USA	24	3.2	0.053		
1-Penten-3-one (1147)					
Europe	2.4	0.34	0.0057	4	2
USA	0.9	0.1	0.002		
1-Octen-3-one (1148)					
Europe	10	1.4	0.024	93	9
USA	0.5	0.07	0.001		190
2-Pentyl-1-buten-3-one (1149)					
Europe	1.2	0.17	0.0029	–	NA
USA	NR	NA	NA		
1-Penten-3-ol (1150)					
Europe	17	2.4	0.040	5300	310
USA	9.1	1.2	0.020		580
1-Hexen-3-ol (1151)					
Europe	4.5	0.64	0.011	+	NA
USA	18	2.3	0.039		
1-Octen-3-ol (1152)					
Europe	2000	290	4.8	15 000	7
USA	180	23	0.39		83

Table 2 (contd)

Substance (No.)	Most recent annual volume (kg)[a]	Intake ('eaters only')[b] µg/day	Intake ('eaters only')[b] µg/kg bw per day	Annual volume in naturally occurring foods (kg)[c]	Consumption ratio[d]
1-Decen-3-ol (1153)					
Europe	NR	NA	NA	–	NA
USA	1.0	0.13	0.0022		
(E,R)-3,7-Dimethyl-1,5,7-octatrien-3-ol (1154)					
Europe	NR	NA	NA	+	NA
USA	45	5.9	0.099		
6-Undecanone (1155)[e]					
Europe	25	3.6	0.059	–	NA
USA	25	3.3	0.055		
2-Methylheptan-3-one (1156)[e]					
Europe	25	3.6	0.059	+	NA
USA	25	3.3	0.055		
Total					
Europe	3900				
USA	1100				

NR, no data reported; NA, not applicable; +, reported to occur naturally in foods (Maarse et al., 1999) but no quantitative data available; –, not reported to occur naturally in foods

[a] From International Organization of the Flavor Industry (1995) and Lucas et al. (1999)
[b] Intake (µg/person per day) calculated as follows: [(annual volume, kg) × (1 × 10^9 µg/kg)/(population × survey correction factor × 365 days)], where population (10%, 'eaters only') = 32 × 10^6 for Europe and 26 × 10^6 for the USA. The correction factor = 0.6 for Europe and 0.8 for the USA, representing the assumption that only 60% and 80% of the annual production volume of the flavour, resepctively, was reported in the poundage surveys. Intake (µg/kg bw per day) calculated as follows: [(µg/person per day)/body weight], where body weight = 60 kg. Slight variations may occur from rounding.
[c] Quantitative data from Stofberg & Grundschober (1987)
[d] Calculated as follows: (annual consumption in food, kg)/(most recently reported volume as a flavouring agent, kg)
[e] Anticipated annual volume in the USA as reported by the Flavor and Extract Manufacturers Association of the United States

Industry, 1995) and 1100 kg in the USA (Lucas et al., 1999) (see Table 2). In Europe and the USA, approximately 73% and 47% of the total annual production volume are accounted for by 6-methyl-5-hepten-2-one (No. 1120) and 1-octen-3-ol (No. 1152), respectively. Production volumes and intake values for each flavouring agent are reported in Table 2.

Twenty-six of the 39 flavouring substances in this group have been reported to occur naturally in foods. They have been detected in a variety of foods, including avocado, raspberry, mango, kiwi fruit, grapefruit juice, orange juice, cocoa, tea, tomato, asparagus, potato chips and rum (Maarse et al., 1999). Quantitative data on natural occurrence and consumption ratios have been reported for 10 of the substances: 6-methyl-5-hepten-2-one (No. 1120); 6,10-dimethyl-5,9-undecadien-2-one (No. 1122); 2,6,10-trimethyl-2,6,10-pentadecatrien-14-one (No. 1123); 3-penten-2-one (No. 1124); 3-octen-2-one (No. 1128); 4-methyl-3-penten-2-one (No. 1131); 1-penten-2-one (No. 1147); 1-octen-3-one (No. 1148); 1-penten-3-ol (No. 1150) and 1-octen-3-ol (No. 1152). The data demonstrate that their intake occurs predominantly from the consumption of traditional foods (i.e., consumption ratio > 1) (Stofberg & Kirschman, 1985; Stofberg & Grundschober, 1987) (see Table 2).

2.3 Biological data

2.3.1 Biochemical data

(a) Hydrolysis

The aliphatic esters, such as 2-pentyl butyrate (No. 1142), are rapidly hydrolysed to the parent secondary alcohol and the component aliphatic carboxylic acids by classes of enzymes recognized as carboxylesterases (Ford & Moran, 1978; Heymann, 1980; Anders, 1989; White, 1990), the most important of which are the β-esterases. Esters of aliphatic secondary alcohols were hydrolysed to their corresponding alcohols and carboxylic acids when incubated with liver homogenates obtained from male Wistar albino rats (Longland et al., 1977) or intestinal mucosal homogenates obtained from pigs (Leegwater & van Straten, 1974b). The ester of the secondary alcohol, 3-methyl-2-pentanol, and acetic acid were completely hydrolysed in intestinal mucosal homogenate within 2 h (Leegwater & van Straten, 1974b). Esters exposed to artificial intestinal fluid and simulated gastric juice were also hydrolysed, but more slowly. Partial hydrolysis (15–40%) was reported 2 h after incubation of 2-propyl butyrate or 4-methyl-2-pentyl acetate with simulated intestinal fluid (Leegwater & van Straten, 1974a). Rat liver homogenates were found to be between 8 and > 25 000 times more efficient than artificial pancreatic juice in hydrolysing a wide variety of aliphatic esters, and hydrolysis in simulated intestinal fluid with pancreatin was much faster than in simulated gastric juice (Longland et al., 1977). The authors concluded that tissue homogenates provide a good model for the situation in vivo. Thus, it is anticipated that the esters (Nos 1142–1146) in this group rapidly hydrolyse to secondary alcohols and carboxylic acids, which are then readily absorbed.

(b) Absorption, distribution, metabolism and excretion

After oral administration, aliphatic secondary alcohols and ketones are absorbed through the gastrointestinal tract and subsequently rapidly eliminated from the blood. Peak blood concentrations normally occur within 1–2 h after dosing (Lehman et al., 1945; Nordmann et al., 1973; Bonte et al., 1981). Although ketones and secondary alcohols are readily interconverted, reduction of the ketones by cytosolic carbonyl reductases (Felsted & Bachur, 1980) is favoured in vivo, yielding the corresponding secondary alcohols, which are excreted in the urine mainly as glucuronic acid conjugates (Kasper & Henton, 1982; Annex 1, reference *138*).

In the case of a methyl ketone, the terminal methyl group may undergo oxidation, eventually yielding α-ketoacid, which is the substrate for further oxidation in the fatty acid pathway and citric acid cycle (Annex 1, reference *138*). In the case of α,β-unsaturated ketones, glutathione may conjugate at the β-position, yielding cysteine and mercapturic acid derivatives, which are also excreted primarily in the urine as the conjugated secondary alcohol (Williams, 1959; Portoghese et al., 1989).

(i) Reduction of aliphatic secondary alcohols and acyclic ketones

Single high doses of a homologous series of aliphatic secondary alcohols and ketones were administered individually by gavage to rabbits, principally to identify the extent of urinary glucuronide metabolite formation. The level of urinary excretion of glucuronic acid conjugates was determined over 24 h. The results showed that secondary alcohols, either administered directly or formed via ketone reduction, are largely excreted as glucuronic acid conjugates (Kamil et al., 1953). The substances, doses, and average urinary output of glucuronide as a percentage of the dose administered are listed in Table 3.

α,β-Unsaturated ketones are also reduced to the corresponding secondary alcohols. Cytoplasmic NADPH-dependent α,β-unsaturated ketone reductase isolated from human liver catalysed the reduction of aliphatic α,β-unsaturated ketones to the corresponding secondary alcohols (Fraser et al., 1967). The resulting alcohols can then be excreted as glucuronic acid conjugates (Williams, 1959). Glucuronic acid conjugates of substances like secondary alcohols are frequently the end-products of metabolism and are excreted in the urine or the bile (Kasper & Henton, 1982).

(ii) α-Oxidation of methylketones

Twenty aliphatic secondary ketones and related esters are methyl ketones (Nos 1119–1124, 1127, 1128, 1130–1132, 1134–1139 and 1149) or are secondary alcohols or related esters that are readily converted to methyl ketones in vivo (Nos 1140 and 1142) (see Table 1). Additional metabolic options are available for these substances. Methyl ketones undergo α-hydroxylation and subsequent oxidation of the terminal methyl group to eventually yield corresponding ketocarboxylic acids (Gabriel et al., 1972). The ketoacids are intermediary metabolites (e.g. α-ketoacids) that undergo oxidative decarboxylation to yield carbon dioxide and simple aliphatic carboxylic acids. The acids may be completely metabolized in the fatty acid pathway and citric acid cycle. The metabolism of α-hydroxyacids and α-ketoacids was recently reviewed (Annex 1, reference *144*).

Table 3. Urinary glucuronic acid conjugate after administration of aliphatic secondary alcohols and ketones to rabbits by gavage

Substance	Dose (mg/kg bw)[a]	Urinary glucuronic acid conjugate (%)
2-Pentanol	740	45
2-Heptanone	950	41
2-Heptanol	960	55
3-Heptanol	960	62
2-Octanol	1100	16

[a] Calculated on the basis of a dose of 25 mmol/3 kg bw

(iii) Glutathione conjugation of α,β-unsaturated ketones

Twenty-six of the substances in this group are α,β-unsaturated ketones (Nos 1124–1139, 1147–1149) or alcohols (Nos 1140, 1141, 1150–1154), which can be oxidized to α,β-unsaturated ketones. In addition to reduction of the carbonyl function and, in some cases, α-oxidation of the methyl group, these substances may also react with free glutathione (GSH) in a nucleophilic Michael-type addition (Portoghese et al., 1989). GSH conjugates can be excreted in the bile or they can be converted enzymically to the mercapturic acid conjugate. It has been shown that α,β-unsaturated carbonyl compounds have an affinity for sulfhydryl groups (Boyland & Chasseaud, 1967, 1968; Portoghese et al., 1989). These substances [e.g., ethyl vinyl ketone (No. 1147)] react with GSH through addition of a nucleophile (Nu^-) to the β-carbon atom of the double-bond. This reaction is catalysed by glutathione-*S*-transferase but can also occur non-enzymatically (see Figure 1) (Esterbauer et al., 1975; Chasseaud, 1979; Portoghese et al., 1989).

The conjugation of glutathione with α,β-unsaturated ketones in vivo is catalysed by glutathione-*S*-tranferases. Up to 13 different forms of this enzyme exist in the

Figure 1. Metabolic fate of aliphatic ketones and secondary alcohols

Reduction of aliphatic ketones (major pathway)

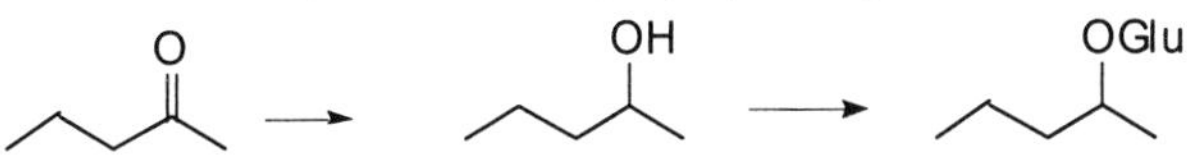

α-Oxidation of methyl ketones

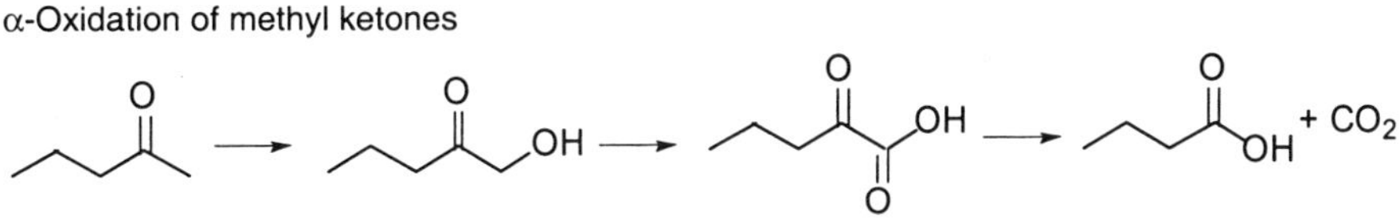

Michael addition of glutathione to α,β-unsaturated ketones

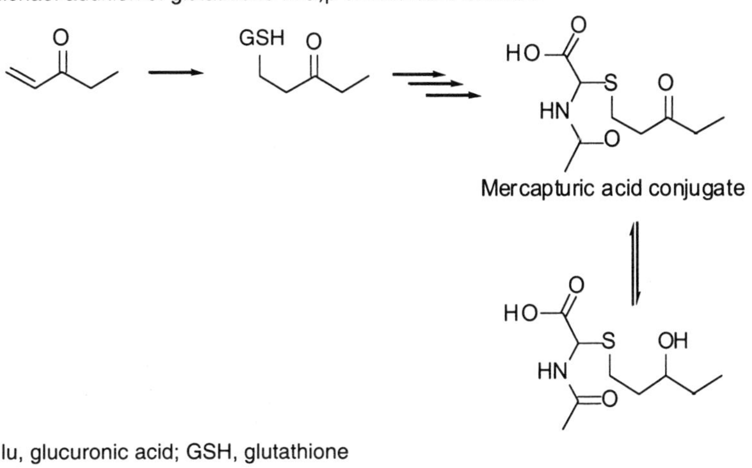

Glu, glucuronic acid; GSH, glutathione

human liver. These isoenzymes have many similarities including subunit size, immunochemical properties, catalytic activities, other binding activities and resistance to sulfhydryl reagents (van der Jagt et al., 1985). They also are identifiable by isoelectric point and haematin binding properties. The conjugation reaction may occur, however, even in the absence of glutathione-*S*-transferase, since many α,β-unsaturated ketones are sufficiently electrophilic to react with glutathione. The reactivities of several α,β-unsaturated ketones for glutathione addition were studied in the absence of glutathione-*S*-transferase (Portoghese et al., 1989). Included in this group were 2-octene-4-one (No. 1129), 4-methyl-3-penten-2-one (No. 1131) and 6-methyl-3,5-heptadien-2-one (No. 1134). It was shown that the rate of reaction was extremely rapid for compounds such as 2-butenone, but that any substitution at the β-position of the double-bond greatly reduced this reactivity. Thus, a methyl substitution at the β-position in 4-methyl-3-penten-2-one (No. 1131) reduced the reactivity towards glutathione addition by 120 000 times as compared with 2-butenone. A similar reduction of relative reactivity was observed with 6-methyl-3,5-heptadien-2-one (No. 1134) (Portoghese et al., 1989).

Typically, two metabolic processes follow conjugation of the α,β-unsaturated ketone. In one, the GSH conjugate undergoes enzyme-catalysed hydrolysis and *N*-acetylation, yielding the corresponding *N*-acetyl-L-cysteine derivative (i.e., mercapturic acid derivative) (Chasseaud, 1976). In the other, the ketone function may be reduced to the corresponding alcohol. The principal metabolite obtained from the urine of rats given 2-propenal by subcutaneous injection was *N*-acetyl-*S*-(3-hydroxypropyl)-L-cysteine (Kaye, 1973). Formation of the mercapturic acid derivative of the corresponding secondary alcohol has been reported with mesityl oxide [4-methyl-3-penten-2-one (No. 1131)] and with ethyl vinyl ketone [1-penten-3-one (No. 1147)] in vitro with rat liver homogenate (Boyland & Chasseaud, 1967; Chasseaud, 1976).

At high cellular concentrations, α,β-unsaturated carbonyl compounds have the potential to react with DNA. A cyclic deoxyguanosine adduct is formed in vitro via Michael addition of the N-1 or exocyclic NH_2 of guanine to the β position of 2-butenone, followed by condensation of the ketone carbonyl with the remaining exocyclic NH_2 or N-1 moiety (Chung et al., 1988). However, these adducts have been found to be formed at a considerably slower pace (24 h at 100 °C) than the glutathione reaction, suggesting that glutathione conjugation takes place in vivo before it can react with DNA (Portoghese et al., 1989).

In addition, the terminal double-bonds may undergo epoxidation, catalysed by microsomal cytochrome P450 isoenzymes (Belluci et al., 1996). In rat liver microsomes, substances such as 1-hexen-3-ol can be epoxidized to yield erythro- and threo-1,2-epoxyhexan-3-ol. These are intermediary metabolites, in that the epoxide function is rapidly hydrolysed to a diol in the presence of microsomal epoxide hydrolase or conjugated with glutathione (Chiappe et al., 1998). This results in formation of erythro- and threo-hexane-1,2,3-triol, a polar metabolite that can be rapidly excreted in the urine (Chiappe et al., 1998).

Analysis of 150 human urine samples revealed the presence of two substances used as flavouring agents. The samples were collected for 2 months on different days from the same volunteers and were analysed by gas chromatography. The urinary profiles varied among individuals but remained constant for the same person over 2 months. Dietary changes during the testing period did not change the urinary profiles. Constituents of normal urine samples were identified by mass spectrometry. Among the numerous compounds from this group of aliphatic secondary alcohols and ketones, 3-penten-2-one (No. 1124) and 4-methyl-3-penten-2-one (No. 1131) were identified as normal constituents of human urine. In a study under the same conditions for compounds present in the urine of 40 patients with diabetes mellitus, one of the several compounds identified as a normal urine constituent was again 3-penten-2-one (No. 1124) (Zlatkis et al., 1973).

A similar type of study was undertaken, in which urinary organic acid metabolites were studied in normal and diabetic C57BL/Ks male mice. The profiles of the diabetic animals showed significantly lower concentrations of certain ketones, including 6-methyl-6-hepten-2-one. These ketones were excreted in low amounts in diabetic animals with depleted fat stores, indicating that their production is related to a deficiency in fat breakdown. However, the concentrations of several other similar ketones, such as 6-methyl-5-hepten-2-one (No. 1120), 3-hepten-2-one (No. 1127) and 3-octen-2-one (No. 1128), showed no difference between normal and diabetic mice (Holland et al., 1984).

(iv) Summary

The studies demonstrate that these 39 aliphatic ketones, secondary alcohols and related esters are metabolized primarily by reduction of the ketone to yield the corresponding secondary alcohol, followed by conjugation with glucuronic acid and excretion in the urine. Other metabolic options include α-oxidation of methyl ketones and glutathione conjugation of the α,β-unsaturated ketones, followed by elimination of the mercapturic acid derivative of the corresponding secondary alcohol. To a minor extent, the double-bond may undergo epoxidation and then hydrolysis or conjugation with glutathione.

Because the α,β-unsaturated carbonyl group is a structural alert for toxicity, the Committee, at previous meetings, devoted considerable attention to the safety of flavouring agents containing this reactive moiety. The Committee concluded at its fifty-seventh meeting (Annex 1, reference *154*) that the presence of cytoprotective processes provides adequate detoxication capacity at the low doses associated with use of such compounds as flavouring agents. With respect to α,β-unsaturated ketones, these cytoprotective processes include reduction of the ketone to the corresponding alcohol (followed by conjugation of the alcohol with glucuronic acid) and conjugation of the glutathione. As reported above, these processes are operative for the aliphatic ketones used as flavouring agents.

2.3.2 Toxicological studies

(a) Acute toxicity

Oral LD_{50} values have been reported for 15 of the 39 substances in this group. In rats, the values ranged from 320 to > 6800 mg/kg bw (Table 4).

(b) Short-term and long-term studies of toxicity

The Committee previously evaluated the structurally related aliphatic ketones 2-heptanone (No. 283), 3-heptanone (No. 285), 4-heptanone (No. 287) and 2,6-dimethyl-4-heptanone (No. 302). The NOEL for 2-heptanone and was 20 mg/kg bw per day (Gaunt et al., 1972), while those of the remaining three aliphatic ketones

Table 4. Studies of acute toxicity of orally administered aliphatic secondary alcohols, ketones and related esters used as flavouring agents

No.	Agent	Species	Sex	LD_{50} (mg/kg bw)	Reference
1120	6-Methyl-5-hepten-2-one	Rat	M,F	4100	Keating (1972)
1120	6-Methyl-5-hepten-2-one	Mouse	M,F	3600	Colaianni (1967)
1121	3,4,5,6-Tetrahydropseudoionone	Mouse	M,F	5200	Moreno (1982)
1121	3,4,5,6-Tetrahydropseudoionone	Rat	M,F	> 5000	Moreno (1977)
1122	6,10-Dimethyl-5,9-undecadien-2-one	Mouse	M,F	> 8600	Colaianni (1967)
1122	6,10-Dimethyl-5,9-undecadien-2-one	Rat	M,F	> 6800	Hoffman (1978)
1122	6,10-Dimethyl-5,9-undecadien-2-one	Rat	M,F	> 5000	Moreno (1975)
1123	2,6,10-Trimethyl-2,6,10-pentadecatrien-14-one	Rat	M,F	> 5000	de Groot et al. (1974)
1124	3-Penten-2-one	Rat	M,F	320	Smyth & Carpenter (1948)
1125	4-Hexen-3-one	Mouse	M,F	780	Oser (1970)
1132	5-Methyl-3-hexen-2-one	Rat	NR	1400	Carpenter et al. (1974)
1133	5-Methyl-2-hepten-4-one	Rat	M F	1400 1400	Reagan & Becci (1984)
1134	6-Methyl-3,5-heptadien-2-one	Mouse	M,F	3200	Colaianni (1967)
1135	(E)-7-Methyl-3-octen-2-one	Rat	M,F	> 2000	Sanders (1996)
1148	1-Octen-3-one	Rat	M F	982 890	Reagan & Becci (1983a)
1149	2-Pentyl-1-buten-3-one	Rat	M F	3700 1800	Reagan & Becci (1983b)
1151	1-Hexen-3-ol	Rat	M,F	440	Til (1977)
1152	1-Octen-3-ol	Rat	M,F	340	Wohl (1974)
1153	1-Decen-3-ol	Rat	M,F	> 1500	Chubb (1995)

M, male; F, female; NR, not reported

were ≥ 1000 mg/kg bw per day (O'Donoghue & Krasavage, 1980; O'Donoghue et al., 1984). The Committee concluded that all four agents were of no safety concern at current levels of intake when used as flavouring agents (Annex 1, reference *138*). The results of short-term studies with repeated dose of two representative aliphatic secondary alcohols and five ketones are summarized in Table 5 and described below.

(i) 5-Methyl-5-hexen-2-one (No. 1119)

A 14-day study to screen for minimum toxicity was conducted, in which a group of five Fischer 344 rats of each sex was maintained on a diet containing 5-methyl-5-hexene-2-one (No. 1119) that provided a target intake of 10 mg/kg bw per day. Four other flavouring agents were tested, with a common control group. The animals

Table 5. Results of short- and long-term studies of toxicity in male and female rats of orally administered aliphatic secondary alcohols, ketones and related esters used as flavouring agents

No.	Substance	No. test groups[a]/ no. per group[b]	Duration (days)	NOEL (mg/kg bw per day)	Reference
1119	5-Methyl-5-hexen-2-one	1/10	14	10[d]	Gill & Van Miller (1987)
1123	2,6,10-Trimethyl-2,6,10-pentadecatrien-14-one	2/10	14	3.5	de Groot et al. (1974)
1125	4-Hexene-3-one	1/46	90	6.6 (F) 6.7 (M)	Shellenberger (1970a,b)
1129	2-Octen-4-one	1/30	90	6.7	Cox et al. (1974a)
1148	1-Octen-3-one	1/30	90	6.7	Cox et al. (1974b)
1152	1-Octen-3-ol	1/28	90	12 (M) 14 (F)	Posternak et al. (1964)
1154	(E,R)-3,7-Dimethyl-1,5,7-octatrien-3-ol	1/14	14	10[d]	Wnorowski (1997)

[a] Does not include control animals
[b] Includes both male and female animals
[c] Performed at either a single or multiple doses that produced no adverse effects. Therefore, this dose level is not a true NOEL but the highest dose tested that produced no adverse effects. The actual NOEL would be higher.

were observed for deaths twice daily and underwent a detailed clinical examination daily. Food consumption was measured on days 7 and 14. No physical signs of toxicity, abnormal body-weight gain, abnormal food consumption or treatment-related effects were observed at necropsy. Statistically significant decreases in absolute and relative liver weights and absolute kidney weight were reported in treated males. As similar changes were found with the four other substances, the authors concluded that the decreases in organ weight were a reflection of higher than normal liver and kidney weights in the control group. This conclusion was supported by the fact that none of the five treated groups showed signs or histopathological changes in these organs (Gill & van Miller, 1987).

(ii) 2,6,10-Trimethyl-2,6,10-pentadecatrien-14-one (No. 1123)

Groups of five weanling rats of each sex were given 2,6,10-trimethylpentadecatrien-14-one in corn oil by gavage at a dose of 0, 0.35 or 3.5 mg/kg bw per day on 6 days per week for 2 weeks. Individual body weights and food intake were recorded on days 1, 7 and 14, and haemoglobin concentration was measured on day 14. All animals survived and were killed at the end of the study. Kidney and liver weights were recorded, and all organs were examined macro-scopically and microscopically. No differences in body-weight gain, food intake or food use efficiency were reported between treated and control groups. Histopathological examination revealed no evidence of alterations to the liver or kidneys. Liver weights were increased in male rats at the lower dose, while kidney weights were increased in male rats at the higher dose. In the absence of histopathological effects, these changes were considered to be of no toxicological significance (deGroot et al., 1974).

(iii) (E,R)-3,7-Dimethyl-1,5,7-octatrien-3-ol (No. 1154)

Groups of seven Sprague-Dawley rats of each sex were given (E,R)-3,7-dimethyl-1,5,7-octatrien-3-ol (No. 1154) in 1 ml of corn oil by gavage at a dose of 10 mg/kg bw per day for 14 consecutive days. Seven vehicle controls of each sex were given 1 ml of corn oil by gavage daily for 14 days. Individual doses were adjusted for body weight on days 1 and 8. All animals were observed daily for signs of gross toxicity and deaths. Weekly measurement of body weights and food consumption revealed no differences between test and control groups. All animals survived to completion of the study. At termination, gross and histopathological examination of the liver and kidneys from all treated animals revealed no lesions that could be associated with administration of the test substance (Wnorowski, 1997).

(iv) 4-Hexen-3-one (No. 1125)

Groups of 23 weanling Sprague-Dawley rats of each sex were maintained on a diet containing 4-hexene-3-one (No. 1125) at a concentration calculated to provide an average daily intake of 6.7 mg/kg bw for males and 6.6 mg/kg bw for females. Weekly measurements of body weight revealed no differences between treated and control groups. Biochemical, haematological and urinary analyses conducted after weeks 6 and 13 also showed no significant differences between treated and control rats. After 90 days, the animals were killed, and a complete necropsy was performed. A number of incidental lesions were reported, the incidences of which were low and randomly distributed among control and treated animals. The authors did not attribute the gross pathological lesions seen in the tissues to treatment with 4-hexen-3-one (Shellenberger, 1970a). In an addendum to the final report, which included the results of histopathological evaluation of tissues obtained from the rats, Shellenberger (1970b) concluded that administration of 4-hexen-3-one had no toxicologically significant effects.

(v) 2-Octen-4-one (No. 1129) and 1-octen-3-one (No. 1148)

In two studies with the same protocol, groups of 15 Wistar rats of each sex were maintained on a diet containing 2-octen-4-one (No. 1129) (Cox, 1974a) or 1-octen-3-one (No. 1148) (Cox, 1974b) at a concentration providing an average target intake of 7.5 mg/kg bw per day, for 90 days. Groups of 15 control rats of each sex were maintained on basal diet. The rats were housed individually and giventheir respective diet and tap water *ad libitum*. Measurements of food consumption throughout the study indicated that the treated groups received 6.7 mg/kg bw per day of 2-octen-4-one or 1-octen-3-one. All rats were monitored daily for survival, behaviour and physical appearance. Weekly measurement of body weight and food consumption revealed no significant differences between treated and control groups. Haematological, blood biochemical and urine analyses performed during weeks 6 and 12 on eight males and eight females from each group showed normal values. At 90 days, complete necropsies were performed. A detailed histopathological examination was undertaken on 23 major organs from eight male and eight female rats, and the livers and kidneys of the remaining seven animals in each group were examined for histopathological changes. No difference was found between treated and control animals in absolute or relative organ weights, and there was no evidence of gross or microscopic alterations with either substance (Cox, 1974a). A sporadic incidence of slight vacuolization of liver cells was observed in both control and treated animals (Cox, 1974b).

(vi) 1-Octen-3-ol (No. 1152)

A group of 14 male and 14 female Charles River rats was maintained on a diet containing 1-octen-3-ol at a concentration calculated to provide an average intake of 12 mg/kg bw per day for males and 14 mg/kg bw per day for females. The rats were given their respective diet and tap water *ad libitum*. All rats were monitored daily for survival, behaviour and physical appearance. Weekly measurements of body weight and food consumption revealed no significant differences between treated and control animals. In general, haematological, blood biochemical and urine analyses performed during weeks 7 and 13 revealed normal values, with the exception of increased haemoglobin and corpuscular values. At necropsy, gross and histopathological examination revealed no evidence of effects attributable to intake of the substance (Posternak, 1964).

(c) Genotoxicity

Seven representative substances in this group have been tested for genotoxicity in vitro. The results of these tests are summarized in Table 6 and described below.

(i) Reverse mutation

Assays for reverse mutation were performed with 6-methyl-5-hepten-2-one (No. 1120), 3-penten-2-one (No.1124), 4-methyl-3-penten-2-one (No. 1131), 6-methyl-3,5-heptadien-2-one (No. 1134), (E)-7-methyl-3-octen-2-one (No. 1135) and 1-decen-3-ol (No. 1153). There was no evidence of mutagenicity at concentrations up to 1500 μg/ml in TA98, TA100, TA1535, TA1537 or TA1538 strains of *Salmonella typhimurium* (Florin et al. 1980; Cheh, 1986; Durward, 1995; Thompson, 1996).

1-Penten-3-one (No. 1147) caused reverse mutation in an assay with preincubation in a single strain of *S. typhimurium* (TA100) in the presence and absence of metabolic activation (Deininger et al., 1990). No increase over spontaneous background reversion rates was reported with 1-penten-3-one at a concentration of 42 μg/plate in the presence of metabolic activation; however, at a concentration of 84 μg/plate, it induced a threefold increase in the rate of reversions. Bacterial toxicity was reduced in the presence of a metabolic activation system, probably due to detoxication by reaction with nucleophilic components in the system, such as glutathione and other SH-group-containing compounds. Similarly, the toxicity was reduced by addition of glutathione, with no effect on mutagenicity (Deininger et al., 1990; Eder et al., 1993).

In order to evaluate the effect of epoxidation of the double-bond on potential mutagenicity, 1-penten-3-one was further evaluated in the presence of SKF 525A, an inhibitor of microsomal monooxygenases, and 1,1,1-trichloropropene-2,3-oxide, an inhibitor of epoxide hydrolase. No mutagenic activity was observed in the presence of 100 μg/ml of SKF 525A, but addition of 1,1,1-trichloropropene-2,3-oxide resulted in an increase in the frequency of reverse mutations in a concentration-dependent manner (Deininger et al., 1990).

Table 6. Studies of genotoxicity in vitro with aliphatic secondary alcohols, ketones and related esters used as flavouring agents

No.	Agent	End-point	Test object	Maximum concentrati.†	Results	Reference
1120	6-Methyl-5-hepten-2-one	Reverse mutation	*S. typhimurium* TA98, TA100, TA1535, TA1537	380 µg/plate	Negative[a]	Florin et al. (1980)
1124	3-Penten-2-one	Reverse mutation	*S. typhimurium* TA98, TA100, TA1535, TA1537	250 µg/plate	Negative[a]	Florin et al. (1980)
1131	4-Methyl-3-penten-2-one	Reverse mutation	*S. typhimurium* TA100	5.8–200 mg/plate	Negative[b]	Cheh (1986)
1134	6-Methyl-3,5-heptadien-2-one	Reverse mutation	*S. typhimurium* TA98, TA100, TA1535, TA1537	370 µg/plate	Negative[a]	Florin et al. (1980)
1135	(E)-7-Methyl-3-octen-2-one	Reverse mutation	*S. typhimurium* TA98, TA100, TA1535, TA1537, TA1538	15–1500 µg/plate	Negative[a]	Thompson (19896)
1147	1-Penten-3-one	Reverse mutation	*S. typhimurium* TA100	42–130 µg/plate	Positive[a,c]	Deininger et al. (1990)
1147	1-Penten-3-one	DNA damage	*E. coli* pQ37	0–4.2 µg/plate	Positive[a]	Deininger et al. (1990)
1153	1-Decen-3-ol	Reverse mutation	*S. typhimurium* TA98, TA100, TA1535, TA1537, TA1538	0.3–1000 µg/plate	Negative[a,d]	Durward (1995)

[a] With and without metabolic activation
[b] Without metabolic activation
[c] Cytotoxic at 50 µg/plate in the absence of metabolic activation
[d] Toxic in some species at 300 µg/plate

(ii) DNA damage

In a non-standardized assay for SOS DNA repair, *Escherichia coli* strain PQ37, with *rfA* and *uvrA* mutations, was incubated with 1-penten-3-one (No. 1147), with and without metabolic activation. *sfiA* gene-linked β-galactosidase activity was used as a measure of the induction of SOS repair. In the absence of metabolic activation, 1-penten-3-one was a weak SOS-inducing agent, with a potential of 0.027 calculated from the linear part of the dose–response curve. Adding metabolic activation did not increase the genotoxicity. The authors concluded that the genotoxicity of 1-penten-3-one is probably a result of its toxicity (Deininger, 1990). As has been pointed out by other authors, it is difficult to detect genotoxic effects of α,β-unsaturated substances in vitro because of their general toxicity to bacteria (Eder et al., 1991).

(iii) Conclusion

Positive results in vitro were obtained only with 1-penten-3-one, for which sufficient metabolic processes are available for detoxication at low levels of intake. These include known conjugation of α,β-unsaturated ketones with glutathione and metabolic conversion of metabolically formed epoxides to dihydrodiols by epoxide hydrolase.

3. REFERENCES

Anders, M.W. (1989) Biotransformation and bioactivation of xenobiotics by the kidney. In: Hutson, D.H., Caldwell, J. & Paulson, G.D., eds, *Intermediary Xenobiotic Metabolism in Animals*, New York: Taylor & Francis, pp. 81–97.

Belluci, G., Chiappe, C., Pucci, L. & Gervasi, P.G. (1996) The mechanism of the oxidation of allylic alcohol to α,β-unsaturated ketones by cytochrome P450. *Chem. Res. Toxicol.*, **9**, 871–874.

Bonte, W., Rudell, E., Sprung, R., Frauenrath, C., Blanke, E., Kupilas, G., Wochinik, J. & Zah, G. (1981) Experimental investigations concerning the blood-analytical detection of small doses of higher aliphatic alcohols in man. *J. Blutalkohol*, **18**, 399–411.

Boyland, E. & Chasseaud, L.F. (1966) Enzyme-catalysed conjugations of glutathione with unsaturated compounds. *Biochem. J.*, **104**, 95–102.

Boyland, E. & Chasseaud, L.F. (1967) Enzyme catalyzing conjugations of glutathione with alpha, beta-unsaturated carbonyl compounds. *Biochem. J.*, **109**, 651–660.

Carpenter, C.P., Weil, C.S. & Smyth, H.F., Jr (1974) Range finding toxicity data: List VIII. *Toxicol. Appl. Pharmacol.*, **28**, 313–319.

Chasseaud, L.F. (1976) Conjugation with glutathione and mercapturic acid excretion. In: Arias, I.R. & Jakoby, W.B., eds, *Glutathione Metabolism and Function*, New York: Raven Press, pp. 77–114.

Chasseaud, L.F. (1979) The role of gluthathione and glutathione-*S*-transferase in the metabolism of chemical carcinogens and other electrophilic agents. *Adv. Cancer Res.*, **29**, 175–274.

Cheh, A.M. (1986) Mutagen production by chlorination of methylated α,β-unsaturated ketones. *Mutat. Res.*, **169**, 1–9.

Chiappe, C., DeRubertis, A., Amato, G. & Gervasi, P.G. (1998) Sterochemistry and biotransformation of 1-hexene and 2-methylh-1-hexene with rat liver microsomes and purified P450s of rats and humans. *Chem. Res. Toxicol.*, **11**, 1487–1493.

Chubb, D.R. (1995) Acute oral toxicity study in the rat (fixed dose procedure). Study No. A/O/41323. Unpublished report from Toxicology Laboratories Ltd, Herefordshire, England. Submitted to WHO by the Flavor and Extract Manufacturers Association of the United States.

Chung, F.L., Roy, K.R. & Hecht, S.S. (1988) A study of reactions of α,β-unsaturated carbonyl compounds with deoxyguanosine. *J. Org. Chem.*, **53**, 14–17.

Colaianni, L.J. (1967) Acute toxicity, eye and skin irritation tests on aromatic compounds. Unpublished report from Roche Chemical Division. Submitted to WHO by the Flavor and Extract Manufacturers Association of the United States.

Cox, G.E., Bailey, D.E. & Morgareidge, K. (1974a) 90-day feeding study in rats with 2-octen-4-one. Unpublished report from FDRL, Inc., Waverly, New York, USA. Submitted to WHO by the Flavor and Extract Manufacturers Association of the United States.

Cox, G.E., Bailey, D.E. & Morgareidge, K. (1974b) 90-day feeding study in rats with 1-octen-3-one. Unpublished report from FDRL, Inc., Waverly, New York, USA. Submitted to WHO by the Flavor and Extract Manufacturers Association of the United States.

Cramer, G.M., Ford, R.A. & Hall, R.L. (1978) Estimation of toxic hazard—A decision tree approach. *Food Cosmet. Toxicol.*, **16**, 255–276.

Deininger, C., Eder, E., Neudecker, T. & Hoffman, C. (1990) Mutagenicity and genotoxicity of ethylvinyl ketone in bacterial tests. *J. Appl. Toxicol.*, **10**, 167–171.

Dietz, F.K., Rodriguez-Giaxola, M., Traiger, G.J., Stella, V.J. & Himmelstein, K.J. (1981) Pharmacokinetics of 2-butanol and its metabolites in the rat. *J. Pharmacokinet. Biopharmacol.*, **9**, 553–576.

Durward, R. (1995) Reverse mutation assay 'Ames test' using Salmonella typhimurium. Unpublished report from Safepharm Laboratories Ltd, Derby, England. Submitted to WHO by the Flavor and Extract Manufacturers Association of the United States.

Eder, E., Hoffman, C. & Deininger, C. (1991) Identification and characterization of deoxyguanosine adducts of methyl vinyl ketone and ethyl vinyl ketone. Genotoxicity of the ketones in the SOS chromotest. *Chem. Res. Toxicol.*. **4**, 50–57.

Eder, E., Scheckenbach, S., Deininger, C. & Hoffman, C. (1993) The possible role of α,β-unsaturated carbonyl compounds in mutagenesis and carcinogenesis. *Toxicol. Lett.*, **67**, 87–103.

Esterbauer, H., Zollner, H. & Scholz, N. (1975) Reaction of glutathione with conjugated carbonyls. *Z. Naturforsch. C*, **30**, 466–473.

Felsted, R.L. & Bachur, N.R. (1980) Ketone reductases. In: Jakoby, W.B., ed., *Enzymatic Basis of Detoxification*, New York: Academic Press, Vol. 1, pp. 281–293.

Florin, I., Rutberg, L., Curvall, M. & Enzell, C.R. (1980) Screening of tobacco smoke constituents for mutagenicity using Ames' test. *Toxicology*, **18**, 219–232.

Ford, D.M. & Moran, E.J. (1978) Preliminary indications of *in vitro* hydrolysis of two flavor chemical esters. Private communication. Submitted to WHO by the Flavor and Extract Manufacturers Association of the United States.

Fraser, I.M., Peters, M.A. & Hardinge, G.M. (1967) Purification and some characteristics of an α,β-unsaturated keone reductases from dog erythrocytes and human liver. *Mol. Pharmacol.*, **3**, 233–247.

Gabriel, M.A., Ilbawi, M. & Al-Khalidi, U.A.S. (1972) The oxidation of acetoin to CO_2 in intact animals and in liver mince preparation. *Comp. Biochem. Physiol.*, **41B**, 493–502.

Gaunt, I.F., Carpanini, F.M.B., Wright, M.G., Grasso, P. & Gangolli, S.D. (1972) Short-term toxicity of methyl amyl ketones in rats. *Food Cosmet. Toxicol.*, **10**, 625–636.

Gill, M.W. & van Miller, J.P. (1987) Fourteen-day dietary minimum toxicity screen (MTS) in albino rats. Unpublished report from Union Carbide Export, Pennsylvania, USA. Submitted to WHO by the Flavor and Extract Manufacturers Association of the United States.

deGroot, A.P., Spanjers, M.T. & van der Heijden, C.A. (1974) Acute and sub-acute oral toxicity studies in rats with five flavour compounds. Unpublished report from Centraal Instituut Voor Voedingsonderzoek TNO, Zeist, Netherlands. Submitted to WHO by the Flavor and Extract Manufacturers Association of the United States.

Heymann, E. (1980) Carboxylesterases and amidases. In: Jakoby, W.B., ed., *Enzymatic Basis of Detoxication*, 2nd Ed., New York: Academic Press, pp. 291-323.

Hoffman, W. (1974) Methylheptenon. Unpublished report from BASF, Ludwigshafen, Germany. Submitted to WHO by the Flavor and Extract Manufacturers Association of the United States.

Hoffman, W. (1978) Acute toxicity study of 2,6-dimethylundecadien-2,6-on-10 (geranylaceton). Unpublished report from BASF, Ludwigshafen, Germany. Submitted to WHO by the Flavor and Extract Manufacturers Association of the United States (in German).

Holland, M.L., Rhodes, G.R., Wiesler, D. & Novotny, M. (1984) Chromatographic profiling of urinary volatile and organic acid metabolites of normal and diabetic C57BL/Ka mice. *J. Chromatogr.*, **306**, 23–37.

Honkanen, E., Karvonen, P. & Virtanen, A.I. (1964) Studies on the transfer of some flavour compounds to milk. *Acta Chem. Scand.*, **18**, 612–618.

International Organization of the Flavor Industry (1995) European inquiry on volume use. Private communication. Submitted to WHO by Flavor and Extract Manufacturers Association of the United States.

van der Jagt, D.L., Hunsaker, L.A., Garcia, K.B. & Royer, R.E. (1985) Isolation and characterization of the multiple gluthathione-*S*-transferases from human liver. *J. Biol. Chem.*, **260**, 11603–11610.

Kasper, C.B. & Henton, D. (1982) Glucuronidation. In: Jakoby, W.B., ed., *Enzymatic Basis of Detoxification*, New York: Academic Press, Vol. 2, pp. 4–36.

Kamil, I.A., Smith, J.N. & Williams, R.T. (1953) Studies in detoxication. 46. The metabolism of aliphatic alcohols; the glucuronic acid conjugation of acyclic aliphatic alcohols. *Biochem. J.*, **53**, 129–136.

Kaye, C.M. (1973) Biosynthesis of mercapturic acids from allyl alcohol, allyl esters and acreolin. *Biochem. J.*, **134**, 1093–1101.

Keating, J.W. (1972) Acute toxicity study in rats and rabbits. Unpublished report from Foster D. Snell, Inc., Elizabeth, New Jersey, USA. Submitted to WHO by the Flavor and Extract Manufacturers Association of the United States.

Leegwater, D.C. & van Straten, S. (1974a) In vitro study on the hydrolysis of twenty-six organic esters by pancreatin. Unpublished report from Centraal Instituut Voor Voedingsonderzoek TNO, Zeist, Netherlands. Submitted to WHO by the Flavor and Extract Manufacturers Association of the United States.

Leegwater, D.C. & van Straten, S. (1974b) In vitro study on the hydrolysis of eight carboxylic esters by intestinal and liver enzymes. Unpublished report from Centraal Instituut Voor Voedingsonderzoek TNO, Zeist, Netherlands. Submitted to WHO by the Flavor and Extract Manufacturers Association of the United States..

Lehman, A.J., Schwerma, H. & Rickards, E. (1945) Isopropyl alcohol: Acquired tolerance in dogs, rate of disappearance from the blood stream in various species, and effects on successive generation of rats. *J. Pharmacol. Exp. Ther.*, **85**, 61–69.

Longland, R.C., Shilling, W.H. & Gangolli, S.D. (1977) The hydrolysis of flavouring estersby artificial gastrointestinal juices and rat tissue preparations. *Toxicology*, **8**, 197–204.

Lucas, C.D., Putnam, J.M. & Hallagan, J.B. (1999) *Flavor and Extract Manufacturers Association of the United States 1995 Poundage and Technical Effects Update Survey*, Washington DC.

Maarse, H., Visscher, C.A., Willemsens, L.C. & Boelens, M.H. (1999) *Volatile Components in Food—Qualitative and Quantitative Data*, Zeist: Centraal Instituut Voor Voedingsonderzioek TNO.

McMahon, R.E. (1982) Alcohols, aldehydes and ketones. in: Jakoby , W.B., Bend, J.R. & Caldwell, J., eds, *Metabolic Basis for Detoxification. Metabolism of Functional Groups*, New York: Academic Press, p. 91.

Moreno, O.M. (1975) Acute toxicity studies in rats rabbits and guinea pigs. Private communication to Research Institute for Fragrance Materials, Hackensack, New Jersey, USA. Submitted to WHO by the Flavor and Extract Manufacturers Association of the United States.

Moreno, O.M.(1977) Acute toxicity studies in rats rabbits and guinea pigs. Private communication to Research Institute for Fragrance Materials, Hackensack, New Jersey, USA. Submitted to WHO by the Flavor and Extract Manufacturers Association of the United States

Moreno, O.M.(1982) Acute toxicity studies in rats rabbits and guinea pigs. Private communication to Research Institute for Fragrance Materials, Hackensack, New Jersey, USA. Submitted to WHO by the Flavor and Extract Manufacturers Association of the United States.

Nordmann, R., Ribière, C., Rouach, H., Béauge, F., Giudicelli, Y. & Nordmann, J. (1973) Metabolic pathways involved in the oxidation of isopropanol into acetone by the intact rat. *Life Sci.*, **13**, 919–932.

O'Donohogue, J.L. & Krasavage, W.J. (1980) 90-day repeated oral administration of five ketones and n-heptane to rats. Private communication. Submitted to WHO by the Flavor and Extract Manufacturers Association of the United States.

O'Donoghue, J.L., Krasavage, W.J., DiVincenzo, G.D. & Katz, G.V. (1984) Further studies on ketone neurotoxicity and interactions. *Toxicol. Appl. Pharmacol.*, **72**, 201–209.

Oser, B.L. (1970) The acute oral toxicity of ten compounds. Unpublished report from FDRL, Inc., Maspeth, New York, USA. Submitted to WHO by the Flavor and Extract Manufacturers Association of the United States.

Portoghese, P.S., Kedziora, G.S., Larson, D.L., Bernard, B.K. & Hall, R.L. (1989) Reactivity of glutathione with α,β-unsaturated ketone flavouring substances. *Food Chem. Toxicol.*, **27**, 773–776.

Posternak, J. (1964) Subacute toxicity (90 days) report on 1-octen-3-ol (amyl vinyl carbinol). Private communication. Submitted to WHO by the Flavor and Extract Manufacturers Association of the United States.

Reagan, E.L. & Becci, P.J. (1983a) Acute oral toxicity study in rats with 1-octene-3-one. Unpublished report from FDRL, Inc., Waverly, New York, USA. Submitted to WHO by the Flavor and Extract Manufacturers Association of the United States.

Reagan, E.L. & Becci, P.J. (1983b) Acute oral toxicity study in rats with 3-methylene-2-octanone. Unpublished report from FDRL, Inc., Waverly, New York, USA. Submitted to WHO by the Flavor and Extract Manufacturers Association of the United States.

Reagan, E.L. & Becci, P.J. (1984) Acute oral LD_{50} study of filbertone in Sprague-Dawley rats. Unpublished report from FDRL, Inc., Waverly, New York, USA. Submitted to WHO by the Flavor and Extract Manufacturers Association of the United States.

Sanders, A. (1996) Acute oral toxicity (limit test) in the rat. Unpublished report from Safepharm Laboratories Ltd, Derby, England. Submitted to WHO by the Flavor and Extract Manufacturers Association of the United States.

Shellenberger, T.E. (1970a) Subacute toxicity evaluation of 2-hexene-4-one with rats. Private communication. Unpublished report from Gulf South Research Institute, New Iberia, Louisiana, USA. Submitted to WHO by the Flavor and Extract Manufacturers Association of the United States.

Shellenberger T.E. (1970b) Histology evaluations: Subacute toxicity evaluation of 2-hexene-4-one with rats. Addendum to the final report. Unpublished report from Gulf South Research Institute, New Iberia, Louisiana, USA. Submitted to WHO by the Flavor and Extract Manufacturers Association of the United States.

Smyth, H.F., Jr & Carpenter, C.P. (1948) Further experience with the range finding test in the industrial toxicology laboratory. *J. Ind. Hyg. Toxicol.*, **30**, 63–68.

Stofberg, J. & Grundschober, F. (1987) Consumption ratio and food predominance of flavoring materials. *Perfumer Flavorist*, **12**, 27.

Stofberg, J. & Kirschman, J.C. (1985) The consumption ratio of flavoring materials: A mechanism for setting priorities for safety evaluation. *Food Chemical Toxicol.*, **23**, 857–860.

Thompson, P.W. (1996) Reverse mutation assay 'Ames test' using Salmonella typhimurium. Unpublished report from Safepharm Laboratories Ltd, Derby, England. Submitted to WHO by the Flavor and Extract Manufacturers Association of the United States.

Til, H.P. (1977) Determination of the acute oral toxicity of 1-vinyl butanol in rats. Unpublished report from Centraal Instituut Voor Voedingsonderzoek TNO, Zeist, Netherlands. Submitted to WHO by the Flavor and Extract Manufacturers Association of the United States.

White, D.A., Heffron, A., Miciak, B., Middleton, B., Knights, S. & Knights, D. (1990) Chemical synthesis of dual radiolabelled cyclandelate and its metabolism in rat hepatocytes and mouse J774 cells. *Xenobiotica*, **20**, 71.

Wnorowski, G. (1997) 14-day repeat dose oral toxicity study: rats. Unpublished report from Product Safety Labs, East Brunswick, New Jersey, USA. Submitted to WHO by the Flavor and Extract Manufacturers Association of the United States.

Wohl, A.J. (1974) Acute oral toxicity of amyl vinyl carbinol. Unpublished report from Foster D. Snell, Inc., Elizabeth, New Jersey, USA. Submitted to WHO by the Flavor and Extract Manufacturers Association of the United States.

Zlatkis, A., Bertsch, W., Lichtenstein, H.A., Tishbee, A., Shunbo, F., Liebich, H.M., Coscia, A.M. & Fleischer, N. (1973) Profile of volatile metabolites in urine by gas chromatography–mass spectrometry. *Anal. Chem.*, **45**, 783–787.

ANNEXES

ANNEX 1

REPORTS AND OTHER DOCUMENTS RESULTING FROM PREVIOUS MEETINGS OF THE JOINT FAO/WHO EXPERT COMMITTEE ON FOOD ADDITIVES

1. **General principles governing the use of food additives** (First report of the Joint FAO/WHO Expert Committee on Food Additives). FAO Nutrition Meetings Report Series, No. 15, 1957; WHO Technical Report Series, No. 129, 1957 (out of print).
2. **Procedures for the testing of intentional food additives to establish their safety for use** (Second report of the Joint FAO/WHO Expert Committee on Food Additives). FAO Nutrition Meetings Report Series, No. 17, 1958; WHO Technical Report Series, No. 144, 1958 (out of print).
3. **Specifications for identity and purity of food additives (antimicrobial preservatives and antioxidants)** (Third report of the Joint FAO/WHO Expert Committee on Food Additives). These specifications were subsequently revised and published as **Specifications for identity and purity of food additives**, Vol. I. **Antimicrobial preservatives and antioxidants**, Rome, Food and Agriculture Organization of the United Nations, 1962 (out of print).
4. **Specifications for identity and purity of food additives (food colours)** (Fourth report of the Joint FAO/WHO Expert Committee on Food Additives). These specifications were subsequently revised and published as **Specifications for identity and purity of food additives**, Vol. II. **Food colours**, Rome, Food and Agriculture Organization of the United Nations, 1963 (out of print).
5. **Evaluation of the carcinogenic hazards of food additives** (Fifth report of the Joint FAO/WHO Expert Committee on Food Additives). FAO Nutrition Meetings Report Series, No. 29, 1961; WHO Technical Report Series, No. 220, 1961 (out of print).
6. **Evaluation of the toxicity of a number of antimicrobials and antioxidants** (Sixth report of the Joint FAO/WHO Expert Committee on Food Additives). FAO Nutrition Meetings Report Series, No. 31, 1962; WHO Technical Report Series, No. 228, 1962 (out of print).
7. **Specifications for the identity and purity of food additives and their toxicological evaluation: emulsifiers, stabilizers, bleaching and maturing agents** (Seventh report of the Joint FAO/WHO Expert Committee on Food Additives). FAO Nutrition Meetings Series, No. 35, 1964; WHO Technical Report Series, No. 281, 1964 (out of print).
8. **Specifications for the identity and purity of food additives and their toxicological evaluation: food colours and some antimicrobials and antioxidants** (Eighth report of the Joint FAO/WHO Expert Committee on Food Additives). FAO Nutrition Meetings Series, No. 38, 1965; WHO Technical Report Series, No. 309, 1965 (out of print).
9. **Specifications for identity and purity and toxicological evaluation of some antimicrobials and antioxidants**. FAO Nutrition Meetings Report Series, No. 38A, 1965; WHO/Food Add/24.65 (out of print).
10. **Specifications for identity and purity and toxicological evaluation of food colours**. FAO Nutrition Meetings Report Series, No. 38B, 1966; WHO/Food Add/66.25.

11. **Specifications for the identity and purity of food additives and their toxicological evaluation: some antimicrobials, antioxidants, emulsifiers, stabilizers, flour treatment agents, acids, and bases** (Ninth report of the Joint FAO/WHO Expert Committee on Food Additives). FAO Nutrition Meetings Series, No. 40, 1966; WHO Technical Report Series, No. 339, 1966 (out of print).
12. **Toxicological evaluation of some antimicrobials, antioxidants, emulsifiers, stabilizers, flour treatment agents, acids, and bases**. FAO Nutrition Meetings Report Series, No. 40A, B, C; WHO/Food Add/67.29.
13. **Specifications for the identity and purity of food additives and their toxicological evaluation: some emulsifiers and stabilizers and certain other substances** (Tenth report of the Joint FAO/WHO Expert Committee on Food Additives). FAO Nutrition Meetings Series, No. 43, 1967; WHO Technical Report Series, No. 373, 1967.
14. **Specifications for the identity and purity of food additives and their toxicological evaluation: some flavouring substances and non nutritive sweetening agents** (Eleventh report of the Joint FAO/WHO Expert Committee on Food Additives). FAO Nutrition Meetings Series, No. 44, 1968; WHO Technical Report Series, No. 383, 1968.
15. **Toxicological evaluation of some flavouring substances and non nutritive sweetening agents**. FAO Nutrition Meetings Report Series, No. 44A, 1968; WHO/Food Add/68.33.
16. **Specifications and criteria for identity and purity of some flavouring substances and non-nutritive sweetening agents**. FAO Nutrition Meetings Report Series, No. 44B, 1969; WHO/Food Add/69.31.
17. **Specifications for the identity and purity of food additives and their toxicological evaluation: some antibiotics** (Twelfth report of the Joint FAO/WHO Expert Committee on Food Additives). FAO Nutrition Meetings Series, No. 45, 1969; WHO Technical Report Series, No. 430, 1969.
18. **Specifications for the identity and purity of some antibiotics**. FAO Nutrition Meetings Series, No. 45A, 1969; WHO/Food Add/69.34.
19. **Specifications for the identity and purity of food additives and their toxicological evaluation: some food colours, emulsifiers, stabilizers, anticaking agents, and certain other substances** (Thirteenth report of the Joint FAO/WHO Expert Committee on Food Additives). FAO Nutrition Meetings Series, No. 46, 1970; WHO Technical Report Series, No. 445, 1970.
20. **Toxicological evaluation of some food colours, emulsifiers, stabilizers, anticaking agents, and certain other substances**. FAO Nutrition Meetings Report Series, No. 46A, 1970; WHO/Food Add/70.36.
21. **Specifications for the identity and purity of some food colours, emulsifiers, stabilizers, anticaking agents, and certain other food additives**. FAO Nutrition Meetings Report Series, No. 46B, 1970; WHO/Food Add/70.37.
22. **Evaluation of food additives: specifications for the identity and purity of food additives and their toxicological evaluation: some extraction solvents and certain other substances; and a review of the technological efficacy of some antimicrobial agents**. (Fourteenth report of the Joint FAO/WHO Expert Committee on Food Additives). FAO Nutrition Meetings Series, No. 48, 1971; WHO Technical Report Series, No. 462, 1971.
23. **Toxicological evaluation of some extraction solvents and certain other substances**. FAO Nutrition Meetings Report Series, No. 48A, 1971; WHO/Food Add/70.39.
24. **Specifications for the identity and purity of some extraction solvents and certain other substances**. FAO Nutrition Meetings Report Series, No. 48B, 1971; WHO/Food Add/70.40.

25. **A review of the technological efficacy of some antimicrobial agents**. FAO Nutrition Meetings Report Series, No. 48C, 1971; WHO/Food Add/70.41.

26. **Evaluation of food additives: some enzymes, modified starches, and certain other substances: Toxicological evaluations and specifications and a review of the technological efficacy of some antioxidants** (Fifteenth report of the Joint FAO/WHO Expert Committee on Food Additives). FAO Nutrition Meetings Series, No. 50, 1972; WHO Technical Report Series, No. 488, 1972.

27. **Toxicological evaluation of some enzymes, modified starches, and certain other substances**. FAO Nutrition Meetings Report Series, No. 50A, 1972; WHO Food Additives Series, No. 1, 1972.

28. **Specifications for the identity and purity of some enzymes and certain other substances**. FAO Nutrition Meetings Report Series, No. 50B, 1972; WHO Food Additives Series, No. 2, 1972.

29. **A review of the technological efficacy of some antioxidants and synergists**. FAO Nutrition Meetings Report Series, No. 50C, 1972; WHO Food Additives Series, No. 3, 1972.

30. **Evaluation of certain food additives and the contaminants mercury, lead, and cadmium** (Sixteenth report of the Joint FAO/WHO Expert Committee on Food Additives). FAO Nutrition Meetings Series, No. 51, 1972; WHO Technical Report Series, No. 505, 1972, and corrigendum.

31. **Evaluation of mercury, lead, cadmium and the food additives amaranth, diethylpyrocarbamate, and octyl gallate**. FAO Nutrition Meetings Report Series, No. 51A, 1972; WHO Food Additives Series, No. 4, 1972.

32. **Toxicological evaluation of certain food additives with a review of general principles and of specifications** (Seventeenth report of the Joint FAO/WHO Expert Committee on Food Additives). FAO Nutrition Meetings Series, No. 53, 1974; WHO Technical Report Series, No. 539, 1974, and corrigendum (out of print).

33. **Toxicological evaluation of some food additives including anticaking agents, antimicrobials, antioxidants, emulsifiers, and thickening agents**. FAO Nutrition Meetings Report Series, No. 53A, 1974; WHO Food Additives Series, No. 5, 1974.

34. **Specifications for identity and purity of thickening agents, anticaking agents, antimicrobials, antioxidants and emulsifiers**. FAO Food and Nutrition Paper, No. 4, 1978.

35. **Evaluation of certain food additives** (Eighteenth report of the Joint FAO/WHO Expert Committee on Food Additives). FAO Nutrition Meetings Series, No. 54, 1974; WHO Technical Report Series, No. 557, 1974, and corrigendum.

36. **Toxicological evaluation of some food colours, enzymes, flavour enhancers, thickening agents, and certain other food additives**. FAO Nutrition Meetings Report Series, No. 54A, 1975; WHO Food Additives Series, No. 6, 1975.

37. **Specifications for the identity and purity of some food colours, enhancers, thickening agents, and certain food additives**. FAO Nutrition Meetings Report Series, No. 54B, 1975; WHO Food Additives Series, No. 7, 1975.

38. **Evaluation of certain food additives: some food colours, thickening agents, smoke condensates, and certain other substances**. (Nineteenth report of the Joint FAO/WHO Expert Committee on Food Additives). FAO Nutrition Meetings Series, No. 55, 1975; WHO Technical Report Series, No. 576, 1975.

39. **Toxicological evaluation of some food colours, thickening agents, and certain other substances**. FAO Nutrition Meetings Report Series, No. 55A, 1975; WHO Food Additives Series, No. 8, 1975.

40. **Specifications for the identity and purity of certain food additives**. FAO Nutrition Meetings Report Series, No. 55B, 1976; WHO Food Additives Series, No. 9, 1976.
41. **Evaluation of certain food additives** (Twentieth report of the Joint FAO/WHO Expert Committee on Food Additives). FAO Food and Nutrition Meetings Series, No. 1, 1976; WHO Technical Report Series, No. 599, 1976.
42. **Toxicological evaluation of certain food additives**. WHO Food Additives Series, No. 10, 1976.
43. **Specifications for the identity and purity of some food additives**. FAO Food and Nutrition Series, No. 1B, 1977; WHO Food Additives Series, No. 11, 1977.
44. **Evaluation of certain food additives** (Twenty-first report of the Joint FAO/WHO Expert Committee on Food Additives). WHO Technical Report Series, No. 617, 1978.
45. **Summary of toxicological data of certain food additives**. WHO Food Additives Series, No. 12, 1977.
46. **Specifications for identity and purity of some food additives, including antioxidant, food colours, thickeners, and others**. FAO Nutrition Meetings Report Series, No. 57, 1977.
47. **Evaluation of certain food additives and contaminants** (Twenty-second report of the Joint FAO/WHO Expert Committee on Food Additives). WHO Technical Report Series, No. 631, 1978.
48. **Summary of toxicological data of certain food additives and contaminants**. WHO Food Additives Series, No. 13, 1978.
49. **Specifications for the identity and purity of certain food additives**. FAO Food and Nutrition Paper, No. 7, 1978.
50. **Evaluation of certain food additives** (Twenty-third report of the Joint FAO/WHO Expert Committee on Food Additives). WHO Technical Report Series, No. 648, 1980, and corrigenda.
51. **Toxicological evaluation of certain food additives**. WHO Food Additives Series, No. 14, 1980.
52. **Specifications for identity and purity of food colours, flavouring agents, and other food additives**. FAO Food and Nutrition Paper, No. 12, 1979.
53. **Evaluation of certain food additives** (Twenty-fourth report of the Joint FAO/WHO Expert Committee on Food Additives). WHO Technical Report Series, No. 653, 1980.
54. **Toxicological evaluation of certain food additives**. WHO Food Additives Series, No. 15, 1980.
55. **Specifications for identity and purity of food additives (sweetening agents, emulsifying agents, and other food additives)**. FAO Food and Nutrition Paper, No. 17, 1980.
56. **Evaluation of certain food additives** (Twenty-fifth report of the Joint FAO/WHO Expert Committee on Food Additives). WHO Technical Report Series, No. 669, 1981.
57. **Toxicological evaluation of certain food additives**. WHO Food Additives Series, No. 16, 1981.
58. **Specifications for identity and purity of food additives (carrier solvents, emulsifiers and stabilizers, enzyme preparations, flavouring agents, food colours, sweetening agents, and other food additives)**. FAO Food and Nutrition Paper, No. 19, 1981.
59. **Evaluation of certain food additives and contaminants** (Twenty-sixth report of the Joint FAO/WHO Expert Committee on Food Additives). WHO Technical Report Series, No. 683, 1982.

60. **Toxicological evaluation of certain food additives**. WHO Food Additives Series, No. 17, 1982.
61. **Specifications for the identity and purity of certain food additives**. FAO Food and Nutrition Paper, No. 25, 1982.
62. **Evaluation of certain food additives and contaminants** (Twenty-seventh report of the Joint FAO/WHO Expert Committee on Food Additives). WHO Technical Report Series, No. 696, 1983, and corrigenda.
63. **Toxicological evaluation of certain food additives and contaminants**. WHO Food Additives Series, No. 18, 1983.
64. **Specifications for the identity and purity of certain food additives**. FAO Food and Nutrition Paper, No. 28, 1983.
65. **Guide to specifications General notices, general methods, identification tests, test solutions, and other reference materials**. FAO Food and Nutrition Paper, No. 5, Rev. 1, 1983.
66. **Evaluation of certain food additives and contaminants** (Twenty-eighth report of the Joint FAO/WHO Expert Committee on Food Additives). WHO Technical Report Series, No. 710, 1984, and corrigendum.
67. **Toxicological evaluation of certain food additives and contaminants**. WHO Food Additives Series, No. 19, 1984.
68. **Specifications for the identity and purity of food colours**. FAO Food and Nutrition Paper, No. 31/1, 1984.
69. **Specifications for the identity and purity of food additives**. FAO Food and Nutrition Paper, No. 31/2, 1984.
70. **Evaluation of certain food additives and contaminants** (Twenty-ninth report of the Joint FAO/WHO Expert Committee on Food Additives). WHO Technical Report Series, No. 733, 1986, and corrigendum.
71. **Specifications for the identity and purity of certain food additives**. FAO Food and Nutrition Paper, No. 34, 1986.
72. **Toxicological evaluation of certain food additives and contaminants**. WHO Food Additives Series, No. 20. Cambridge University Press, 1987.
73. **Evaluation of certain food additives and contaminants** (Thirtieth report of the Joint FAO/WHO Expert Committee on Food Additives). WHO Technical Report Series, No. 751, 1987.
74. **Toxicological evaluation of certain food additives and contaminants**. WHO Food Additives Series, No. 21. Cambridge University Press, 1987.
75. **Specifications for the identity and purity of certain food additives**. FAO Food and Nutrition Paper, No. 37, 1986.
76. **Principles for the safety assessment of food additives and contaminants in food**. WHO Environmental Health Criteria, No. 70. Geneva, World Health Organization, 1987 (out of print). The full text is available electronically at www.who.int/pcs.
77. **Evaluation of certain food additives and contaminants** (Thirty-first report of the Joint FAO/WHO Expert Committee on Food Additives). WHO Technical Report Series, No. 759, 1987 and corrigendum.
78. **Toxicological evaluation of certain food additives**. WHO Food Additives Series, No. 22. Cambridge University Press, 1988.
79. **Specifications for the identity and purity of certain food additives**. FAO Food and Nutrition Paper, No. 38, 1988.
80. **Evaluation of certain veterinary drug residues in food** (Thirty-second report of the Joint FAO/WHO Expert Committee on Food Additives). WHO Technical Report Series, No. 763, 1988.

81. **Toxicological evaluation of certain veterinary drug residues in food**. WHO Food Additives Series, No. 23. Cambridge University Press, 1988.
82. **Residues of some veterinary drugs in animals and foods**. FAO Food and Nutrition Paper, No. 41, 1988.
83. **Evaluation of certain food additives and contaminants** (Thirty-third report of the Joint FAO/WHO Expert Committee on Food Additives). WHO Technical Report Series, No. 776, 1989.
84. **Toxicological evaluation of certain food additives and contaminants**. WHO Food Additives Series, No. 24. Cambridge University Press, 1989.
85. **Evaluation of certain veterinary drug residues in food** (Thirty-fourth report of the Joint FAO/WHO Expert Committee on Food Additives). WHO Technical Report Series, No. 788, 1989.
86. **Toxicological evaluation of certain veterinary drug residues in food**. WHO Food Additives Series, No. 25, 1990.
87. **Residues of some veterinary drugs in animals and foods**. FAO Food and Nutrition Paper, No. 41/2, 1990.
88. **Evaluation of certain food additives and contaminants** (Thirty-fifth report of the Joint FAO/WHO Expert Committee on Food Additives). WHO Technical Report Series, No. 789, 1990, and corrigenda.
89. **Toxicological evaluation of certain food additives and contaminants**. WHO Food Additives Series, No. 26, 1990.
90. **Specifications for identity and purity of certain food additives**. FAO Food and Nutrition Paper, No. 49, 1990.
91. **Evaluation of certain veterinary drug residues in food** (Thirty-sixth report of the Joint FAO/WHO Expert Committee on Food Additives). WHO Technical Report Series, No. 799, 1990.
92. **Toxicological evaluation of certain veterinary drug residues in food**. WHO Food Additives Series, No. 27, 1991.
93. **Residues of some veterinary drugs in animals and foods**. FAO Food and Nutrition Paper, No. 41/3, 1991.
94. **Evaluation of certain food additives and contaminants** (Thirty-seventh report of the Joint FAO/WHO Expert Committee on Food Additives). WHO Technical Report Series, No. 806, 1991, and corrigenda.
95. **Toxicological evaluation of certain food additives and contaminants**. WHO Food Additives Series, No. 28, 1991.
96. **Compendium of Food Additive Specifications**. Joint FAO/WHO Expert Committee on Food Additives (JECFA). Combined specifications from 1st through the 37th Meetings, 1956-1990. FAO, 1992 (2 volumes).
97. **Evaluation of certain veterinary drug residues in food** (Thirty-eighth report of the Joint FAO/WHO Expert Committee on Food Additives). WHO Technical Report Series, No. 815, 1991.
98. **Toxicological evaluation of certain veterinary residues in food**. WHO Food Additives Series, No. 29, 1991.
99. **Residues of some veterinary drugs in animals and foods**. FAO Food and Nutrition Paper, No. 41/4, 1991.
100. **Guide to specifications—General notices, general analytical techniques, identification tests, test solutions, and other reference materials**. FAO Food and Nutrition Paper, No. 5, Ref. 2, 1991.
101. **Evaluation of certain food additives and naturally occurring toxicants** (Thirty-ninth report of the Joint FAO/WHO Expert Committee on Food Additives). WHO Technical Report Series No. 828, 1992.
102. **Toxicological evaluation of certain food additives and naturally occurring toxicants**. WHO Food Additive Series, No. 30, 1993.

103. **Compendium of food additive specifications: Addendum 1**. FAO Food and Nutrition Paper, No. 52, 1992.
104. **Evaluation of certain veterinary drug residues in food** (Fortieth report of the Joint FAO/WHO Expert Committee on Food Additives). WHO Technical Report Series, No. 832, 1993.
105. **Toxicological evaluation of certain veterinary drug residues in food**. WHO Food Additives Series, No. 31, 1993.
106. **Residues of some veterinary drugs in animals and foods**. FAO Food and Nutrition Paper, No. 41/5, 1993.
107. **Evaluation of certain food additives and contaminants** (Forty-first report of the Joint FAO/WHO Expert Committee on Food Additives). WHO Technical Report Series, No. 837, 1993.
108. **Toxicological evaluation of certain food additives and contaminants**. WHO Food Additives Series, No. 32, 1993.
109. **Compendium of food additive specifications: Addendum 2**. FAO Food and Nutrition Paper, No. 52, Add. 2, 1993.
110. **Evaluation of certain veterinary drug residues in food** (Forty-second report of the Joint FAO/WHO Expert Committee on Food Additives). WHO Technical Report Series, No. 851, 1995.
111. **Toxicological evaluation of certain veterinary drug residues in food**. WHO Food Additives Series, No. 33, 1994.
112. **Residues of some veterinary drugs in animals and foods**. FAO Food and Nutrition Paper, No. 41/6, 1994.
113. **Evaluation of certain veterinary drug residues in food** (Forty-third report of the Joint FAO/WHO Expert Committee on Food Additives). WHO Technical Report Series, No. 855, 1995, and corrigendum.
114. **Toxicological evaluation of certain veterinary drug residues in food**. WHO Food Additives Series, No. 34, 1995.
115. **Residues of some veterinary drugs in animals and foods**. FAO Food and Nutrition Paper, No. 41/7, 1995.
116. **Evaluation of certain food additives and contaminants** (Forty-fourth report of the Joint FAO/WHO Expert Committee on Food Additives). WHO Technical Report Series, No. 859, 1995.
117. **Toxicological evaluation of certain food additives and contaminants**. WHO Food Additives Series, No. 35, 1996.
118. **Compendium of food additive specifications: Addendum 3**. FAO Food and Nutrition Paper, No. 52, Add. 3, 1995.
119. **Evaluation of certain veterinary drug residues in food** (Forty-fifth report of the Joint FAO/WHO Expert Committee on Food Additives). WHO Technical Report Series, No. 864, 1996.
120. **Toxicological evaluation of certain veterinary drug residues in food**. WHO Food Additives Series, No. 36, 1996.
121. **Residues of some veterinary drugs in animals and foods**. FAO Food and Nutrition Paper, No. 41/8, 1996.
122. **Evaluation of certain food additives and contaminants** (Forty-sixth report of the Joint FAO/WHO Expert Committee on Food Additives). WHO Technical Report Series, No. 868, 1997.
123. **Toxicological evaluation of certain food additives**. WHO Food Additives Series, No. 37, 1996.
124. **Compendium of food additive specifications, addendum 4**. FAO Food and Nutrition Paper, No. 52, Add. 4, 1996.

125. **Evaluation of certain veterinary drug residues in food** (Forty-seventh report of the Joint FAO/WHO Expert Committee on Food Additives). WHO Technical Report Series, No. 876, 1998.
126. **Toxicological evaluation of certain veterinary drug residues in food**. WHO Food Additives Series, No. 38, 1996.
127. **Residues of some veterinary drugs in animals and foods**. FAO Food and Nutrition Paper, No. 41/9, 1997.
128. **Evaluation of certain veterinary drug residues in food** (Forty-eighth report of the Joint FAO/WHO Expert Committee on Food Additives). WHO Technical Report Series, No. 879, 1998.
129. **Toxicological evaluation of certain veterinary drug residues in food.** WHO Food Additives Series, No. 39, 1997.
130. **Residues of some veterinary drugs in animals and foods**. FAO Food and Nutrition Paper, No. 41/10, 1998.
131. **Evaluation of certain food additives and contaminants** (Forty-ninth report of the Joint FAO/WHO Expert Committee on Food Additives). WHO Technical Report Series, No. 884, 1999.
132. **Safety evaluation of certain food additives and contaminants**. WHO Food Additives Series, No. 40, 1998.
133. **Compendium of food additive specifications: Addendum 5**. FAO Food and Nutrition Paper, No. 52, Add. 5, 1997.
134. **Evaluation of certain veterinary drug residues in food** (Fiftieth report of the Joint FAO/WHO Expert Committee on Food Additives). WHO Technical Report Series, No. 888, 1999.
135. **Toxicological evaluation of certain veterinary drug residues in food.** WHO Food Additives Series, No. 41, 1998.
136. **Residues of some veterinary drugs in animals and foods**. FAO Food and Nutrition Paper, No. 41/11, 1999.
137. **Evaluation of certain food additives** (Fifty-first report of the Joint FAO/WHO Expert Committee on Food Additives). WHO Technical Report Series, No. 891, 2000.
138. **Safety evaluation of certain food additives**. WHO Food Additives Series, No. 42, 1999.
139. **Compendium of food additive specifications, addendum 6**. FAO Food and Nutrition Paper, No. 52, Add. 6, 1998.
140. **Evaluation of certain veterinary drug residues in food** (Fifty-second report of the Joint FAO/WHO Expert Committee on Food Additives). WHO Technical Report Series, No. 893, 1999.
141. **Toxicological evaluation of certain veterinary drug residues in food**. WHO Food Additives Series, No. 43, 2000
142. **Residues of some veterinary drugs in animals and foods**. FAO Food and Nutrition Paper, No. 41/12, 2000.
143. **Evaluation of certain food additives and contaminants** (Fifty-third report of the Joint FAO/WHO Expert Committee on Food Additives). WHO Technical Report Series, 2000.
144. **Safety evaluation of certain food additives and contaminants**. WHO Food Additives Series, No. 44, 2000.
145. **Compendium of food additive specifications, addendum 7**. FAO Food and Nutrition Paper, No. 52, Add. 7, 1999.
146. **Evaluation of certain veterinary drug residues in food** (Fifty-fourth report of the Joint FAO/WHO Expert Committee on Food Additives). WHO Technical Report Series, 2000.

147. **Toxicological evaluation of certain veterinary drug residues in food**. WHO Food Additives Series, No. 45, 2000.
148. **Residues of some veterinary drugs in animals and foods**. FAO Food and Nutrition Paper, No. 41/13, 2000.
149. **Evaluation of certain food additives and contaminants** (Fifty-fifth report of the Joint FAO/WHO Expert Committee on Food Additives). WHO Technical Report Series No. 901, 2001.
150. **Safety evaluation of certain food additives and contaminants**. WHO Food Additives Series, No. 46, 2001.
151. **Compendium of food additive specifications: Addendum 8**. FAO Food and Nutrition Paper, No. 52, Add. 8, 2000.
152. **Evaluation of certain mycotoxins in food** (Fifty-sixth report of the Joint FAO/WHO Expert Committee on Food Additives). WHO Technical Report Series No. 906, 2002.
153. **Safety evaluation of certain mycotoxins in food**. WHO Food Additives Series, No. 47/FAO Food and Nutrition Paper 74, 2001.
154. **Evaluation of certain food additives and contaminants** (Fifty-seventh report of the Joint FAO/WHO Expert Committee on Food Additives). WHO Technical Report Series No. 909, 2002.
155. **Safety evaluation of certain food additives and contaminants**. WHO Food Additives Series, No. 48, 2002.
156. **Compendium of food additive specifications: Addendum 9**. FAO Food and Nutrition Paper, No. 52, Add. 9, 2001.
157. **Evaluation of certain veterinary drug residues in food** (Fifty-eighth report of the Joint FAO/WHO Expert Committee on Food Additives). WHO Technical Report Series No. 911, 2002.
158. **Toxicological evaluation of certain veterinary drug residues in food** WHO Food Additives Series, No. 49, 2002.
159. **Residues of some veterinary drugs in animals and foods**. FAO Food and Nutrition Paper, No. 41/14, 2002.
160. **Evaluation of certain food additives** (Fifty-ninth report of the Joint FAO/WHO Expert Committee on Food Additives). WHO Technical Report Series No. 913, 2002.

ANNEX 2

ABBREVIATIONS USED IN THE MONOGRAPHS

ADI	acceptable daily intake
ALDH	alcohol dehydrogenase
ALT	alanine aminotransferase
AST	aspartate aminotransferase
ATP	adenosine triphosphate
AUC	area under the curve (concentration–time)
Aus	Australia
bw	body weight
c_{max}	maximum plasma concentration
CI	confidence interval
DMBA	7,12-dimethylbenz[*a*]anthracene
F	female
F_0	parental generation
F_1	first filial generation
F_2	second filial generation
FAO	Food and Agricultural Organization of the United Nations
GEMS	Global Environemtnal Monitoring System
GGT	γ-glutamyl transpeptidase
GLP	good laboratory practice
GMP	good manufacturing practice
GSFA	General Standards for Food Safety
GSH	glutathione
IDDM	insulin-dependent diabetes mellitus
Ig	immunoglobulin
IPCS	International Programme on Chemical Safety
JECFA	Joint FAO/WHO Expert Committee on Food Additives
LD_{50}	median lethal dose
LOQ	limit of quantification
M	male
ND	not detected
NDMA	*N*-nitrosodimethylamine
NHL	non-Hodgkin lymphoma
NOEL	no-observed-effect level
NPRO	*N*-nitrosoproline
NR	not reported
NZ	New Zealand
OECD	Organisation for Economic Co-operation and Development
OR	odds ratio
ppm	parts per million

S9	9000 × *g* microsomal frcation of rat liver
SIDS	sudden infant death syndrome
TNF	tumour necrosis factor
UK	United Kingdom
WHO	World Health Organization
w/w	weight for weight

ANNEX 3

Joint FAO/WHO Expert Committee on Food Additives

Geneva, 4–13 June 2002

Members

Dr J. Alexander, Professor and Director, Department of Food Toxicology, Division of Environmental Medicine, Norwegian Institute of Public Health, Oslo, Norway

Dr Junshi Chen, Chinese Centers for Disease Control and Prevention, Institute of Nutrition and Food Safety, Beijing, China

Dr S.M. Dagher. Professor, American University of Beirut, Beirut, Lebanon

Dr D.G. Hattan, Senior Toxicologist, Office of Food Additives Safety, Center for Food Safety and Applied Nutrition, Food and Drug Administration, College Park, MD, USA

Dr Y. Kawamura, Section Chief, Division of Food Additives, National Institute of Health Sciences, Tokyo, Japan (*FAO Consultant*)

Dr A.G.A.C. Knaap, Center for Substances and Risk Assessment, National Institute of Public Health and the Environment, Bilthoven, Netherlands

Dr P.M. Kuznesof, Senior Chemist, Office of Food Additive Safety, Center for Food Safety and Applied Nutrition, Food and Drug Administration, College Park, MD, USA (*Joint Rapporteur*)

Dr J.C. Larsen, Head of Division, Division of Gene Technology and Biochemistry, Institute of Food Safety and Nutrition, Danish Veterinary and Food Administration, Ministry of Food, Agriculture and Fisheries, Søborg, Denmark (*Chairman*)

Mrs I. Meyland, Senior Scientific Adviser, Institute of Food Research and Nutrition, Danish Veterinary and Food Administration, Ministry of Food, Agriculture and Fisheries, Søborg, Denmark (*Vice-chairman*)

Dr G. Pascal, Scientific Director for Human Nutrition and Food Safety, National Institute for Agricultural Research, Paris, France (*Joint Rapporteur*)

Dr M.V. Rao, Head of Chemistry Unit, Food and Environment Laboratory, Dubai Municipality, Dubai, United Arab Emirates

Ms E. Vavasour, Food Directorate, Health Canada, Ottawa, Ontario, Canada

Professor R. Walker, Emeritus Professor of Food Science, School of Biomedical and Life Sciences, University of Surrey, Guildford, Surrey, England

Mrs H. Wallin, Senior Food Control Officer, National Food Agency, Helsinki, Finland

Dr D.B. Whitehouse, Consultant, Food Regulatory Affairs, Bowdon, Cheshire, England

Secretariat

Dr P.J. Abbott, Australia New Zealand Food Authority, Canberra, ACT, Australia (*WHO Temporary Adviser*)

Dr D. Benford, Food Standards Agency, London, England (*WHO Temporary Adviser*)

Prof dr P. van den Brandt, Department of Epidemiology, Maastricht University, Maastricht, Netherlands (*WHO Temporary Adviser*)

Dr C.E. Fisher, Hatfield, Herts, England (*FAO Consultant*)

Ms T.L. Hambridge, Australia New Zealand Food Authority, Canberra, ACT, Australia (*WHO Temporary Adviser*)

Mr E.F.F. Hecker, Chairman of the Codex Committee on Food Additives and Contaminants, Deputy Director of the Department of Food and Veterinary Affairs, Ministry of Agriculture, Nature Management and Fisheries, The Hague, Netherlands

Dr J.L. Herrman, Scientist, International Programme on Chemical Safety, World Health Organization, Geneva, Switzerland (*Joint Secretary*)

Mrs E. Heseltine, Communication in Science, Lajarthe, Saint-Léon-sur-Vézère, France *(Editor)*

Dr T. Inoue, Director, Biological Safety Research Centre, National Institute of Health Sciences, Tokyo, Japan (*WHO Temporary Adviser*)

Dr Taiyi Jin, Department of Occupational Health and Toxicology, Public Health School, Fudan University, Shanghai, China (*WHO Temporary Adviser*)

Dr C.A. Lawrie, Food Standards Agency, London, England (*FAO Consultant*)

Ms C. LeClercq, National Research Institute for Food and Nutrition Research, Rome, Italy (*FAO Consultant*)

Mr M. Luetzow, Food Quality and Standards Service, Food and Nutrition Division, Food and Agriculture Organization of the United Nations, Rome, Italy (*Joint Secretary*)

Professor J.R. Lupton, Regents Professor and University Faculty Fellow, Faculty of Nutrition, Texas A&M University, College Station, TX, USA (*WHO Temporary Adviser*)

Dr A. Mattia, Division of Petition Review, Office of Food Additive Safety, Center for Food Safety and Applied Nutrition, Food and Drug Administration, College Park, MD, USA (*WHO Temporary Adviser*)

Dr G. Moy, Food Safety Programme, World Health Organization, Geneva, Switzerland

Dr I.C. Munro, CanTox Health Sciences International, Mississauga, Ontario, Canada (*WHO Temporary Adviser*)

Dr A. Nishikawa, Section Chief, Division of Pathology, National Institute of Health Sciences, Tokyo, Japan (*WHO Temporary Adviser*)

Dr S.W. Page, International Programme on Chemical Safety, World Health Organization, Geneva, Switzerland

Professor A.G. Renwick, Clinical Pharmacology Group, University of Southampton, Southampton, England (*WHO Temporary Adviser*)

Dr S. Resnik, Department of Industry, Faculty of Exact and Natural Sciences, University of Buenos Aires, Buenos Aires, Argentina (*FAO Consultant*)

Dr O. Sabzevari, Associate Professor in Molecular Toxicology and Drug Metabolism, Department of Toxicology and Pharmacology, Faculty of Pharmacy, Tehran University of Medical Sciences, Tehran, Iran (*WHO Temporary Adviser*)

Prof I.G.Sipes, Professor and Head, Department of Pharmacology, College of Medicine, University of Arizona, Tucson, Arizona, USA (*WHO Temporary Adviser*)

Dr G.J.A. Speijers, Section Public Health of the Centre for Substances & Risk Assessment, National Institute of Public Health and Environmental Protection, Bilthoven, Netherlands (*WHO Temporary Adviser*)

Dr C. Tohyama, Director, Environmental Health Sciences Division, National Institute for Environmental Studies, Tsukuba, Japan (*WHO Temporary Adviser*)

Dr P.J.P. Verger, Scientific Directorate on Human Nutrition and Food Safety, National Institute for Agricultural Research, Paris, France (*FAO Consultant*)

Professor G. Williams, Professor of Pathology, Director, Environmental Pathology and Toxicology, New York Medical College, Valhalla, NY, USA (*WHO Temporary Adviser*)

ANNEX 4

ACCEPTABLE DAILY INTAKES, OTHER TOXICOLOGICAL INFORMATION, AND INFORMATION ON SPECIFICATIONS

Food additives evaluated toxicologically

Food additive	Specifications[a]	Acceptable daily intake (ADI) in mg/kg bw and other toxicological recommendations
Alitame	R	0–1 mg/kg bw[b]
Cross-linked sodium carboxy cellulose	N	Included in the group ADI 'not specified'[c] with methyl other modified celluloses
Mineral oils (low- and medium-viscosity)		
Class I[d]	R	0–10 mg/kg bw
Class II[e] and Class III[f]	R	0–0.01 mg/kg bw (group ADI) (temporary)[g]
Nitrite	S	0–0.07 mg/kg bw (expressed as nitrite ion)
Nitrate	S	0–3.7 mg/kg bw (expressed as nitrate ion)
Salatrim (short- and long-chain acyltriglyceride molecules)	R	Adequate information was not available to evaluate safety and nutritional effects.

[a] N, new specifications prepared; R, existing specifications revised; S, existing specifications were not considered.
[b] ADI was established at the forty-sixth meeting (WHO Technical Report Series, No. 868, 1997). Evaluation of a 90-day study of teolerance in patients with diabetes was postponed.
[c] ADI 'not specified' is used to refer to a food substance of very low toxicity which, on the basis of the available data (chemical, biochemical, toxicological and other) and the total dietary intake of the substance arising from its use at the levels necessary to achieve the desired effects and from its acceptable background levels in food, does not, in the opinion of the Committee, represent a hazard to health. For that reason, and for the reasons stated in the individual evaluations, the establishment of an ADI expressed in numerical form is not deemed necessary. An additive that meets this criterion must be used within the bounds of good manufacturing practice, i.e. it should be technologically efficacious and should be used at the lowest level necessary to achieve its effect, it should not conceal food of inferior quality or adulterated food, and it should not create a nutritional imbalance.
[d] Including P70(H) oil.
[e] Including N70(H) and N70(A) oils.
[f] Including P15(H), N15(H), and N10(A) oils
[g] See Annex 3.

Food additive considered for specifications only

Food additive	Specifications[a]
Amyloglucosidase from *Aspergillus niger*, var.	R

[a]R, existing specifications revised

Substances evaluated by the Procedure for the Safety Evaluation of Flavouring Agents

A. *Alicyclic primary alcohols, aldehydes, acids and related esters*

Flavouring agent	No.	Specifications[a]	Conclusions based on current intake
Cyclohexanecarboxylic acid	961	N	No safety concern
Methyl cyclohexanecarboxylate	962	N	No safety concern
Ethyl cyclohexanecarboxylate	963	N	No safety concern
Cyclohexaneethyl acetate	964	N	No safety concern
Cyclohexaneacetic acid	965	N	No safety concern
Ethyl cyclohexanepropionate	966	N	No safety concern
2,2,3-Trimethylcyclopent-3-en-1-yl acetaldehyde	967	N	No safety concern
cis-5-Isopropenyl-*cis*-2-methylcyclopentan-1-carboxaldehyde	968	N	No safety concern
Campholene acetate	969	N	No safety concern
α-Campholenic alcohol	970	N	No safety concern
para-Menth-1-en-9-al	971	N	No safety concern
1-*para*-Menthen-9-yl acetate	972	N	No safety concern
para-Mentha-1,8-dien-7-al	973	N	No safety concern
para-Mentha-1,8-dien-7-ol	974	N	No safety concern
para -Mentha-1,8-dien-7-yl acetate	975	N	No safety concern
1,2,5,6-Tetrahydrocuminic acid	976	N	No safety concern
2,6,6-Trimethylcyclohexa-1,3-dienyl methanal	977	N	No safety concern
2,6,6-Trimethyl-1-cyclohexen-1-acetaldehyde	978	N	No safety concern
2,6,6-Trimethyl-1&2-cyclohexen-1-carboxaldehyde	979	N	No safety concern
2-Formyl-6,6-dimethylbicyclo[3.1.1]hept-2-ene (Myrtenal)	980	N	No safety concern
Myrtenol	981	N	No safety concern
Myrtenyl acetate	982	N	No safety concern
6,6-Myrtenyl formate	983	N	No safety concern
Santalol (α and β)	984	N	No safety concern
Santalyl acetate (α and β)	985	N	No safety concern
10-Hydroxymethylene-2-pinene	986	N	No safety concern

[a]N, new specifications prepared

B. *Phenethyl alcohol, aldehyde, acid and related acetals and esters*

Flavouring agent	No.	Specifications[a]	Conclusions based on current intake
Phenethyl alcohol	987	N	No safety concern
Phenethyl formate	988	N	No safety concern
Phenethyl acetate	989	N	No safety concern
Phenethyl propionate	990	N	No safety concern
Phenethyl butyrate	991	N	No safety concern
Phenethyl isobutyrate	992	N	No safety concern
Phenethyl 2-methylbutyrate	993	N	No safety concern
Phenethyl isovalerate	994	N	No safety concern
Phenethyl hexanoate	995	N	No safety concern
Phenethyl octanoate	996	N	No safety concern
Phenethyl tiglate	997	N	No safety concern
Phenethyl senecioate	998	N	No safety concern
Phenethyl phenylacetate	999	N	No safety concern
Acetaldehyde phenethyl propyl acetal	1000	N	No safety concern
Acetaldehyde butyl phenethyl acetal	1001	N	No safety concern
Phenylacetaldehyde	1002	N	No safety concern
Phenylacetaldehyde dimethyl acetal	1003	N	No safety concern

Flavouring agent	No.	Specifications[a]	Conclusions based on current intake
Phenylacetaldehyde glyceryl acetal	1004	N	No safety concern
Phenylacetic acid	1007	N	No safety concern
Methyl phenylacetate	1008	N	No safety concern
Ethyl phenylacetate	1009	N	No safety concern
Propyl phenylacetate	1010	N	No safety concern
Isopropyl phenylacetate	1011	N	No safety concern
Butyl phenylacetate	1012	N	No safety concern
Isobutyl phenylacetate	1013	N	No safety concern
Isoamyl phenylacetate	1014	N	No safety concern
Hexyl phenylacetate	1015	N	No safety concern
3-Hexenyl phenylacetate	1016	N	No safety concern
Octyl phenylacetate	1017	N	No safety concern
Rhodinyl phenylacetate	1018	N	No safety concern
Linalyl phenylacetate	1019	N	No safety concern
Geranyl phenylacetate	1020	N	No safety concern
Citronellyl phenylacetate	1021	N	No safety concern
Santalyl phenylacetate (α and β)	1022	N	No safety concern
para-Tolylacetaldehyde	1023	N	No safety concern
para-Isopropylphenylacetaldehyde	1024	N	No safety concern
Methyl *para*-*tert*-butylphenylacetate	1025	N	No safety concern
Phenoxyacetic acid	1026	N	No safety concern
Ethyl (*para*-tolyloxy)acetate	1027	N	No safety concern
2-Phenoxyethyl isobutyrate	1028	N	No safety concern
Sodium 2-(4-methoxyphenoxy)propanoate	1029	N	No safety concern

[a]N, new specifications prepared

C. Sulfur-containing heterocyclic compounds

Flavouring agent	No.	Specifications[a]	Conclusions based on current intake
Thiamine hydrochloride	1030	N	No safety concern
4-Methyl-5-thiazoleethanol	1031	N	No safety concern
Thiazole	1032	N	No safety concern
2-(1-Methylpropyl)thiazole	1033	N	No safety concern
2-Isobutylthiazole	1034	N	No safety concern
4,5-Dimethylthiazole	1035	N	No safety concern
2,4,5-Trimethylthiazole	1036	N	No safety concern
2-Isopropyl-4-methylthiazole	1037	N	No safety concern
4-Methyl-5-vinylthiazole	1038	N	No safety concern
2,4-Dimethyl-5-vinylthiazole	1039	N	No safety concern
Benzothiazole	1040	N	No safety concern
2-Acetylthiazole	1041	N	No safety concern
2-Propionylthiazole	1042	N	No safety concern
4-Methylthiazole	1043	N	No safety concern
2-Ethyl-4-methylthiazole	1044	N	No safety concern
4,5-Dimethyl-2-isobutyl-3-thiazoline	1045	N	No safety concern
2-Isobutyl-4,6-dimethyldihydro-1,3,5-dithiazine and 4-isobutyl-2,6-dimethyldihydro-1,3,5-dithiazine (mixture)	1046	N	No safety concern
2-Isopropyl-4,6-dimethyl and 4-isopropyl-2,6-dimethyldihydro-1,3,5-dithiazine (mixture)	1047	N	No safety concern
2,4,6-Triisobutyl-5,6-dihydro-4*H*-1,3,5-dithiazine	1048	N	No safety concern
2,4,6-Trimethyldihydro-4*H*-1,3,5-dithiazine	1049	N	No safety concern
5-Methyl-2-thiophenecarboxaldehyde	1050	N	No safety concern
3-Acetyl-2,5-dimethylthiophene	1051	N	No safety concern

Flavouring agent	No.	Specifications[a]	Conclusions based on current intake
2-Thienylmercaptan	1052	N	No safety concern
2-Thienyl disulfide	1053	N	No safety concern
2-Ethoxythiazole	1056	N	No safety concern
2-Methyl-5-methoxythiazole	1057	N	No safety concern
4,5-Dimethyl-2-ethyl-3-thiazoline	1058	N	No safety concern
2-(2-Butyl)-4,5-dimethyl-3-thiazoline	1059	N	No safety concern

[a]N, new specifications prepared

D. Sulfur-substituted furan derivatives

Flavouring agent	No.	Specifications[a]	Conclusions based on current intake
2-Methyl-3-furanthiol	1060	N	No safety concern
2-Methyl-3-(methylthio)furan	1061	N	No safety concern
2-Methyl-5-(methylthio)furan	1062	N	No safety concern
2,5-Dimethyl-3-furanthiol	1063	N	No safety concern
Methyl 2-methyl-3-furyl disulfide	1064	N	No safety concern
Propyl 2-methyl-3-furyl disulfide	1065	N	No safety concern
Bis(2-methyl-3-furyl) disulfide	1066	N	No safety concern
Bis(2,5-dimethyl-3-furyl) disulfide	1067	N	No safety concern
Bis(2-methyl-3-furyl) tetrasulfide	1068	N	No safety concern
2-Ethanethoic acid, *S*-(2-methyl-3-furanyl) ester	1069	N	No safety concern
2,5-Dimethyl-3-furan thioisovalerate	1070	N	No safety concern
2,5-Dimethyl-3-thiofuroylfuran	1071	N	No safety concern
Furfuryl mercaptan	1072	N	No safety concern
S-Furfuryl thioformate	1073	N	No safety concern
S-Furfuryl thioacetate	1074	N	No safety concern
S-Furfuryl thiopropionate	1075	N	No safety concern
Furfuryl methyl sulfide	1076	N	No safety concern
Furfuryl isopropyl sulfide	1077	N	No safety concern
Methyl furfuryl disulfide	1078	N	No safety concern
Propyl furfuryl disulfide	1079	N	No safety concern
2,2'-(Thiodimethylene)difuran	1080	N	No safety concern
2,2'-(Dithiodimethylene)difuran	1081	N	No safety concern
2-Methyl-3-, 5- or 6-(furfurylthio)pyrazine	1082	N	No safety concern
S-Methyl thiofuroate	1083	N	No safety concern
4-[(2-Furanmethyl)thio]-2-pentanone	1084	N	No safety concern
3-[(2-Methyl-3-furyl)thio]-4-heptanone	1085	N	No safety concern
2,6-Dimethyl-3-[(2-methyl-3-furyl)thio]-4-heptanone	1086	N	No safety concern
4-[(2-Methyl-3-furyl)thio]-5-nonanone	1087	N	No safety concern
Ethyl 3-(furfurylthio)propionate	1088	N	No safety concern
2-Methyl-3-thioacetoxy-4,5-dihydrofuran	1089	N	No safety concern
2-Methyl-3-tetrahydrofuranthiol	1090	N	No safety concern
2,5-Dimethyl-3-tetrahydrofuranthiol, *cis* and *trans* isomers	1091	N	No safety concern
2,5-Dimethyl-3-thioacetoxy-tetrahydrofuran, *cis* and *trans* isomers	1092	N	No safety concern

[a]N, new specifications prepared

E. *Alicyclic ketones, secondary alcohols and related esters*

Flavouring agent	No.	Specifications[a]	Conclusions based on current intake
3,3,5-Trimethyl cyclohexanol	1099	N	No safety concern
Cyclohexanone	1100	N	No safety concern
Cyclopentanone	1101	N	No safety concern
2-Methylcyclohexanone	1102	N	No safety concern
3-Methylcyclohexanone	1103	N	No safety concern
4-Methylcyclohexanone	1104	N	No safety concern
1-Methyl-1-cyclopenten-3-one	1105	N	No safety concern
2-Hexylidene cyclopentanone	1106	N	No safety concern
3-Methyl-2-cyclohexen-1-one	1107	N	No safety concern
2,2,6-Trimethylcyclohexanone	1108	N	No safety concern
2-*sec*-Butylcyclohexanone	1109	N	No safety concern
4-Isopropyl-2-cyclohexenone	1110	N	No safety concern
Tetramethylethylcyclohexenone (mixture of isomers)	1111	N	No safety concern
Isophorone	1112	N	No safety concern
3-Methyl-5-propyl-2-cyclohexen-1-one	1113	N	No safety concern
3-Methyl-2-(2-pentenyl)-2-cyclopenten-1-one	1114	N	No safety concern
Isojasmone	1115	N	No safety concern
(*E*)-2-(2-Octenyl)cyclopentanone	1116	N	No safety concern
2-(3,7-Dimethyl-2,6-octadienyl)cyclopentanone	1117	N	No safety concern

[a]N, new specifications prepared

F. *Aliphatic secondary alcohols, ketones and related esters*

Flavouring agent	No.	Specifications[a]	Conclusions based on current intake
3-Decanone	1118	N	No safety concern
5-Methyl-5-hexen-2-one	1119	N	No safety concern
6-Methyl-5-hepten-2-one	1120	N	No safety concern
3,4,5,6-Tetrahydropseudoionone	1121	N	No safety concern
6,10-Dimethyl-5,9-undecadien-2-one	1122	N	No safety concern
2,6,10-Trimethyl-2,6,10-pentadecatrien-14-one	1123	N	No safety concern
3-Penten-2-one	1124	N	No safety concern
4-Hexen-3-one	1125	N	No safety concern
2-Hepten-4-one	1126	N	No safety concern
3-Hepten-2-one	1127	N	No safety concern
3-Octen-2-one	1128	N	No safety concern
2-Octen-4-one	1129	N	No safety concern
3-Decen-2-one	1130	N	No safety concern
4-Methyl-3-penten-2-one	1131	N	No safety concern
5-Methyl-3-hexen-2-one	1132	N	No safety concern
5-Methyl-2-hepten-4-one	1133	N	No safety concern
6-Methyl-3,5-heptadien-2-one	1134	N	No safety concern
(*E*)-7-Methyl-3-octen-2-one	1135	N	No safety concern
3-Nonen-2-one	1136	N	No safety concern
(*E*) and (Z)-4,8-Dimethyl-3,7-nonadien-2-one	1137	N	No safety concern
(*E*)-6-Methyl-3-hepten-2-one	1138	N	No safety concern
(*E*,*E*)-3,5-Octadien-2-one	1139	N	No safety concern
3-Octen-2-ol	1140	N	No safety concern
(*E*)-2-Octen-4-ol	1141	N	No safety concern
2-Pentyl butyrate	1142	N	No safety concern
(+/–)Heptan-3-yl acetate	1143	N	No safety concern
(+/–)Heptan-2-yl butyrate	1144	N	No safety concern
(+/–)Nonan-3-yl acetate	1145	N	No safety concern
2-Pentyl acetate	1146	N	No safety concern
1-Penten-3-one	1147	N	No safety concern

Flavouring agent	No.	Specifications[a]	Conclusions based on current intake
1-Octen-3-one	1148	N	No safety concern
2-Pentyl-1-buten-3-one	1149	N	No safety concern
1-Penten-3-ol	1150	N	No safety concern
1-Hexen-3-ol	1151	N	No safety concern
1-Octen-3-ol	1152	N	No safety concern
1-Decen-3-ol	1153	N	No safety concern
(*E*,R)-3,7-Dimethyl-1,5,7-octatrien-3-ol	1154	N	No safety concern
6-Undecanone	1155	N	No safety concern
2-Methylheptan-3-one	1156	N	No safety concern

[a]N, new specifications prepared

Flavouring agents considered for specifications only

No.	Flavouring agent	Specifications[a]
111	Lauric acid	R
113	Myristic acid	R
115	Palmitic acid	R
116	Stearic acid	R
182	Isoamyl laurate	R
310	Isopropyl isovalerate	R
390	γ-Ionone	R
455	Butyl sulfide	R
476	Ethyl 3-methylthiopropionate	R
483	Ethyl thioacetate	R
490	Allyl thiopropionate	R
492	Methylthio 2-(acetyloxy)propionate	R
493	Methylthio 2-(propionyloxy) propionate	R
540	1,6-Hexanedithiol	R
542	1,9-Nonanedithiol	R
551	2-Mercaptopropionic acid	R
553	Ethyl 3-mercaptopropionate	R
564	Dimethyl disulfide	R
566	Propyl disulfide	R
601	Ethyl 3-hydroxyhexanoate	R
602	Ethyl 3-oxohexanoate	R
606	Laevulinic acid	R
608	Butyl levulinate	R
609	1,4-Nonanediol diacetate	R
614	Diethyl malonate	R
616	Dimethyl succinate	R
617	Diethyl succinate	R
622	Diethyl tartrate	R
624	Diethyl sebacate	R
625	Dibutyl sebacate	R
626	Ethylene brassylate	R
627	Aconitic acid	R
642	3-Phenylpropyl hexanoate	R
678	α-Amylcinnamyl isovalerate	R
729	Dihydroxyacetophenone	R
745	5-Methylfurfural	R
752	2-Phenyl-3-carbethoxyfuran	R
814	α-Methylphenethyl butyrate	R
817	4-(*para*-Tolyl)-2-butanone	R

No.	Flavouring agent	Specifications[a]
847	Benzyl 2,3-dimethylcrotonate	R
861	Glyceryl tribenzoate	R
862	Propylene glycol dibenzoate	S
866	Tolualdehydes (mixed *ortho*, *meta* and *para*)	R
870	Butyl *para*-hydroxybenzoate	S
872	Anisyl formate	R
910	3-Oxohexanoic acid diglyceride	R
911	3-Oxooctanoic acid glyceride	R
914	3-Oxodecanoic acid glyceride	R
915	3-Oxododecanoic acid glyceride	R
916	3-Oxotetradecanoic acid glyceride	R
917	3-Oxohexadecanoic acid glyceride	R
918	Glyceryl monostearate	R,T
919	Glyceryl monooleate	R,T
921	Glyceryl tripropanoate	S
922	Tributyrin	S
923	Glycerol 5-hydroxydecanoate	S,T
924	Glycerol 5-hydroxydodecanoate	S,T
926	Propylene glycol stearate	R
937	Pyruvaldehyde	S,T
943	Acetaldehyde ethyl cis-3-hexenyl acetal	R
954	Ethyl vanillin propylene glycol acetal	R
955	4-Hydroxybenzyl alcohol	N
956	4-Hydroxybenzaldehyde	N
957	4-Hydroxybenzoic acid	N
958	2-Hydroxybenzoic acid	N
959	4-Hydroxy-3-methoxy benzoic acid	N
960	Vanillin erythro- and threo-butan-2,3-diol acetal	N

[a]R, existing specifications revised; S, existing specifications were maintained; T, the existing, new, or revised specifications are tentative and new information is required.

Flavouring agents for which confirmation of flavour use was required

The Committee could not complete its evaluation of 18 substances at its fifty-seventh meeting, pending confirmation that they were in current use as flavouring agents. The Committee received information from the flavour industry on these substances, and concluded that flavour uses have been established for 16 of them. The conclusions of the Committee are summarized in the following table.

No.	Flavouring agent	Evaluation based on current intake as a flavouring agent
850	Benzoic acid[a]	No safety concern
861	Glyceryl tribenzoate	No safety concern
862	Propylene glycol dibenzoate	No safety concern
870	Butyl-*para*-hydroxybenzoate	No safety concern
909	Glycerol[b]	Evaluation not finalized, pending definition of 'flavouring agent'
914	3-Oxodecanoic acid glyceride	No safety concern
915	3-Oxododecanoic acid glyceride	No safety concern
916	3-Oxotetradecanoic acid glyceride	No safety concern
917	3-Oxohexadecanoic acid glyceride	No safety concern
918	Glyceryl monostearate[c]	No safety concern

No.	Flavouring agent	Evaluation based on current intake as a flavouring agent
919	Glyceryl monooleate[c]	No safety concern
920	Triacetin[d]	No safety concern
921	Glyceryl tripropionate (Tripropionin)	No safety concern
922	Tributyrin	No safety concern
923	Glycerol 5-hydroxydecanoate	No safety concern
924	Glycerol 5-hydroxydodecanoate	No safety concern
925	Propylene glycol[e]	Evaluation not finalized, pending definition of 'flavouring agent'
926	Propylene glycol stearate[f]	No safety concern

ANNEX 5

CORRIGENDUM

Food Additive Series 48: Safety evaluation of certain food additives and contaminants, 2002

page 451: *Add as co-author of the first draft of the monograph on polychlorinated dibenzodioxins, polychlorinated dibenzofurans and coplanar polychlorinated biphenyls:*

H. Tryphonas, Bureau of Chemical Safety, Food Directorate, Health Products and Food Branch, Health Canada, Ottawa, Ontario, Canada